Interdisciplinary Applied Mathematics

Volume 48

More information about this series at http://www.springer.com/series/1390

Alain Bensoussan

Estimation and Control of Dynamical Systems

Springer

Alain Bensoussan
International Center for Risk
and Decision Analysis
Jindal School of Management
University of Texas at Dallas
Richardson, TX, USA

Department of Systems Engineering
and Engineering Management
City University of Hong Kong
Kowloon, Hong Kong

ISSN 0939-6047 ISSN 2196-9973 (electronic)
Interdisciplinary Applied Mathematics
ISBN 978-3-030-09236-8 ISBN 978-3-319-75456-7 (eBook)
https://doi.org/10.1007/978-3-319-75456-7

Mathematics Subject Classification: 49-XX, 49-L XX

Softcover re-print of the Hardcover 1st edition 2018

Printed on acid-free paper

This Springer imprint is published by the registered company Springer International Publishing AG part of Springer Nature.
The registered company address is: Gewerbestrasse 11, 6330 Cham, Switzerland

Contents

Chapter 1
Introduction

Dynamical systems are systems that evolve with respect to time. They represent the overwhelming majority of systems. The objective of scientists is first to model them as accurately as possible in order to have optimal knowledge of their evolution and to design controls of the evolution in order to accomplish certain objectives. A significant evolution in the modeling and theory of dynamical systems occurred in the late 1950s and early 1960s, resulting in what is called the *modern theory of dynamical systems and control theory.* A key element is the **state** representation of dynamical systems, also called the **internal** representation. Previously, dynamical systems were modeled by an **external** representation, also called an *input–output* relation. It is the idea of a black box. In such a framework, one can see only how the system reacts to inputs. This approach is particularly meaningful when there is no knowledge model available, justifying the terminology **black box.** For an engineer, who is not so interested in a full knowledge of a system but rather in its behavior when it receives stimuli, the external representation is a useful approach. We shall not develop it here, since it has clearly been superseded by the modern theory. Let us simply recall a few basic elements. Suppose a system can receive m inputs, represented by a time-dependent vector $v(t) \in R^m$. Suppose we observe outputs $z(t) \in R^p$. In the simplest case, in which the system reacts linearly, the input–output relation can be written as

$$z(t) = \int_{t_0}^{t} h(t, \tau)v(\tau)d\tau,$$

in which t_0 is an initial time, and the matrix function $h(t, \tau)$ characterizes the reaction function. All the information of interest for the engineer is included in that function. The stationary case corresponds to $t_0 = 0$ and

$$h(t, \tau) = h(t - \tau),$$

A. Bensoussan, *Estimation and Control of Dynamical Systems*, Interdisciplinary Applied Mathematics 48, https://doi.org/10.1007/978-3-319-75456-7_1

so in fact, z reduces to the convolution product $z = h * v$:

$$z(t) = \int_0^t h(t-\tau)v(\tau)d\tau.$$

It is natural to introduce Laplace transforms $V(s)$, $Z(s)$, $H(s)$ defined by

$$V(s) = \int_0^{+\infty} v(t)\exp(-st)\,dt,$$

in which s is complex. This integral must be well defined, so there may be possible restrictions on s. We define similarly $H(s)$, $Z(s)$. We obtain immediately

$$Z(s) = H(s)V(s).$$

The matrix $H(s)$ is called the *transfer function.* Much work has been devoted to transfer functions that completely characterize linear stationary systems described by an external representation.

The internal representation introduces the very important concept of *state of the system.* The idea is reminiscent of that of a *knowledge model*, in which one models as acccurately as possible all the components of the system, starting with the physical laws that are involved. For instance, if a dynamical system is a rocket in flight, then one will use the laws of mechanics and possibly more advanced physical and chemical laws to describe the propulsion. The state will be the position and velocity of the center of the rocket, and there may be additional state variables.

Of course, such models are not linear, but a linearization can be made around a predefined profile, provided the system remains close to that profile.

An external representation can be always obtained from an internal representation. An interesting question is the reverse. Can one define an internal representation from a given external representation? One may think that this is a problem of model identification or calibration, which is the case when one prescribes the model and fits parameters to satisfy the input–output relationship. In addition, in systems dynamics, a new idea was that the state may not have a physical meaning. The state is then purely a mathematical object. This approach, called realization theory, was intensely developed in the 1960s for linear systems. The current tendency, however, is to rely on knowledge models, in which the concept of state remains predominant, but the state has a physical meaning. Mathematically, once the internal representation has been made precise, it makes no difference whether the state has a physical meaning or not. But the situation is completely different from the applications point of view. See [13, 25] for basic references in the domain.

Since the 1960s, control theory has evolved considerably, and it has been applied successfully in many fields. Space exploration, which was a very important domain of application in modern control theory in the early phase, has given way to new domains such as economics and management science. The remarkable expansion of the field of mathematical finance is a good example. Networks in telecommunications and computer science, along with applications to robotics, provide other directions of development. We currently see the need to control very large systems, such as a national electric grid. In a totally different context, control of nanosystems has emerged as a field that may offer a rejuvenation of the original external representation of systems. Differential games, motivated originally by defense problems, has seen a considerable expansion due to competition in economics and management science. Here the role of information has been overwhelming, and completely new problems such as principal–agent, asymmetry of information, and incentives have appeared, presenting considerable difficulties but also yielding quite surprising new results and ideas.

I lived through this evolution with a certain sense of bewilderment. In a nutshell, control theory has been extremely successful. At the same time, it has metamorphosed in many directions that are less and less similar. One consequence is that recent books in control theory have been rather specialized, and the risk is that the core of control theory may be lost. Of course, old books still exist, but because they are old, they appear obsolete and not used by students. It is in this context that this book was conceived. I wished to write a book that could serve as a bridge between the core concepts and the more modern ramifications. This book is not an encyclopedia. In fact, an encyclopedia exists, see [1], having been published in 2015. But an encyclopedia is a collection of contributions and not the work of a single author. The book represents my experience over the past half-century. Even though the period is long, the knowledge of one person is limited. Another limitation is that a book cannot be of unlimited length, which introduces restrictions in the choice of topics. Several large areas are not addressed at all, such as the general theory of control of partially observable systems (only the linear-quadratic case is presented), the control of distributed parameter systems, and the recent and very interesting theory of mean field games, to mention only those closest to my particular interests. In contrast to what may be found in the above-mentioned encyclopedia, engineering applications are not considered here, but applications to finance are.

This book is also biased toward stochastic control, but it does not consider adaptive control, for instance. It is my belief that stochastic control is the core of control theory. One can find excellent books on stochastic control—such as [22, 29, 37, 44, 49, 55], among the most important—and I have relied on them in this work. They represent more focused visions; in particular, they address recent progress in stochastic control, with limited connection to the past. For instance, all that concerns linear-quadratic systems is seen as a particular case of general nonlinear systems. But students should know that they can be approached directly, and one does not need the general theory to study them. It is important to learn about the evolution of control theory and not just the latest results. Such a historical point of view represents the spirit of this book, and I hope that students will find such an approach helpful. Another important point is the connection between estimation and control. Estimation is a part of control for stochastic systems, and can also be viewed as a control problem. It is necessary to understand the link between the two. For instance, the separation principle belongs to the core knowledge. It is rarely discussed in more recent works, certainly because it is not a common case. Conceptually, however, it is important, because one may apply the separation principle as an approximation. At any rate, the reader more interested in control than in estimation can easily omit the corresponding chapters, at least on a first reading. We insist on the fact that this book is no substitute for the numerous works that have appeared previously, but it provides a somewhat comprehensive approach in a single text. In addition, since the current refinements of stochastic control can be very technical, I have preferred to use formal arguments rather than to strive for absolute rigor when such an approach seemed useful. I have tried to emphasize concepts and intuition rather than full rigor.

So to sum up, this book fills some gaps between classical and more recent, but specialized, developments. It targets scientists who want to use control theory in their research and are interested in understanding the intuition behind the results. We have tried our best to avoid relying on the reader's background. In particular, what is needed in the theory of stochastic processes is recalled. I hope that this book will help those developing more advanced and relevant theory and therefore will be useful to doctoral students.

I thank warmly Laurent Mertz, who has read a draft version. Some comments from reviewers have been useful and are appreciated. Special thanks to Springer, which plays such a great role in publishing scientific books. All comments from readers are welcome.

Chapter 2
State Representation of Linear Dynamical Systems

2.1 General Description

2.1.1 The Model: Internal Representation

We have indicated in the introduction the two possible representations of dynamical systems. We consider here the situation in which the input–output relationship is obtained through the evolution of a state variable $x(t) \in R^n$, as follows:

$$\begin{aligned} \frac{dx}{dt} &= F(t)x(t) + G(t)v(t), \\ z(t) &= H(t)x(t). \end{aligned} \tag{2.1.1}$$

We use indifferently the notation $\frac{dx}{dt}$ or $\dot{x}$. The input is $v(t)$ and the output $z(t)$. Although everything is deterministic and the model is known, we do not have access to the state $x(t)$, because in (2.1.1), neither the initial time has been defined (not necessarily at time 0) nor the initial state given. It may be quite common that we do not know it. Therefore, even though we can control through $v(.)$, we have no access to the state $x(t)$. On the other hand, we know the output $z(t)$. Naturally,

$$F(t) \in \mathcal{L}(R^n; R^n), \quad G(t) \in \mathcal{L}(R^m; R^n), \quad H(t) \in \mathcal{L}(R^n; R^p),$$

where $\mathcal{L}(R^n; R^n)$ represents the vector space of linear maps from R^n to R^n. It is, of course, the set of $n \times n$ matrices.

We will write the external representation below, see Section 2.1.3, after having recalled useful properties of matrices. At this stage it is important to emphasize that the input $v(t)$ and the output $z(t)$ are the only quantities that can be observed at any

A. Bensoussan, *Estimation and Control of Dynamical Systems*, Interdisciplinary Applied Mathematics 48, https://doi.org/10.1007/978-3-319-75456-7_2

time t. The input is a decision of the operator using actuators, while the output is observed through sensors. In this framework, there is no access to the state $x(t)$, although the model is completely deterministic. What is lacking is the initial state.

2.1.2 Fundamental Matrix

The differential equation

$$\begin{aligned} \frac{dx}{dt} &= F(t)x(t), \quad t > t_0 \\ x(t_0) &= \xi \end{aligned} \tag{2.1.2}$$

can be solved explicitly using the fundamental matrix $\Phi(t, t_0)$ as follows:

$$x(t) = \Phi(t, t_0)\xi. \tag{2.1.3}$$

This matrix is the solution of the differential equation

$$\begin{aligned} \frac{\partial}{\partial t}\Phi(t, t_0) &= F(t)\Phi(t, t_0) \\ \Phi(t_0, t_0) &= I. \end{aligned} \tag{2.1.4}$$

In the stationary case, one has $F(t) = F$ and $\Phi(t, t_0) = \exp F(t - t_0)$. This matrix is invertible (see in particular Liouville's theorem below), and it is convenient to write

$$\Phi(t, \tau)^{-1} = \Phi(\tau, t), \quad t > \tau. \tag{2.1.5}$$

We have the semigroup property

$$\Phi(t, s) = \Phi(t, \tau)\Phi(\tau, s), \ \forall t, s, \tau, \tag{2.1.6}$$

which comes from the differential equation when $t > \tau > s$, and the notation for the inverse otherwise (see 2.1.5).

The matrix $\Phi(t, t_0)$ is also differentiable with respect to the second argument, and one has

$$\frac{\partial}{\partial t_0}\Phi(t, t_0) = -\Phi(t, t_0)F(t_0). \tag{2.1.7}$$

EXERCISE 2.1. Prove (2.1.7) by considering the differential equation

$$-\frac{dp}{dt} = F^*(t)p(t)$$

and using the fact that $p^*(t)x(t)$ is constant.

A useful result is Liouville's theorem.

THEOREM 2.1. *det* $\Phi(t, t_0) = \exp \int_{t_0}^{t} tr\, F(s)ds.$

EXERCISE 2.2. Prove Liouville's theorem when $n = 2$.

EXERCISE 2.3. Prove (2.1.7) using the invertibility of $\Phi(t, t_0)$.

We shall also use the following important property of matrices. Let $A \in \mathcal{L}(R^n; R^m)$, $m \geq n$. We have

$$\text{Range } A = (\text{Ker } A^*)^{\perp}, \tag{2.1.8}$$

which means

$$\{y|\ y = Ax,\ x \in R^n\} \equiv \{y|\ y^*z = 0, \forall z \text{ such that } A^*z = 0\}.$$

We give a proof for the convenience of the reader.

PROOF. If M is a subspace of R^m, then $M = M^{\perp\perp}$, so it is equivalent to prove that

$$\text{Ker } A^* = (\text{Range } A)^{\perp}.$$

If $z \in R^m$ satisfies $A^*z = 0$, then $z^*y = 0$ for $y = Ax$. Conversely, if z satisfies $z^*Ax = 0,\ \forall x$, then $A^*z = 0$. This proves the result. □

2.1.3 External Representation

We can use the fundamental matrix to give explicitly the solution of (2.1.1). If $x(t_0) = \xi$, then for $t > t_0$ one can write

$$x(t) = \Phi(t, t_0)\xi + \int_{t_0}^{t} \Phi(t, s)G(s)v(s)ds. \tag{2.1.9}$$

The presence of ξ makes $x(t)$ an affine functional of $v(.)$ (standard notation for functions). To get a linear system, we have to take $\xi = 0$. We can then write

$$z(t) = \int_{t_0}^{t} H(t)\Phi(t, s)G(s)v(s)ds, \tag{2.1.10}$$

which is the external representation of Chapter 1 with

$$h(t, \tau) = H(t)\Phi(t, \tau)G(\tau). \tag{2.1.11}$$

2.1.4 Stationary Case

If $F(t) = F$, things simplify. It is easy to check that

$$\Phi(t, t_0) = \exp F(t - t_0) \tag{2.1.12}$$
$$= \sum_{k=0}^{+\infty} F^k \frac{(t - t_0)^k}{k!}.$$

Therefore, if $G(t) = G,\; H(t) = H$, we get

$$h(t, \tau) = h(t - \tau) = H \exp F(t - \tau)G, \tag{2.1.13}$$

and thus

$$h(t) = H \;\exp Ft\; G. \tag{2.1.14}$$

We introduced in the introduction the transfer function $H(s)$, with a complex argument s, $H(s) = \int_0^{+\infty} h(t)\exp(-st)\,dt$, not to be confused, of course, with the function of time $H(t)$ defined in (2.1.10). The transfer function is given by

$$H(s) = H(sI - F)^{-1}G, \tag{2.1.15}$$

provided the inverse is defined and $\exp(F - sI)t \to 0$ as $t \to +\infty$.

2.2 Controllability

Consider the linear dynamical system

$$\frac{dx}{dt} = F(t)x(t) + G(t)v(t) \tag{2.2.1}$$
$$x(\tau) = \xi$$

with matrices $F(t) \in \mathcal{L}(R^n; R^n)$, $G(t) \in \mathcal{L}(R^m; R^n)$.

We say that a system is *completely controllable* at time τ if $\forall\; \eta \in R^n$, we can find some $t > \tau$ and a control $v(s)\; \tau < s < t$ such that $x(t) = \eta$. This condition means that we can find a control $v(s),\; \tau < s < t$ such that

$$\eta - \Phi(t, \tau)\xi = \int_\tau^t \Phi(t, s)G(s)v(s)ds. \tag{2.2.2}$$

Let us introduce the controllability matrix

$$\mathcal{C}(t, \tau) = \int_\tau^t \Phi(t, s)G(s)G^*(s)\Phi^*(t, s)ds. \tag{2.2.3}$$

It is clear that if for some $t > \tau$, $\eta - \Phi(t,\tau)\xi$ is in the range of $C(t,\tau)$, then the system is completely controllable at time τ. Indeed, there will exist $t > \tau$ and $\zeta \in R^n$ such that

$$\eta - \Phi(t,\tau)\xi = C(t,\tau)\zeta. \tag{2.2.4}$$

Then

$$v(s) = G^*(s)\Phi^*(t,s)\zeta \tag{2.2.5}$$

satisfies the property (2.2.2). What is interesting is that the converse is true. So we can state the following result.

PROPOSITION 2.1. *The system (2.2.1) is completely controllable at time τ if and only if for all $\eta \in R^n$, there exists $t > \tau$ such that $\eta - \Phi(t,\tau)\xi$ is in the range of $C(t,\tau)$.*

PROOF. It remains to prove that if there exists a control $v(s)$, $\tau < s < t$, such that (2.2.2) holds, then $\eta - \Phi(t,\tau)\xi$ is in the range of $C(t,\tau)$. Suppose that is not the case. Then from (2.1.8) it follows that $\eta - \Phi(t,\tau)\xi$ is not in $(\text{Ker } C(t,\tau))^{\perp}$. Therefore, there exists $\zeta \in \text{Ker } C(t,\tau)$ such that

$$\zeta^*(\eta - \Phi(t,\tau)\xi) \neq 0. \tag{2.2.6}$$

Since $\zeta \in \text{Ker } C(t,\tau)$, we have

$$\left(\int_\tau^t \Phi(t,s)G(s)G^*(s)\Phi^*(t,s)ds\right)\zeta = 0,$$

but then, multiplying on the left by ζ^*, we obtain easily

$$\zeta^*\Phi(t,s)G(s) = 0 \forall s \in (\tau,t)$$

and also

$$\int_\tau^t \zeta^*\Phi(t,s)G(s)v(s)ds = 0,$$

which contradicts (2.2.6), taking account of (2.2.2). This completes the proof. □

REMARK 2.1. If a system is completely controllable at τ, this does not mean that $C(t,\tau)$ is invertible for some t. Of course, if this is the case, the system is completely controllable at τ.

Let us establish a property of the control defined by formulas (2.2.4), (2.2.5).

PROPOSITION 2.2. *Assume that $C(t, \tau)$ is positive definite. Then the control defined by formulas (2.2.4), (2.2.5) achieves the transfer from (ξ, τ) to (η, t) with minimum energy.*

PROOF. Consider controls $v(.)$ that achieve the transfer from (ξ, τ) to (η, t). They satisfy

$$\eta - \Phi(t, \tau)\xi = \int_\tau^t \Phi(t, s)G(s)v(s)ds,$$

and let

$$u(s) = G^*(s)\Phi^*(t, s)C(t, \tau)^{-1}(\eta - \Phi(t, \tau)\xi).$$

If we set $\tilde{v}(s) = v(s) - u(s)$, we get

$$\int_\tau^t \Phi(t, s)G(s)\tilde{v}(s)ds = 0$$

and also

$$\int_\tau^t u(s)^*\tilde{v}(s)ds = (\eta - \Phi(t, \tau)\xi)^*C(t, \tau)^{-1}\int_\tau^t \Phi(t, s)G(s)\tilde{v}(s)ds = 0.$$

Hence

$$\int_\tau^t |v(s)|^2ds = \int_\tau^t |u(s)|^2ds + \int_\tau^t |\tilde{v}(s)|^2ds \geq \int_\tau^t |u(s)|^2ds.$$

This proves that $u(.)$ achieves the transfer with minimum energy. □

Consider now the stationary case, for which we have the following theorem.

THEOREM 2.2. *A stationary linear dynamical system is completely controllable (i.e., completely controllable at every time τ) if and only if for the matrix $[G, FG, \cdots, F^{n-1}G] \in \mathcal{L}(R^{mn}; R^n)$, one has*

$$\text{Rank}\,[G, FG, \cdots, F^{n-1}G] = n. \tag{2.2.7}$$

PROOF. Consider the matrix $W \in \mathcal{L}(R^n; R^n)$ defined by

$$W = [G, FG, \cdots, F^{n-1}G][G, FG, \cdots, F^{n-1}G]^*.$$

Let us show that $\forall t, \tau$ with $t > \tau$, the matrices $C(t, \tau)$ and W have the same range. It is sufficient to show that they have the same kernel, since they are both symmetric matrices. Now if $x \neq 0$ lies in the kernel of $C(t, \tau)$, we have

$$x^*[\int_\tau^t \exp F(t-s)\,GG^* \exp F^*(t-s)\,ds]x = 0.$$

Therefore,

$$G^* \exp(F^*\lambda)\, x = 0, \forall\lambda \in (0, t-\tau).$$

This implies, on taking $\lambda = 0$ in the function and its derivatives in λ,

$$G^*x = G^*F^*x = \cdots = G^*F^{*(n-1)}x = 0,$$

and hence

$$x^*Wx = 0,$$

which implies that x lies in the kernel of W. Conversely, assume that x lies in the kernel of W. We use the Cayley–Hamilton theorem, which claims that for $k \geq n$, F^k is a linear combination of $I, F, \cdots, F^{n-1}$. Therefore, also

$$\exp(Fs) = \sum_{i=0}^{n-1} \alpha_i(s) F^i,$$

where the $\alpha_i(s)$ are finite.

From

$$x^*G = x^*FG = \cdots, x^*F^{n-1}G = 0,$$

we deduce

$$x^* \exp Fs\, G = 0, \;\forall s > 0$$

which implies that $x \in \mathrm{Kerl}\, C(t, \tau)\; \forall t > \tau$.

Now, the system is completely controllable at time τ if $\forall\eta \in R^n$, $\eta - \exp F(t-\tau)\,\xi$ belongs to the range of $C(t, \tau)$, and hence to the range of W. It follows that the range of W is n. Since it is a symmetric square matrix, its kernel is 0 and the matrix W is invertible. Therefore,

$$x^*[G, FG, \cdots, F^{n-1}G] = 0$$

implies $x = 0$. This leads to the rank condition. □

EXERCISE 2.4. Find $\alpha_i(s)$ in the expression of $\exp(Fs)$.

DEFINITION 2.1. It is convenient to say that the pair (F, G) is controllable when the rank condition is satisfied or when the stationary linear dynamical system

$$\frac{dx}{dt} = Fx(t) + Gv(t)$$

is completely controllable.

2.3 Stability

2.3.1 *Definition*

Consider the dynamical system

$$\frac{dx}{dt} = F(t)x(t) \tag{2.3.1}$$

with $||F(t)||$ bounded.

We recall that $||F|| = \sup_{x \neq 0} \frac{|Fx|}{|x|}$. Using the fundamental matrix, we can write

$$x(t) = \Phi(t, t_0)\xi.$$

We can state the following definition.

DEFINITION 2.2. The system (2.3.1) is asymptotically stable if

$$\forall t_0,\ ||\Phi(t, t_0)|| \to 0, \text{ as } t \to +\infty.$$

It is uniformly asymptotically stable if $\forall t_0, \forall \epsilon,\ \exists T_\varepsilon$ independent of t_0 such that

$$||\Phi(t, t_0)|| < \epsilon, \forall t > t_0 + T_\epsilon.$$

We have the following important result.

THEOREM 2.3. *The system (2.3.1) is uniformly asymptotically stable if and only if it is exponentially stable, i.e., if* $\exists$ *two constants* c, K *such that*

$$||\Phi(t, \tau)|| \leq K \exp(-c(t - \tau)),\ \forall t > \tau. \tag{2.3.2}$$

PROOF. It is easy to check that if the system is exponentially stable, it is uniformly asymptotically stable. Conversely, suppose that the system is uniformly asymptotically stable. We can then find a number T such that

$$||\Phi(t + T, t)|| < \frac{1}{2},\ \forall t$$

We can then write, for $t - \tau > T$,

$$\Phi(t, \tau) = \Phi(t, \tau+[\frac{t-\tau}{T}]T)\Phi(\tau+[\frac{t-\tau}{T}]T, \tau+([\frac{t-\tau}{T}]-1)T)\cdots\Phi(\tau+T, \tau), \tag{2.3.3}$$

in which

$$[\frac{t-\tau}{T}] = \text{integer part of } \frac{t-\tau}{T}.$$

We note that

$$[\frac{t-\tau}{T}]T < t-\tau \leq [\frac{t-\tau}{T}]T + T$$

Also, if $t-\tau \leq T$, we have

$$||\Phi(t,\tau)|| \leq \sup_{\tau \leq s \leq \tau+T} ||\Phi(s,\tau)|| \leq k, \tag{2.3.4}$$

using the fact that $F(t)$ is bounded. Therefore, from (2.3.3) and (2.3.4) we obtain

$$\begin{aligned} ||\Phi(t,\tau)|| &\leq k\,(\frac{1}{2})^{[\frac{t-\tau}{T}]} \leq k(\frac{1}{2})^{\frac{t-\tau}{T}-1} \\ &= 2k\exp(-\frac{t-\tau}{T}\log 2). \end{aligned}$$

This is valid if $t-\tau > T$, since $1 \leq 2\exp(-\frac{t-\tau}{T}\log 2)$ if $t-\tau \leq T$, and using (2.3.4) again, we have in general

$$||\Phi(t,\tau)|| \leq 2k\exp(-\frac{t-\tau}{T}\log 2),$$

On setting $c = \dfrac{\log 2}{T}$ and $K = 2k$, the result (2.3.2) follows. □

THEOREM 2.4. *The system (2.3.1) is uniformly asymptotically stable if and only if there exist two symmetric matrices* $P(t)$, $Q(t)$ *such that*

$$\alpha I \leq P(t) \leq \beta I \quad \gamma I \leq Q(t) \leq \delta I$$

and

$$\frac{dP}{dt} + P(t)F(t) + F^*(t)P(t) + Q(t) = 0. \tag{2.3.5}$$

PROOF. We can replace uniformly asymptotically stable by exponentially stable, using Theorem 2.3. We first check that the condition is sufficient. Indeed,

$$\begin{aligned} \frac{d}{dt}x^*(t)P(t)x(t) &= -x^*(t)Q(t)x(t) \leq -\gamma||x(t)||^2 \\ &\leq -\frac{\gamma}{\beta}x^*(t)P(t)x(t). \end{aligned}$$

Therefore, for $t > \tau$ we have

$$x^*(t)P(t)x(t) \leq x^*(\tau)P(\tau)x(\tau)\exp-\frac{\gamma}{\beta}(t-\tau),$$

whence also

$$\alpha||x(t)||^2 \leq \beta||x(\tau)||^2 \exp(-\frac{\gamma}{\beta}(t-\tau)),$$

which implies

$$||\Phi(t,\tau)|| \leq \sqrt{\frac{\beta}{\alpha}} \exp -\frac{\gamma}{2\beta}(t-\tau),$$

which proves exponential stability. We prove next that the condition is necessary. We can take an arbitrary symmetric matrix $Q(t)$ satisfying $\gamma I \leq Q(t) \leq \delta I$. We define the matrix $P(t)$ by the formula

$$P(t) = \int_t^{+\infty} \Phi^*(\tau,t)Q(\tau)\Phi(\tau,t)d\tau.$$

By the exponential stability property this integral is well defined. Moreover, by differentiating, we get the relation (2.3.5). Next, we have

$$\begin{aligned} x^*P(t)x &\leq \delta \int_t^{+\infty} |\Phi(\tau,t)x|^2 d\tau \\ &\leq \delta K^2 \int_t^{+\infty} \exp(-2c(\tau-t))d\tau \; ||x||^2 \\ &= \frac{\delta K^2}{2c}||x||^2, \end{aligned}$$

and this proves that

$$||P(t)|| \leq \beta = \frac{\delta K^2}{2c}.$$

On the other hand,

$$x^*P(t)x \geq \gamma \int_t^{t+T} |\Phi(\tau,t)x|^2 d\tau.$$

Now consider $\Phi(t,\tau)$, which is the inverse of $\Phi(t,\tau)$. We have

$$\frac{d}{d\tau}\Phi(t,\tau) = -\Phi(t,\tau)F(\tau),$$

and therefore,

$$||\Phi(t,\tau)|| \leq c_T, \; t \leq \tau \leq t+T.$$

Since

$$x = \Phi(t,\tau)\Phi(\tau,t)x,$$

we get

$$|x| \leq c_T |\Phi(\tau, t)x|, \ \forall t \leq \tau \leq t + T.$$

Collecting the above results, we obtain

$$x^* P(t)x \geq \frac{\gamma T}{c_T^2} |x|^2,$$

which implies

$$P(t) \geq \alpha I, \ \alpha = \frac{\gamma T}{c_T^2},$$

and this concludes the proof. □

We have been considering above the stability of a given dynamical system. One of the great interests in controlling dynamical systems lies in the possibility of stabilizing them. We will develop this idea in the stationary case. In particular, we shall see that if the pair (F, G) is controllable, then there exists a matrix $K \in \mathcal{L}(R^n; R^m)$ such that $F - GK$ is exponentially stable, i.e., the system

$$\frac{dx}{dt} = (F - GK)x(t)$$

is exponentially stable. This evolution is the result of controlling the linear dynamical system

$$\frac{dx}{dt} = Fx(t) + Gv(t) \tag{2.3.6}$$

with a feedback law $v(t) = -Kx(t)$. The concept of feedback is essential in systems dynamics and control theory. There are several approaches to prove the stabilizability property through a feedback. We shall present a constructive one (we construct explicitly the feedback) that also uses basic ideas of optimal control theory, to be developed in following chapters

2.3.2 *Stabilizability*

For stability, the assumption of controllability can be replaced by a more general one. Consider the linear system (2.3.6) with an initial condition

$$x(0) = h, \tag{2.3.7}$$

and define the set of controls

$$\mathcal{V}(h) = \{v(.)|v(.) \in L^2(0, +\infty; R^m),\ x(.) \in L^2(0, +\infty; R^n)\}. \tag{2.3.8}$$

We introduce the following definition.

DEFINITION 2.3. The system (2.3.6) is stabilizable if $\forall h \in R^n$, the set $\mathcal{V}(h)$ is not empty.

If a system is controllable, then it is stabilizable. Indeed, from the definition of controllability, for all h there exists a control that brings the state to 0, in a finite time T. We then extend the control by 0 beyond T. The state remains 0, and thus $\mathcal{V}(h)$ is not empty. We define next

$$J_h(v(.)) = \int_0^{+\infty} |x(t)|^2 dt + \int_0^{+\infty} |v(t)|^2 dt,$$

and by the stabilizability property, we have $J_h(v(.)) < +\infty$ for some control $v(.)$. Therefore,

$$\chi(h) = \inf_{v(.)} J_h(v(.)) < +\infty, \tag{2.3.9}$$

and we can state the following theorem.

THEOREM 2.5. *Assume that the system (2.3.6) is stabilizable. Then we can write*

$$\chi(h) = h^* P h, \tag{2.3.10}$$

where P is the unique symmetric positive definite matrix solution of

$$PF + F^* P - PGG^* P + I = 0. \tag{2.3.11}$$

Moreover, the matrix $F - GK$, with $K = -G^* P$, is exponentially stable.

PROOF. We note that

$$\chi(h) = \inf_{v(.) \in \mathcal{V}(h)} J_h(v(.)),$$

and for all $v(.) \in \mathcal{V}(h)$, we have $x(t) \to 0$ as $t \to \infty$. Indeed, we have the formula

$$|x(T)|^2 = 2 \int_0^T x(t). \frac{dx}{dt}(t) dt + |h|^2,$$

and since the functions $x(.)$ and $\frac{dx}{dt}(.)$ are in $L^2(0, +\infty; R^n)$, we see that $|x(T)|$ has a limit as $T \to +\infty$. This limit is necessarily 0, since otherwise, $x(.)$ could not be in $L^2(0, +\infty; R^n)$. The functional $J_h(v(.))$ is strictly convex and goes to $+\infty$ as the norm of $v(.)$ in $L^2(0, +\infty; R^m)$ goes to $+\infty$. Therefore, there exists a unique optimal control that minimizes $J_h(v(.))$. Let us call it $u(.)$, and the corresponding state is called $y(.)$. Then $u(.)$ depends linearly on h. This is because $u(.)$ is the unique solution of the necessary and sufficient condition of optimality (Euler condition)

$$\int_0^{+\infty} (y^*(t)z(t) + u^*(t)v(t))dt = 0$$

for all $v(.)$ in $\mathcal{V}(0)$, and $z(.)$ is the corresponding state. From this necessary and sufficient condition, it is immediate that $u(.)$ depends linearly on h. But then $\chi(h)$ is quadratic, by the definition of $J_h(u(.))$. It can be written as (2.3.10) for some a symmetric matrix P. Clearly $P \geq 0$,and if $Ph = 0$, then $\chi(h) = 0$. Thus $u(.) = 0$, $y(.) = 0$, and necessarily $h = 0$. It follows that the rank of P is n, and P is invertible.

Let v_0 be arbitrary in R^m. Consider the function $y_0(t)$ defined by

$$\frac{dy_0}{dt} = Fy_0(t) + Gv_0,$$
$$y_0(0) = h.$$

Let $\delta > 0$ and consider $v(.) \in \mathcal{V}(y_0(\delta))$. Define the control

$$v_\delta(t) = \left| \begin{array}{ll} v_0, & 0 < t < \delta, \\ v(t-\delta), & t > \delta, \end{array} \right.$$

and consider the state associated with $v_\delta(.)$, called $y_\delta(.)$. If we define $\tilde{y}_\delta(.)$ by

$$\frac{d\tilde{y}_\delta}{dt} = F\tilde{y}_\delta(t) + Gv(t),$$
$$\tilde{y}_\delta(0) = y_0(\delta),$$

then one can check easily that

$$y_\delta(t) = \left| \begin{array}{ll} y_0(t), & 0 < t < \delta, \\ \tilde{y}_\delta(t-\delta), & t > \delta. \end{array} \right.$$

Then

$$J_h(v_\delta(.)) = \delta|v_0|^2 + \int_0^\delta |y_0(t)|^2 dt + J_{y_0(\delta)}(v(.)),$$

and therefore, $v_\delta(.) \in \mathcal{V}(h)$. It follows that

$$\chi(h) \leq \delta|v_0|^2 + \int_0^\delta |y_0(t)|^2 dt + J_{y_0(\delta)}(v(.)),$$

and since $v(.)$ is arbitrary in $\mathcal{V}(y_0(\delta))$, we can assert that

$$\chi(h) \leq \delta|v_0|^2 + \int_0^\delta |y_0(t)|^2 dt + \chi(y_0(\delta)),$$

or

$$h^* P h \leq \delta|v_0|^2 + \int_0^\delta |y_0(t)|^2 dt + (y_0(\delta))^* P y_0.(\delta)$$

Now we have

$$y_0(\delta) = h + \int_0^\delta (F y_0(t) + G v_0) dt,$$

and hence

$$h^* P h \leq \delta|v_0|^2 + \int_0^\delta |y_0(t)|^2 dt + h^* P h + 2h^* P \int_0^\delta (F y_0(t) + G v_0) dt +$$
$$+ (\int_0^\delta (F y_0(t) + G v_0) dt)^* P (\int_0^\delta (F y_0(t) + G v_0) dt).$$

On canceling $h^* P h$, dividing by δ, and letting $\delta \to 0$, we obtain

$$|v_0|^2 + |h|^2 + 2h^* P(Fh + G v_0) \geq 0.$$

Recalling that v_0 is arbitrary, we can use

$$v_0 = -G^* P h$$

and obtain

$$h^*(PF + F^* P - PGG^* P + I) h \geq 0. \tag{2.3.12}$$

Next, recalling that $u(.)$, $y(.)$ denote the optimal control and state, we consider $u_\delta(t) = u(t+\delta)$, for which the corresponding state is simply $y_\delta(t) = y(t+\delta)$. We have

$$\begin{aligned}\chi(h) &= J_h(u(.)) \\ &= \int_0^\delta |u(t)|^2 dt + \int_0^\delta |y(t)|^2 dt + J_{y(\delta)}(u_\delta(t)),\end{aligned}$$

and therefore

$$\chi(h) \geq \int_0^\delta |u(t)|^2 dt + \int_0^\delta |y(t)|^2 dt + \chi(y(\delta)).$$

Thus

$$h^*Ph \geq \int_0^\delta [|u(t)|^2 + |y(t)|^2 + 2h^*P(Fy(t) + Gu(t))]dt + h^*Ph$$
$$+ (\int_0^\delta (Fy(t) + Gu(t))dt)^* P(\int_0^\delta (Fy(t) + Gu(t))dt);$$

hence

$$0 \geq \int_0^\delta [|u(t)|^2 + |y(t)|^2 + 2h^*P(Fy(t) + Gu(t))]dt.$$

Now one checks easily that

$$|u(t)|^2 + 2h^*PGu(t) \geq -h^*PGG^*Ph,$$

and therefore, we can state that

$$0 \geq -h^*PGG^*Ph\,\delta + \int_0^\delta (|y(t)|^2 + 2h^*PFy(t))dt.$$

Dividing by δ and letting $\delta \to 0$, we obtain easily

$$h^*(PF + F^*P - PGG^*P + I)h \leq 0,$$

and comparing with (2.3.12), we obtain (2.3.11), since h is arbitrary. Now let Q be a matrix that is symmetric positive semidefinite and let T be fixed. Consider $v(.) \in L^2(0, T; R^m)$ and the linear dynamical system (2.3.6). We have

$$\frac{d}{dt}\, x(t)^*Qx(t) + |x(t)|^2 + |v(t)|^2 = 2v(t)^*G^*Qx(t)$$
$$+|v(t)|^2 + |x(t)|^2 + 2x(t)^*FQx(t).$$

If Q is a solution of (2.3.11), we have

$$\frac{d}{dt}\, x(t)^*Qx(t) + |x(t)|^2 + |v(t)|^2 = |v(t) + G^*Qx(t)|^2. \tag{2.3.13}$$

Integrating between 0 and T and then letting $T \to +\infty$, recalling that $x(T) \to 0$, we obtain

$$h^*Qh \leq J_h(v(.)).$$

Hence since $v(.)$ is arbitrary in $\mathcal{V}(h)$, we deduce $h^*Qh \leq \chi(h) = h^*Ph$.

However, we may consider the control $v(.)$ defined by the feedback

$$v(t) = -G^*Qx(t),$$

in which case $x(t)$ is the solution of the differential equation

$$\frac{dx}{dt} = (F - GG^* Q)x(t),$$
$$x(0) = h.$$

From (2.3.13) we can state that

$$\frac{d}{dt}x(t)^* Qx(t) + |x(t)|^2 + |v(t)|^2 = 0,$$

and integrating between 0 and T, it follows that

$$\int_0^T (|x(t)|^2 + |v(t)|^2)dt \le h^* Qh.$$

But then we can let $T \uparrow +\infty$ and thus also

$$\int_0^{+\infty} (|x(t)|^2 + |v(t)|^2)dt \le h^* Qh.$$

Hence $v(.) \in \mathcal{V}(h)$ and

$$J_h(v(.)) \le h^* Qh.$$

Since by definition, $J_h(v(.)) \ge h^* Ph$, we obtain $h^* Ph \le h^* Qh$. Hence $h^* Ph = h^* Qh$. Since h is arbitrary, we have $Q = P$, and uniqueness is established.

If we consider the linear system

$$\frac{dy}{dt} = (F - GG^* P)y(t),$$
$$y(0) = h,$$

and set $u(t) = -G^* Py(t)$, we obtain easily that

$$J_h(u(.)) = h^* Ph,$$

and thus $u(.)$ is optimal. Moreover, by direct computation and (2.3.11), we have

$$\frac{d}{dt}y^*(t)Py(t) \le -|y(t)|^2 \le -\frac{1}{\beta}y^*(t)Py(t)$$

with β such that $||P|| \le \beta$. It follows that

$$y^*(t)Py(t) \le h^* Ph \, \exp -\frac{t}{\beta},$$

which implies

$$\alpha |y(t)|^2 \leq \beta |h|^2 \exp -\frac{t}{\beta},$$

proving that $F - GG^*P$ is exponentially stable. This concludes the proof. □

2.4 Observability

2.4.1 Definition

We consider the linear system

$$\begin{aligned}\frac{dx}{dt} &= F(t)x(t) + G(t)v(t), \\ z(t) &= Hx(t).\end{aligned} \tag{2.4.1}$$

We do not have access to the state, since we do not know the initial condition. We have access to the input $v(.)$ and to the output $z(.)$. Can we recover the state? We introduce the following definition.

DEFINITION 2.4. The system (2.4.1) is observable at time t if there exists $\tau < t$ such that $x(t)$ is uniquely defined by the knowledge of $v(s), z(s), s \in (\tau, t)$.

Since for $s < t$, one has

$$x(t) = \Phi(t,s)x(s) + \int_s^t \Phi(t,\theta)G(\theta)v(\theta)d\theta,$$

on applying the matrix $\Phi(s,t)$, the inverse of $\Phi(t,s)$, we get

$$x(s) = \Phi(s,t)x(t) - \int_s^t \Phi(s,\theta)G(\theta)v(\theta)d\theta,$$

and thus

$$z(s) + H(s)\int_s^t \Phi(s,\tau)G(\tau)v(\theta)d\tau = H(s)\Phi(s,t)x(t). \tag{2.4.2}$$

The left-hand side is known for all $s \in (\tau, t)$, and the unknown is $x(t)$. We introduce the <u>observability matrix</u>

$$O(t,\tau) = \int_\tau^t \Phi^*(s,t)H^*(s)H(s)\Phi(s,t)ds. \tag{2.4.3}$$

We have the following result.

THEOREM 2.6. *The system (2.4.1) is observable at time t if and only if there exists $\tau < t$ such that $O(t,\tau)$ is positive definite.*

PROOF. Define

$$z_1(s) = z(s) + H(s)\int_s^t \Phi(s,\tau)G(\tau)v(\theta)d\tau.$$

We deduce from (2.4.2) that

$$O(t,\tau)x(t) = \int_\tau^t \Phi^*(s,t)H^*(s)z_1(s)ds, \tag{2.4.4}$$

and the right-hand side of (2.4.4) is known for all τ. If the system is observable at time t, this equation must have a unique solution for some τ. This implies $\text{Ker}\, O(t,\tau) = \{0\}$, and since $O(t,\tau)$ is symmetric, its range is R^n. Hence $O(t,\tau)$ is positive definite. The converse is clearly true, whence the result. □

REMARK 2.2. Since $\Phi^*(s,t)$ satisfies

$$\frac{d}{dt}\Phi^*(s,t) = -F^*(t)\Phi^*(s,t),$$

it can be interpreted as the fundamental matrix of $-F^*(t)$. So the observability matrix of the pair $F(t), H(t)$ is identical to the controllability matrix of the pair $-F^*(t), -H^*(t)$. ■

Consider now the stationary case. The observability of the pair (F,H) is equivalent to the controllability of the pair $(-F^*, -H^*)$. This means that the rank of the matrix $[-H^*, (-F^*)(-H^*), \cdots, (-F^*)^{n-1}(-H^*)]$ is n. This is equivalent to

$$x^*[-H^*, (-F^*)(-H^*), \cdots, (-F^*)^{n-1}(-H^*)] = 0 \Rightarrow x = 0,$$

which also means that

$$x^*H^* = 0, \cdots, x^*(F^*)^{n-1}H^* = 0 \Rightarrow x = 0.$$

Hence

$$\text{Rank}\,[H^*, F^*H^*, \cdots, (F^*)^{n-1}H^*] = n,$$

which is also equivalent to

$$\text{Rank}\begin{bmatrix} H \\ HF \\ \vdots \\ HF^{n-1} \end{bmatrix} = n.$$

Likewise, the observability of the pair (F,H) is equivalent to the controllability of the pair (F^*, H^*).

2.4.2 Observers

Consider the system

$$\frac{dx}{dt} = Fx(t) + Gv(t). \tag{2.4.5}$$
$$z(t) = Hx(t)$$

An observer is a linear dynamic system with inputs the pair $v(.), z(.)$ whose state evolution is known and can become arbitrarily close to $x(t)$. It can be constructed as follows:

$$\frac{d\hat{x}}{dt} = F\hat{x}(t) + Gv(t) + K(z(t) - H\hat{x}(t)), \tag{2.4.6}$$

with arbitrary initial condition. The matrix $K \in \mathcal{L}(R^p; R^n)$ is to be chosen and is called the gain of the observer. Let us set $\tilde{x}(t) = x(t) - \hat{x}(t)$. Then on combining (2.4.1) and (2.4.6), we obtain

$$\frac{d\tilde{x}}{dt} = (F - KH)\tilde{x}(t).$$

Let us assume that the pair (F, H) is observable. Then the pair (F^*, H^*) is controllable. It follows from Theorem 2.5 that there exists a symmetric positive definite matrix P such that

$$PF^* + FP - PH^*HP + I = 0 \tag{2.4.7}$$

and the matrix $F^* - H^*HP$ is exponentially stable. It follows that the transpose $F - PH^*H$ is exponentially stable. We can thus take $K = PH^*$ to construct the observer (2.4.6). We may recall that to obtain the matrix P, we need a weaker assumption than that (F^*, H^*) is controllable. We need (F^*, H^*) to be stabilizable. By analogy with the correspondence between controllability and observability, we shall use the following definition.

DEFINITION 2.5. The pair F, H is detectable if the pair (F^*, H^*) is stabilizable. So detectability is sufficient to obtain an observer.

Chapter 3
Optimal Control of Linear Dynamical Systems

We have seen in the previous chapter that a control can be used to bring the state of a system to a given value or to stabilize the system. In dealing with stability we have also used a method that consists in solving an optimal control problem. Optimal control represents an essential branch of control theory. We shall present the general theory later on. In the case of linear systems, the results can be obtained by ad hoc techniques that are useful to know.

3.1 Finite Horizon Problem

The general formulation of the finite horizon problem is the following. The evolution of the state $x(t)$ is described by

$$\frac{dx}{dt} = F(t)x(t) + G(t)v(t) + f(t), \tag{3.1.1}$$
$$x(0) = x_0.$$

It is a slight extension of (2.1.1). We want to minimize the payoff

$$J(v(.)) = \int_0^T (x(t)^*M(t)x(t) + v(t)^*N(t)v(t) + 2m(t)^*x(t) + 2n(t)^*v(t))dt + \\ + \; x(T)^*M_Tx(T) + 2m_T^*x(T). \tag{3.1.2}$$

The control $v(.)$ belongs to $L^2(0, T; R^m)$. We assume that $M(.)$, M_T are symmetric positive semidefinite $n \times n$ matrices, and $N(.)$ is an $n \times n$ symmetric positive definite matrix, with bounded inverse $N^{-1}(.)$. All functions of time entering in the definition of the state evolution and of the payoff are bounded.

A. Bensoussan, *Estimation and Control of Dynamical Systems*, Interdisciplinary Applied Mathematics 48, https://doi.org/10.1007/978-3-319-75456-7_3

3.1.1 Solution of the Problem

We first describe the solution. It is obtained through a feedback rule. We have already encountered this concept in the study of stability. It is an essential concept of control theory. We first solve the backward Riccati equation

$$\frac{d\pi}{dt} + \pi(t)F(t) + F(t)^*\pi(t) - \pi(t)G(t)N^{-1}(t)G^*(t)\pi(t) + M(t) = 0,$$
$$\pi(T) = M_T. \tag{3.1.3}$$

We then solve the affine equation

$$\frac{dr}{dt} + (F(t)^* - \pi(t)G(t)N^{-1}(t)G^*(t))r(t) + \pi(t)(f(t)$$
$$-G(t)N^{-1}(t)n(t)) + m(t) = 0,$$
$$r(T) = m_T. \tag{3.1.4}$$

We can then define the feedback rule

$$\hat{v}(x,t) = -N^{-1}(t)[n(t) + G^*(t)(\pi(t)x + r(t))]. \tag{3.1.5}$$

Using this feedback rule in the state equation (3.1.1), we obtain the differential equation

$$\frac{dy}{dt} = (F(t) - G(t)N^{-1}(t)G^*(t)\pi(t))y(t) + f(t) - G(t)N^{-1}(t)(n(t) + G^*(t)r(t)),$$
$$y(0) = x_0. \tag{3.1.6}$$

Then the function $y(t)$ is the optimal state, and the optimal control is given by

$$u(t) = \hat{v}(y(t),t) \tag{3.1.7}$$

We state the following theorem.

THEOREM 3.1. *The optimal control $u(.)$ of the problem (3.1.1), (3.1.2) is uniquely defined by the feedback rule (3.1.7).*

3.1.2 Proof of Theorem

If we consider a control $v(.)$ and the corresponding state $x(.)$, then we can define $\mu(.)$ by the formula

$$v(t) = \Lambda(t)x(t) + \lambda(t) + \mu(t), \tag{3.1.8}$$

in which $\Lambda(t), \lambda(t)$ will be defined below. We can see this equality as a change of unknown function and $\mu(.)$ as a new control. The idea is to try to obtain an optimal value of the new control that is 0. Note that the trajectory

$$\frac{dy}{dt} = (F(t) + G(t)\Lambda(t))y(t) + f(t) + G(t)\lambda(t),$$
$$y(0) = x_0, \tag{3.1.9}$$

corresponds to a value of $\mu(t) = 0$. The corresponding control is

$$u(t) = \Lambda(t)y(t) + \lambda(t). \tag{3.1.10}$$

For a general $\mu(.)$, the state $x(.)$ becomes the solution of

$$\frac{dx}{dt} = (F(t) + G(t)\Lambda(t))x(t) + f(t) + G(t)\lambda(t) + G(t)\mu(t), \tag{3.1.11}$$
$$x(0) = x_0.$$

It follows that

$$x(t) = y(t) + \tilde{x}(t),$$
$$v(t) = u(t) + \Lambda(t)\tilde{x}(t) + \mu(t),$$

with $\tilde{x}(t)$ the solution of

$$\frac{d\tilde{x}}{dt} = (F(t) + G(t)\Lambda(t))\tilde{x}(t) + G(t)\mu(t),$$
$$\tilde{x}(0) = 0.$$

By an easy but tedious calculation we see next that

$$\begin{aligned} J(v(.)) &= J(u(.)) + \qquad (3.1.12)\\ &+ \int_0^T [\tilde{x}^* M\tilde{x} + (\Lambda\tilde{x} + \mu)^* N(\Lambda\tilde{x} + \mu)]dt + \tilde{x}(T)^* M_T \tilde{x}(T)\\ &+ 2\int_0^T [\tilde{x}^*(My + m) + (\Lambda\tilde{x} + \mu)^*(Nu + n)]dt\\ &+ 2\tilde{x}(T)^*(M_T y(T) + m_T). \end{aligned}$$

If we can choose the functions $\Lambda(.), \lambda(.)$ so that the mixed term is identically 0 for all $v(.)$, then it will clearly follow that the control $u(.)$ defined by formulas (3.1.9), (3.1.10) is optimal. We show that this is possible. We set

$$X = \int_0^T [\tilde{x}^*(My + m) + (\Lambda\tilde{x} + \mu)^*(Nu + n)]dt + \tilde{x}(T)^*(M_T y(T) + m_T),$$

and we introduce two new functions $\pi(t), r(t)$, where $\pi(t)$ is a symmetric matrix, with only the conditions

$$\pi(T) = M_T,\ r(T) = m_T.$$

We assume differentiability of the functions $\pi(t), r(t)$. We set next

$$p(t) = \pi(t)y(t) + r(t).$$

We then compute

$$\begin{aligned}\frac{d}{dt}\tilde{x}(t)^* p(t) &= [\tilde{x}^*(F^* + \Lambda^* G^*) + \mu^* G^*]p + \\ &+ \tilde{x}^*[\frac{d\pi}{dt}y + \pi((F + G\Lambda)y + f + G\lambda) + \frac{dr}{dt}].\end{aligned}$$

Note that

$$\begin{aligned}\tilde{x}(T)^*(M_T y(T) + m_T) &= \tilde{x}(T)^* p(T) \\ &= \int_0^T \frac{d}{dt}\tilde{x}(t)^* p(t)\, dt,\end{aligned}$$

and collecting results, we can write

$$\begin{aligned}X = \int_0^T &[\tilde{x}^*\{M y + m + \Lambda^*(Nu + n) + (F^* + \Lambda^* G^*)p + \\ &+ \frac{d\pi}{dt}y + \pi((F + G\Lambda)y + f + G\lambda) + \frac{dr}{dt}\} + \mu^*(Nu + n + G^* p)]dt.\end{aligned}$$

In order to have $X = 0$ for all $v(.)$, and hence all $\mu(.)$, we must have

$$\begin{aligned}M y + m + \Lambda^*(Nu + n) + (F^* + \Lambda^* G^*)p + \frac{d\pi}{dt}y + \pi((F + G\Lambda)y + f + G\lambda) \\ + \frac{dr}{dt} = 0, \\ Nu + n + G^* p = 0.\end{aligned} \tag{3.1.13}$$

Combining these, we get

$$\begin{aligned}M y + m + F^*(\pi y + r) + \frac{d\pi}{dt}y + \pi((F + G\Lambda)y + f + G\lambda) + \frac{dr}{dt} = 0, \\ N(\Lambda y + \lambda) + n + G^*(\pi y + r) = 0,\end{aligned}$$

and this can be achieved by expressing that in these two relations, the coefficients of y are 0, and the terms independent of y are also 0. We get four conditions, for which we have four unknowns Λ, π, λ, r. We get

$$M + F^*\pi + \frac{d\pi}{dt} + \pi(F + G\Lambda) = 0,$$

$$m + F^*r + \pi(f + G\lambda) + \frac{dr}{dt} = 0,$$

$$N\Lambda + G^*\pi = 0,$$

$$N\lambda + n + G^*r = 0.$$

Necessarily,

$$\Lambda(t) = -N^{-1}(t)G^*(t)\pi(t),$$
$$\lambda(t) = -N^{-1}(t)(n(t) + G^*(t)r(t)),$$

and $\pi(.), r(.)$ must satisfy (3.1.3), (3.1.4). So the matching can be completed, and with this choice of $\Lambda(.), \lambda(.)$ we have $X = 0$. Therefore,

$$J(v(.)) = J(u(.)) + \int_0^T [\tilde{x}^* M\,\tilde{x} + (\Lambda\tilde{x} + \mu)^* N(\Lambda\tilde{x} + \mu)]dt + \tilde{x}(T)^* M_T \tilde{x}(T),$$

and $u(.)$ is optimal, corresponding to an optimal choice of $\mu(.) = 0$. The proof will have been completed if we can guarantee that the solution of the Riccati equation (3.1.3) exists. We will discuss this issue in the following lemma. ■

LEMMA 3.1. *The Riccati equation (3.1.3) has a unique solution that is symmetric positive semidefinite.*

PROOF. We consider a simpler control problem without the first-order terms. So the state evolution is defined by

$$\frac{dx}{dt} = F(t)x(t) + G(t)v(t), \tag{3.1.14}$$
$$x(0) = x_0,$$

and the payoff is defined by

$$J(v(.)) = \int_0^T (x(t)^* M(t)x(t) + v(t)^* N(t)v(t))dt +$$
$$+ x(T)^* M_T x(T). \tag{3.1.15}$$

Suppose the Riccati equation has a solution, which is necessarily symmetric, but we do not assume that it is positive semidefinite. Nevertheless, the proof of Theorem 3.1 shows that the control problem (3.1.14), (3.1.15) has a solution defined by the feedback

$$u(t) = -N^{-1}(t)G^*(t)\pi(t)y(t),$$

and the optimal state $y(t)$ satisfies

$$\begin{aligned}\frac{dy}{dt} &= (F(t) - G(t)N^{-1}(t)G^*(t)\pi(t))y(t), \\ y(0) &= x_0.\end{aligned} \tag{3.1.16}$$

But then

$$J(u(.)) = \int_0^T y(t)^*(M(t) + \pi(t)G(t)N^{-1}(t)G^*(t)\pi(t))y(t)dt + y(T)^* M_T y(T).$$

We can compute the derivative

$$\frac{d}{dt}y(t)^*\pi(t)y(t) = -y(t)^*(M(t) + \pi(t)G(t)N^{-1}(t)G^*(t)\pi(t))y(t),$$

and therefore,

$$J(u(.)) = x_0^*\pi(0)x_0 \geq 0.$$

Now $J(u(.)) \leq J(0)$ and

$$J(0) = x_0^*[\int_0^T \Phi^*(t,0)M(t)\Phi(t,0)dt + \Phi^*(T,0)M_T\Phi(T,0)]x_0,$$

where $\Phi(t,s)$ is the fundamental matrix of $F(t)$. Since x_0 is arbitrary, we have proven

$$0 \leq \pi(0) \leq \int_0^T \Phi^*(t,0)M(t)\Phi(t,0)dt + \Phi^*(T,0)M_T\Phi(T,0).$$

In fact, considering a control problem on (s,T) instead of $(0,T)$, we can check that

$$0 \leq \pi(s) \leq \int_s^T \Phi^*(t,s)M(t)\Phi(t,s)dt + \Phi^*(T,s)M_T\Phi(T,s). \tag{3.1.17}$$

So we see that if the Riccati equation has a solution, it is bounded. Since the Riccati equation has a solution on an interval $(T-\epsilon, T]$, by the classical existence result of local solutions of differential equations, it does not blow up at $T-\epsilon$, and therefore it can be extended outside the interval. Clearly, it can be extended to 0, and thus it has a global solution. This completes the proof. □

We can then compute the optimal value $J(u(.))$ of the functional (3.1.2).

EXERCISE 3.1. Show first that

$$
\begin{aligned}
J(u(.)) = \int_0^T & [y(t)^*(M(t) + \pi(t)G(t)N^{-1}(t)G^*(t)\pi(t))y(t) + \\
& + 2y(t)^*(\pi(t)G(t)N^{-1}(t)G^*(t)r(t) + m(t)) \\
& + r^*(t)G(t)N^{-1}(t)G^*(t)r(t) - n^*(t)N^{-1}(t)n(t)]dt \\
& + y(T)^*M_T y(T) + 2m_T^* y(T);
\end{aligned}
$$

then define

$$\Phi(t) = y(t)^*\pi(t)y(t) + 2r(t)^*y(t) + \rho(t)$$

with

$$\rho(t) = \int_t^T (2r^*(s)f(s) - (G^*(s)r(s) + n(s))^*N^{-1}(s)(G^*(s)r(s) + n(s)))ds.$$

Show that

$$
\begin{aligned}
J(u(.)) &= \Phi(0) = \\
&= x_0^*\pi(0)x_0 + 2r(0)^*x_0 + \rho(0).
\end{aligned}
$$

3.2 Infinite Horizon Problem

We have already considered an infinite horizon control problem in the study of stability in Theorem 2.5. We have seen that the solution is linked to an algebraic Riccati equation (3.1.9). Here we shall consider a more general problem. We consider again the linear dynamical system

$$
\begin{aligned}
\frac{dx}{dt} &= Fx(t) + Gv(t), \\
x(0) &= h,
\end{aligned}
\tag{3.2.1}
$$

and the cost functional

$$J_h(v(.)) = \int_0^{+\infty} (|Hx(t)|^2 + v^*(t)Nv(t))dt, \tag{3.2.2}$$

in which N is symmetric positive definite. In fact, N could be taken to be the identity. The interesting new feature with respect to the situation of Theorem 2.5 is the presence of H. We introduce

$$\mathcal{V}(h) = \{v(.) \in L^2(0, +\infty; R^m)|\ Hx(.) \in L^2(0, +\infty; R^p)\},$$

along with the following definition.

DEFINITION 3.1. The pair (F, G) is H stabilizable if $\forall h$, $\mathcal{V}(h)$ is not empty.

This definition generalizes that of stabilizability, which corresponds to $H = I$. If (F, G) is stabilizable, it is H stabilizable for all H. We then define

$$\chi(h) = \inf_{v(.)\in\mathcal{V}(h)} J_h(v(.)).$$

We have the following result.

THEOREM 3.2. *We assume that the pair (F, G) is H stabilizable. Then the function $\chi(h)$ can be written as*

$$\chi(h) = h^* P h,$$

where P is a symmetric positive semidefinite matrix solution of

$$PF + F^*P - PGN^{-1}G^*P + H^*H = 0. \tag{3.2.3}$$

The matrix P is the minimum symmetric positive semidefinite lower solution of (3.2.3), i.e., $P \le Q$ for every symmetric positive semidefinite Q satisfying

$$QF + F^*Q - QGN^{-1}G^*Q + H^*H \le 0. \tag{3.2.4}$$

There exists a unique optimal control defined by the feedback

$$u(t) = -N^{-1}G^*Py(t), \tag{3.2.5}$$

in which $y(t)$ is the optimal state solution of

$$\frac{dy}{dt} = (F - GN^{-1}G^*P)y, \tag{3.2.6}$$
$$y(0) = h.$$

PROOF. Many steps are similar to the case $H = I$; see Theorem 2.5 of Chapter 2. We first have

$$\chi(h) = \inf_{v(.)} J_h(v(.)),$$

and the optimal control exists. We can proceed as in Theorem 2.5 to show that $\chi(h)$ is quadratic and can be written as

$$\chi(h) = h^* P h$$

with P symmetric and positive semidefinite. As in Theorem 2.5, we can check that P is a solution of (3.2.3). This time, the solution is not unique. However, let Q

be any symmetric lower solution of (3.2.3), i.e., one that satisfies (3.2.4). Then as for (3.1.12), we have

$$\frac{d}{dt}x(t)^*Qx(t)+|Hx(t)|^2+v^*(t)Nv(t)\le(v(t)+N^{-1}G^*Qx(t))^*N(v(t)$$
$$+N^{-1}G^*Qx(t)). \tag{3.2.7}$$

In particular, we take a control such that

$$v(t)=-N^{-1}G^*Qx(t),$$

which means that $x(t)$ is the solution of

$$\begin{aligned}\frac{dx}{dt}&=(F-GN^{-1}G^*Q)x,\\ x(0)&=h.\end{aligned} \tag{3.2.8}$$

From (3.2.7) and integrating over $(0, T)$, it follows that

$$\int_0^T(|Hx(t)|^2+v^*(t)Nv(t))dt\le h^*Qh.$$

Therefore, the control is in $\mathcal{V}(h)$, and letting $T\uparrow+\infty$, we get

$$\chi(h)\le J_h(v(.))\le h^*Qh.$$

Hence $P\le Q$. Therefore, P is the minimum lower solution. This concludes the proof. □

REMARK 3.1. The optimal state is not stable, and the matrix $F-GN^{-1}G^*P$ is not necessarily exponentially stable.

We can complete the study of the structure of the set of solutions of the algebraic Riccati equation (3.2.3) when

$$\text{the pair } F, G \text{ is stabilizable.} \tag{3.2.9}$$

We have the following result.

PROPOSITION 3.1. *We assume (3.2.9). Consider the set of symmetric positive semidefinite matrices Q that satisfy*

$$QF+F^*Q-QGN^{-1}G^*Q+H^*H\ge0. \tag{3.2.10}$$

Such matrices Q are called upper solutions of (3.2.3). This set has a maximum element $\bar{P}\ge Q$, $\forall Q$ satisfying (3.2.10). Moreover, $\bar{P}$ is a solution of the algebraic Riccati equation (3.2.3).

PROOF. We introduce the set of controls

$$\overline{\mathcal{V}}(h) = \{v(.) \in L^2(0,\infty; R^m) |\, x(.) \in L^2(0,\infty; R^n)\}.$$

Thanks to the assumption, this set is not empty for all h. We can then consider

$$\bar{\chi}(h) = \inf_{v(.)\in\overline{\mathcal{V}}(h)} J_h(v(.)). \tag{3.2.11}$$

This function is finite and positive. Moreover, $\bar{\chi}(h) \geq \chi(h), \forall h$. However, the infimum is not necessarily attained in (3.2.11). We have shown that it is attained in the larger set $\mathcal{V}_h$. So we cannot claim that

$$\bar{\chi}(h) = h^*\bar{P}h \tag{3.2.12}$$

as we did in Theorems 2.5 and 3.2. We proceed with an approximation. Let us introduce a penalized functional

$$J_h^\epsilon(v(.)) = J_h(v(.)) + \epsilon \int_0^{+\infty} |x(t)|^2 dt.$$

Clearly,

$$\inf_{v(.)\in\overline{\mathcal{V}}(h)} J_h^\epsilon(v(.)) = \inf_{v(.)} J_h^\epsilon(v(.)),$$

and thus for the problem $J_h^\epsilon(v(.))$ we are in the situation of Theorem 2.5. Consequently,

$$\inf_{v(.)} J_h^\epsilon(v(.)) = h^*P^\epsilon h,$$

with P^ϵ the unique solution of

$$P^\epsilon F + F^*P^\epsilon - P^\epsilon GN^{-1}G^*P^\epsilon + H^*H + \epsilon I = 0.$$

The sequence P^ϵ is monotonically decreasing (as ϵ decreases to 0) and bounded below by 0. So it converges to $\bar{P}$. Necessarily $\bar{P}$ is a solution of (3.2.3). Moreover,

$$\inf_{v(.)\in\overline{\mathcal{V}}(h)} J_h^\epsilon(v(.)) \downarrow \inf_{v(.)\in\overline{\mathcal{V}}(h)} J_h(v(.)), \text{ as } \epsilon \downarrow 0.$$

We obtain (3.3.1). Consider now a matrix Q satisfying (3.2.10). For all $v(.) \in L^2(0,T; R^m)$, we can write as for (3.2.7),

$$\frac{d}{dt}x(t)^*Qx(t)+|Hx(t)|^2+v^*(t)Nv(t) \geq (v(t)+N^{-1}G^*Qx(t))^*N(v(t)+N^{-1}G^*Qx(t)).$$

We take $v(.) \in \overline{\mathcal{V}}(h)$. We now integrate the preceding inequality between 0 and T, and let $T \uparrow +\infty$. Note that $x(T) \to 0$. We obtain

$$h^* Q h \leq J_h(v(.)) \leq J_h^\epsilon(v(.)).$$

Since $v(.)$ is arbitrary in $\overline{\mathcal{V}}(h)$, we deduce $h^* Q h \leq h^* P^\epsilon h$. Therefore, also $Q \leq \bar{P}$. This completes the proof. □

Let us also state the following result

PROPOSITION 3.2. *Assume that F, G are stabilizable. If there exists a matrix Q that is a solution of the algebraic Riccati equation (3.2.3) such that the solution of (3.2.8) satisfies $x(t) \to 0$ as $t \uparrow +\infty$, then $\bar{P} = Q$, and thus it is unique.*

PROOF. Consider the system (3.2.8) and set

$$v(t) = -N^{-1} G^* Q x(t).$$

We can check that

$$\frac{d}{dt} x(t)^* Q x(t) + |Hx(t)|^2 + v^*(t) N v(t) = 0.$$

Therefore, using $x(t) \to 0$ as $t \uparrow +\infty$, we get immediately

$$h^* Q h = J_h(v(.)).$$

On the other hand, $x(t) \to 0$ as $t \uparrow +\infty$ implies also exponential stability, whence we have $x(.) \in L^2(0, +\infty; R^n)$. Therefore, $v(.) \in \overline{\mathcal{V}}(h)$. Consequently,

$$h^* Q h \geq \bar{\chi}(h)$$

and $Q \geq \bar{P}$. However, $\bar{P}$ is the maximum upper solution of (3.2.3). Therefore, $Q = \bar{P}$, and such a matrix is necessarily unique. This concludes the proof. □

Let us give additional results for the algebraic Riccati equation (3.2.3). We first introduce the following definition.

DEFINITION 3.2. The pair (F, H) is observable if

$$\begin{aligned} \frac{dx}{dt} &= Fx(t) \\ x(0) &= h \end{aligned} \tag{3.2.13}$$

implies

$$Hx(t) = 0, \forall t \Rightarrow h = 0. \tag{3.2.14}$$

We can then state the following result.

PROPOSITION 3.3. *Assume that (F, G) is H stabilizable and that (F, H) is observable. Then the minimum solution P of the algebraic Riccati equation (3.2.3) is positive definite. If (F, G) is stabilizable and (F, H) is observable, then the algebraic Riccati equation has a unique solution.*

PROOF. Let us check that P is positive definite. Indeed, consider the optimal control of (3.2.16), (3.2.17), namely

$$u(t) = -N^{-1}G^*Py(t),$$

with $y(t)$ the solution of

$$\begin{aligned} \frac{dy}{dt} &= (F - GN^{-1}G^*P)y(t), \\ y(0) &= h. \end{aligned} \tag{3.2.15}$$

Moreover, we have the relation

$$h^*Ph = \int_0^{+\infty} (|Hy(t)|^2 + |u(t)|^2)dt.$$

If P is not positive definite, then there exists $h \neq 0$ such that $h^*Ph = 0$. Necessarily $u(t) = 0$ and $Hy(t) = 0$, since

$$\begin{aligned} \frac{dy}{dt} &= Fy(t), \\ y(0) &= h, \end{aligned}$$

Since F, H is observable, we must have $h = 0$, and this is a contradiction. We also have

$$h^*Ph = \int_0^T (|Hy(t)|^2 + |u(t)|^2)dt + y^*(T)Py(T).$$

Therefore, $y^*(T)Py(T) \downarrow 0$ as $T \uparrow +\infty$. Since P is positive definite, it follows that $y(T) \to 0$. Now if we assume that (F, G) is stabilizable, we can refer to Proposition 3.2 to assert that $P = \bar{P}$, and thus the solution is unique. This concludes the proof. □

EXERCISE 3.2. Study the problem

$$\begin{aligned} \frac{dx}{dt} &= Fx(t) + Gv(t), \\ x(0) &= h, \end{aligned} \tag{3.2.16}$$

and the cost functional

$$J_h(v(.)) = \int_0^{+\infty} \exp -\alpha t \; (|Hx(t)|^2 + v^*(t)Nv(t))dt. \tag{3.2.17}$$

Assume that $\alpha > 2||F||$. Show that for all $v(.)$ such that

$$\int_0^{+\infty} \exp -\alpha t \; |v(t)|^2 dt \; < +\infty,$$

one has

$$\int_0^{+\infty} \exp -\alpha t \; |x(t)|^2 dt \; < +\infty.$$

Solve the optimal control problem (3.2.16), (3.2.17). What Riccati equation is obtained?

3.3 Positivity

3.3.1 Positive Real Lemma

Positive systems (one says also passive systems) are systems such that positive inputs lead to positive outputs. This is not a precise definition, just an intuitive idea. Such systems arise naturally when one wants to guarantee a quality of service that should occur automatically, for instance a minimum temperature in a heating system. We shall define the concept mathematically below. We consider the following linear system:

$$\begin{aligned} \frac{dx}{dt} &= Fx(t) + Gv(t), \\ z(t) &= Hx(t) + Dv(t). \end{aligned} \tag{3.3.1}$$

We do not specify an initial condition. Note also that we have included a direct impact of the input on the output (not just indirect through the state). In this section, D can be 0, but in the next section it must be different from 0. We assume $p = m$, and hence the dimension of the vector of inputs is equal to the dimension of the vector of outputs.

The transfer function of this system is (recall that s is a complex argument)

$$H(s) = D + H(sI - F)^{-1}G.$$

There is a relation between the condition

$$H(i\omega) + H^*(-i\omega) \geq 0$$

and the property

$$\int_{-\infty}^{+\infty} z^*(t)v(t)dt \geq 0, \ \forall v(.).$$

This is related to Parseval's theorem, but we do not elaborate further, since we will consider this property as the definition of positivity. One can interpret the left-hand side of the inequality as an energy. We shall assume that

$$(F, G) \text{ is controllable,} \tag{3.3.2}$$

and we use the following definition.

DEFINITION 3.3. The system (3.3.1) is positive if for all $v(.) \in L^2(-\infty, 0; R^m)$ such that $x(.) \in L^2(-\infty, 0; R^n)$, one has

$$\int_{-\infty}^{0} z^*(t)v(t)dt \geq 0.$$

From the condition (3.3.2), there exist controls $v(.) \in L^2(-\infty, 0; R^m)$ such that $x(.) \in L^2(-\infty, 0; R^n)$. Note that the choice of $(-\infty, 0)$ cannot be replaced by $(0, +\infty)$, because the value at 0 is a target, and the origin at $-\infty$ is 0. The "positive real lemma," due to Kalman, Popov, and Yakubovitch, see [26, 46], is the following.

THEOREM 3.3. *We assume (3.3.2). The system (3.3.1) is positive if and only if there exist matrices P, Q, R, S, where P, Q, R are symmetric and*

$$P \geq 0, \qquad \begin{pmatrix} Q & S \\ S^* & R \end{pmatrix} \geq 0, \tag{3.3.3}$$

$$\begin{aligned} PF + F^*P &= -Q, \\ H^* - PG &= S, \\ D + D^* &= R. \end{aligned} \tag{3.3.4}$$

PROOF. Let us check that the condition is sufficient. We compute

$$\begin{aligned} \frac{d}{dt}x^*(t)Px(t) &= (v^*(t)G^* + x^*(t)F^*)Px(t) + x^*(t)P(Fx(t) + Gv(t)) \\ &= -x^*(t)Qx(t) + 2x^*(t)(H^* - S)v(t) \\ &= -x^*(t)Qx(t) - 2x^*(t)Sv(t) + 2z^*(t)v(t) - v^*(t)Rv(t), \end{aligned}$$

and from the second condition (3.3.3), we get

$$\frac{d}{dt}x^*(t)Px(t) \leq 2z^*(t)v(t),$$

and we obtain

$$x^*(0)Px(0) \leq 2\int_{-\infty}^{0} z^*(t)v(t)dt,$$

which implies the positivity.

Let us prove that the condition is necessary. For $h \in R^n$, we introduce the set

$$\mathcal{V}(h) = \{v(.) \in L^2(-\infty; R^m)|\ x(.) \in L^2(-\infty; R^n),\ x(0) = h\}.$$

By the assumption (3.3.2), this set is not empty. We define on $\mathcal{V}(h)$ the functional

$$J_h(v(.)) = 2\int_{-\infty}^{0} z^*(t)v(t)dt,$$

which is always positive. This function is convex in $\mathcal{V}(h)$. Indeed, if $v_1(.), v_2(.)$ belong to $\mathcal{V}(h)$, then $\theta v_1(.) + (1-\theta)v_2(.)$ belongs to $\mathcal{V}(h)$. The corresponding $z(.)$ is $\theta z_1(.) + (1-\theta)z_2(.)$. Hence

$$J_h(\theta v_1(.) + (1-\theta)v_2(.)) - \theta J_h(v_1(.)) - (1-\theta)J_h(v_2(.)) =$$
$$-2\theta(1-\theta)\int_{-\infty}^{0} (z_1(t) - z_2(t))^*(v_1(t) - v_2(t))dt \leq 0$$

by the positivity property, since $z_1(.) - z_2(.)$ is the output corresponding to $v_1(.) - v_2(.)$. We then set

$$\chi(h) = \inf_{v(.)\in\mathcal{V}(h)} J_h(v(.)).$$

This function is well defined and positive. We want to show that

$$\chi(h) = h^*Ph. \tag{3.3.5}$$

We face the difficulty that the infimum in the definition of $\chi(h)$ may not be attained, since $\mathcal{V}(h)$ is not closed in $L^2(-\infty; R^m)$. We then proceed with an approximation as follows: Define

$$J_h^\epsilon(v(.)) = J_h(v(.)) + \epsilon\int_{-\infty}^{0} (|v(t)|^2 + |x(t)|^2)dt.$$

We consider

$$\chi^{\epsilon}(h) = \inf_{v(.)\in\mathcal{V}(h)} J_h^{\epsilon}(v(.)).$$

Then the infimum is attained at a single point, since $J_h^{\epsilon}(v(.))$ is strictly convex. Let $u^{\epsilon}(.)$ be this minimum, and let $y^{\epsilon}(.),\ z^{\epsilon}(.)$ be the corresponding state and output. They satisfy the necessary and sufficient condition of optimality

$$\int_{-\infty}^{0} (\tilde{z}(t)^* u^{\epsilon}(t) + \tilde{v}(t)^*(z^{\epsilon}(t) + \epsilon u^{\epsilon}(t)) + \epsilon \tilde{x}(t)^* y^{\epsilon}(t))dt = 0,$$

in which $\tilde{v}(t)$ is any element of $\mathcal{V}(0)$ and $\tilde{x}(t), \tilde{z}(t)$ are the corresponding state and output. This relation defines uniquely $u^{\epsilon}(.)$, and we see that it is a linear function of h. Also $y^{\epsilon}(t)$ and $z^{\epsilon}(t)$ are linear functions of h, and $\chi^{\epsilon}(h)$ is a quadratic function of h. We can thus write

$$\chi^{\epsilon}(h) = h^* P^{\epsilon} h.$$

Now, $\chi^{\epsilon}(h) \downarrow \chi(h)$, and the sequence of symmetric positive matrices P^{ϵ} also converges to P. Hence (3.3.5) is valid. Once P is defined, we define Q, S, R by the formulas (3.3.4). It remains to see that

$$\mathcal{P} = \begin{pmatrix} Q & S \\ S^* & R \end{pmatrix} \geq 0. \tag{3.3.6}$$

By a calculation already made to check the sufficiency, we have, for all $v(.) \in L^2(-\infty; R^m)$ such that $x(.) \in L^2(-\infty; R^n)$,

$$\begin{aligned} x^*(0)Px(0) + \int_{-\infty}^{0} \begin{pmatrix} x(t) \\ v(t) \end{pmatrix}^* \mathcal{P} \begin{pmatrix} x(t) \\ v(t) \end{pmatrix} dt &= 2\int_{-\infty}^{0} z^*(t)v(t)dt \qquad (3.3.7) \\ &\geq x^*(0)Px(0), \end{aligned}$$

since $v(.) \in \mathcal{V}(x(0))$. Therefore,

$$\int_{-\infty}^{0} \begin{pmatrix} x(t) \\ v(t) \end{pmatrix}^* \mathcal{P} \begin{pmatrix} x(t) \\ v(t) \end{pmatrix} dt \geq 0$$

for all $v(.) \in L^2(-\infty; R^m)$ such that $x(.) \in L^2(-\infty; R^n)$.

Let now $h \in R^n$, and consider $v^{\epsilon}(.) \in \mathcal{V}(h)$ such that

$$J_h(v^{\epsilon}(.)) \leq h^* Ph + \epsilon^2.$$

Since

$$J_h(v^\epsilon(.)) = h^*Ph + \int_{-\infty}^{0} \begin{pmatrix} x^\epsilon(t) \\ v^\epsilon(t) \end{pmatrix}^* \mathcal{P} \begin{pmatrix} x^\epsilon(t) \\ v^\epsilon(t) \end{pmatrix} dt,$$

where $x^\epsilon(.)$ corresponds to $v^\epsilon(.)$, we can state that

$$\int_{-\infty}^{0} \begin{pmatrix} x^\epsilon(t) \\ v^\epsilon(t) \end{pmatrix}^* \mathcal{P} \begin{pmatrix} x^\epsilon(t) \\ v^\epsilon(t) \end{pmatrix} dt \leq \epsilon^2. \tag{3.3.8}$$

We take next v arbitrary. We define

$$\tilde{v}^\epsilon(t) = \begin{vmatrix} v, & \forall t \in (-\epsilon, 0), \\ v^\epsilon(t+\epsilon), & \forall t < -\epsilon, \end{vmatrix}$$

and the corresponding state $\tilde{x}^\epsilon(t)$ satisfies

$$\tilde{x}^\epsilon(t) = x^\epsilon(t+\epsilon), \ \forall t < -\epsilon,$$

which means that

$$\begin{aligned} \frac{d}{dt}\tilde{x}^\epsilon &= F\tilde{x}^\epsilon(t) + Gv, \ \forall t \in (-\epsilon, 0), \\ \tilde{x}^\epsilon(-\epsilon) &= h. \end{aligned}$$

Next

$$\begin{aligned} 0 &\leq \int_{-\infty}^{0} \begin{pmatrix} \tilde{x}^\epsilon(t) \\ \tilde{v}^\epsilon(t) \end{pmatrix}^* \mathcal{P} \begin{pmatrix} \tilde{x}^\epsilon(t) \\ \tilde{v}^\epsilon(t) \end{pmatrix} dt = \\ &= \int_{-\infty}^{-\epsilon} \begin{pmatrix} x^\epsilon(t+\epsilon) \\ v^\epsilon(t+\epsilon) \end{pmatrix}^* \mathcal{P} \begin{pmatrix} x^\epsilon(t+\epsilon) \\ v^\epsilon(t+\epsilon) \end{pmatrix} dt + \int_{-\epsilon}^{0} \begin{pmatrix} \tilde{x}^\epsilon(t) \\ v \end{pmatrix}^* \mathcal{P} \begin{pmatrix} \tilde{x}^\epsilon(t) \\ v \end{pmatrix} dt \\ &\leq \epsilon^2 + \int_{-\epsilon}^{0} \begin{pmatrix} \tilde{x}^\epsilon(t) \\ v \end{pmatrix}^* \mathcal{P} \begin{pmatrix} \tilde{x}^\epsilon(t) \\ v \end{pmatrix} dt. \end{aligned}$$

By dividing by ϵ and letting ϵ tend to 0, we obtain easily

$$0 \leq \begin{pmatrix} h \\ v \end{pmatrix}^* \mathcal{P} \begin{pmatrix} h \\ v \end{pmatrix},$$

and the property (3.3.6) has been proven. This completes the proof. □

3.3.2 Characterization of P

In this section we assume that

$$R \text{ is invertible.} \tag{3.3.9}$$

We shall begin by proving a lemma.

LEMMA 3.2. *Assume (3.3.9). The conditions (3.3.3), (3.3.4) are equivalent to*

$$\begin{gathered} P, Q \text{ symmetric, positive semidefinite,} \\ PF + F^*P + (H - G^*P)^*R^{-1}(H - G^*P) \leq 0, \\ Q = -(PF + F^*P) \quad S = H^* - GP. \end{gathered} \tag{3.3.10}$$

PROOF. Assume (3.3.3), (3.3.4). Then for every pair $\xi, \eta \in R^n$, we have

$$-\xi^*(PF + F^*P)\xi + 2\xi^*(H^* - PG)\eta + \eta^*R\eta \geq 0.$$

We can minimize in η, taking account of the fact that R is invertible. We obtain immediately the second condition (3.3.10). We can proceed backward and check that (3.3.10) implies (3.3.10). This completes the proof. □

Consider now the Riccati equation

$$PF + F^*P + (H - G^*P)^*R^{-1}(H - G^*P) = 0 \tag{3.3.11}$$

and the set of lower solutions of (3.3.11), which means the set of symmetric positive semidefinite matrices such that

$$\Pi F + F^*\Pi + (H - G^*\Pi)^*R^{-1}(H - G^*\Pi) \leq 0. \tag{3.3.12}$$

We want to characterize the matrix identified in Theorem 3.3 by (3.3.5) as the largest lower solution, and to show also that it is a solution, i.e., that it satisfies (3.3.11). From the positive real lemma, Theorem 3.3, and Lemma 3.2, we know that it is a lower solution.

PROPOSITION 3.4. *We assume (3.3.2) and (3.3.9). The matrix P defined by (3.3.5) is the maximum lower solution of (3.3.11) and is a solution of (3.3.11).*

PROOF. Let Π be symmetric and positive semidefinite satisfying (3.3.12). Then for every control $v(.) \in \mathcal{V}(h)$ and the corresponding state $x(.)$, we have

$$\begin{aligned} \frac{d}{dt}x(t)^*\Pi x(t) &= 2x(t)^*\Pi(Fx(t) + Gv(t)) \\ &\leq -x(t)^*(H - G^*\Pi)^*R^{-1}(H - G^*\Pi)x(t) + 2x(t)^*\Pi Gv(t) \end{aligned}$$

$$
\begin{aligned}
&= v(t)^* R v(t) + 2x(t)^* H^* v(t) \\
&- (v(t) + R^{-1}(H - G^* P)x(t))^* R(v(t) + R^{-1}(H - G^* P)x(t)) \\
&\leq 2z^*(t)v(t).
\end{aligned}
$$

Therefore, integrating between $-\infty$ and 0, we get

$$
h^* \Pi h \leq J_h(v(.)) \,, \forall v(.) \in \mathcal{V}(h).
$$

It follows that

$$
h^* \Pi h \leq \chi(h) = h^* P h,
$$

which implies indeed $\Pi \leq P$. Therefore, P is the maximum lower solution. Let us check that it is a solution. We recall the calculation (3.3.7), which implies that for all $v(.) \in \mathcal{V}(h)$, we have

$$
\begin{aligned}
h^* P h + \int_{-\infty}^{0} \begin{pmatrix} x(t) \\ v(t) \end{pmatrix}^* \mathcal{P} \begin{pmatrix} x(t) \\ v(t) \end{pmatrix} dt &= 2 \int_{-\infty}^{0} z^*(t) v(t) dt \\
&= J_h(v(.)).
\end{aligned}
$$

We would like to apply this to the optimal control, but it does not exist. So we consider the regularized problem $J_h^\epsilon(v(.))$, and the matrix P^ϵ such that

$$
h^* P^\epsilon h = \inf_{v(.) \in \mathcal{V}(h)} J_h^\epsilon(v(.)).
$$

We define

$$
Q^\epsilon = -(P^\epsilon F + F^* P^\epsilon), \quad S^\epsilon = H^* - P^\epsilon G
$$

and

$$
\mathcal{P}^\epsilon = \begin{pmatrix} Q^\epsilon & S^\epsilon \\ S^{\epsilon *} & R \end{pmatrix}.
$$

Then a similar calculation shows that

$$
h^* P^\epsilon h + \int_{-\infty}^{0} \begin{pmatrix} x(t) \\ v(t) \end{pmatrix}^* (\mathcal{P}^\epsilon + \epsilon I) \begin{pmatrix} x(t) \\ v(t) \end{pmatrix} dt = J_h^\epsilon(v(.)).
$$

We have here an optimal control $u^\epsilon(.)$, and the corresponding state is denoted by $y^\epsilon(.)$. Applying the preceding relation to $v(.) = u^\epsilon(.)$ and noting that $J_h^\epsilon(u^\epsilon(.)) = h^* P^\epsilon h$, we deduce that

$$
\int_{-\infty}^{0} \begin{pmatrix} y^\epsilon(t) \\ u^\epsilon(t) \end{pmatrix}^* (\mathcal{P}^\epsilon + \epsilon I) \begin{pmatrix} y^\epsilon(t) \\ u^\epsilon(t) \end{pmatrix} dt = 0,
$$

which implies

$$(\mathcal{P}^\epsilon + \epsilon I)\begin{pmatrix} y^\epsilon(t) \\ u^\epsilon(t) \end{pmatrix} = 0, \forall t,$$

and hence

$$(Q^\epsilon + \epsilon I)y^\epsilon(t) + S^\epsilon u^\epsilon(t) = 0,$$
$$(S^\epsilon)^* y^\epsilon(t) + (R + \epsilon I)u^\epsilon(t) = 0.$$

Therefore, also

$$(Q^\epsilon + \epsilon I)y^\epsilon(t) = S^\epsilon(R + \epsilon I)^{-1}(S^\epsilon)^* y^\epsilon(t).$$

Applying this relation at time $t = 0$ yields, since h is arbitrary,

$$Q^\epsilon + \epsilon I = S^\epsilon(R + \epsilon I)^{-1}(S^\epsilon)^*,$$

but we know that $P^\epsilon \downarrow P$; hence $Q^\epsilon \to Q,\ S^\epsilon \to S$ as $\epsilon \downarrow 0$. It follows that

$$Q = SR^{-1}S^*,$$

and P is a solution of (3.3.11). The proof is complete. □

EXERCISE 3.3. Consider the linear system

$$\frac{dx}{dt} = -Fx(t) - Gv(t),\ t \geq 0,$$
$$x(0) = h,$$

and assume that the pair $(-F, -G)$ is stabilizable. Define the payoff

$$J_h(v(.)) = \int_0^{+\infty} (2x^*(t)H^*v(t) + v(t)^* Rv(t))dt$$

with R symmetric and positive definite. Show that the existence of a symmetric positive semidefinite lower solution of (3.3.11) is equivalent to

$$\inf_{v(.)\in\mathcal{V}(h)} J_h(v(.)) \geq 0,$$

in which

$$\mathcal{V}(h) = \{v(.) \in L^2(0,\infty; R^m)|\ x(.) \in L^2(0,\infty; R^n)\}.$$

Chapter 4
Estimation Theory

Estimation is very much related to control theory. We have already seen many connections, for example in the presentation of controllability and observability. In practice also, the engineer will have to mix estimation and decision-making. There is, however, a current tendency to consider estimation as a branch of statistics, rather than as a part of control theory. We believe that students should learn both. This is the motivation of this chapter and the next. Of course, they can be perused rapidly on a first reading.

4.1 Deterministic Approach

Consider the relation

$$z = Hx + \epsilon, \tag{4.1.1}$$

in which x is a vector of parameters, $x \in R^n$, and z is a vector of observations, $z \in R^p$. We may interpret x as the state of a system that is not a dynamical system. So it does not change with time. We look at Hx as the output. However, there is an error, ϵ. So the observation is not exactly Hx. We do not know the error, and we do not have any information about its nature. Later on, we shall consider it a random variable, but not here. This is why we call our present approach deterministic. We know the matrix H. We assume $p \geq n$ and that the matrix H has rank n. Estimating x from z means that we define a function $x(z)$ that is *as close as possible* to x. The **least squares estimator** simply minimizes, with respect to x, the distance

$$J(x) = \frac{1}{2}|z - Hx|^2.$$

A. Bensoussan, *Estimation and Control of Dynamical Systems*, Interdisciplinary Applied Mathematics 48, https://doi.org/10.1007/978-3-319-75456-7_4

Since

$$DJ(x) = -(z - Hx)^* H,$$

we get a unique solution

$$\hat{x}(z) = (H^* H)^{-1} H^* z. \tag{4.1.2}$$

If we define the error $\tilde{x}$ and the residual r by

$$\tilde{x} = x - \hat{x},$$
$$r = z - Hx,$$

then we can express them in terms of ϵ by the formulas

$$\tilde{x} = -(H^* H)^{-1} H^* \epsilon, \tag{4.1.3}$$
$$r = (I - H(H^* H)^{-1} H^*)\epsilon.$$

4.2 Bayesian Approach

4.2.1 Definition

In general, we have some knowledge about the error ϵ. It is convenient to consider it a random variable. Therefore, z is random. So it has a probability distribution depending on the value of x. The Bayesian approach consists in assuming that z has a probability distribution indexed by x. We shall assume for the sake of simplification that the distribution has a density. Hence we start with a function $f(z|x)$ that represents the dependence of the probability density on z with respect to x. Although x is deterministic, it is convenient to call $f(z|x)$ the conditional probability density of z, given x. Note that we use the same notation for the random variable z and the argument of its conditional probability density. This probability density is also called the **likelihood function.** It plays an essential role in the theory of estimation. In the Bayesian approach, one then randomizes x in a *subjective* manner. We consider that x is random variable, with probability density $\mu(x)$. We call it the *prior probability density* of x. It is subjective in the sense that it corresponds to our belief. In fact, x is deterministic, but we do not know its value.

We next consider that the two random variables x, z have a joint probability density given by the formula $f(z|x)\mu(x)$. The marginal probability density of z is

$$\hat{f}(z) = \int f(z|x)\mu(x)dx.$$

Bayes's rule defines the *posterior probability density* of x, given z, by the formula

$$\hat{\mu}(x|z) = \frac{f(z|x)\mu(x)}{\hat{f}(z)},$$

which can be written in a more symmetric way as follows:

$$f(z|x)\mu(x) = \hat{\mu}(x|z)\hat{f}(z).$$

We consider $\hat{\mu}(x|z)$ the conditional probability of x, given z.

4.2.2 Examples

EXERCISE 4.1. We take

$$f(z|x) = \frac{1}{(\sigma\sqrt{2\pi})^p}\prod_{i=1}^{p}\exp -\frac{1}{2}\frac{(z_i - x)^2}{\sigma^2}, \tag{4.2.1}$$

which means that the components of z are independent random variables with mean x and variance σ^2. If the prior probability of x is $\mu(x)$, show that the posterior is

$$\hat{\mu}(x|z) = \frac{\exp -\frac{p}{2}\frac{(x-\bar{z})^2}{\sigma^2}\,\mu(x)}{\int \exp -\frac{p}{2}\frac{(\xi-\bar{z})^2}{\sigma^2}\,\mu(\xi)d\xi},$$

in which

$$\bar{z} = \frac{\sum_{i=1}^{p} z_i}{p}. \tag{4.2.2}$$

We call $\bar{z}$ the empirical mean. Indeed, we are observing a sample $z_1, \cdots, z_p$ of independent realizations of a Gaussian random variable with mean x that is not known and variance σ^2 that is known. We observe also that in the posterior probability density, z appears only through $\bar{z}$. We say that $\bar{z}$ is a **sufficient statistic**. We do not need to know all the values of z, only the empirical mean.

We now consider another example.

EXERCISE 4.2. Assume

$$f(z|x) = \prod_{i=1}^{p}\exp -x\frac{x^{z_i}}{z_i!}. \tag{4.2.3}$$

This time, the components of z are independent realizations of a Poisson random variable with mean x. Check that

$$\hat{\mu}(x|z) = \frac{\exp - px\, x^{p\bar{z}} \mu(x)}{\int \exp - p\xi\, \xi^{p\bar{z}}\, \mu(\xi) d\xi},$$

and therefore, $\bar{z}$ is again a sufficient statistic.

4.3 Good Estimators

4.3.1 Properties

Suppose as before that an observation $z \in R^p$ has a probability density $f(z|x)$ when the parameter has value $x \in R^n$. Assume that the prior probability density of x is $\mu(x)$. So the joint probability distribution of the pair x, z is $f(z|x)\mu(x)$.

An estimator is simply a function $\hat{x}(z)$. To select a good estimator, one has to work out its properties. Defining the error

$$\tilde{x} = x - \hat{x}(z), \tag{4.3.1}$$

we shall say that the estimator is unbiased if

$$E\tilde{x} = 0. \tag{4.3.2}$$

EXERCISE 4.3. Suppose that $\hat{x}(z)$ satisfies

$$\int \hat{x}(z) f(z|x) dz = x \;\forall x. \tag{4.3.3}$$

Then the estimator is unbiased. We say that the estimator is **absolutely unbiased.**

Bayes's rule has defined a posterior probability density $\hat{\mu}(x|z))$. A natural estimator of x is the mean with respect to the probability density $\hat{\mu}(x|z))$, namely

$$\hat{x}(z) = \int x \hat{\mu}(x|z) dx. \tag{4.3.4}$$

EXERCISE 4.4. Show that the estimator (4.3.4) is unbiased.

A second important property is consistency. This means that as p (the size of the space of observations, which is equivalent to the number of observations) increases, the accuracy of the approximation given by the estimator improves. Mathematically, we say that the estimator is consistent whenever

$$\hat{x}(z) \to x, \text{ in probability, as } p \to +\infty. \tag{4.3.5}$$

We finally define efficiency. We say that $\hat{x}(z)$ is efficient if it is unbiased and has the smallest covariance among all unbiased estimators. This means that

$$\int\int (x-\hat{x}(z))(x-\hat{x}(z))^* f(z|x)\mu(x)dxdz \leq \int\int (x-\varphi(z))(x-\varphi(z))^* f(z|x)\mu(x)dxdz \tag{4.3.6}$$

for every unbiased estimator $\varphi(z)$. This definition can be reinforced into being **absolutely efficient** by requiring

$$\int (x-\hat{x}(z))(x-\hat{x}(z))^* f(z|x)dz \leq \int (x-\varphi(z))(x-\varphi(z))^* f(z|x)dz, \ \forall x \tag{4.3.7}$$

and for every absolutely unbiased estimator $\varphi(z)$.

EXERCISE 4.5. Show that a biased estimator cannot be efficient.

4.3.2 *The Cramér–Rao Inequality*

It is possible to find a lower bound for the covariance of an absolutely unbiased estimator. We introduce the Fisher information matrix

$$J = J(x) = \int [(D_x \log f(z|x))^* D_x \log f(z|x)] f(z|x)dz, \tag{4.3.8}$$

in which the expectation is to be taken with respect to the joint probability density $f(z|x)\mu(x)$. Note that in the notation (4.3.8), $D_x \log f(z|x)$ is a line matrix, and therefore, $(D_x \log f(z|x))^*$ is a column vector.

We note that

$$J = -\int D_x^2 \log f(z|x) f(z|x)dz. \tag{4.3.9}$$

We have the following theorem.

THEOREM 4.1. *Assume that $J(x)$ is invertible for all x. We have the inequality (Cramér–Rao)*

$$\int (x-\hat{x}(z))(x-\hat{x}(z))^* f(z|x)dz \geq J(x)^{-1}, \forall x \tag{4.3.10}$$

and for all absolutely unbiased estimator $\hat{x}(z)$. There is equality only whenever

$$(D_x \log f(z|x))^* = J(x)(\hat{x}(z)-x), \tag{4.3.11}$$

which is satisfied when

$$(D_x \log f(z|x))^* = c(x)(\hat{x}(z)-x), \tag{4.3.12}$$

where $c(x)$ is a constant.

PROOF. We have

$$x = \int \hat{x}(z) f(z|x) dz$$

and

$$1 = \int f(z|x) dz.$$

We deduce by differentiating with respect to x ttha

$$\int (\hat{x}(z) - x) D_x f(z|x) dz = I$$

or

$$\int (\hat{x}(z) - x) D_x log\ f(z|x)\ f(z|x) dz = I.$$

Consider arbitrary $a, b \in R^n$. We deduce that

$$a^* b = \int a^* (\hat{x}(z) - x) D_x log\ f(z|x) b\ f(z|x) dz.$$

Hence by the Cauchy–Schwarz inequality, we have

$$\begin{aligned}(a^* b)^2 &\leq \int (a^*(\hat{x}(z) - x))^2 f(z|x) dz \int (D_x log\ f(z|x) b)^2 f(z|x) dz \\ &= a^* \int (x - \hat{x}(z))(x - \hat{x}(z))^* f(z|x) dz\, a\, b^* J b.\end{aligned}$$

By taking $b = J^{-1} a$, we obtain

$$a^* J^{-1} a \leq a^* \int (x - \hat{x}(z))(x - \hat{x}(z))^* f(z|x) dz\, a,$$

and the result (4.3.10) follows immediately.

To study the case of equality, we simplify the notation by writing

$$X = D_x log\ f(z|x), \quad Y = \hat{x}(z) - x.$$

Integration with respect to the density $f(z|x)$ yields what is called the expected value and is denoted by E. Note that X is a row vector and Y a column vector of size n. We recall that

$$EYX = I, \tag{4.3.13}$$

and the equality in (4.3.10) means that

$$EYY^* = (EX^*X)^{-1}. \tag{4.3.14}$$

We need to prove that these relations imply

$$X^* = (EX^*X)\,Y. \tag{4.3.15}$$

First from (4.3.14), we have

$$EYY^*EX^*X = I.$$

It follows that for $a, b \in R^n$,

$$a^*EYY^*EX^*X\,b = a^*b,$$

and if we take

$$a = EX^*X\,b,$$

we obtain

$$a^*EYY^*a = b^*EX^*X\,b = a^*b,$$

which is interpreted as

$$\begin{aligned} E(a^*Y)^2 &= E(Xb)^2 = a^*b \\ &= a^*EYX\,b \\ &= E(a^*Y)(Xb). \end{aligned}$$

But then

$$E\frac{a^*Y}{\sqrt{E(a^*Y)^2}}\frac{Xb}{\sqrt{E(Xb)^2}} = 1,$$

which implies

$$\frac{a^*Y}{\sqrt{E(a^*Y)^2}} = \frac{Xb}{\sqrt{E(Xb)^2}},$$

and since the denominators are equal, we obtain

$$a^*Y = Xb.$$

Therefore,

$$b^*EX^*X\,Y = b^*X^*,$$

which is the result (4.3.15).

Now if $X^* = cY$, then

$$EX^*X = cEYX = cI,$$

and thus (4.3.15) is satisfied. This completes the proof. □

COROLLARY 4.1. *If the estimator $\hat{x}(z)$ satisfies*

$$\int (x - \hat{x}(z))(x - \hat{x}(z))^* f(z|x)dz = J(x)^{-1}, \tag{4.3.16}$$

then it is absolutely efficient.

EXERCISE 4.6. Suppose the components z_i of z are independent Poisson variables, with parameter x. Show that

$$\bar{z} = \frac{\sum_{i=1}^{p} z_i}{p}$$

is an absolutely efficient estimator of x. Denoting by E the result of integration with respect to the density $f(z|x)$, show that

$$E(x - \bar{z})^2 = J\,\frac{x^2}{p^2}$$

and $J = \dfrac{p}{x}$.

4.4 Minimum Mean Square Estimator

4.4.1 Definition

We consider the pair x, z of random variables in R^n, R^p and assume that the joint probability density is $f(z, x)$. The observation z is known, and we want to define an estimator of x that is a function $\hat{x}(z)$. A natural way to proceed is to minimize the mean square error

$$E|x - \varphi(z)|^2 \tag{4.4.1}$$

among all measurable functions $\varphi : R^p \to R^n$. Note that we assume implicitly $E|x|^2 < +\infty$, and we want $E|\varphi(z)|^2 < +\infty$. In fact, $\hat{x}(z)$ exists and is uniquely defined. It is simply the conditional expectation $E[x|z]$. The conditional expectation is the projection of x in the sense of the Hilbert space $L^2(\Omega, \mathcal{A}, P; R^n)$ on the closed subspace of measurable functions of z. It satisfies the Euler optimality condition

$$E[(x - \hat{x}(z))^*\varphi(z)] = 0, \ \forall \varphi(z) \in L^2(\Omega, \mathcal{A}, P; R^n). \tag{4.4.2}$$

It is called the *minimum mean square estimator*, or MMSE for short.

4.4.2 Properties

We can write

$$E[(x - \hat{x}(z))_i \varphi_j(z)] = 0,$$

and hence also

$$E[(x - \hat{x}(z))\varphi(z)^*] = 0, \ \forall \varphi(z) \in L^2(\Omega, \mathcal{A}, P; R^n). \tag{4.4.3}$$

EXERCISE 4.7. Show that $\hat{x}(z)$ satisfies

$$E[(x-\varphi(z))(x-\varphi(z))^*] \geq E[(x-\hat{x}(z))(x-\hat{x}(z))^*], \forall \varphi(z) \in L^2(\Omega, \mathcal{A}, P; R^n).$$

■

We can define

$$f(x|z) = \frac{f(z,x)}{f(z)}, \tag{4.4.4}$$

where $f(z)$ is the marginal probability density of z,

$$f(z) = \int f(z,x)dx,$$

and we call $f(x|z)$ the conditional probability density of x, given z. Then we have the following result.

PROPOSITION 4.1. *The conditional expectation (MMSE) $\hat{x}(z)$ is given by*

$$\hat{x}(z) = \frac{\int x f(z,x)dx}{f(z)}.$$

PROOF. The proposition is an easy consequence of the Euler optimality condition. □

We have also the following.

PROPOSITION 4.2. *If an estimator $\hat{x}(z)$ is absolutely efficient, then it coincides with the MMSE.*

PROOF. Indeed, if

$$\int (x - \hat{x}(z))(x - \hat{x}(z))^* f(z|x)dz \leq \int (x - \varphi(z))(x - \varphi(z))^* f(z|x)dz, \ \forall x,$$

then

$$\int\int (x-\hat{x}(z))(x-\hat{x}(z))^* f(z,x)dzdx \leq \int\int (x-\varphi(z))(x-\varphi(z))^* f(z,x)dzdx,$$

and the result follows. □

EXERCISE 4.8. Let $n = 1$ and

$$f(z, x) = \frac{1}{(\sqrt{2\pi\sigma^2})^p} \exp -\frac{1}{2} \frac{\sum_{i=1}^{p}(z_i - x)^2}{\sigma^2} \mu(x).$$

Show that

$$\bar{z} = \frac{\sum_{i=1}^{p} z_i}{p}$$

is the MMSE.

Hints: Show that the Fisher information number is $J(x) = \frac{p}{\sigma^2}$ and prove (4.3.16).

4.4.3 *MMSE for Gaussian Variables*

We assume that the pair z, x is Gaussian. To simplify, we assume that the means are 0. Let us introduce next the variance and covariance matrices

$$Q_z = Ezz^*,$$
$$Q_x = Exx^*,$$
$$Q_{xz} = Exz^*.$$

Note that

$$Q_{zx} = Ezx^* = Q_{xz}^*$$

and set

$$Q = \begin{pmatrix} Q_x & Q_{xz} \\ Q_{zx} & Q_z \end{pmatrix}.$$

There is a density only when the matrix Q is invertible. In that case, we have the formula

$$f(z, x) = \frac{1}{(2\pi)^{\frac{n+p}{2}}} \frac{1}{(\det Q)^{\frac{1}{2}}} \exp -\frac{1}{2}(x^* z^*) Q^{-1} \begin{pmatrix} x \\ z \end{pmatrix}. \tag{4.4.5}$$

We do not need this strong assumption to define the MMSE. We need only for Q_z to be invertible. We have the following result.

PROPOSITION 4.3. *Assume that Q_z is invertible. Then the MMSE for Gaussian vectors with 0 mean is given by*

$$E[x|z] = \hat{\Lambda} z = Q_{xz} Q_z^{-1} z. \tag{4.4.6}$$

PROOF. Set

$$\hat{\epsilon} = x - Q_{xz} Q_z^{-1} z.$$

Obviously,

$$E \hat{\epsilon} z^* = 0,$$

and thus $\hat{\epsilon}$ and z are not correlated. Since they are Gaussian, they are independent. The orthogonality property (4.4.3) is satisfied, with

$$\hat{x}(z) = \hat{\Lambda} z,$$

which completes the proof. □

If the joint probability density exists (see (4.4.5)), then we can define the conditional probability of x given z by the formula

$$f(x|z) = \frac{f(z, x)}{f(z)}.$$

We now introduce the matrix

$$P = Q_x - Q_{xz} Q_z^{-1} Q_{zx}, \tag{4.4.7}$$

and then we have the following.

PROPOSITION 4.4. *Assume that P is invertible. Then*

$$f(x|z) = \frac{1}{(2\pi)^{\frac{n}{2}} (det P)^{\frac{1}{2}}} \exp -\frac{1}{2} (x - \hat{\Lambda} z)^* P^{-1} (x - \hat{\Lambda} z).$$

PROOF. Although a direct calculation is possible whenever Q is invertible, the fastest way to prove the result is to use the characteristic function (Fourier transform) and show that

$$\int \exp i\lambda^* x \; f(x|z) dx = \exp i\lambda^* \hat{\Lambda} z \; \exp -\frac{1}{2} \lambda^* P \lambda, \; \forall \lambda \in R^n. \tag{4.4.8}$$

However,

$$\begin{aligned} \int \exp i\lambda^* x \; f(x|z) dx &= \exp i\lambda^* \hat{\Lambda} z \int \exp i\lambda^* (x - \hat{\Lambda} z) \; f(x|z) dx \\ &= \exp i\lambda^* \hat{\Lambda} z \int \exp i\lambda^* \hat{\epsilon} \; f(x|z) dx. \end{aligned}$$

But $\hat{\epsilon}$ is independent of z. Therefore, for every bounded measurable function $\varphi(z)$, we have

$$\int\int \exp i\lambda^* \hat{\epsilon} \varphi(z) f(z,x) dz dx = \int\int \exp i\lambda^* \hat{\epsilon} f(z,x) dz dx \int \varphi(z) f(z) dz,$$

and since $\varphi(z)$ is arbitrary, we get

$$\int \exp i\lambda^* \hat{\epsilon} f(z,x) dx = f(z) \int\int \exp i\lambda^* \hat{\epsilon} f(z,x) dz dx,$$

whence

$$\int \exp i\lambda^* \hat{\epsilon}\ f(x|z) dx = \int\int \exp i\lambda^* \hat{\epsilon} f(z,x) dz dx = E \exp i\lambda^* \hat{\epsilon}.$$

But $\hat{\epsilon}$ is a Gaussian variable with mean 0 and covariance

$$\begin{aligned} E\hat{\epsilon}\hat{\epsilon}^* &= E(x - Q_{xz}Q_z^{-1}z)(x - Q_{xz}Q_z^{-1}z)^* \\ &= P, \end{aligned}$$

and thus

$$\int \exp i\lambda^* \hat{\epsilon}\ f(x|z) dx = \exp -\frac{1}{2}\lambda^* P \lambda,$$

and the result (4.4.8) follows. □

REMARK 4.1. We do not need the invertibility of P to claim that the conditional probability distribution of x, given z, is a Gaussian with mean $\hat{\Lambda} z$ and covariance matrix P. ■

Suppose now that z, x are not zero-mean and that the expected values are $\bar{z}, \bar{x}$. Then the MMSE of x is defined by

$$\hat{x}(z) = \bar{x} + Q_{xz}Q_z^{-1}(z - \bar{z}). \tag{4.4.9}$$

4.5 Minimum Variance Linear Estimator

4.5.1 Definition

If the pair z, x is not Gaussian, then the MMSE is not linear. However, it is possible to define a best linear estimator, in the sense of minimizing the variance of the error. Suppose, for simplicity's sake, that the random variables have 0 mean, and let Q be their joint covariance

$$Q = \begin{pmatrix} Q_x & Q_{xz} \\ Q_{zx} & Q_z \end{pmatrix}. \tag{4.5.1}$$

We define the best linear estimate as follows: find $\Lambda \in L(R^p; R^n)$ to minimize

$$\mathcal{J}(\Lambda) = E[(x - \Lambda z)^*(x - \Lambda z)]. \tag{4.5.2}$$

Define the error

$$\epsilon_\Lambda = x - \Lambda z$$

and the covariance matrix of the error

$$\Gamma_\Lambda = E\,\epsilon_\Lambda \epsilon_\Lambda^*.$$

Then one checks easily that

$$\mathcal{J}(\Lambda) = \text{tr}\,\Gamma_\Lambda. \tag{4.5.3}$$

We have seen in Proposition 4.3 that if Q_z is invertible, then the unique solution of the problem (4.5.2) is given by

$$\hat{\Lambda} = Q_{xz} Q_z^{-1}.$$

We will give in the next section a more general necessary and sufficient condition for optimality.

4.5.2 Necessary and Sufficient Condition

PROPOSITION 4.5. *The following properties are equivalent:*

(1) $\text{tr}\,\Gamma_{\hat{\Lambda}} \leq \text{tr}\,\Gamma_\Lambda$;
(2) $\Gamma_{\hat{\Lambda}} \leq \Gamma_\Lambda$;
(3) $E\epsilon_{\hat{\Lambda}} z^* = 0$;
(4) $\hat{\Lambda} Q_z = Q_{xz}$.

PROOF. If the first property is true, then one has $\mathcal{J}(\hat{\Lambda}) \leq \mathcal{J}(\Lambda)$. We may consequently write

$$\mathcal{J}(\hat{\Lambda}) \leq \mathcal{J}(\hat{\Lambda} + \theta\Lambda), \forall \Lambda, \forall \theta \in R,$$

which implies

$$E[(x - \hat{\Lambda} z)^* \Lambda z] = 0$$

and

$$E\text{tr}[\Lambda z (x - \hat{\Lambda} z)^*] = 0$$

or

$$\text{tr}[\Lambda (Q_{zx} - Q_z \hat{\Lambda}^*)] = 0, \ \forall \Lambda.$$

By taking $\Lambda = Q_{xz} - \hat{\Lambda} Q_z$, we get

$$\operatorname{tr}[(Q_{xz} - \hat{\Lambda} Q_z)(Q_{zx} - Q_z \hat{\Lambda}^*)] = 0,$$

and therefore,

$$Q_{xz} - \hat{\Lambda} Q_z = 0,$$

which is the fourth assertion. This is the same as the third assertion. Finally,

$$\Gamma_\Lambda = Q_x + \Lambda Q_z \Lambda - \Lambda Q_{zx} - Q_{xz} \Lambda^*,$$

which can be written as

$$\Gamma_\Lambda = \Gamma_{\hat{\Lambda}} + (\Lambda - \hat{\Lambda}) Q_z (\Lambda - \hat{\Lambda})^* \geq \Gamma_{\hat{\Lambda}},$$

which is the second assertion.

Conversely if the second assertion holds, then the first assertion holds, which implies also the third and fourth assertions. ∎

In the preceding formulation, the unknown is a map Λ. Naturally, the map $\hat{\Lambda}$ also minimizes $\psi^* \Gamma_\Lambda \psi$ for all $\psi \in R^n$. But we may notice that by writing

$$v = \Lambda^* \psi \in R^p,$$

we have

$$\begin{aligned} \tilde{J}_\psi(v) &= \psi^* \Gamma_\Lambda \psi \\ &= \psi^* Q_x \psi + v^* Q_z v - 2\psi^* Q_{xz} v, \end{aligned} \tag{4.5.4}$$

and the minimum is achieved with

$$\hat{v} = \hat{\Lambda}^* \psi. \tag{4.5.5}$$

4.5.3 Least Squares Estimator

In this section we assume that Q is invertible. We also assume, for the sake of simplification, that the means are 0. A least squares estimator is the statistic $\hat{x}(z)$ that minimizes

$$J_z(\varphi) = (\varphi^* \; z^*) Q^{-1} \begin{pmatrix} \varphi \\ z \end{pmatrix}. \tag{4.5.6}$$

From the assumption, Q_z is also invertible. We then have the following.

PROPOSITION 4.6. *The least squares estimator coincides with the minimum variance linear estimator*

$$\hat{x}(z) = Q_{xz} Q_z^{-1} z.$$

PROOF. One can give an explicit expression for Q^{-1}. One can check that

$$Q^{-1} = \begin{pmatrix} A & B \\ B^* & C \end{pmatrix}$$

with

$$\begin{aligned} A &= P^{-1}, \; P = Q_x - Q_{xz} Q_z^{-1} Q_{zx}, \\ B &= -A Q_{xz} Q_z^{-1}, \\ C &= (Q_z - Q_{zx} Q_x^{-1} Q_{xz})^{-1}, \end{aligned}$$

and also the expression

$$(\varphi^* \; z^*) Q^{-1} \begin{pmatrix} \varphi \\ z \end{pmatrix} = (\varphi - \hat{x}(z))^* P^{-1} (\varphi - \hat{x}(z)) + z^* Q_z^{-1} z, \tag{4.5.7}$$

which proves the assertion. □

Recalling the problem $\tilde{J}_\psi(v)$, we call the two problems $J_z(\varphi)$, $\tilde{J}_\psi(v)$ **dual problems.** The unknowns φ and v are in R^n and R^p respectively. The solutions are

$$\hat{x}(z) = \hat{\Lambda} z, \quad \hat{v}(\psi) = \hat{\Lambda}^* \psi.$$

The first problem is referred to as the least squares optimization problem. The second problem is referred to as the minimum variance optimization problem. It is important to note that these are not equivalent in terms of assumptions. For the first problem, we assume Q to be invertible, and for the second problem, only Q_z is assumed to be invertible. The concept of duality plays an essential role in the theory of optimization and in control theory.

4.5.4 A Particular Structure

We have considered in Section 4.1 the model

$$z = Hx + \epsilon \tag{4.5.8}$$

in a deterministic context. We can now consider the same model in a stochastic context. We assume that x, ϵ are random variables that are uncorrelated and have 0 mean (for simplicity) and covariance matrices Q_x and R. We deduce easily

$$\begin{aligned} Q_z &= H Q_x H^* + R, \\ Q_{zx} &= H Q_x. \end{aligned}$$

If R is positive definite, then Q_z is invertible. We can then define the minimum variance linear estimator

$$\hat{x}(z) = \hat{\Lambda} z = Q_x H^* (H Q_x H^* + R)^{-1}, \tag{4.5.9}$$

and the covariance matrix of the error is

$$P = Q_x - Q_x H^* (H Q_x H^* + R)^{-1} H Q_x. \tag{4.5.10}$$

EXERCISE 4.9. Show that if Q_x is invertible, then

$$P = (Q_x^{-1} + H^* R^{-1} H)^{-1},$$
$$\hat{\Lambda} = P H^* R^{-1}.$$

We then have

$$Q = \begin{pmatrix} Q_x & Q_x H^* \\ H Q_x & H Q_x H^* + R \end{pmatrix},$$

and thus Q is invertible as soon as Q_x is invertible. We then have

$$Q^{-1} = \begin{pmatrix} P^{-1} & -H^* R^{-1} \\ -R^{-1} H & R^{-1} \end{pmatrix}.$$

It follows that the dual problems are defined as follows:

$$J_z(\varphi) = \varphi^* Q_x^{-1} \varphi + (z - H\varphi)^* R^{-1} (z - H\varphi),$$
$$\tilde{J}_\psi(v) = (\psi - H^* v)^* Q_x (\psi - H^* v) + v^* R v.$$

4.6 The Maximum Likelihood Method

4.6.1 Definition

Consider the likelihood function $f(z|x)$. A natural idea to obtain an estimator is to maximize the likelihood function in x. This is equivalent to maximizing the logarithm of the likelihood, called the log likelihood. So we define the maximum likelihood estimator by the system of equations

$$D_x \log f(z|x) = 0, \tag{4.6.1}$$

called the *likelihood equations.*

EXERCISE 4.10. Consider a sample of p independent variables z_i from a Gaussian probability density $N(x, \sigma^2)$. Show that the maximum likelihood estimator is again

$$\hat{x}(z) = \bar{z} = \frac{\sum_{i=1}^{p} z_i}{p}.$$

4.6.2 *Properties*

We begin with a proposition.

PROPOSITION 4.7. *If there is an absolutely efficient estimator (in the sense that it achieves the minimum in the Cramér–Rao inequality), then it is also the maximum likelihood estimator.*

PROOF. From Theorem 4.1, we know that there is equality in the Cramér–Rao inequality whenever

$$(D_x \log f(z|x))^* = J(x)(\hat{x}(z) - x).$$

Therefore, the left-hand side vanishes at $x = \hat{x}(z)$, which proves that $\hat{x}(z)$ is the maximum likelihood estimator. ■

We next study consistency. To be precise, we consider a sample of size M of independent realizations of the observation z. In fact, z is changed into

$$z = \begin{pmatrix} z_1 \\ \vdots \\ z_M \end{pmatrix},$$

and the size of this vector is Mp. Since the random variables $z_1, \cdots, z_M$ are independent, we have

$$f(z|x) = \prod_{i=1}^{M} f(z_i|x),$$

with a slight abuse of notation, and we shall notate in the same way the likelihood of the sample and the likelihood of each realization.

We then define the function

$$s^M(x) = D_x \log f(z|x) = \sum_{i=1}^{M} D_x \log f(z_i|x),$$

called the **score**, and the estimator is $\hat{x}^M(z)$, such that

$$s^M(\hat{x}^M(z)) = 0.$$

□

EXERCISE 4.11. Show that

$$Es^M(x) = 0,$$
$$Es^M(x)(s^M(x))^* = MJ(x).$$

From the central limit theorem, we can assert that

$$\frac{s^M(x)}{\sqrt{M}} \sim N(0, J(x)), \tag{4.6.2}$$

where $N(0, J(x))$ represents the Gaussian distribution in R^n with mean 0 and covariance matrix $J(x)$, the Fisher information matrix. Next, from Taylor's expansion, we have

$$s^M(x) = \sum_{i=1}^{M} \int_0^1 D_{x^2}^2 \log f(z_i|\hat{x}^M + \lambda(x - \hat{x}^M))d\lambda \; (x - \hat{x}^M).$$

Now we proceed formally to claim that

$$\frac{1}{M}\sum_{i=1}^{M} \int_0^1 D_{x^2}^2 \log f(z_i|\hat{x}^M + \lambda(x - \hat{x}^M))d\lambda \to E[D_{x^2}^2 \log f(z|x)] = -J(x)|.$$

Assuming $J(x)$ invertible, we can write

$$\hat{x}^M - x \sim J(x)^{-1}\frac{s^M(x)}{M}.$$

Therefore,

$$\sqrt{M}(\hat{x}^M - x) \sim J(x)^{-1}\frac{s^M(x)}{\sqrt{M}},$$

and from (4.6.2) we can state that

$$\sqrt{M}(\hat{x}^M - x) \sim N(0, J(x)^{-1}).$$

This allows us to claim that the maximum likelihood estimate is asymptotically both **consistent** and **efficient**.

We shall give a complete proof of this result for the interested reader in the appendix to this chapter.

4.6.3 Maximum Posterior Probability Estimators

If we know the prior probability density $\mu(x)$ of x, we can replace the likelihood by the joint probability density $f(z|x)\mu(x)$. So the likelihood equations (4.6.1) are replaced by

$$D_x \log f(z|x) + D_x \log \mu(x) = 0. \tag{4.6.3}$$

EXERCISE 4.12. Consider the situation of Exercise 4.10, and assume that the prior of x is a Gaussian with mean 0 and variance $\bar{\sigma}^2$. Show that the maximum posterior probability estimate is given by

$$\hat{x}(z) = \frac{\bar{\sigma}^2}{\sigma^2 + p\bar{\sigma}^2} \sum_{i=1}^{p} z_i.$$

The importance of the knowledge of a prior probability density becomes less when $\bar{\sigma}^2$ or p is large. We see that $\hat{x}(z)$ becomes close to $\bar{z}$, which is the case when there is no prior probability density.

4.7 Dynamic Models

4.7.1 Fixed Parameter

In dynamic models, we make observations sequentially. So the observation vector is a collection of observations at a sequence of times. We consider here the situation in which the parameter $x \in R^n$ does not evolve with time. On the other hand, we make an observation at time $i = 1, \cdots$, denoted by $z_i \in R^d$. Each observation is related to the parameter by the relation

$$z_i = H_i x + \epsilon_i, \tag{4.7.1}$$

in which ϵ_i is a sequence of independent random variables in R^d, with 0 mean and covariance matrix R_i. We assume that R_i is invertible.

Suppose we have a sequence of observations up to $i = k$. So at time k, the observation is

$$z^k = \begin{bmatrix} z_1 \\ \vdots \\ z_k \end{bmatrix},$$

and $z^k \in R^{dk}$. The variable x is independent of $\epsilon_1, \cdots, \epsilon_k$ and has mean $\bar{x}$ and covariance matrix P_0. This model at the same time extends and is a particular case of the model (4.5.8). The fact that we have replaced the notation Q_x by P_0 will be explained below.

Since we have not expressed the joint probability density, but simply the means and variances, the only estimators we can think of are the least squares estimator and the minimum variance linear estimator, which are the same, except that the assumptions for existence are not identical. We have seen in Exercise (4.9) that if P_0 is invertible, then the least squares problem is well defined. In the present case, it amounts to minimizing in $\varphi \in R^n$ the functional

$$J_{z^k}(\varphi) = (\varphi - \bar{x})^* P_0^{-1}(\varphi - \bar{x}) + \sum_{i=1}^{k}(z_i - H_i\varphi)^* R_i^{-1}(z_i - H_i\varphi). \tag{4.7.2}$$

It is easy to compute the gradient and to prove the following result.

EXERCISE 4.13. We denote the optimal φ by $\hat{x}_k$. Verify the formula

$$\hat{x}_k = (P_0^{-1} + \sum_{i=1}^{k} H_i^* R_i^{-1} H_i)^{-1}(P_0^{-1}\bar{x} + \sum_{i=1}^{k} H_i^* R_i^{-1} z_i) \tag{4.7.3}$$

and that the error covariance matrix is given by

$$\begin{aligned} P_k &= E(x - \hat{x}_k)(x - \hat{x}_k)^* \\ &= (P_0^{-1} + \sum_{i=1}^{k} H_i^* R_i^{-1} H_i)^{-1}. \end{aligned} \tag{4.7.4}$$

4.7.2 Recursive Formulas

Expressions (4.7.3) and (4.7.4) are not recursive. We can give recursive formulas

$$\begin{aligned} P_k^{-1} &= P_0^{-1} + \sum_{i=1}^{k} H_i^* R_i^{-1} H_i \\ &= P_{k-1}^{-1} + H_k^* R_k^{-1} H_k. \end{aligned}$$

EXERCISE 4.14. Show that

$$P_k = P_{k-1} - P_{k-1}H_k^*(H_k P_{k-1} H_k^* + R_k)^{-1} H_k P_{k-1}. \tag{4.7.5}$$

This formula is clearly recursive, starting with P_0, which explains the notation. Moreover, it defines a sequence of matrices that is well defined even when P_0 is not invertible. Show next that

$$\hat{x}_k = \hat{x}_{k-1} + K_k(z_k - H_k\hat{x}_{k-1}) \tag{4.7.6}$$

with

$$\begin{aligned} K_k &= P_k H_k^* R_k^{-1} \\ &= P_{k-1} H_k^* (H_k P_{k-1} H_k^* + R_k)^{-1}. \end{aligned} \tag{4.7.7}$$

The sequence $\hat{x}_k$ is obtained in a recursive manner, starting with $\hat{x}_0 = \bar{x}$. It is well defined even when the matrices P_k are not invertible.

4.7.3 Dual Formulation

If we look at formula (4.5.4), then we have $Q_x = P_0$,

$$Q_z = \begin{pmatrix} H_1 P_0 H_1^* & 0 & \cdots & 0 \\ 0 & H_2 P_0 H_2^* & 0 & \cdots \\ \cdots & \cdots & \cdots & \cdots \\ 0 & 0 & \cdots & H_k P_0 H_k^* \end{pmatrix}$$

with $Q_z \in \mathcal{L}(R^{dk}; R^{dk})$ and

$$Q_{xz} = (P_0 H_1^* \; P_0 H_2^* \cdots\cdots P_0 H_k^*) \in \mathcal{L}(R^{dk}; R^n).$$

Taking $v^k \in R^{dk}$ and $\psi \in R^n$, the dual problem is defined by

$$\tilde{J}_\psi(v^k) = \psi^* Q_x \psi - 2\psi^* Q_{xz} v^k + (v^k)^* Q_z v^k.$$

Writing

$$v^k = \begin{pmatrix} v_1 \\ \vdots \\ v_k \end{pmatrix},$$

we can obtain easily the explicit equation

$$\tilde{J}_\psi(v^k) = (\psi - \sum_{i=1}^k H_i^* v_i)^* P_0 (\psi - \sum_{i=1}^k H_i^* v_i) + \sum_{i=1}^k v_i^* R_i v_i. \tag{4.7.8}$$

The duality is expressed as follows: the solution (4.7.3) can be written

$$\hat{x}_k = \bar{x} + \sum_{i=1}^k \hat{\Lambda}_i^{(k)} (z_i - H_i \bar{x}) \tag{4.7.9}$$

with

$$\hat{\Lambda}_i^{(k)} = P_k H_i^* R_i^{-1}. \tag{4.7.10}$$

The operator $\hat{\Lambda}^{(k)}$ maps R^{dk} into R^n. The solution of the dual problem is

$$\hat{v}^k = (\hat{\Lambda}^{(k)})^* \psi,$$

whence

$$\hat{v}_i^k = R_i^{-1} H_i P_k \psi. \tag{4.7.11}$$

EXERCISE 4.15. Prove directly that (4.7.11) is the solution of the dual problem.

4.7.4 *The Gaussian Case*

Suppose now that x, z^k are Gaussian. In that case, the minimum-variance linear estimator coincides with the conditional mean, which is the MMSE. So we have

$$\hat{x}_k = E[x|z^k].$$

In fact, we can compute the conditional mean easily and recover the formulas (4.7.5), (4.7.6), (4.7.7). The idea is to introduce the process

$$\nu_k = z_k - H_k \hat{x}_{k-1}, \tag{4.7.12}$$

which is called the **innovation process.** We have the following.

PROPOSITION 4.8. *The innovation process is Gaussian, consisting of independent random variables such that*

$$\begin{aligned} E\nu_k &= 0, \\ E\nu_k \nu_k^* &= H_k P_{k-1} H_k^* + R_k, \end{aligned} \tag{4.7.13}$$

in which

$$P_k = E(x - \hat{x}_k)(x - \hat{x}_k)^*.$$

PROOF. The fact that it is Gaussian follows from the definition, where we recall that $\hat{x}_k$ is itself Gaussian, since it depends linearly on x, z^k. Clearly, ν_k is adapted to z^k. We now introduce the error

$$\tilde{x}_k = x - \hat{x}_k.$$

From the properties of the conditional mean, $\tilde{x}_k$ is independent of z^k. Therefore,

$$\nu_k = H_k \tilde{x}_{k-1} + \epsilon_k$$

is independent of z^{k-1}, since both $\tilde{x}_{k-1}$ and ϵ_k are independent of z^{k-1}. This implies also that ν_k is independent of ν_l $l < k$. Clearly, ν_k has mean 0.

Next, ϵ_k and $\tilde{x}_{k-1}$ are independent. Therefore, the second formula (4.7.13) follows, using the definition of P_k. □

The next idea is to notice that the σ-algebra generated by z^k is equivalent to the σ-algebra generated by z^{k-1} and ν_k. Therefore, we can write

$$\hat{x}_k = E(x|z^{k-1}, \nu_k),$$

and by the linearity of the conditional mean, we have

$$\hat{x}_k = \hat{x}_{k-1} + K_k \nu_k.$$

for a convenient *gain* K_k. The error $\tilde{x}_k$ can be written as

$$\begin{aligned}\tilde{x}_k &= \tilde{x}_{k-1} - K_k \nu_k \\ &= (I - K_k H_k)\tilde{x}_{k-1} - K_k \epsilon_k.\end{aligned}$$

Since ϵ_k is independent of $\tilde{x}_{k-1}$, it follows that

$$\begin{aligned}P_k &= (I - K_k H_k) P_{k-1} (I - K_k H_k)^* + K_k R_k K_k^* \\ &= P_{k-1} - K_k H_k P_{k-1} - P_{k-1} H_k^* K_k^* + K_k (H_k P_{k-1} H_k^* + R_k) K_k^*.\end{aligned}$$

Therefore, we also have that

$$\begin{aligned}P_k = P_{k-1} &- P_{k-1} H_k^* (H_k P_{k-1} H_k^* + R_k)^{-1} H_k P_{k-1} + \\ &+ (K_k - P_{k-1} H_k^* (H_k P_{k-1} H_k^* + R_k)^{-1})(H_k P_{k-1} H_k^* + R_k) \\ &(K_k - P_{k-1} H_k^* (H_k P_{k-1} H_k^* + R_k)^{-1})^*.\end{aligned}$$

Since P_k must be the minimum covariance matrix, it follows necessarily that K_k is given by formula (4.7.7). ■

4.7.5 The Kalman Filter in Discrete Time

We now allow the parameter x to also evolve with time. In this context, one should think of x not as a parameter, but as the state of a dynamical system. The objective is to estimate the state as it evolves. We thus consider the following model:

$$\begin{aligned}x_{i+1} &= F_i x_i + f_i + w_i, \; i = 1, \cdots, \qquad (4.7.14) \\ x_1 &= \xi, \\ z_i &= H_i x_i + h_i + \epsilon_i.\end{aligned}$$

We have $x_i \in R^n$, $z_i \in R^d$.

We assume that ξ is Gaussian with mean $\bar{\xi}$ and covariance matrix P_0. The sequence w_i is a sequence of independent Gaussian variables with mean 0 and covariance matrix Q_i. Similarly, ϵ_i is a sequence of independent Gaussian variables with mean 0 and covariance matrix R_i. The triple ξ, w_i, ϵ_i are mutually independent. The vectors f_i, h_i are deterministic. We define as in the previous section

$$z^k = (z_1, \cdots, z_k),$$

and we look for the conditional mean (also the MMSE)

$$\hat{x}_k = E(x_k|z^k).$$

Because all variables are Gaussian, we know that the conditional mean depends linearly on the observation. We can naturally use the equivalence with the minimum variance linear estimator, and also make use of the dual problem. However, the shortest route is to use the innovation process. So we limit our analysis to this approach.

We need to introduce another estimator,

$$\hat{x}_k^- = E(x_k|z^{k-1}).$$

When the state does not evolve, as in the previous section, this is simply $\hat{x}_{k-1}$, but not here. When $k = 1$, there is no observation, and clearly $\hat{x}_1^- = \bar{\xi}$.

We define the errors

$$\tilde{x}_k = x_k - \hat{x}_k, \; \tilde{x}_k^- = x_k - \hat{x}_k^-,$$

and the corresponding covariance matrices

$$P_k = E\tilde{x}_k\tilde{x}_k^*, \; P_k^- = E\tilde{x}_k^-\tilde{x}_k^{-*}.$$

We have the following result.

THEOREM 4.2. *We have the relations*

$$\begin{aligned} \hat{x}_{k+1}^- &= F_k\hat{x}_k + f_k, \; \hat{x}_1^- = \bar{\xi}, \\ \hat{x}_k &= \hat{x}_k^- + P_k^- H_k^*(H_k P_k^- H_k^* + R_k)^{-1}(z_k - H_k\hat{x}_k^- - h_k), \end{aligned} \tag{4.7.15}$$

$$\begin{aligned} P_{k+1}^- &= Q_k + F_k P_k F_k^*, \\ P_k &= P_k^- - P_k^- H_k^*(H_k P_k^- H_k^* + R_k)^{-1} H_k P_k^-, \\ P_1^- &= P_0. \end{aligned} \tag{4.7.16}$$

PROOF. We introduce the innovation process

$$\nu_k = z_k - H_k\hat{x}_k^- - h_k,$$

and again it is a sequence of Gaussian variables; ν_k is z^k measurable and independent of z^{k-1}; hence they are mutually independent. Since

$$\nu_k = H_k \tilde{x}_k^- + \epsilon_k$$

and ϵ_k is independent of z^{k-1}, it is independent of $\tilde{x}_k^-$. Therefore,

$$\begin{aligned} E\nu_k &= 0, \\ E\nu_k \nu_k^* &= H_k P_k^- H_k^* + R_k. \end{aligned}$$

We then check as in the previous section that

$$\hat{x}_k = \hat{x}_k^- + K_k \nu_k$$

for a convenient K_k. We next have

$$\begin{aligned} \tilde{x}_k &= \tilde{x}_k^- - K_k \nu_k \\ &= (I - K_k H_k)\tilde{x}_k^- - K_k \epsilon_k. \end{aligned}$$

We can then proceed as in the previous section to define the optimal K_k. This concludes the proof. □

We can obtain a system of relations for $\hat{x}_k^-$, P_k^- as follows:

$$\begin{aligned} \hat{x}_{k+1}^- &= F_k \hat{x}_k^- + f_k + \\ &+ F_k P_k^- H_k^* (H_k P_k^- H_k^* + R_k)^{-1} (z_k - H_k \hat{x}_k^- - h_k), \\ \hat{x}_1^- &= \bar{\xi}, \end{aligned} \tag{4.7.17}$$

$$\begin{aligned} P_{k+1}^- &= Q_k + F_k P_k^- F_k^* \\ &\quad - F_k P_k^- H_k^* (H_k P_k^- H_k^* + R_k)^{-1} H_k P_k^- F_k^*, \\ P_1^- &= P_0. \end{aligned} \tag{4.7.18}$$

4.8 Appendix

4.8.1 Preliminaries

We want to study the asymptotic properties of the maximum likelihood estimate (MLE). Since we follow [24] with some modifications, we will use their notation. Namely, the parameter x will be replaced by θ. So we shall consider a probability density $f(z, \theta)$, in which the parameter θ is in R^k. We assume necessary smoothness, without stating it explicitly. The random variable z is in R^d. We call θ_0 the true value, which we want to estimate. In fact, the probability density of z is $f(z, \theta_0)$, and $f(z, \theta)$ refers to the function. We proceed with a sample of independent random variables $z^1, \cdots, z^M$. The joint probability density of this sample is, of course,

$$\prod_{j=1}^{M} f(z^j, \theta_0).$$

We assume, to simplify technicalities, that we have an open bounded subset of R^k, denoted by Θ, such that $\theta_0 \in \Theta$. We set

$$Z_M(u) = \prod_{j=1}^{M} \frac{f(z^j, \theta_0 + u)}{f(z^j, \theta_0)}. \tag{4.8.1}$$

A maximum likelihood estimate (MLE) $\hat{\theta}_M = \theta_0 + \hat{u}_M$ satisfies

$$Z_M(\hat{u}_M) = \sup_{\{u|\theta_0+u\in\Theta\}} Z_M(u). \tag{4.8.2}$$

Since Θ is open, we cannot guarantee the existence of a maximum. Thus, we will postulate its existence and derive properties of an MLE.

In the following, we shall not write explicitly the constraint $\theta_0 + u \in \Theta$ unless it is useful to do so.

The *consistency* of $\hat{\theta}_M$ is the property that $\hat{\theta}_M \to \theta_0$ as $M \to \infty$. We may have consistency a.s. or in probability. We shall use the observation

$$\begin{aligned}
\{|\hat{\theta}_M - \theta_0| > \gamma\} &= \{|\hat{u}_M| > \gamma\} \\
&= \left\{ \sup_{|u|>\gamma} Z_M(u)| \geq \sup_{|u|\leq\gamma} Z_M(u) \right\} \\
&\subset \left\{ \sup_{|u|>\gamma} Z_M(u)| \geq Z_M(0) \right\} \\
&= \left\{ \sup_{|u|>\gamma} Z_M(u)| \geq 1 \right\},
\end{aligned} \tag{4.8.3}$$

which will be very useful in obtaining estimates.

4.8.2 Consistency

Consider the quantities

$$\pi_\theta(\gamma) = \inf_{\substack{\theta' \in \Theta \\ |\theta' - \theta| \geq \gamma}} \int_{R^d} (f^{\frac{1}{2}}(z, \theta') - f^{\frac{1}{2}}(z, \theta))^2 dz, \tag{4.8.4}$$

$$\varpi_\theta^2(\delta) = \int_{R^d} \sup_{|\theta'-\theta|\leq\delta} (f^{\frac{1}{2}}(z,\theta') - f^{\frac{1}{2}}(z,\theta))^2 dz. \tag{4.8.5}$$

We have the following.

THEOREM 4.3. *Assume*

$$\forall\theta \in \Theta, \gamma > 0,\ \pi_\theta(\gamma) > 0, \tag{4.8.6}$$

$$\forall\theta \in \bar{\Theta}\ \lim_{\delta\to 0} \varpi_\theta(\delta) = 0. \tag{4.8.7}$$

Then

$$\hat{\theta}_M \to \theta_0 \text{ as } M \to \infty, \text{ a.s.} \tag{4.8.8}$$

REMARK 4.2. The first assumption simply means that for a pair of elements $\theta, \theta' \in \Theta$ such that $|\theta - \theta'| \geq \gamma$, one has necessarily

$$f(z,\theta) \neq f(z,\theta'), \text{ on a set of positive measure.}$$

We begin with a lemma.

LEMMA 4.1. *The property (4.8.7) implies the stronger property*

$$\lim_{\delta\to 0} \sup_{\theta\in\bar{\Theta}} \varpi_\theta(\delta) = 0. \tag{4.8.9}$$

PROOF. If the property is not true, then there exists a sequence $\theta_n \in \bar{\Theta}$, $\delta_n \to 0$, such that

$$\varpi_{\theta_n}(\delta_n) \geq \beta > 0.$$

We can assume that for a subsequence, we have

$$\theta_n \to \theta^*.$$

We then use the inequality

$$\varpi_{\theta_n}(\delta_n) \leq \varpi_{\theta^*}(|\theta^* - \theta_n|) + \varpi_{\theta^*}(|\theta^* - \theta_n| + \delta_n),$$

and from (4.8.7) we get necessarily $\varpi_{\theta_n}(\delta_n) \to 0$, which contradicts the assumption. □

PROOF OF THEOREM 4.3. Consider u_0 such that $|u_0| \geq \gamma$ and $\theta_0 + u_0 \in \bar{\Theta}$. Such points exist for γ sufficiently small, since Θ is open and $\theta_0 \in \Theta$. Let Γ_0 be the sphere with center $\theta_0 + u_0$ and radius δ. We estimate

$$E \sup_{\{u|\theta_0+u\in\Gamma_0\}} Z_M^{\frac{1}{2}}(u) = E \sup_{\Gamma_0} Z_M^{\frac{1}{2}}(u).$$

We write

$$Z_M^{\frac{1}{2}}(u) = \prod_{j=1}^{M} f^{-\frac{1}{2}}(z^j, \theta_0) \prod_{j=1}^{M} f^{\frac{1}{2}}(z^j, \theta_0 + u)$$

$$\leq \prod_{j=1}^{M} f^{-\frac{1}{2}}(z^j, \theta_0) \prod_{j=1}^{M} (f^{\frac{1}{2}}(z^j, \theta_0 + u_0) + |f^{\frac{1}{2}}(z^j, \theta_0 + u) - f^{\frac{1}{2}}(z^j, \theta_0 + u_0)|).$$

Hence

$$\sup_{\Gamma_0} Z_M^{\frac{1}{2}}(u) \leq \prod_{j=1}^{M} f^{-\frac{1}{2}}(z^j, \theta_0) \prod_{j=1}^{M} (f^{\frac{1}{2}}(z^j, \theta_0 + u_0) + \sup_{\Gamma_0} |f^{\frac{1}{2}}(z^j, \theta_0 + u) - f^{\frac{1}{2}}(z^j, \theta_0 + u_0)|).$$

Therefore,

$$E \sup_{\Gamma_0} Z_M^{\frac{1}{2}}(u) \leq (X_{\Gamma_0})^M$$

with

$$X_{\Gamma_0} = \int_{R^d} f^{\frac{1}{2}}(z, \theta_0) f^{\frac{1}{2}}(z, \theta_0 + u_0) dz + + \int_{R^d} f^{\frac{1}{2}}(z, \theta_0) \sup_{\Gamma_0} |f^{\frac{1}{2}}(z, \theta_0 + u) - f^{\frac{1}{2}}(z, \theta_0 + u_0)| dz.$$

We then use

$$\int_{R^d} f^{\frac{1}{2}}(z, \theta_0) f^{\frac{1}{2}}(z, \theta_0 + u_0) dz = 1 - \frac{1}{2} \int_{R^d} (f^{\frac{1}{2}}(z, \theta_0) - f^{\frac{1}{2}}(z, \theta_0 + u_0))^2 dz$$
$$\leq 1 - \frac{1}{2} \pi_{\theta_0}(\gamma).$$

Next,

$$\int_{R^d} f^{\frac{1}{2}}(z, \theta_0) \sup_{\Gamma_0} |f^{\frac{1}{2}}(z, \theta_0 + u) - f^{\frac{1}{2}}(z, \theta_0 + u_0)| dz \leq$$
$$\sqrt{\int_{R^d} \sup_{\Gamma_0} |f^{\frac{1}{2}}(z, \theta_0 + u) - f^{\frac{1}{2}}(z, \theta_0 + u_0)|^2 dz} = \varpi_{\theta_0 + u_0}(\delta).$$

Collecting results, we can write

$$X_{\Gamma_0} \leq 1 - \frac{1}{2}\pi_{\theta_0}(\gamma) + \varpi_{\theta_0+u_0}(\delta).$$

Therefore,

$$E \sup_{\Gamma_0} Z_M^{\frac{1}{2}}(u) \leq \left(1 - \frac{1}{2}\pi_{\theta_0}(\gamma) + \varpi_{\theta_0+u_0}(\delta)\right)^M$$
$$\leq \exp -M(\frac{1}{2}\pi_{\theta_0}(\gamma) - \varpi_{\theta_0+u_0}(\delta)), \tag{4.8.10}$$

in which we have used the elementary inequality

$$a + 1 \leq \exp a \ , \forall a$$

applied with $1 + a > 0$.

For each vector u such that $|u| \geq \gamma$ and $\theta_0 + u \in \bar{\Theta}$, we consider the ball with center $\theta_0 + u$ of radius δ. We obtain a covering of the set $\{\theta_0 + u \in \bar{\Theta}, |u| \geq \gamma\}$. Since this set is compact, we obtain a finite covering of this set by balls Γ_j with center $\theta_0 + u_j$ and radius δ, with $j = 1, \cdots J$. Hence

$$\sup_{|u|>\gamma} Z_M^{\frac{1}{2}}(u)| \leq \sum_{j=1}^{J} \sup_{\Gamma_j} Z_M^{\frac{1}{2}}(u)$$

and

$$E \sup_{|u|>\gamma} Z_M^{\frac{1}{2}}(u)| \leq \sum_{j=1}^{J} \exp -M(\frac{1}{2}\pi_{\theta_0}(\gamma) - \varpi_{\theta_0+u_j}(\delta))$$
$$\leq J \exp -M(\frac{1}{2}\pi_{\theta_0}(\gamma) - \sup_{\theta\in\bar{\Theta}} \varpi_\theta(\delta)).$$

From the property (4.8.9) we can choose δ sufficiently small that

$$\sup_{\theta\in\bar{\Theta}} \varpi_\theta(\delta) \leq \frac{1}{4}\pi_{\theta_0}(\gamma),$$

and hence

$$E \sup_{|u|>\gamma} Z_M^{\frac{1}{2}}(u)| \leq J \ \exp -M(\frac{1}{4}\pi_{\theta_0}(\gamma)).$$

Now from (4.8.3), we have

$$P\left(\{|\hat{\theta}_M - \theta_0| > \gamma\}\right) \leq P\left(\left\{\sup_{|u|>\gamma} Z_M(u)| \geq 1\right\}\right)$$
$$= P\left(\left\{\sup_{|u|>\gamma} Z_M^{\frac{1}{2}}(u)| \geq 1\right\}\right)$$
$$\leq E \sup_{|u|>\gamma} Z_M^{\frac{1}{2}}(u)| \leq J \exp -M(\frac{1}{4}\pi_{\theta_0}(\gamma)).$$

It follows that

$$P\left(\cup_{M\geq M_0}\{|\hat{\theta}_M - \theta_0| > \gamma\}\right) \leq \sum_{M\geq M_0} P\left(\{|\hat{\theta}_M - \theta_0| > \gamma\}\right)$$
$$\leq J \sum_{M\geq M_0} \exp -M(\frac{1}{4}\pi_{\theta_0}(\gamma))$$
$$= J \frac{\exp -M_0(\frac{1}{4}\pi_{\theta_0}(\gamma))}{1 - \exp -(\frac{1}{4}\pi_{\theta_0}(\gamma))} \to 0, \text{ as } M_0 \to +\infty.$$

Therefore,

$$P\left(\cap_{M_0=1}^{\infty} \cup_{M\geq M_0} \{|\hat{\theta}_M - \theta_0| > \gamma\}\right) = 0.$$

Since γ is arbitrary, the sequence $\hat{\theta}_M$ almost surely cannot have an accumulation point different from θ_0, which implies (4.8.8) and completes the proof. ■

4.8.3 Asymptotic Normality

Let us consider the log likelihood

$$L(z, \theta) = \log f(z, \theta).$$

The MLE maximizes

$$\sum_{j=1}^{M} L(z^j, \theta)$$

over θ in the open domain Θ. Therefore, we can write

$$\sum_{j=1}^{M} D_\theta L(z^j, \hat{\theta}_M) = 0. \tag{4.8.11}$$

We assume some regularity on the derivatives of the log likelihood function. Namely, there exists δ, $\frac{1}{2} < \delta < 1$, such that

$$E|D_\theta L(z,\theta_0)|^{1+\delta} < +\infty, \tag{4.8.12}$$

$$E||\, D_\theta^2 L(z,\theta_0)||^{1+\delta} < +\infty. \tag{4.8.13}$$

Next, we define

$$R(z,\theta) = \sup_{\theta' \in \Theta} \frac{||\, D_\theta^2 L(z,\theta') - D^2 L(z,\theta)||}{|\theta' - \theta|^\delta} \tag{4.8.14}$$

and assume

$$ER(z,\theta_0) < +\infty. \tag{4.8.15}$$

Consider the Fisher information matrix

$$I(\theta_0) = -ED_\theta^2 L(z,\theta_0). \tag{4.8.16}$$

It is well known that

$$I(\theta_0) = E\left(D_\theta L(z,\theta_0)(D_\theta L(z,\theta_0))^*\right) \geq 0.$$

We assume

$$I(\theta_0) \text{ invertible} \tag{4.8.17}$$

and state the main result.

THEOREM 4.4. *We use the assumptions of Theorem 4.3 and (4.8.12), (4.8.13), (4.8.15), (4.8.17). We then have the property*

$$\sqrt{M}(\hat{\theta}_M - \theta_0) \to N(0, I^{-1}(\theta_0)), \tag{4.8.18}$$

the convergence being in law, and the limit Gaussian having mean 0 *and covariance matrix* $I^{-1}(\theta_0)$.

Since Theorem 4.3 holds, we know that $\hat{\theta}_M - \theta_0 \to 0$ a.s. We will prove a stronger result.

LEMMA 4.2. *We have*

$$M^{\frac{\delta}{1+\delta}}(\hat{\theta}_M - \theta_0) \to 0, \; a.s. \tag{4.8.19}$$

PROOF. We note that $\frac{\delta}{1+\delta} < \frac{1}{2}$, since $\delta < 1$. From (4.8.11) we can write

$$\sum_{j=1}^{M} D_\theta L(z^j, \theta_0) = \sum_{j=1}^{M} (D_\theta L(z^j, \theta_0) - D_\theta L(z^j, \hat{\theta}_M))$$

$$= -\sum_{j=1}^{M} \int_0^1 D_\theta^2 L(z^j, \theta_0 + \lambda(\hat{\theta}_M - \theta_0)) d\lambda \, (\hat{\theta}_M - \theta_0)$$

$$= -\sum_{j=1}^{M} \int_0^1 D_\theta^2 L(z^j, \theta_0)(\hat{\theta}_M - \theta_0) -$$

$$-\sum_{j=1}^{M} \int_0^1 \left(D_\theta^2 L(z^j, \theta_0 + \lambda(\hat{\theta}_M - \theta_0)) - D_\theta^2 L(z^j, \theta_0) \right) d\lambda \, (\hat{\theta}_M - \theta_0).$$

It follows that

$$\sum_{j=1}^{M} D_\theta L(z^j, \theta_0) = M \, I(\theta_0)(\hat{\theta}_M - \theta_0)$$

$$-\sum_{j=1}^{M} \left(D_\theta^2 L(z^j, \theta_0) - E D_\theta^2 L(z^j, \theta_0) \right) (\hat{\theta}_M - \theta_0)$$

$$-\sum_{j=1}^{M} \int_0^1 \left(D_\theta^2 L(z^j, \theta_0 + \lambda(\hat{\theta}_M - \theta_0)) - D_\theta^2 L(z^j, \theta_0) \right) d\lambda \, (\hat{\theta}_M - \theta_0).$$

This is an equality between vectors in R^k. We multiply on the left by the line vector

$$\frac{(\hat{\theta}_M - \theta_0)^*}{|\hat{\theta}_M - \theta_0|}$$

and obtain

$$\frac{M}{|\hat{\theta}_M - \theta_0|} (\hat{\theta}_M - \theta_0)^* I(\theta_0)(\hat{\theta}_M - \theta_0) = \frac{(\hat{\theta}_M - \theta_0)^* \sum_{j=1}^{M} D_\theta L(z^j, \theta_0)}{|\hat{\theta}_M - \theta_0|} +$$

$$+ \frac{(\hat{\theta}_M - \theta_0)^* \sum_{j=1}^{M} \left(D_\theta^2 L(z^j, \theta_0) - E D_\theta^2 L(z^j, \theta_0) \right) (\hat{\theta}_M - \theta_0)}{|\hat{\theta}_M - \theta_0|} +$$

$$+ \frac{(\hat{\theta}_M - \theta_0)^* \sum_{j=1}^{M} \int_0^1 \left(D_\theta^2 L(z^j, \theta_0 + \lambda(\hat{\theta}_M - \theta_0)) - D_\theta^2 L(z^j, \theta_0) \right) d\lambda \, (\hat{\theta}_M - \theta_0)}{|\hat{\theta}_M - \theta_0|}.$$

Since $I(\theta_0)$ is invertible, we deduce that

$$\alpha M|\hat{\theta}_M - \theta_0| \leq |\sum_{j=1}^{M} D_\theta L(z^j, \theta_0)| + \tag{4.8.20}$$

$$+||\sum_{j=1}^{M}\left(D_\theta^2 L(z^j, \theta_0) - ED_\theta^2 L(z^j, \theta_0)\right)|||\hat{\theta}_M - \theta_0| + (\sum_{j=1}^{M} R(z^j, \theta_0))|\hat{\theta}_M - \theta_0|^{1+\delta},$$

where $\alpha > 0$. We set

$$\mathcal{R}_M = \frac{\sum_{j=1}^M R(z^j, \theta_0)}{M}.$$

Since the random variables $R(z^j, \theta_0)$ are independent i.i.d. and $ER(z^j, \theta_0) = ER(z, \theta_0) < +\infty$, we can refer to Kolmogorov's strong law of large numbers to claim that

$$\mathcal{R}_M \to ER(z, \theta_0), \text{ a.s.} \tag{4.8.21}$$

Therefore,

$$\mathcal{R}_M|\hat{\theta}_M - \theta_0|^\delta \to 0, \text{ a.s.} \tag{4.8.22}$$

From the assumptions (4.8.12), (4.8.13), we can assert that

$$\frac{\sum_{j=1}^M D_\theta L(z^j, \theta_0)}{M^{\frac{1}{1+\delta}}} \to 0, \quad \frac{\sum_{j=1}^M \left(D_\theta^2 L(z^j, \theta_0) - ED_\theta^2 L(z^j, \theta_0)\right)}{M^{\frac{1}{1+\delta}}} \to 0, \text{ a.s.} \tag{4.8.23}$$

This follows from a result of Marcinkiewicz, whose proof can be found in Loève [35]. The result is the following: let $\xi_1, \cdots \xi_n, \cdots$ be independent identically distributed random variables such that $E|\xi_n|^{1+\delta} < \infty$, $0 \leq \delta < 1$. Then

$$\frac{\sum_{j=1}^n (\xi_j - E\xi_j)}{n^{\frac{1}{1+\delta}}} \to 0, \text{ a.s.}$$

From (4.8.20) we can write

$$M^{\frac{\delta}{1+\delta}}|\hat{\theta}_M - \theta_0|(\alpha - \mathcal{R}_M|\hat{\theta}_M - \theta_0|^\delta) \leq \left|\frac{\sum_{j=1}^M D_\theta L(z^j, \theta_0)}{M^{\frac{1}{1+\delta}}}\right|$$

$$+\left|\frac{\sum_{j=1}^M \left(D_\theta^2 L(z^j, \theta_0) - ED_\theta^2 L(z^j, \theta_0)\right)}{M^{\frac{1}{1+\delta}}}\right|,$$

and the result (4.8.19) follows immediately. □

We turn now to the proof of Theorem 4.4.

PROOF. We write now

$$\sqrt{M}\, I(\theta_0)(\hat{\theta}_M - \theta_0) = \frac{1}{\sqrt{M}} \sum_{j=1}^{M} D_\theta L(z^j, \theta_0) + \Gamma_1^M + \Gamma_2^M$$

with

$$\Gamma_1^M = \frac{1}{\sqrt{M}} \sum_{j=1}^{M} \left(D_\theta^2 L(z^j, \theta_0) - E D_\theta^2 L(z^j, \theta_0) \right) (\hat{\theta}_M - \theta_0)$$

and

$$\Gamma_2^M = \frac{1}{\sqrt{M}} \sum_{j=1}^{M} \int_0^1 \left(D_\theta^2 L(z^j, \theta_0 + \lambda(\hat{\theta}_M - \theta_0)) - D_\theta^2 L(z^j, \theta_0) \right) d\lambda\, (\hat{\theta}_M - \theta_0).$$

Then

$$|\Gamma_2^M| \leq \frac{1}{\sqrt{M}} \sum_{j=1}^{M} R(z^j, \theta_0) |\hat{\theta}_M - \theta_0|^{1+\delta} =$$
$$= \mathcal{R}_M \sqrt{M} |\hat{\theta}_M - \theta_0|^{1+\delta}.$$

Since $\delta > \frac{1}{2}$, we have $M^{\frac{1}{2(1+\delta)}} (\hat{\theta}_M - \theta_0) \to 0$, a.s. Since $\mathcal{R}_M$is a.s. bounded, we get $\Gamma_2^M \to 0$ a.s. Next we write

$$\Gamma_1^M = \frac{\sum_{j=1}^M \left(D_\theta^2 L(z^j, \theta_0) - E D_\theta^2 L(z^j, \theta_0) \right)}{M^{\frac{1}{1+\delta}}} M^{\frac{1-\delta}{2(1+\delta)}} (\hat{\theta}_M - \theta_0).$$

Since $\frac{1-\delta}{2} < \delta$, we can assert that $M^{\frac{1-\delta}{2(1+\delta)}} (\hat{\theta}_M - \theta_0) \to 0$ a.s. Thanks also to the second part of (4.8.23), we can conclude that $\Gamma_1^M \to 0$ a.s.

Furthermore, it is standard that

$$\frac{1}{\sqrt{M}} \sum_{j=1}^{M} D_\theta L(z^j, \theta_0) \to N(0, I(\theta_0))$$

in law. Hence

$$\frac{1}{\sqrt{M}} I(\theta_0)^{-1} \sum_{j=1}^{M} D_\theta L(z^j, \theta_0) \to N(0, (I(\theta_0))^{-1})$$

in law. This implies the result (4.8.18). ■

Chapter 5
Further Techniques of Estimation

We continue the study of estimation. The contents of this chapter are more of interest to statisticians and can be skipped by readers having control theory as a priority.

5.1 Generalized Linear Models

Generalized linear models (GLM) are a generalization of the classical linear Gaussian model

$$z = Hx + h + \epsilon, \; x \in R^n, \; z \in R^d, \tag{5.1.1}$$

in which x is the variable of interest, H is a given matrix in $\mathcal{L}(R^n; R^d)$, and ϵ is a Gaussian variable in R^d with mean 0 and variance R. The major generalization is in giving up the Gaussian assumption. However, this prevents us from writing the observation as in (5.1.1). We have to work with the conditional probability density $f(z|x)$. We shall assume in the general presentation for the sake of simplicity that $d = 1$.

We consider a real random variable, still denoted by z, whose range is not necessarily $(-\infty, +\infty)$ and assume that it has a probability density defined on the range depending on a parameter θ, also a scalar. This parameter is called the **canonical parameter**. We recall that we use the same notation for the random variable z and the argument of the probability density, also called the likelihood. So we call $f(z, \theta)$ the likelihood of the random variable z. If z is a discrete probability distribution, we keep the notation to represent the probabilities of specific values of z. We consider the log likelihood $L(z, \theta) = \log\, f(z, \theta)$, defined on the range of values of the random variable z.

A. Bensoussan, *Estimation and Control of Dynamical Systems*, Interdisciplinary Applied Mathematics 48, https://doi.org/10.1007/978-3-319-75456-7_5

The basic assumption is that this function has the form

$$L(z,\theta) = z\theta - b(\theta) + c(z). \tag{5.1.2}$$

Because the function $L(z,\theta)$ is a log likelihood, we shall see that the function $b(\theta)$ cannot be arbitrary. If it is smooth, we shall show that it is necessarily convex. First of all, we check the relation

$$\mu = b'(\theta), \tag{5.1.3}$$

in which μ is the mean of z. Indeed,

$$E\,\frac{\partial L(z,\theta)}{\partial\theta} = 0,$$

from which (5.1.3) follows immediately. Assume that the function $b'(\theta)$ is invertible on the range of values of z. Therefore, the canonical parameter can be expressed in terms of the mean. Next, from the relation

$$\frac{\partial^2 L(z,\theta)}{\partial\theta^2} + (\frac{\partial L(z,\theta)}{\partial\theta})^2 = \frac{1}{f(z,\theta)}\frac{\partial^2 f(z,\theta)}{\partial\theta^2}, \tag{5.1.4}$$

we have

$$E[\frac{\partial^2 L(z,\theta)}{\partial\theta^2} + (\frac{\partial L(z,\theta)}{\partial\theta})^2] = 0$$

and

$$V = E(z-\mu)^2 = b''(\theta), \tag{5.1.5}$$

which proves the convexity of $b(\theta)$. Also,

$$\frac{\partial^3 L}{\partial\theta^3} + (\frac{\partial L}{\partial\theta})^3 + 3\frac{\partial L}{\partial\theta}\frac{\partial^2 L}{\partial\theta^2} = \frac{1}{f}\frac{\partial^3 f(z,\theta)}{\partial\theta^3},$$

from which we deduce

$$E[\frac{\partial^3 L}{\partial\theta^3} + (\frac{\partial L}{\partial\theta})^3] = 0,$$

and hence

$$E(z-\mu)^3 = b'''(\theta). \tag{5.1.6}$$

The next ingredient of GLM models is the **link function**. It connects the canonical parameter to the variables of interest $x \in R^n$, and uses the mean as an intermediary. We express the link by the relation

$$h^* x = g(\mu), \tag{5.1.7}$$

where $h^* \in \mathcal{L}(R^n; R)$ is equivalent to the matrix H in formula (5.1.1), since $d = 1$, with the constant term taken as 0. The function g is a link function, defined on the range of values of z. If g is invertible on its domain, we can express the mean as a function of h^*x, and also the canonical parameter as a function of the variables of interest, by inverting the relation

$$h^*x = g(b'(\theta)). \tag{5.1.8}$$

These considerations form the core of the GLM models, up to some extensions. We will review the case of vector observations in Section 5.4.

5.2 Examples

5.2.1 The Gaussian Distribution

Let

$$f(z) = \frac{1}{\sqrt{2\pi\sigma^2}} \exp -\frac{(z-\mu)^2}{2\sigma^2}$$

and

$$\log f(z) = -\frac{(z-\mu)^2}{2\sigma^2} - \log \sqrt{2\pi}\sigma.$$

If we set $\theta = \frac{\mu}{\sigma^2}$ and $b(\theta) = \frac{1}{2}\sigma^2\theta^2$, considering σ as a fixed constant, we can write

$$\log f(z) = L(z, \theta)$$

with $L(z, \theta)$ defined by (5.1.2), in which

$$c(z) = -\frac{1}{2}\frac{z^2}{\sigma^2} - \log \sqrt{2\pi}\sigma.$$

5.2.2 The Exponential Distribution

Let

$$f(z) = \frac{1}{\mu} \exp -\frac{z}{\mu}$$

for $z \in R^+$. On the range we have

$$\log f(z) = -\log \mu - \frac{z}{\mu},$$

which can be written as $L(z, \theta)$ with

$$\theta = -\frac{1}{\mu} \quad b(\theta) = -\log -\theta \quad c(z) = 0. \tag{5.2.1}$$

The function $b(\theta)$ is defined on R^-.

5.2.3 The Poisson Distribution

We have

$$f(z) = \exp -\mu \, \frac{\mu^z}{z!}$$

with z an integer. Of course $f(z)$ is not a density. We keep this notation for convenience. Therefore,

$$\begin{aligned} \log f(z) &= z \log \mu - \mu - \log z! \\ &= L(z, \theta), \end{aligned}$$

with $\theta = \log \mu$, $b(\theta) = \exp \theta$, $c(z) = -\log z!$.

5.2.4 The Binomial Distribution

Let

$$f(z) = C_q^z \pi^z (1 - \pi)^{q-z}$$

with z running from 0 to q. So

$$\begin{aligned} \log f(z) &= z \log \frac{\pi}{1 - \pi} + q \log(1 - \pi) + \log C_q^z \\ &= L(z, \theta), \end{aligned}$$

with

$$\theta = \log \frac{\pi}{1 - \pi}, \; b(\theta) = q \log(1 + \exp \theta), \; c(z) = \log C_q^z.$$

We have

$$\mu = q\pi = q\frac{\exp\theta}{1+\exp\theta}.$$

5.2.5 The Gamma Distribution

We have

$$f(z) = \frac{1}{\beta^{\alpha}\Gamma(\alpha)}z^{\alpha-1}\exp -\frac{z}{\beta}.$$

The range is R^{+}, and α, β are positive parameters. We have

$$\begin{aligned}\log f(z) &= -\frac{z}{\beta} - \alpha\log\beta + (\alpha-1)\log z - \log\Gamma(\alpha)\\ &= L(z,\theta),\end{aligned}$$

with

$$\theta = -\frac{1}{\beta},\ b(\theta) = -\alpha\log -\theta,\ c(z) = (\alpha-1)\log z - \log\Gamma(\alpha).$$

The function $b(\theta)$ is defined on R^{-}.

We consider α a given number. Note that $\mu = \alpha\beta = -\frac{\alpha}{\theta}$. For $\alpha = 1$, we recover the exponential distribution.

5.2.6 The Weibull Distribution

The Weibull distribution is defined by

$$\varphi(y) = \frac{k}{\lambda}(\frac{y}{\lambda})^{k-1}\exp -(\frac{y}{\lambda})^{k}$$

over the range $y > 0$. We have used a different notation $\varphi(y)$ instead of $f(z)$ intentionally, because this distribution does not satisfy the assumptions of GLM.

If $k = 1$, it reduces to the exponential distribution with mean λ. The mean is given by

$$\mu = \lambda\Gamma(1+\frac{1}{k}).$$

The parameter $k \geq 1$ is called the shape parameter, and λ is called the scale parameter. Suppose λ is linked to the variables of interest x by the relation

$$\lambda = h^{*}x. \tag{5.2.2}$$

Can we estimate x by observing the random variable y? The answer is yes, because we can associate to y a random variable, observable when y is observed, that belongs to the GLM family. This is done by defining

$$z = y^k,$$

for which the density is given by

$$f(z) = \frac{1}{\lambda^k} \exp - \frac{z}{\lambda^k},$$

defined on R^+. Thus it is an exponential distribution that belongs to the GLM family with

$$\theta = -\frac{1}{\lambda^k}, \quad b(\theta) = -\log -\theta.$$

The function $b(\theta)$ is defined on R^-. If now we introduce the link function

$$g(\mu) = \mu^{\frac{1}{k}},$$

we can write

$$h^* x = g(b'(\theta)) = (-\frac{1}{\theta})^{\frac{1}{k}},$$

and we are in the general framework described in the introduction.

5.2.7 Nonlinear Gaussian Model

Consider the model

$$z = \varphi(h^* x) + \epsilon, \tag{5.2.3}$$

in which ϵ is Gaussian with mean 0 and variance σ^2. We assume that φ is invertible. It belongs to the GLM family with

$$b(\theta) = \frac{\sigma^2 \theta^2}{2}, \quad g(\mu) = \varphi^{-1}(\mu). \tag{5.2.4}$$

5.2.8 Canonical Links

A link function $g(.)$ is **canonical** if

$$g(b'(\theta)) = c\theta, \tag{5.2.5}$$

where c is a constant. Therefore, the following link functions are canonical for the GLM models indicated in parentheses:

$$g(\mu) = \mu, \quad \text{(Gaussian)}$$

$$g(\mu) = -\frac{1}{\mu}, \quad \text{(Exponential)}$$

$$g(\mu) = \log \mu, \quad \text{(Poisson)}$$

$$g(\mu) = \log \frac{\mu}{q - \mu}, \quad \text{(Binomial)}$$

$$g(\mu) = -\frac{\alpha}{\mu}, \quad \text{(Gamma)}$$

Note that in the first case, the constant is σ^2. In the other cases, the constant is 1. For the Weibull distribution, discussed in Section 5.2.6, the link function for the exponential variable $z = y^k$ is not canonical, since

$$g(b'(\theta)) = (-\frac{1}{\theta})^{\frac{1}{k}},$$

As in the nonlinear Gaussian case, the link function is not canonical. For canonical link functions, we have simply

$$c\theta = h^* x \tag{5.2.6}$$

and also

$$g'(b'(\theta))b''(\theta) = c. \tag{5.2.7}$$

REMARK. By changing h into $\frac{h}{c}$, it is always possible to take $c = 1$, which is the more common definition of canonical link.

5.3 MLE for Generalized Linear Models

5.3.1 Statement of the Problem and Notation

Consider now a sequence of independent random variables $z^1, \cdots, z^M$ that follow GLM distributions with canonical parameters $\theta^1, \cdots, \theta^M$. We continue to assume that these variables are scalar for the sake of simplicity. So the canonical parameters are also scalar. These canonical parameters are linked to the variables of interest x by the relations

$$(h^j)^* x = g(b'(\theta^j)), \tag{5.3.1}$$

So the link function $g(.)$ and the function $b(\theta)$ are identical for all the variables. We define the functions $\mu^j(x)$ by solving

$$(h^j)^* x = g(\mu^j(x)), \tag{5.3.2}$$

which is possible, since g is invertible. Similarly, we define the functions $\theta^j(x)$ by solving

$$\mu^j(x) = b'(\theta^j(x)) \tag{5.3.3}$$

and

$$V^j(x) = b''(\theta^j(x)). \tag{5.3.4}$$

Recalling the function

$$f(z,\theta) = \exp(z\theta - b(\theta) + c(z)), \tag{5.3.5}$$

we see that the probability density of the variable z^j is $f(z, \theta^j(x_0))$, in which we have denoted by x_0 the true value of the parameter. We note that the variables z^j are not identically distributed, which introduces a slight difficulty with respect to the MLE developed in the previous section. The log likelihood function is $L(z, \theta^j(x))$, where $L(z,\theta) = \log f(z,\theta)$. The joint probability density of the sample $z^1, \cdots, z^M$ is

$$\prod_{j=1}^{M} f(z^j, \theta^j(x_0)). \tag{5.3.6}$$

The MLE is obtained by maximizing the function of x

$$\prod_{j=1}^{M} f(z^j, \theta^j(x)). \tag{5.3.7}$$

As we have done for the MLE in general, we shall assume, to simplify technicalities, that we know a bounded convex open domain, denoted by $\mathcal{X}$, and

$$x_0 \in \mathcal{X}.$$

So we maximize the function (5.3.7) on $\mathcal{X}$, and we assume that such a maximum exists, which we denote by $\hat{x}_M$. To prove the asymptotic properties of $\hat{x}_M$, we shall adapt the methods used for the MLE in general. We first introduce some notation. From formula (5.3.1), we can write

$$\theta^j(x) = \varphi((h^j)^* x), \tag{5.3.8}$$

with

$$\varphi(\eta) = b'^{-1}(g^{-1}(\eta)). \tag{5.3.9}$$

Hence

$$D_x\theta^j(x) = \varphi'((h^j)^*x)h^j,$$

and then

$$\begin{aligned} D_xL(z,\theta^j(x)) &= (z - b'(\theta^j(x)))D_x\theta^j(x) \\ &= (z-\mu^j(x))\varphi'((h^j)^*x)h^j \\ &= (z - b'(\varphi((h^j)^*x)))\varphi'((h^j)^*x)h^j \\ &= [z\varphi'((h^j)^*x) - (bo\varphi)'((h^j)^*x)]h^j. \end{aligned} \tag{5.3.10}$$

Therefore,

$$D_x^2L(z,\theta^j(x)) = [z\varphi''((h^j)^*x) - (bo\varphi)''((h^j)^*x)]h^j(h^j)^*. \tag{5.3.11}$$

Note that

$$\varphi'(\eta) = \frac{1}{g'(b'(\varphi(\eta)))b''(\varphi(\eta))},$$

and therefore,

$$\varphi'((h^j)^*x) = \frac{1}{g'(b'(\theta^j(x)))b''(\theta^j(x))}.$$

Recalling

$$\mu^j(x) = b'(\theta^j(x)),\ \ V^j(x) = b''(\theta^j(x)),$$

we have

$$\varphi'((h^j)^*x) = \frac{1}{g'(\mu^j(x))V^j(x)}.$$

It is convenient to introduce the weights

$$W^j(x) = \frac{1}{(g')^2(\mu^j(x))V^j(x)}.$$

Therefore,

$$\varphi'((h^j)^*x) = g'(\mu^j(x))W^j(x). \tag{5.3.12}$$

Also, we have

$$\mu^j(x) = b'o\varphi((h^j)^*x).$$

Next, we note that

$$b'o\varphi(\eta)\varphi''(\eta) - (bo\varphi)''(\eta) = -b''(\varphi(\eta))(\varphi')^2(\eta).$$

Therefore, from (5.3.11) we get

$$
\begin{aligned}
D_x^2 L(z,\theta^j(x)) &= [(z-\mu^j(x))\varphi''((h^j)^*x) - V^j(x)(\varphi')^2((h^j)^*x)]h^j(h^j)^* \\
&= [(z-\mu^j(x))\varphi''((h^j)^*x) - W^j(x)]h^j(h^j)^*. \qquad (5.3.13)
\end{aligned}
$$

Considering the true value of the parameter x_0, we note that

$$ED_xL(z^j,\theta^j(x_0)) = 0,$$

whence

$$Ez^j = \mu^j(x_0) \qquad (5.3.14)$$

and

$$ED_x^2L(z^j,\theta^j(x_0)) = -W^j(x_0)h^j(h^j)^*. \qquad (5.3.15)$$

Also,

$$
\begin{aligned}
ED_x^2L(z^j,\theta^j(x_0)) &= -E(D_xL(z^j,\theta^j(x_0)))(D_xL(z^j,\theta^j(x_0)))^* \\
&= -E(z-\mu^j(x_0))^2(\varphi'((h^j)^*x_0))^2h^j(h^j)^*.
\end{aligned}
$$

It follows that

$$
\begin{aligned}
W^j(x_0) &= E((z-\mu^j(x_0)))^2(\varphi'((h^j)^*x_0))^2 \\
&= E(z-\mu^j(x_0))^2\frac{1}{(g'(\mu^j(x_0))V^j(x_0))^2}, \\
\frac{1}{(g')^2(\mu^j(x_0))V^j(x_0)} &= E(z-\mu^j(x_0))^2\frac{1}{(g'(\mu^j(x_0))V^j(x_0))^2},
\end{aligned}
$$

which implies the interpretation

$$E(z-\mu^j(x_0))^2 = V^j(x_0). \qquad (5.3.16)$$

We recover of course (5.1.5).

5.3.2 Examples

In the Gaussian case, we have $g(\mu)=\mu$ and $b(\theta)=\frac{1}{2}\sigma\theta^2$. We get easily

$$D_xL(Z;\,x) = \frac{1}{\sigma}\sum_{j=1}^{M}(z^j-(h^j)^*x)h^j, \qquad (5.3.17)$$

and thus $\hat{x}_M$ satisfies

$$\sum_{j=1}^{M} z^j h^j = \sum_{j=1}^{M} h^j (h^j)^* \hat{x}_M, \tag{5.3.18}$$

and this system has one and only one solution, provided the matrix $\sum_{j=1}^{M} h^j (h^j)^*$ (which belongs to $\mathcal{L}(R^n; R^n)$), is invertible.

Let us consider the Weibull distribution case, with known shape k; see Section 5.2.6. We have for the variables $z^j = (y^j)^k$, $g(\mu) = \mu^{\frac{1}{k}}$, $b(\theta) = -\log(-\theta)$. Hence

$$\mu^j(x) = ((h^j)^* x)^k, \quad \theta^j(x) = -\frac{1}{\mu^j(x)}, \tag{5.3.19}$$

$$V^j(x) = (\mu^j(x))^2, \quad W^j(x) = \frac{k^2}{(\mu^j(x))^{\frac{2}{k}}}.$$

We obtain the system

$$\sum_{j=1}^{M} (\frac{z^j}{\mu^j(\hat{x}_M)} - 1) \frac{h^j}{(h^i)^* \hat{x}_M} = 0. \tag{5.3.20}$$

Let us finally consider the nonlinear Gaussian case; see Section 5.2.7. We have

$$g(\mu) = \varphi^{-1}(\mu), \quad b(\theta) = \frac{\sigma^2 \theta^2}{2};$$

hence

$$\mu^j(x) = \varphi((h^j)^* x), \quad \theta^j(x) = \frac{\mu^j(x)}{\sigma^2}, \tag{5.3.21}$$

$$V^j(x) = \sigma^2, \quad W^j(x) = \frac{(\varphi'((h^j)^* x))^2}{\sigma^2}.$$

We obtain the system

$$\sum_{j=1}^{M} (z^j - \varphi((h^j)^* \hat{x}_M)) \varphi'((h^j)^* \hat{x}_M) h^j = 0. \tag{5.3.22}$$

5.3.3 Consistency

We consider the issue of consistency. We will need an assumption of linear independence of the vectors h^j. More precisely, let us consider for $i = 1, \cdots$ the $n \times n$ matrix

$$H^i = \begin{pmatrix} (h^{(i-1)n+1})^* \\ \vdots \\ (h^{in})^* \end{pmatrix}.$$

We assume that

$$H^i \text{ is invertible } \forall i \tag{5.3.23}$$

and

$$||H^i||,\ ||(H^i)^{-1}|| \leq C. \tag{5.3.24}$$

We shall need the following assumption:

$$\int_R \sup_{\theta\in\Theta} f(z,\theta)dz < \infty,\ \int_R z^2 \sup_{\theta\in\Theta} f(z,\theta)dz < \infty \forall\Theta \text{ compact interval.} \tag{5.3.25}$$

THEOREM 5.1. *We consider the GLM defined by (5.3.5), where* $b(\theta)$ *is* C^2 *and strictly convex. We assume that the link function* $g(\mu)$ *has an inverse and is* C^1. *We also assume (5.3.23), (5.3.24), and (5.3.25). Then we have the consistency property*

$$\hat{x}_M \rightarrow x_0 \ a.s. \tag{5.3.26}$$

PROOF. We will operate as in Theorem 4.3. We need to obtain properties similar to (4.8.4) and (4.8.5). Consider the functions $\theta^j(x)$ and define vector functions (with values in R^n)

$$\bar{\theta}^i(x) = \begin{pmatrix} \theta^{(i-1)n+1}(x) \\ \vdots \\ \theta^{in}(x) \end{pmatrix}.$$

We next define a sequence of probability densities in R^n, depending on the vector $\bar{\theta}^i(x)$, given by the formula

$$\bar{f}(\bar{z},\bar{\theta}^i(x)) = \prod_{l=1}^{n} f(\bar{z}_l, \theta^{(i-1)n+l}(x)), \tag{5.3.27}$$

where the argument $\bar{z}$ is in R^n, and the $\bar{z}_l$ are in the range $l = 1,\ldots,n$. We define the random vector $\bar{z}^i \in R^n$ by

$$\bar{z}^i = \begin{pmatrix} z^{(i-1)n+1} \\ \vdots \\ z^{in} \end{pmatrix}.$$

We notice that the sequence of scalar random variables $z^1, \cdots, z^{nM}$ is equivalent to the sequence of vector random variables $\bar{z}^1, \cdots, \bar{z}^M$. We first consider the random function

$$Z_{nM}(u) = \prod_{j=1}^{nM} \frac{f(z^j, \theta^j(x_0+u))}{f(z^j, \theta^j(x_0))},$$

and we can write

$$\begin{aligned} Z_{nM}(u) &= \bar{Z}_M(u) \\ &= \prod_{i=1}^{M} \frac{\bar{f}(\bar{z}^i, \bar{\theta}^i(x_0+u))}{\bar{f}(\bar{z}^i, \bar{\theta}^i(x_0))}. \end{aligned}$$

Let u_0 with $|u_0| \geq \gamma$ and $x_0 + u_0 \in \bar{\mathcal{X}}$. We consider the sphere with center $x_0 + u_0$ and radius δ. We call it Γ_0. We shall estimate

$$E \sup_{\Gamma_0} \bar{Z}_M^{\frac{1}{2}}(u) = E \sup_{\{u | x_0 + u \in \Gamma_0\}} \bar{Z}_M^{\frac{1}{2}}(u).$$

Writing

$$\begin{aligned} \sup_{\Gamma_0} \bar{Z}_M^{\frac{1}{2}}(u) \leq \prod_{i=1}^{M} [&\bar{f}^{-\frac{1}{2}}(\bar{z}^i, \bar{\theta}^i(x_0)) \bar{f}^{\frac{1}{2}}(\bar{z}^i, \bar{\theta}^i(x_0+u_0)) + \\ &+ \bar{f}^{-\frac{1}{2}}(\bar{z}^i, \bar{\theta}^i(x_0)) \sup_{\Gamma_0} |\bar{f}^{\frac{1}{2}}(\bar{z}^i, \bar{\theta}^i(x_0+u)) - \bar{f}^{\frac{1}{2}}(\bar{z}^i, \bar{\theta}^i(x_0+u_0))|], \end{aligned}$$

we get

$$E \sup_{\Gamma_0} \bar{Z}_M^{\frac{1}{2}}(u) \leq \prod_{i=1}^{M} X_{\Gamma_0}^i,$$

with

$$\begin{aligned} \bar{X}_{\Gamma_0}^i = \int_{R^n} &\bar{f}^{\frac{1}{2}}(\bar{z}, \bar{\theta}^i(x_0)) [\bar{f}^{\frac{1}{2}}(\bar{z}, \bar{\theta}^i(x_0+u_0)) + \\ &+ \sup_{\Gamma_0} |\bar{f}^{\frac{1}{2}}(\bar{z}, \bar{\theta}^i(x_0+u)) - \bar{f}^{\frac{1}{2}}(\bar{z}, \bar{\theta}^i(x_0+u_0))|] d\bar{z}. \end{aligned}$$

We first have

$$\int_{R^n} \bar{f}^{\frac{1}{2}}(\bar{z}, \bar{\theta}^i(x_0)) \bar{f}^{\frac{1}{2}}(\bar{z}, \bar{\theta}^i(x_0+u_0)) d\bar{z} = 1 -$$

$$-\frac{1}{2} \int_{R^n} |\bar{f}^{\frac{1}{2}}(\bar{z}, \bar{\theta}^i(x_0+u_0)) - \bar{f}^{\frac{1}{2}}(\bar{z}, \bar{\theta}^i(x_0))|^2 d\bar{z}.$$

Next,

$$\int_{R^n} |\bar{f}^{\frac{1}{2}}(\bar{z}, \bar{\theta}^i(x_0 + u_0)) - \bar{f}^{\frac{1}{2}}(\bar{z}, \bar{\theta}^i(x_0))|^2 d\bar{z} \geq \tag{5.3.28}$$

$$\inf_{\left\{\begin{array}{c} x \in \mathcal{X} \\ |x - x_0| \geq \gamma \end{array}\right\}} \int_{R^n} |\bar{f}^{\frac{1}{2}}(\bar{z}, \bar{\theta}^i(x)) - \bar{f}^{\frac{1}{2}}(\bar{z}, \bar{\theta}^i(x_0))|^2 d\bar{z}.$$

We recall the relations

$$(h^{(i-1)n+l})^* x = g(b'(\theta^{(i-1)n+l}(x))).$$

For $x \in \mathcal{X}$, it follows from the properties of the functions g and b, and from the assumption (5.3.24), that $\theta^{(i-1)n+l}(x)$ is bounded, so $\bar{\theta}^i(x)$ lies in a compact set $\bar{\Theta}^n$ of R^n, where $\bar{\Theta}$ is a compact interval. Since

$$(h^{(i-1)n+l})^*(x - x_0) = g(b'(\theta^{(i-1)n+l}(x))) - g(b'(\theta^{(i-1)n+l}(x_0))),$$

we deduce easily, using the fact that g is C^1, that b is C^2, and the bounds on the arguments

$$|(h^{(i-1)n+l})^*(x - x_0)| \leq c|\theta^{(i-1)n+l}(x) - \theta^{(i-1)n+l}(x_0)|$$

This can also be written as

$$|H^i(x - x_0)| \leq c|\bar{\theta}^i(x) - \bar{\theta}^i(x_0)|.$$

From the assumptions (5.3.23), (5.3.24), we obtain also

$$|x - x_0| \leq \rho|H^i(x - x_0)| \leq c\rho|\bar{\theta}^i(x) - \bar{\theta}^i(x_0)|.$$

Therefore, for $x \in \mathcal{X}$ and $|x - x_0| \geq \gamma$, we get

$$|\bar{\theta}^i(x) - \bar{\theta}^i(x_0)| \geq \beta = \frac{\gamma}{c\rho}. \tag{5.3.29}$$

Collecting results, we obtain

$$\int_{R^n} |\bar{f}^{\frac{1}{2}}(\bar{z}, \bar{\theta}^i(x_0 + u_0)) - \bar{f}^{\frac{1}{2}}(\bar{z}, \bar{\theta}^i(x_0))|^2 d\bar{z} \geq$$

$$\inf_{\left\{\begin{array}{c} \bar{\theta}, \bar{\theta}' \in \bar{\Theta}^n \\ |\bar{\theta} - \bar{\theta}'| \geq \beta \end{array}\right\}} \int_{R^n} |\bar{f}^{\frac{1}{2}}(\bar{z}, \bar{\theta}) - \bar{f}^{\frac{1}{2}}(\bar{z}, \bar{\theta}')|^2 d\bar{z} = \pi(\beta).$$

And we claim that $\pi(\beta) > 0$ for $\beta > 0$. Since $\bar{\Theta}^n$ is compact, it is easy to check that if we had $\pi(\beta) = 0$, then there would exist $\bar{\theta}$ and $\bar{\theta}'$ in $\bar{\Theta}^n$ such that

$$|\bar{\theta} - \bar{\theta}'| \geq \beta,\ \bar{f}(\bar{z}, \bar{\theta}) = \bar{f}(\bar{z}, \bar{\theta}'),\ \forall \bar{z} \in R^n.$$

Since

$$\bar{f}(\bar{z}, \bar{\theta}) = \exp \sum_{l=1}^{n} (\bar{z}_l \bar{\theta}_l - b(\bar{\theta}_l) + c(\bar{z}_l)),$$

we need to have

$$\sum_{l=1}^{n} (\bar{z}_l \bar{\theta}_l - b(\bar{\theta}_l)) = \sum_{l=1}^{n} (\bar{z}_l \bar{\theta}_l' - b(\bar{\theta}_l'))$$

for every real $\bar{z}_l$, $l = 1, \cdots n$. Suppose there were l_0 such that $\bar{\theta}_{l_0} \neq \bar{\theta}_{l_0}'$. We take

$$\bar{z}_l = \frac{b(\bar{\theta}_l) - b(\bar{\theta}_l')}{\bar{\theta}_l - \bar{\theta}_l'} \text{ if } \bar{\theta}_l - \bar{\theta}_l' \neq 0,$$

$$\bar{z}_{l_0} = 0,\ \bar{z}_l \text{ arbitrary if } \bar{\theta}_l - \bar{\theta}_l' = 0.$$

It clearly follows that $b(\bar{\theta}_{l_0}) = b(\bar{\theta}_{l_0}')$, and from the invertibility of the function b, we obtain $\bar{\theta}_{l_0} = \bar{\theta}_{l_0}'$, which is a contradiction. Hence $\bar{\theta}_l = \bar{\theta}_{l'}$, which is impossible. Finally, we have obtained

$$\int_{R^n} \bar{f}^{\frac{1}{2}}(\bar{z}, \bar{\theta}^i(x_0)) \bar{f}^{\frac{1}{2}}(\bar{z}, \bar{\theta}^i(x_0 + u_0)) d\bar{z} \leq 1 - \frac{1}{2}\pi(\beta). \tag{5.3.30}$$

We next write

$$\int_{R^n} \bar{f}^{\frac{1}{2}}(\bar{z}, \bar{\theta}^i(x_0)) \sup_{\Gamma_0} |\bar{f}^{\frac{1}{2}}(\bar{z}, \bar{\theta}^i(x_0 + u)) - \bar{f}^{\frac{1}{2}}(\bar{z}, \bar{\theta}^i(x_0 + u_0))| d\bar{z} \leq$$

$$\sqrt{\int_{R^n} \sup_{\Gamma_0} |\bar{f}^{\frac{1}{2}}(\bar{z}, \bar{\theta}^i(x_0 + u)) - \bar{f}^{\frac{1}{2}}(\bar{z}, \bar{\theta}^i(x_0 + u_0))|^2 d\bar{z}} \leq$$

$$\sqrt{\int_{R^n} \sup_{\left\{ \begin{array}{c} x \in \mathcal{X} \\ |x' - x| \leq \delta \end{array} \right\}} |\bar{f}^{\frac{1}{2}}(\bar{z}, \bar{\theta}^i(x')) - \bar{f}^{\frac{1}{2}}(\bar{z}, \bar{\theta}^i(x))|^2 d\bar{z}} \leq$$

$$\sqrt{\int_{R^n} \sup_{\left\{ \begin{array}{c} \bar{\theta} \in \bar{\Theta}^n \\ |\bar{\theta}' - \bar{\theta}| \leq c(\delta) \end{array} \right\}} |\bar{f}^{\frac{1}{2}}(\bar{z}, \bar{\theta}') - \bar{f}^{\frac{1}{2}}(\bar{z}, \bar{\theta})|^2 d\bar{z}} = \varpi(\delta),$$

and we see easily using (5.3.25) that $\varpi(\delta) \to 0$ as $\delta \to 0$. Therefore, we have obtained, recalling (5.3.30), that

$$\bar{X}^i_{\Gamma_0} \leq 1 - \frac{1}{2}\pi(\beta) + \varpi(\delta)$$

and

$$E \sup_{\Gamma_0} Z^{\frac{1}{2}}_{nM}(u) \leq \exp -M(\frac{1}{2}\pi(\beta) - \varpi(\delta)). \tag{5.3.31}$$

We next write

$$Z^{\frac{1}{2}}_{M}(u) = \bar{Z}^{\frac{1}{2}}_{n\left[\frac{M}{n}\right]}(u) \prod_{j=n\left[\frac{M}{n}\right]+1}^{M} \frac{f(z^j, \theta^j(x_0+u))}{f(z^j, \theta^j(x_0)}.$$

Therefore,

$$E \sup_{\Gamma_0} Z^{\frac{1}{2}}_{M}(u) \leq E \sup_{\Gamma_0} \bar{Z}^{\frac{1}{2}}_{n\left[\frac{M}{n}\right]}(u) \prod_{i=n\left[\frac{M}{n}\right]+1}^{M} \int_R \sup_{\Gamma_0} f^{\frac{1}{2}}(z, \theta^j(x_0) f^{\frac{1}{2}}(z, \theta^j(x_0+u)dz$$

$$\leq E \sup_{\Gamma_0} \bar{Z}^{\frac{1}{2}}_{n\left[\frac{M}{n}\right]}(u) \left(\sqrt{\int_R \sup_{\theta \in \bar{\Theta}} f(z, \theta)dz}\right)^n,$$

where we have used the fact that

$$\sqrt{\int_R \sup_{\theta \in \bar{\Theta}} f(z, \theta)dz} > 1.$$

Thanks to (5.3.25), we can assert that

$$E \sup_{\Gamma_0} Z^{\frac{1}{2}}_{M}(u) \leq C_n \exp -\left[\frac{M}{n}\right](\frac{1}{2}\pi(\beta) - \varpi(\delta)).$$

We are then exactly in the situation of the MLE; see Theorem 4.3. We cover the set $\{x = x_0 + u | \, |u| \geq \gamma$ and $x_0 + u \in \mathcal{X}\}$ with a finite number J of balls similar to Γ_0, with δ chosen such that $\varpi(\delta) \leq \frac{1}{4}\pi(\beta)$. We obtain

$$E \sup_{|u| \geq \gamma} Z^{\frac{1}{2}}_{M}(u) \leq JC_n \exp -\left[\frac{M}{n}\right](\frac{1}{4}\pi(\beta))$$

and

$$P\left(\{\hat{x}_M - x_0| > \gamma\}\right) \leq J \exp -\left[\frac{M}{n}\right]\frac{1}{4}\pi(\frac{\gamma}{c\rho}).$$

As in Theorem 4.3, we deduce

$$\hat{x}_M \to x_0 \text{ a.s. as } M \to +\infty,$$

which concludes the proof. □

5.3.4 Further Consistency Estimates

Our objective is to prove convergence results as follows:

$$M^\beta(\hat{x}_M - x_0) \to 0, \text{ a.s. and in } L^q, \forall 1 \le q < \infty, \forall \beta < \frac{1}{2}. \tag{5.3.32}$$

We cannot use the method of Lemma 3.2. This is because the Marcinkiewicz theorem used in that lemma necessitates that the variables be independent, which is not the case. We shall proceed differently, following ideas of [24].

It is convenient also to introduce the following notation. We know that the true value of the parameter is x_0. However, we may define the probability that the true value is any given value x. We call it P_x. So the true probability is $P = P_{x_0}$. With probability P, the variables z^j are independent and have a marginal density $f(z, \theta^j(x))$. We shall assume that

$$\sup_{x\in\mathcal{X}} \sup_j E_x|z^j - \mu^j(x)|^m < +\infty, \; m > n. \tag{5.3.33}$$

We have the following result.

PROPOSITION 5.1. *We make the assumptions of Theorem 5.1 and (5.3.33). Then the property (5.3.32) holds.*

PROOF. We begin with preliminary estimates. We define

$$Z_M(u) = \frac{\prod_{j=1}^M f(z^j, \theta^j(x_0 + \frac{u}{M^\beta}))}{\prod_{j=1}^M f(z^j, \theta^j(x_0))}, \tag{5.3.34}$$

and we will consider vectors u such that $x_0 + \frac{u}{M^\beta} \in \mathcal{X}$. We have used for $Z_M(u)$ the same notation as in Theorem 5.1, but there is no risk of confusion. We recover the notation of Theorem 5.1 by taking $\beta = 0$. Let u, v be such that $x_0 + \frac{u}{M^\beta}, x_0 + \frac{v}{M^\beta} \in \mathcal{X}$. We want to estimate

$$E|Z_M^{\frac{1}{m}}(u) - Z_M^{\frac{1}{m}}(v)|^m = E|\sum_{i=1}^n \int_0^1 (v_i - u_i)\frac{\partial}{\partial u_i} Z_M^{\frac{1}{m}}(u + \lambda(v-u))d\lambda|^m.$$

From Minkowski's inequality, we have

$$E^{\frac{1}{m}}|\sum_{i=1}^{n}\int_0^1 (v_i - u_i)\frac{\partial}{\partial u_i}Z_M^{\frac{1}{m}}(u+\lambda(v-u))d\lambda|^m \leq$$

$$\sum_{i=1}^{n} E^{\frac{1}{m}}|\int_0^1 (v_i - u_i)\frac{\partial}{\partial u_i}Z_M^{\frac{1}{m}}(u+\lambda(v-u))d\lambda|^m \leq$$

$$\sum_{i=1}^{n}|v_i - u_i|E^{\frac{1}{m}}\int_0^1 |\frac{\partial}{\partial u_i}Z_M^{\frac{1}{m}}(u+\lambda(v-u))|^m d\lambda \leq$$

$$|v-u|\int_0^1 \sum_{i=1}^{n} E^{\frac{1}{m}}|\frac{\partial}{\partial u_i}Z_M^{\frac{1}{m}}(u+\lambda(v-u))|^m d\lambda.$$

Therefore,

$$E|Z_M^{\frac{1}{m}}(u) - Z_M^{\frac{1}{m}}(v)|^m \leq |v-u|^m \int_0^1 (\sum_{i=1}^{n} E^{\frac{1}{m}}|\frac{\partial}{\partial u_i}Z_M^{\frac{1}{m}}(u+\lambda(v-u))|^m)^m d\lambda.$$

Using the inequality

$$(\sum_{i=1}^{n} a_i)^m \leq n^{m-1}\sum_{i=1}^{n} a_i^m \tag{5.3.35}$$

for numbers $a_i > 0$, we can assert finally

$$E|Z_M^{\frac{1}{m}}(u) - Z_M^{\frac{1}{m}}(v)|^m \leq |v-u|^m n^{m-1}\sum_{i=1}^{n}\int_0^1 E|\frac{\partial}{\partial u_i}Z_M^{\frac{1}{m}}(u+\lambda(v-u))|^m d\lambda. \tag{5.3.36}$$

To pursue the estimation, we consider

$$\frac{\partial}{\partial u_i}Z_M^{\frac{1}{m}}(u) = \frac{1}{mM^\beta}Z_M^{\frac{1}{m}}(u)\sum_{j=1}^{M}\frac{\partial}{\partial x_i}\log f(z^j, \theta^j(x_0 + \frac{u}{M^\beta})).$$

Therefore,

$$E|\frac{\partial}{\partial u_i}Z_M^{\frac{1}{m}}(u)|^m = \frac{1}{m^m M^{m\beta}}E_{x_0+\frac{u}{M^\beta}}|\sum_{j=1}^{M}\frac{\partial}{\partial x_i}\log f(z^j, \theta^j(x_0 + \frac{u}{M^\beta}))|^m. \tag{5.3.37}$$

Note that in (5.3.37), we take on the right-hand side the expected value with respect to the probability P_x, with $x = x_0 + \frac{u}{M^\beta}$. We note also that

$$E_x \frac{\partial}{\partial x_i} \log f(z^j, \theta^j(x)) = 0.$$

We then use the Marcinkiewicz–Zygmund inequality, which we now recall. Let $\xi_1, \cdots \xi_M$ be independent random variables with 0 mean. Then

$$E|\sum_{j=1}^M \xi_j|^m \leq C_m E(\sum_{j=1}^M |\xi_j|^2)^{\frac{m}{2}},$$

and from (5.3.35), we deduce

$$E|\sum_{j=1}^M \xi_j|^m \leq C_m M^{\frac{m}{2}-1} E \sum_{j=1}^M |\xi_j|^m.$$

Applying this inequality to (5.3.37), we get

$$E|\frac{\partial}{\partial u_i} Z_M^{\frac{1}{m}}(u)|^m \leq \frac{C_m}{m^m} \frac{1}{M^{1-m(\frac{1}{2}-\beta)}} \sum_{j=1}^M E_{x_0+\frac{u}{M^\beta}} |\frac{\partial}{\partial x_i} \log f(z^j, \theta^j(x_0 + \frac{u}{M^\beta}))|^m. \tag{5.3.38}$$

But as is easily seen,

$$E_x|\frac{\partial}{\partial x_i} \log f(z^j, \theta^j(x))|^m = E_x|z^j - \mu^j(x)|^m |W^j(x) g'(\mu^j(x)) h_i^j|^m.$$

From the assumptions, in particular (5.3.33), we get, using C_m as a generic constant depending only on m and the compact set $\bar{\mathcal{X}}$,

$$\sup_{j, x \in \mathcal{X}} E_x|\frac{\partial}{\partial x_i} \log f(z^j, \theta^j(x))|^m \leq C_m.$$

Hence also

$$\sup_{\left\{u,v \left| \begin{array}{l} x_0 + \frac{u}{M^\beta} \in \mathcal{X} \\ x_0 + \frac{v}{M^\beta} \in \mathcal{X} \end{array} \right. \right\}} \frac{E|Z_M^{\frac{1}{m}}(u) - Z_M^{\frac{1}{m}}(v)|^m}{|v-u|^m} \leq C_m n^{m-1} M^{m(\frac{1}{2}-\beta)}. \tag{5.3.39}$$

Note that

$$E|Z_M^{\frac{1}{m}}(u)|^m = E Z_M(u) = 1 \leq C_m n^{m-1} M^{m(\frac{1}{2}-\beta)}.$$

We can then use a result on the uniform continuity of stochastic processes—see [24], appendix I, Theorem 19—to claim that for $m > n$,

$$E \sup_{\left\{u,v \middle| \begin{array}{l} x_0 + \frac{u}{M^\beta}, x_0 + \frac{v}{M^\beta} \in \mathcal{X} \\ |u|,\ |v| \le L,\ |u-v| \le h \end{array}\right\}} |Z_M^{\frac{1}{m}}(u) - Z_M^{\frac{1}{m}}(v)| \le B_{n,m}|L|^{\frac{n}{m}} h^{\frac{m-n}{m}} M^{\frac{1}{2}-\beta}, \tag{5.3.40}$$

where $B_{n,m}$ is a constant depending only on n, m and the compact set $\overline{\mathcal{X}}$. We proceed with another estimate. Consider

$$EZ_M^{\frac{1}{2}}(u) = \prod_{j=1}^{M} \int f^{\frac{1}{2}}(z, \theta^j(x_0)) f^{\frac{1}{2}}(z, \theta^j())dz.$$

We want to prove the estimate

$$EZ_M^{\frac{1}{2}}(u) \le \exp(-|u|^2 \alpha_n M^{1-2\beta}),\ \forall u \text{ such that } x_0 + \frac{u}{M^\beta} \in \mathcal{X}, \tag{5.3.41}$$

where the constant α_n is strictly positive and depends only on n and the compact set $\bar{\mathcal{X}}$. We introduce the probability $\bar{f}(\bar{z}, \bar{\theta}^i(x))$ defined in (5.3.27) and the random function

$$\bar{Z}_M(u) = \frac{\prod_{i=1}^{M} \bar{f}(\bar{z}^i, \bar{\theta}^i(x_0 + \frac{u}{M^\beta}))}{\prod_{i=1}^{M} \bar{f}(\bar{z}^i, \bar{\theta}^i(x_0))}$$

with the notation of Theorem 5.1. We recall that

$$\bar{Z}_M(u) = Z_{nm}(u).$$

Since

$$EZ_M^{\frac{1}{2}}(u) \le E\bar{Z}^{\frac{1}{2}}_{n\left[\frac{M}{n}\right]}(u) \left(\sqrt{\int_R \sup_{\theta \in \bar{\Theta}} f(z, \theta) dz}\right)^n,$$

It is sufficient to estimate $E\bar{Z}_M(u)$. We recall that $\bar{\theta}^i(x) \in \bar{\Theta}^n$, $\forall x \in \mathcal{X}$, where $\bar{\Theta}$ is a compact interval. We have

$$E\bar{Z}_M^{\frac{1}{2}}(u) = \prod_{i=1}^{m} [1 - \frac{1}{2} \int_{R^n} (\bar{f}^{\frac{1}{2}}(\bar{z}, \bar{\theta}^i(x_0 + \frac{u}{M^\beta})) - \bar{f}^{\frac{1}{2}}(\bar{z}, \bar{\theta}^i(x_0))^2 d\bar{z}]. \tag{5.3.42}$$

We are going to check that

$$\inf_{\bar{\theta}, \bar{\theta}' \in \bar{\Theta}^n} \frac{\int_{R^n} (\bar{f}^{\frac{1}{2}}(\bar{z}, \bar{\theta}) - \bar{f}^{\frac{1}{2}}(\bar{z}, \bar{\theta}'))^2 d\bar{z}}{|\bar{\theta} - \bar{\theta}'|^2} = \beta > 0. \tag{5.3.43}$$

Suppose that (5.3.43) is not true. Then considering a minimizing sequence $\bar{\theta}_k$, $\bar{\theta}'_k$, we must have $\bar{\theta}_k$-$\bar{\theta}'_k \rightarrow 0$. Indeed, if $\bar{\theta}_k - \bar{\theta}'_k$ has an accumulation point that is not 0, we may assume, since $\bar{\Theta}^n$ is compact, by taking a subsequence, that $\bar{\theta}_k \rightarrow \bar{\theta}$, $\bar{\theta}'_k \rightarrow \bar{\theta}'$, and $|\bar{\theta}$-$\bar{\theta}'|\neq 0$. By continuity,

$$\bar{f}^{\frac{1}{2}}(\bar{z},\bar{\theta}_k) - \bar{f}^{\frac{1}{2}}(\bar{z},\bar{\theta}'_k) \rightarrow \bar{f}^{\frac{1}{2}}(\bar{z},\bar{\theta}) - \bar{f}^{\frac{1}{2}}(\bar{z},\bar{\theta}')$$

and

$$\int_{R^n} (\bar{f}^{\frac{1}{2}}(\bar{z},\bar{\theta}) - \bar{f}^{\frac{1}{2}}(\bar{z},\bar{\theta}'))^2 d\bar{z} > 0,$$

as we have seen in the proof of Theorem 5.1. This will lead to a contradiction, since

$$\frac{\int_{R^n} (\bar{f}^{\frac{1}{2}}(\bar{z},\bar{\theta}_k) - \bar{f}^{\frac{1}{2}}(\bar{z},\bar{\theta}'_k))^2 d\bar{z}}{|\bar{\theta}_k - \bar{\theta}'_k|^2} \rightarrow 0. \tag{5.3.44}$$

Now we can write

$$\bar{f}^{\frac{1}{2}}(\bar{z},\bar{\theta}'_k) - \bar{f}^{\frac{1}{2}}(\bar{z},\bar{\theta}_k) = \frac{1}{2}\bar{f}^{\frac{1}{2}}(\bar{z},\bar{\theta}_k) D_{\bar{\theta}}\bar{L}\,(\bar{z},\bar{\theta}_k)(\bar{\theta}'_k - \bar{\theta}_k) +$$

$$+\frac{1}{2}\int_0^1\int_0^1 \lambda \bar{f}^{\frac{1}{2}}(\bar{z},\bar{\theta}_k + \lambda\mu(\bar{\theta}'_k - \bar{\theta}_k)) \Big[(\bar{\theta}'_k - \bar{\theta}_k)^* D^2_{\bar{\theta}}\bar{L}\,(\bar{z},\bar{\theta}_k + \lambda\mu(\bar{\theta}'_k - \bar{\theta}_k))(\bar{\theta}'_k - \bar{\theta}_k) +$$
$$+ \frac{1}{2}((\bar{\theta}'_k - \bar{\theta}_k)^* D_{\bar{\theta}}\bar{L}\,(\bar{z},\bar{\theta}_k + \lambda\mu(\bar{\theta}'_k - \bar{\theta}_k)))^2 \Big] d\mu d\lambda.$$

Recalling

$$D_{\bar{\theta}}\bar{L}\,(\bar{z},\bar{\theta}) = \begin{pmatrix} \bar{z}_1 - b'(\bar{\theta}_1) \\ \vdots \\ \bar{z}_n - b'(\bar{\theta}_n) \end{pmatrix},$$

$$(D^2_{\bar{\theta}}\bar{L}\,(\bar{z},\bar{\theta}))_{ll'} = -b''(\bar{\theta}_l)\delta_{ll'},$$

we get

$$\bar{f}^{\frac{1}{2}}(\bar{z},\bar{\theta}'_k) - \bar{f}^{\frac{1}{2}}(\bar{z},\bar{\theta}_k) - \frac{1}{2}\bar{f}^{\frac{1}{2}}(\bar{z},\bar{\theta}_k)\sum_{l=1}^n (\bar{z}_l - b'(\bar{\theta}_{kl}))(\bar{\theta}'_{kl} - \bar{\theta}_{kl})$$

$$= +\frac{1}{2}\int_0^1\int_0^1 \lambda \bar{f}^{\frac{1}{2}}(\bar{z},\bar{\theta}_k + \lambda\mu(\bar{\theta}'_k - \bar{\theta}_k)) \Big[-\sum_{l=1}^n b''(\bar{\theta}_{kl} + \lambda\mu(\bar{\theta}'_{kl} - \bar{\theta}_{kl}))(\bar{\theta}'_{kl} - \bar{\theta}_{kl})^2 +$$

$$+\frac{1}{2}(\sum_{l=1}^n (\bar{z}_l - b'(\bar{\theta}_{kl}))(\bar{\theta}'_{kl} - \bar{\theta}_{kl}))^2 \Big],$$

and thus

$$\frac{\bar{f}^{\frac{1}{2}}(\bar{z},\bar{\theta}'_k)-\bar{f}^{\frac{1}{2}}(\bar{z},\bar{\theta}_k)-\frac{1}{2}\bar{f}^{\frac{1}{2}}(\bar{z},\bar{\theta}_k)\sum_{l=1}^n(\bar{z}_l-b'(\bar{\theta}_{kl}))(\bar{\theta}'_{kl}-\bar{\theta}_{kl})}{|\bar{\theta}_k-\bar{\theta}'_k|}\rightarrow 0, \forall \bar{z}.$$

We can also bound this function by a fixed function that is square integrable. From Lebesgue's theorem, we obtain easily

$$\frac{\int_{R^n}(\bar{f}^{\frac{1}{2}}(\bar{z},\bar{\theta}_k)-\bar{f}^{\frac{1}{2}}(\bar{z},\bar{\theta}'_k))^2d\bar{z}}{|\bar{\theta}_k-\bar{\theta}'_k|^2}$$
$$-\frac{1}{4}\frac{\int_{R^n}\bar{f}(\bar{z},\bar{\theta}_k)(\sum_{l=1}^n(\bar{z}_l-b'(\bar{\theta}_{kl}))(\bar{\theta}'_{kl}-\bar{\theta}_{kl}))^2d\bar{z}}{|\bar{\theta}_k-\bar{\theta}'_k|^2}\rightarrow 0, \text{ as } k\rightarrow\infty.$$

However,

$$\int_{R^n}\bar{f}(\bar{z},\ \bar{\theta}_k)(\sum_{l=1}^n(\bar{z}_l-b'(\bar{\theta}_{kl}))(\bar{\theta}'_{kl}-\bar{\theta}_{kl}))^2d\bar{z}$$
$$=\sum_{l=1}^n\int_{R^n}\bar{f}(\bar{z},\bar{\theta}_k)(\bar{z}_l-b'(\bar{\theta}_{kl}))^2d\bar{z}(\bar{\theta}'_{kl}-\bar{\theta}_{kl}))^2.$$

Since by (5.1.5) we have

$$\int_R(z-b'(\theta))^2f(z,\theta)dz=b''(\theta)$$

and

$$\int_{R^n}\bar{f}(\bar{z},\bar{\theta}_k)(\bar{z}_l-b'(\bar{\theta}_{kl}))^2d\bar{z}=\int_R(z-b'(\theta_{kl}))^2f(z,\theta_{kl})dz=b''(\theta_{kl})\geq c>0,$$

we deduce

$$\int_{R^n}\bar{f}(\bar{z},\bar{\theta}_k)(\sum_{l=1}^n(\bar{z}_l-b'(\bar{\theta}_{kl}))(\bar{\theta}'_{kl}-\bar{\theta}_{kl}))^2d\bar{z}\geq c|\bar{\theta}_k-\bar{\theta}'_k|^2.$$

We obtain again a contradiction with (5.3.44). Therefore, (5.3.43) is established. Now from the property (5.3.29) (see the proof of Theorem 5.1), we have

$$|\bar{\theta}^i(x_0+\frac{u}{M^\beta})-\bar{\theta}^i(x_0)|\geq\frac{1}{c\rho}\frac{|u|}{M^\beta}.$$

Combining this inequality with (5.3.43) yields

$$\int_{R^n} (\bar{f}^{\frac{1}{2}}(\bar{z}, \bar{\theta}^i(x_0 + \frac{u}{M^\beta})) - \bar{f}^{\frac{1}{2}}(\bar{z}, \bar{\theta}^i(x_0))^2 d\bar{z} \geq \frac{\beta}{c^2\rho^2} \frac{|u|^2}{M^{2\beta}}.$$

Hence from (5.3.42), we obtain

$$E\bar{Z}_M^{\frac{1}{2}}(u) \leq (1 - \delta\frac{|u|^2}{M^{2\beta}})^M.$$

As in Theorem 5.1, we conclude that (5.3.41) holds for an appropriate constant $\alpha_n > 0$.

We next estimate

$$P(\{|\hat{x}_M - x_0|M^\beta > \gamma) \leq P(\sup_{|u|>\gamma} Z_M(u) \geq 1). \tag{5.3.45}$$

Let

$$\Gamma_r = \{u|\, x_0 + \frac{u}{M^\beta} \in \mathcal{X},\ \gamma + r \leq |u| < \gamma + r + 1\},$$

so that

$$P(\sup_{|u|>\gamma} Z_M(u) \geq 1) \leq \sum_{r=0}^{+\infty} P(\sup_{u\in\Gamma_r} Z_M(u) \geq 1). \tag{5.3.46}$$

We next estimate $P(\sup_{u\in\Gamma_r} Z_M(u) \geq 1)$. We note that $\Gamma_r \subset [-(\gamma + r + 1), \gamma + r + 1]^n$. We subdivide this cube into cubes of diameter h. The number of such cubes is

$$N = B_n\frac{(\gamma + r + 1)^n}{h^n},$$

where B_n is a generic constant depending only on n. We consider all the cubes of diameter h that have a nonempty intersection with Γ_r. The number of such cubes is $N' < N$. We obtain a covering of Γ_r with nonoverlapping cubes of diameter h. We call these cubes Γ_r^j, $j = 1, \cdots N'$. Let u_r^j be a point in Γ_r^j that also belongs to Γ_r.

Let $\hat{u}_r$ be the point that maximizes $Z_M(u)$ over Γ_r. This point belongs to one and only one of the small cubes, say $\Gamma_r^{j_r}$. We can assert that

$$\{\sup_{u\in\Gamma_r} Z_M(u) \geq 1\} = \{\sup_{u\in\Gamma_r} Z_M^{\frac{1}{m}}(u) \geq 1\} = \{Z_M^{\frac{1}{m}}(\hat{u}_r) \geq 1\}$$

$$\subset\{Z_M^{\frac{1}{m}}(u_r^{j_r}) \geq \frac{1}{2}\} \cup \{|Z_M^{\frac{1}{m}}(\hat{u}_r) - Z_M^{\frac{1}{m}}(u_r^{j_r})| \geq \frac{1}{2}\}$$

$$\subset \{Z_M^{\frac{1}{m}}(u_r^{j_r}) \geq \frac{1}{2}\} \cup \{ \sup_{\substack{|u-v| \leq h \\ u,v \in \Gamma_r}} |Z_M^{\frac{1}{m}}(u) - Z_M^{\frac{1}{m}}(v)| \geq \frac{1}{2}\}$$

$$\subset \cup_{j=1}^{N'} \{Z_M^{\frac{1}{m}}(u_r^{j}) \geq \frac{1}{2}\} \cup \{ \sup_{\substack{|u-v| \leq h \\ u,v \in \Gamma_r}} |Z_M^{\frac{1}{m}}(u) - Z_M^{\frac{1}{m}}(v)| \geq \frac{1}{2}\}.$$

Therefore,

$$P(\{\sup_{u\in\Gamma_r} Z_M(u) \geq 1\} \leq \sum_{j=1}^{N'} P(\{Z_M^{\frac{1}{m}}(u_r^{j}) \geq \frac{1}{2}\}) + P(\{ \sup_{\substack{|u-v| \leq h \\ u,v \in \Gamma_r}} |Z_M^{\frac{1}{m}}(u) - Z_M^{\frac{1}{m}}(v)| \geq \frac{1}{2}\}). \tag{5.3.47}$$

Now

$$\{Z_M^{\frac{1}{m}}(u_r^{j}) \geq \frac{1}{2}\} = \{Z_M^{\frac{1}{2}}(u_r^{j}) \geq (\frac{1}{2})^{\frac{m}{2}}\},$$

and hence

$$P(\{Z_M^{\frac{1}{m}}(u_r^{j}) \geq \frac{1}{2}\}) \leq 2^{\frac{m}{2}} E Z_M^{\frac{1}{2}}(u_r^{j}),$$

and from (5.3.41), we obtain

$$\begin{aligned} E Z_M^{\frac{1}{2}}(u_r^{j}) &\leq \exp(-|u_r^{j}|^2 \alpha_n M^{1-2\beta}) \\ &\leq \exp(-(\gamma + r)^2 \alpha_n M^{1-2\beta}). \end{aligned}$$

Next, from (5.3.40) we have

$$\begin{aligned} P(\{ \sup_{\substack{|u-v| \leq h \\ u,v \in \Gamma_r}} |Z_M^{\frac{1}{m}}(u) - Z_M^{\frac{1}{m}}(v)| \geq \frac{1}{2}\}) &\leq 2E \sup_{\substack{|u-v| \leq h \\ u,v \in \Gamma_r}} |Z_M^{\frac{1}{m}}(u) - Z_M^{\frac{1}{m}}(v)| \\ &\leq B_{n,m}|\gamma + r + 1|^{\frac{n}{m}} h^{\frac{m-n}{m}} M^{\frac{1}{2}-\beta}. \end{aligned}$$

Therefore, from (5.3.47) we obtain

$$\begin{aligned} P(\{\sup_{u\in\Gamma_r} Z_M(u) \geq 1\} &\leq N \exp(-(\gamma+r)^2 \alpha_n M^{1-2\beta}) + B_{n,m}|\gamma+r+1|^{\frac{n}{m}} h^{\frac{m-n}{m}} M^{\frac{1}{2}-\beta} \\ &\leq B_n(\frac{\gamma + r + 1}{h})^n \exp(-(\gamma + r)^2 \alpha_n M^{1-2\beta}) + B_{n,m}|\gamma + r + 1|^{\frac{n}{m}} h^{\frac{m-n}{m}} M^{\frac{1}{2}-\beta}, \end{aligned}$$

where B_n, $B_{n,m}$ are generic constants. So far, h has not been fixed. We now choose h such that

$$\frac{\exp(-(\gamma+r)^2\alpha_n M^{1-2\beta})}{h^n} = h^{\frac{m-n}{m}} M^{\frac{1}{2}-\beta},$$

which means that

$$h = \frac{\exp(-\frac{(\gamma+r)^2\alpha_n M^{1-2\beta}}{1+n(1-\frac{1}{m})})}{M^{\frac{\frac{1}{2}-\beta}{1+n(1-\frac{1}{m})}}}.$$

With this choice, we can state that

$$P(\{\sup_{u\in\Gamma_r} Z_M(u) \geq 1\} \leq B_{n,m}(\gamma+r+1)^n M^{\frac{nm(\frac{1}{2}-\beta)}{nm+m-n}} \exp -\frac{(\gamma+r)^2\alpha_n M^{1-2\beta}(m-n)}{m-n+mn}.$$

By changing the constant α_n, we have also

$$P(\{\sup_{u\in\Gamma_r} Z_M(u) \geq 1\} \leq B_{n,m} M^{\frac{nm(\frac{1}{2}-\beta)}{nm+m-n}} \exp -\frac{(\gamma+r)^2\alpha_n M^{1-2\beta}(m-n)}{m-n+mn}.$$

Hence from (5.3.46), it follows easily that

$$P(\{|\hat{x}_M - x_0|M^\beta > \gamma) \leq B_{n,m} M^{\frac{nm(\frac{1}{2}-\beta)}{nm+m-n}} \exp -\frac{\gamma^2\alpha_n M^{1-2\beta}(m-n)}{m-n+mn},$$

and by changing α_n again, we get

$$P(\{|\hat{x}_M - x_0|M^\beta > \gamma) \leq B_{n,m} \exp -\frac{\gamma^2\alpha_n M^{1-2\beta}(m-n)}{m-n+mn}. \tag{5.3.48}$$

As in Theorem 4.3, we deduce

$$|\hat{x}_M - x_0|M^\beta \to 0, \text{ a.s. as } M \to +\infty. \tag{5.3.49}$$

To show that the variable tends to 0 in L^q, we can write

$$E(|\hat{x}_M - x_0|M^\beta)^q = \sum_{r=0}^{\infty} E[(|\hat{x}_M - x_0|M^\beta)^q \mathbb{1}_{r\leq|\hat{x}_M-x_0|M^\beta<r+1}]$$

$$\leq \sum_{r=1}^{\infty}(r+1)^q P(\{|\hat{x}_M - x_0|M^\beta > r) + E[(|\hat{x}_M - x_0|M^\beta)^q \mathbb{1}_{|\hat{x}_M-x_0|M^\beta<1}]$$

$$\leq B\sum_{r=1}^{\infty}(r+1)^q \exp -\frac{r^2\alpha_n M^{1-2\beta}(m-n)}{m-n+mn} + E[(|\hat{x}_M - x_0|M^\beta)^q \mathbb{1}_{|\hat{x}_M-x_0|M^\beta<1}],$$

and both terms tend to 0 as $M \to +\infty$. Therefore,

$$E(|\hat{x}_M - x_0|M^\beta)^q \to 0, \text{ as } M \to \infty$$

The proof is complete. □

5.3.5 Asymptotic Normality

Our objective now is to prove a result similar to that of Theorem 4.4, namely asymptotic normality. Unfortunately, the method does not work, since we cannot invoke Lemma 3.2, due to the fact that the Marcinkiewicz theorem used in that lemma necessitates that the variables be independent. We shall proceed differently, making use of Proposition 5.1. Define the matrix

$$\Pi_M(x) = \sum_{j=1}^{M} W^j(x) h^j (h^j)^*. \tag{5.3.50}$$

From the assumptions of Theorem 5.1, it is easy to check that $W^j(x_0) \geq \alpha > 0$, and hence for $M > n$, we have

$$\sum_{j=1}^{M} W^j(x_0) h^j (h^j)^* \geq \alpha \sum_{j=1}^{M} h^j (h^j)^* \geq \alpha \sum_{j=1}^{\left[\frac{M}{n}\right]n} h^j (h^j)^*$$

$$= \alpha \sum_{i=1}^{\left[\frac{M}{n}\right]} (H^i)^* H^i \geq \alpha c \left[\frac{M}{n}\right] I \geq \alpha c (\frac{M}{n} - 1) I.$$

Therefore,

$$\frac{\Pi_M(x_0)}{M} \geq \alpha c (\frac{1}{n} - \frac{1}{M}) I,$$

whence

$$\frac{\Pi_M(x_0)}{M} \geq \alpha c \frac{1}{2n} I \text{ , } \forall M \geq 2n. \tag{5.3.51}$$

With these preliminaries out of the way, our objective is to prove the following asymptotic normality result.

THEOREM 5.2. *We make the assumptions of Theorem 5.1, (5.3.33), and*

$$\varphi(\eta) \text{ is } C^2, \tag{5.3.52}$$

$$|W^j(x') - W^j(x)| \leq c|x' - x|^\beta 0 < \beta \leq 1, \forall x, x' \in \mathcal{X}, \tag{5.3.53}$$

$$|\varphi''(x') - \varphi''(x)| \leq c|x' - x|^\beta 0 < \beta \leq 1, \forall x, x' \in \mathcal{X}.$$

We then have

$$(\Pi_M(x_0))^{\frac{1}{2}}(\hat{x}_M - x_0) \to N(0, I), \tag{5.3.54}$$

the convergence being in law, in which $N(0, I)$ *represents the Gaussian law in* R^n, *with mean* 0 *and covariance matrix the identity.*

PROOF. Since $\hat{x}_M$ maximizes the likelihood in an open domain, we have

$$\sum_{j=1}^{M} D_x L(z^j, \theta^j(\hat{x}_M)) = 0.$$

We can then write

$$\begin{aligned}
\sum_{j=1}^{M} D_x L(z^j, \theta^j(x_0)) &= \sum_{j=1}^{M}(D_x L(z^j, \theta^j(x_0)) - D_x L(z^j, \theta^j(\hat{x}_M))) \\
&= -\sum_{j=1}^{M} \int_0^1 D_x^2 L(z^j, \theta^j(x_0 + \lambda(\hat{x}_M - x_0)))d\lambda(\hat{x}_M - x_0) \\
&= -\sum_{j=1}^{M} D_x^2 L(z^j, \theta^j(x_0))(\hat{x}_M - x_0) \\
&\quad - [\sum_{j=1}^{M} \int_0^1 (D_x^2 L(z^j, \theta^j(x_0 + \lambda(\hat{x}_M - x_0))) \\
&\quad - D_x^2 L(z^j, \theta^j(x_0)))d\lambda](\hat{x}_M - x_0).
\end{aligned}$$

Recalling (5.3.15), we get

$$\begin{aligned}
\Pi_M(x_0)(\hat{x}_M - x_0) = &\sum_{j=1}^{M} D_x L(z^j, \theta^j(x_0)) + \\
&+ [\sum_{j=1}^{M}(D_x^2 L(z^j, \theta^j(x_0)) - E(D_x^2 L(z^j, \theta^j(x_0))))](\hat{x}_M - x_0) \\
&+ [\sum_{j=1}^{M} \int_0^1 (D_x^2 L(z^j, \theta^j(x_0 + \lambda(\hat{x}_M - x_0))) \\
&- D_x^2 L(z^j, \theta^j(x_0)))d\lambda](\hat{x}_M - x_0).
\end{aligned}$$

We can then write, recalling the formulas for $D_x L(z^j,\theta^j(x_0))$ and $(D_x^2 L(z^j,\theta^j(x_0))$,

$$\Pi_M^{\frac{1}{2}}(x_0)(\hat{x}_M - x_0) = \Pi_M^{-\frac{1}{2}}(x_0)\sum_{j=1}^M W^j(x_0)g'(\mu^j(x_0))(z^j - \mu^j(x_0))h^j +$$

$$+\Pi_M^{-\frac{1}{2}}(x_0)\sum_{j=1}^M (z^j - \mu^j(x_0))\varphi''((h^j)^*(x_0))d\lambda h^j(h^j)^*(\hat{x}_M - x_0) + \tag{5.3.55}$$

$$+\Pi_M^{-\frac{1}{2}}(x_0)\left(\sum_{j=1}^M \{z^j \int_0^1 (\varphi''((h^j)^*(x_0+\lambda(\hat{x}_M - x_0))) - \varphi''((h^j)^* x_0))d\lambda + \right.$$

$$- \int_0^1 (\mu^j(x_0+\lambda(\hat{x}_M - x_0))\varphi''((h^j)^*(x_0+\lambda(\hat{x}_M - x_0))) - \mu^j(x_0)\varphi''((h^j)^* x_0))d\lambda \tag{5.3.56}$$

$$\left. - \int_0^1 (W^j(x_0+\lambda(\hat{x}_M - x_0)) - W^j(x_0))d\lambda\} h^j(h^j)^*\right)\Pi_M^{-\frac{1}{2}}(x_0)\Pi_M^{\frac{1}{2}}(x_0)(\hat{x}_M - x_0).$$

Define

$$\Gamma_M = \Pi_M^{-\frac{1}{2}}(x_0)\left(\sum_{j=1}^M \{z^j \int_0^1 (\varphi''((h^j)^*(x_0 + \lambda(\hat{x}_M - x_0))) - \varphi''((h^j)^* x_0))d\lambda + \right.$$

$$- \int_0^1 (\mu^j(x_0 + \lambda(\hat{x}_M - x_0))\varphi''((h^j)^*(x_0 + \lambda(\hat{x}_M - x_0))) - \mu^j(x_0)\varphi''((h^j)^* x_0))d\lambda$$

$$\left. - \int_0^1 (W^j(x_0 + \lambda(\hat{x}_M - x_0)) - W^j(x_0))d\lambda\} h^j(h^j)^*\right)\Pi_M^{-\frac{1}{2}}(x_0).$$

Noting that

$$D_x \mu^j(x) = \frac{1}{g'(\mu^j(x))}h^j,$$

we conclude that $\mu^j(x)$ is C^1. From the assumption (5.3.53), we can state that

$$||\Gamma_M|| \leq C(\sum_{j=1}^M \frac{|z^j|}{M} + 1)|\hat{x}_M - x_0|^\beta.$$

Set

$$X_{3M} = \Gamma_M \Pi_M^{\frac{1}{2}}(x_0)(\hat{x}_M - x_0),$$

which is the third term on the right-hand side of (5.3.55). Using (5.3.51), we have

$$|X_{3M}| \leq C(\sum_{j=1}^{M} \frac{|z^j|}{M} + 1) M^{\frac{1}{2}} |\hat{x}_M - x_0|^{1+\beta}.$$

The variable $\sum_{j=1}^{M} \frac{|z^j|}{M}$ is bounded in L^2. Let $1 < \delta < 2$. We have

$$E|X_{3M}|^{\delta} \leq C \left([(E(\sum_{j=1}^{M} \frac{|z^j|}{M})^2)^{\frac{\delta}{2}} + 1](E(M^{\frac{1}{2(1+\beta)}} |\hat{x}_M - x_0|)^{\frac{2\delta(1+\beta)}{2-\delta}})^{\frac{2-\delta}{2}} \right)$$

$$\leq C(E(M^{\frac{1}{2(1+\beta)}} |\hat{x}_M - x_0|)^{\frac{2\delta(1+\beta)}{2-\delta}})^{\frac{2-\delta}{2}}.$$

But from Proposition 5.1, we have

$$E(M^{\frac{1}{2(1+\beta)}} |\hat{x}_M - x_0|)^{\frac{2\delta(1+\beta)}{2-\delta}} \to 0,$$

and hence

$$X_{3M} \to 0, \text{ in } L^{\delta}. \tag{5.3.57}$$

Consider next

$$X_{2M} = \Pi_M^{-\frac{1}{2}}(x_0) \sum_{j=1}^{M} (z^j - \mu^j(x_0)) \varphi''((h^j)^*(x_0)) d\lambda h^j (h^j)^* (\hat{x}_M - x_0).$$

We have

$$|X_{2M}| \leq ||\Pi_M^{-\frac{1}{2}}(x_0) \sum_{j=1}^{M} (z^j - \mu^j(x_0)) \varphi''((h^j)^*(x_0)) d\lambda h^j (h^j)^*|| \, |\hat{x}_M - x_0|.$$

For $1 < \delta < 2$, we obtain

$$E|X_{2M}|^{\delta} \leq \left(E||\Pi_M^{-\frac{1}{2}}(x_0) \sum_{j=1}^{M} (z^j - \mu^j(x_0)) \varphi''((h^j)^*(x_0)) d\lambda h^j (h^j)^*||^2 \right)^{\frac{\delta}{2}}$$

$$(E|\hat{x}_M - x_0|^{\frac{2\delta}{2-\delta}})^{\frac{2-\delta}{2}}.$$

We can take $||A||^2 = \text{tr}\, A^*A$ as the norm of a matrix A. Therefore,

$$E||\Pi_M^{-\frac{1}{2}}(x_0)\sum_{j=1}^{M}(z^j - \mu^j(x_0))\varphi''((h^j)^*(x_0))d\lambda h^j(h^j)^*||^2 =$$

$$\sum_{j=1}^{M} V^j(x_0)(\varphi''((h^j)^*(x_0)))^2 \text{tr} h^j(h^j)^*\Pi_M^{-1}(x_0)h^j(h^j)^* \le C,$$

and thus also

$$X_{2M} \to 0, \text{ in } L^\delta. \tag{5.3.58}$$

Setting finally

$$X_{1M} = \Pi_M^{-\frac{1}{2}}(x_0)\sum_{j=1}^{M} W^j(x_0)g'(\mu^j(x_0))(z^j - \mu^j(x_0))h^j,$$

we want to prove that

$$X_{1M} \to N(0, I) \tag{5.3.59}$$

for the convergence in law. What we must prove is

$$E\exp i\lambda^* X_{1M} \to \exp -\frac{1}{2}|\lambda|^2 \tag{5.3.60}$$

for all $\lambda \in R^n$, with $i = \sqrt{-1}$. From the independence of variables z^j, we have

$$\begin{aligned} E\exp i\lambda^* X_{1M} &= \prod_{j=1}^{M} E\exp i\lambda^*(\Pi^M(x_0))^{-\frac{1}{2}}h^j(z^j - \mu^j(x_0))W^j(x_0)g'(\mu^j(x_0)) \\ &= \prod_{j=1}^{M} E\exp i\chi_j^M. \end{aligned} \tag{5.3.61}$$

Write

$$E\exp i\chi_j^M = 1 - a_j^M.$$

Since $E\chi_j^M = 0$, we can write

$$a_j^M = E\,[(\chi_j^M)^2 \int_0^1\int_0^1 u\,\exp(iuv\chi_j^M)dudv].$$

Therefore,

$$|a_j^M| \leq \frac{1}{2} E\,(\chi_j^M)^2,$$

and from (5.3.51), we get

$$|E\,(\chi_j^M)^2| \leq \frac{C}{M},$$

whence also

$$\begin{aligned} |(a_j^M)^k| = |a_j^M|^k &\leq \frac{1}{2^k}(E\,(\chi_j^M)^2)^k \\ &= \frac{1}{2^k}(\frac{C}{M})^k. \end{aligned}$$

Since M is large, we can assume that $|a_j^M| < 1, \forall j$. We use the definition

$$\log(1-\eta) = -\sum_{k=1}^{\infty} \frac{\eta^k}{k}$$

for $|\eta| < 1$, and therefore,

$$\log E \exp i \chi_j^M = -\sum_{k=1}^{\infty} \frac{(a_j^M)^k}{k}.$$

Next, we have

$$\begin{aligned} \log E \exp i\lambda^* X_{1M} &= \sum_{j=1}^{M} \log E \exp i \chi_j^M \\ &= -\sum_{j=1}^{M}\sum_{k=1}^{\infty} \frac{(a_j^M)^k}{k}. \end{aligned}$$

We first consider

$$\begin{aligned} |\sum_{j=1}^{M}\sum_{k=2}^{\infty} \frac{(a_j^M)^k}{k}| &\leq \sum_{j=1}^{M}\sum_{k=2}^{\infty} \frac{|(a_j^M)^k|}{k} \\ &\leq \sum_{j=1}^{M}\sum_{k=2}^{\infty} \frac{1}{k2^k}(\frac{C}{M})^k \end{aligned}$$

$$\leq \sum_{k=2}^{\infty} \frac{1}{k2^k} \frac{C^k}{M^{k-1}}$$

$$= \frac{C^2}{4M} \sum_{k=0}^{\infty} \frac{1}{(k+2)2^k} (\frac{C}{M})^k.$$

This implies

$$\sum_{j=1}^{M} \sum_{k=2}^{\infty} \frac{(a_j^M)^k}{k} \to 0, \text{ as } M \to +\infty. \tag{5.3.62}$$

Next,

$$\sum_{j=1}^{M} a_j^M = \sum_{j=1}^{M} E\,(\chi_j^M)^2 \int_0^1 \int_0^1 u \, \exp(iuv\chi_j^M) du dv$$

$$= \frac{1}{2} \sum_{j=1}^{M} E\,(\chi_j^M)^2 + \gamma^M,$$

with

$$\gamma^M = \sum_{j=1}^{M} E\,(\chi_j^M)^2 \int_0^1 \int_0^1 u\,(\exp(iuv\chi_j^M) - 1) du dv$$

$$= \sum_{j=1}^{M} E\,(\chi_j^M)^3 \int_0^1 \int_0^1 \int_0^1 iu^2 v \, \exp(iuvw\chi_j^M) du dv dw.$$

Therefore,

$$|\gamma^M| \leq \frac{1}{6} \sum_{j=1}^{M} |E\,(\chi_j^M)^3|$$

$$= \frac{1}{6} \sum_{j=1}^{M} |(\lambda^*(\Pi^M(x_0))^{-\frac{1}{2}} h^j)^3 (W^j(x_0) g'(\mu^j(x_0)))^3 E(z^j - \mu^j(x_0))^3|,$$

and from (5.1.6), we have

$$|\gamma^M| \leq \frac{1}{6} \sum_{j=1}^{M} |(\lambda^*(\Pi^M(x_0))^{-\frac{1}{2}} h^j)^3 (W^j(x_0) g'(\mu^j(x_0)))^3 b'''(\theta^j(x_0))|,$$

and from the assumptions, we obtain

$$|\gamma^M| \leq \frac{C}{\sqrt{M}}.$$

Finally,

$$\begin{aligned}\frac{1}{2}\sum_{j=1}^{M} E\,(\chi_j^M)^2 &= \frac{1}{2}\sum_{j=1}^{M}(\lambda^*(\Pi^M(x_0))^{-\frac{1}{2}}h^j)^2 V^j(x_0)(W^j(x_0)g'(\mu^j(x_0)))^2 \\ &= \frac{1}{2}\sum_{j=1}^{M}(\lambda^*(\Pi^M(x_0))^{-\frac{1}{2}}h^j)^2 W^j(x_0),\end{aligned}$$

from which it follows immediately that

$$\frac{1}{2}\sum_{j=1}^{M} E\,(\chi_j^M)^2 = \frac{1}{2}|\lambda|^2. \tag{5.3.63}$$

Therefore,

$$\sum_{j=1}^{M} a_j^M \to \frac{1}{2}|\lambda|^2.$$

Collecting results, we obtain

$$\log E \exp i\lambda^* X_{1M} \to -\frac{1}{2}|\lambda|^2,$$

which is equivalent to (5.3.59).

Finally, going back to (5.3.55), we have

$$\Pi_M^{\frac{1}{2}}(x_0)(\hat{x}_M - x_0) = X_{1M} + X_{2M} + X_{3M},$$

and X_{1M} converges in law to $N(0, I)$ and $X_{2M} + X_{3M} \to 0$ in L^δ. This implies easily (5.3.54) and completes the proof. □

5.4 The Vector Case

5.4.1 Notation and Preliminaries

In the preceding sections, we have considered that the observation z is a scalar, whereas the unknown parameter x is a vector in R^n. At the beginning, see (5.1.1), we recalled the classical linear model, in which the observation is a vector in R^d. So for the sake of completeness, we return here to the vector case. We begin with a density $f(z, \theta)$ and the log likelihood $L(z, \theta) = \log f(z, \theta)$, with

$$L(z, \theta) = z^* \Sigma^{-1} \theta - b(\theta) + c(z), \tag{5.4.1}$$

in which $z \in R^d$, $\theta \in R^d$, and Σ is a symmetric invertible $d \times d$ matrix. The function $b : R^d \to R$ will satisfy properties given below. The parameter θ is the canonical parameter. We again let z denote the random variable, whose probability density is $f(z, \theta)$, to save notation. We have

$$Ez = \mu = \Sigma D_\theta b(\theta) \tag{5.4.2}$$

and

$$E(z - \mu)(z - \mu)^* = V = \Sigma D_\theta^2 b(\theta) \Sigma. \tag{5.4.3}$$

We relate the unknown variable x to the mean, via the link function, so we write

$$g(\mu) = Hx + h, \tag{5.4.4}$$

in which $H \in \mathcal{L}(R^n; R^d)$ and $g : R^d \to R^d$. So the canonical parameter is linked to the unknown variable x by the relation

$$g(\Sigma D_\theta b(\theta)) = Hx + h. \tag{5.4.5}$$

We shall assume that the map $\theta \to g(\Sigma D_\theta b(\theta))$ is invertible and thus define the function $\varphi(\eta)$ by solving

$$g(\Sigma D_\theta b(\varphi(\eta))) = \eta, \forall \eta \in R^d. \tag{5.4.6}$$

So the canonical parameter is linked to the unknown variable by the relation

$$\theta = \theta(x) = \varphi(Hx + h). \tag{5.4.7}$$

Hence

$$D\theta(x) = D\varphi(Hx + h)H.$$

We set, by abuse of notation,

$$L(z, x) = L(z, \theta(x)). \tag{5.4.8}$$

We compute easily

$$D_x L(z, x) = H^* D\varphi^*(Hx + h)(\Sigma^{-1} z - D_\theta b(\theta(x))). \tag{5.4.9}$$

In order to differentiate a second time in x, it is convenient to write the preceding relation as follows:

$$D_x L(z, x) = H^* \sum_{k=1}^{d} ((\Sigma^{-1} z)_k - D_{\theta_k} b(\theta(x))) D\varphi_k(Hx + h), \tag{5.4.10}$$

which implies

$$\begin{aligned} D_x^2 L(z, x) = H^*[&-D\varphi^*(Hx + h) D^2 b(\theta(x)) D\varphi(Hx + h)+ \\ &+\sum_{k=1}^{d} ((\Sigma^{-1} z)_k - D_{\theta_k} b(\theta(x))) D^2 \varphi_k(Hx + h)]H. \end{aligned} \tag{5.4.11}$$

From the relation (5.4.6) we can write

$$I = Dg(\mu) \Sigma D^2 b(\varphi(\eta)) D\varphi(\eta),$$

and hence

$$\Sigma^{-1} D\varphi(\eta) = (\Sigma D^2 b(\varphi(\eta)) \Sigma)^{-1} (Dg(\mu))^{-1}, \tag{5.4.12}$$

with

$$\mu = \Sigma D_\theta b(\varphi(\eta)).$$

Using $\eta = Hx + h$, we get

$$\begin{aligned} \theta(x) &= \varphi(Hx + h), \\ \mu(x) &= \Sigma D_\theta b(\theta(x)), \\ V(x) &= \Sigma D^2 b(\theta(x)) \Sigma, \end{aligned} \tag{5.4.13}$$

whence

$$\Sigma^{-1} D\varphi(Hx + h) = V(x)^{-1} (Dg(\mu(x)))^{-1}. \tag{5.4.14}$$

Therefore, using (5.4.9), we can write

$$D_x L(z, x) = H^*((Dg(\mu(x)))^*)^{-1} V(x)^{-1}(z - \mu(x)). \tag{5.4.15}$$

We introduce the weight matrix $W(x)$ defined by

$$W(x) = ((Dg(\mu(x)))^*)^{-1} V(x)^{-1} (Dg(\mu(x)))^{-1}. \tag{5.4.16}$$

So we can write

$$D_x L(z, x) = H^* W(x) Dg(\mu(x))(z - \mu(x)). \tag{5.4.17}$$

Using

$$E D_x^2 L(z, x) = -E D_x L(z, x)(E D_x L(z, x))^*,$$

we get

$$E D_x^2 L(z, x) = -H^* W(x) Dg(\mu(x)) V(x) (Dg(\mu(x)))^* W(x) H,$$

and from (5.4.16), it follows that

$$E D_x^2 L(z, x) = -H^* W(x) H. \tag{5.4.18}$$

We can summarize results as follows:

PROPOSITION 5.2. *We assume that $b(\theta)$ is C^2 on R^d and $D^2 b(\theta)$ is strictly positive on compacts sets. We assume that the link function $g(\mu)$ is C^1 from R^d to R^d and is invertible with bounded inverse on compact sets. The function $\theta \to g(\Sigma D_\theta b(\theta))$ is then invertible with inverse $\varphi(\eta)$ C^1 on compact sets. The canonical parameter, mean, and variance are expressed as functions of x by formulas (5.4.13). The weight matrix function is then defined by formula (5.4.16) and is continuous on compact sets. We can then express the gradient of the log likelihood with respect to x by formula (5.4.17), and we have formula (5.4.18).*

5.4.2 MLE Estimate

We consider now a sequence of independent random variables $z^1, \cdots, z^M$. Each of them follows a GLM distribution. We assume that the unknown variable x is the same for all random variables. However, the matrix H and vector h vary from experiment to experiment, so we have a sequence H^j, h^j. Similarly, we have functions $b^j(\theta)$, $g^j(\mu)$ and matrices Σ^j depending on the experiment, with identical properties. So we get functions $\theta^j(x)$, $\mu^j(x)$, $V^j(x)$, $W^j(x)$, and thus

$$D_x L^j(z,x) = (H^j)^* W^j(x) Dg^j(\mu^j(x))(z - \mu^j(x)).$$

The MLE $\hat{x}_M$, if it exists, is the solution of the following system of nonlinear equations:

$$\sum_{j=1}^{M} (H^j)^* W^j(\hat{x}_M) Dg^j(\mu^j(\hat{x}_M))(z^j - \mu^j(\hat{x}_M)) = 0. \tag{5.4.19}$$

We propose to solve (5.4.19) by an iterative method, adapted from Newton's method. We shall define a sequence $\hat{x}_M^k$, written $\hat{x}^k$ to save notation, as follows:

$$\sum_{j=1}^{M} (H^j)^* W^j(\hat{x}^k)(H^j \hat{x}^{k+1} + h^j) = \sum_{j=1}^{M} (H^j)^* W^j(\hat{x}^k)(H^j \hat{x}^k + h^j) +$$
$$+ \sum_{j=1}^{M} (H^j)^* W^j(\hat{x}^k) Dg^j(\mu^j(\hat{x}^k))(z^j - \mu^j(\hat{x}^k)).$$

It is clear that if this iteration converges, then the limit point is a solution of (5.4.19). Noting that

$$H^j \hat{x}^k + h^j = g^j(\mu^j(\hat{x}^k)),$$

we can rewrite the iteration as follows:

$$\sum_{j=1}^{M} (H^j)^* W^j(\hat{x}^k)(H^j \hat{x}^{k+1} + h^j) = \sum_{j=1}^{M} (H^j)^* W^j(\hat{x}^k)[g^j(\mu^j(\hat{x}^k)) +$$
$$+ Dg^j(\mu^j(\hat{x}^k))(z^j - \mu^j(\hat{x}^k))].$$

Assuming $\sum_{j=1}^{M} (H^j)^* W^j(\hat{x}^k) H^j$ to be invertible, we get the following iteration:

$$\hat{x}^{k+1} = (\sum_{j=1}^{M} (H^j)^* W^j(\hat{x}^k) H^j)^{-1} \left(\sum_{j=1}^{M} (H^j)^* W^j(\hat{x}^k)[-h^j + g^j(\mu^j(\hat{x}^k)) \right. \tag{5.4.20}$$
$$\left. + Dg^j(\mu^j(\hat{x}^k))(z^j - \mu^j(\hat{x}^k))] \right).$$

5.4.3 The Gaussian Case

The Gaussian case corresponds to the model

$$z^j = H^j x + h^j + \epsilon^j, \tag{5.4.21}$$

where the variables ϵ^j are independent Gaussian with mean 0 and covariance matrix Σ^j. Indeed, if we take in the model (5.4.1),

$$b^j(\theta) = \frac{1}{2}\theta^*(\Sigma^j)^{-1}\theta,\ c^j(z) = -\frac{1}{2}z^*(\Sigma^j)^{-1}z - \frac{d}{2}\log(2\pi|\Sigma^j|)|,$$

$$g^j(\mu) = \mu,$$

then we have

$$g^j(\Sigma^j D_\theta b^j(\theta)) = \theta,$$

and hence $\varphi(\eta) = \eta$. It follows that

$$\theta^j(x) = H^j x + h^j,$$

$$\mu^j(x) = H^j x + h^j,$$

$$V^j(x) = \Sigma^j,$$

$$W^j(x) = (\Sigma^j)^{-1}.$$

We clearly have

$$f^j(z,x) = \frac{\exp -\frac{1}{2}(z - (H^j x + h^j))^*(\Sigma^j)^{-1}(z - (H^j x + h^j))}{(2\pi|\Sigma^j|)^{\frac{d}{2}}},$$

which is equivalent to (5.4.21). The system (5.4.19) becomes

$$\sum_{j=1}^{M}(H^j)^*(\Sigma^j)^{-1}(z^j - H^j\hat{x}_M - h^j) = 0, \tag{5.4.22}$$

and therefore, $\hat{x}_M$ is given explicitly by the formula

$$\hat{x}_M = (\sum_{j=1}^{M}(H^j)^*(\Sigma^j)^{-1}H^j)^{-1}\sum_{j=1}^{M}(H^j)^*(\Sigma^j)^{-1}(z^j - h^j). \tag{5.4.23}$$

We can check immediately that the iteration (5.4.20) leads to $\hat{x}^k = \hat{x}_M, \forall k$.

5.4.4 Recursivity

It is well known that formula (5.4.23), although nonrecursive in M, can be given a recursive form. Similarly, the iterative algorithm (5.4.20) is not recursive in M. To get a recursive formula we define a sequence $\hat{x}_j$ as follows. Given $\hat{x}_j$, define $\hat{\mu}_j$ by

$$g(\hat{\mu}_j) = H^j \hat{x}_j + h^j, \tag{5.4.24}$$

then $\hat{\theta}_j$ by

$$(\Sigma^j)^{-1} \hat{\mu}_j = Db(\hat{\theta}_j), \tag{5.4.25}$$

and next $\hat{V}_j$ by

$$\hat{V}_j = \Sigma^j D^2 b(\hat{\theta}_j) \Sigma^j. \tag{5.4.26}$$

Finally, $\hat{W}_j$ is given by

$$\hat{W}_j = ((Dg(\hat{\mu}_j))^*)^{-1} \hat{V}_j^{-1} (Dg(\hat{\mu}_j))^{-1}. \tag{5.4.27}$$

We write

$$\hat{x}_{M+1} = (\sum_{j=1}^{M} (H^j)^* \hat{W}_j H^j)^{-1} \left(\sum_{j=1}^{M} (H^j)^* \hat{W}_j [-h^j + g^j(\hat{\mu}_j) \right. \tag{5.4.28}$$
$$\left. + Dg^j(\hat{\mu}_j)(z^j - \hat{\mu}_j)] \right).$$

We can then give a recursive form to this formula. We set

$$P_M = (\sum_{j=1}^{M} (H^j)^* \hat{W}_j H^j)^{-1}. \tag{5.4.29}$$

Then

$$P_M^{-1} = P_{M-1}^{-1} + (H^M)^* \hat{W}_M H^M,$$

from which we obtain easily the following recursive relation:

$$P_M = P_{M-1} - P_{M-1}(H^M)^* (H^M P_{M-1} (H^M)^* + \hat{W}_M^{-1})^{-1} H^M P_{M-1}. \tag{5.4.30}$$

Let us introduce the *corrected* observation

$$\hat{\zeta}_M = g^M(\hat{\mu}_M) + Dg^M(\hat{\mu}_M)(z^M - \hat{\mu}_M). \tag{5.4.31}$$

Then from (5.4.28), we get

$$\begin{aligned} P_M^{-1}\hat{x}_{M+1} &= \sum_{j=1}^{M}(H^j)^*\hat{W}_j(-h^j + \hat{\zeta}_j) \\ &= P_{M-1}^{-1}\hat{x}_M + (H^M)^*\hat{W}_M(-h^M + \hat{\zeta}_M). \end{aligned}$$

and hence

$$\hat{x}_{M+1} = P_M P_{M-1}^{-1}\hat{x}_M + P_M(H^M)^*\hat{W}_M(-h^M + \hat{\zeta}_M).$$

Using

$$P_M P_{M-1}^{-1} = I - P_M(H^M)^*\hat{W}_M H^M,$$

we obtain

$$\hat{x}_{M+1} = \hat{x}_M + P_M(H^M)^*\hat{W}_M(\hat{\zeta}_M - H^M\hat{x}_M - h^M), \tag{5.4.32}$$

or equivalently,

$$\hat{x}_{M+1} = \hat{x}_M + P_M(H^M)^*\hat{W}_M Dg(\hat{\mu}_M)(z^M - \hat{\mu}_M). \tag{5.4.33}$$

In such a recursive algorithm, the initial condition $\hat{x}_1$ is arbitrary and corresponds to the best prior estimate of the parameter x, . without any observation. We then define the values of $\hat{\mu}_1$, $\hat{\theta}_1$, $\hat{V}_1$, $\hat{W}_1$ by formulas (5.4.24),(5.4.25), (5.4.26), (5.4.27), and

$$P_1 = ((H^1)^*\hat{W}_1 H^1)^{-1}. \tag{5.4.34}$$

5.4.5 *Examples*

5.4.5.1 The Binomial Distribution

The observation takes only the finite values $0, 1, \cdots, q$. We take $d = 1$, $\Sigma = 1$,

$$b(\theta) = q\log(1 + \exp\theta),\ \ c(z) = \log C_q^z,$$

$$g(\mu) = \log\frac{\mu}{q - \mu},\ \ 0 < \mu < q,$$

and so noting that

$$\pi = \frac{\exp \theta}{1 + \exp \theta},$$

we get easily

$$f(z, \theta) = C_q^z \pi^z (1 - \pi)^{q-z}.$$

We take

$$Hx + h = h^* x,$$

with the abuse of notation as regards h. So

$$\mu(x) = q \frac{\exp h^* x}{1 + \exp h^* x}, \; \theta(x) = h^* x,$$

$$V(x) = q \frac{\exp h^* x}{(1 + \exp h^* x)^2}, \; W(x) = V(x),$$

so $W(x) g'(\mu(x)) = 1$. Therefore, the maximum likelihood estimator is the solution of the system of nonlinear equations

$$\sum_{j=1}^{M} h^j (z^j - q \frac{\exp h^* \hat{x}_M}{1 + \exp h^* \hat{x}_M}) = 0; \tag{5.4.35}$$

see (5.4.19). The recursive algorithm (5.4.33) reduces to

$$\hat{x}_{M+1} = \hat{x}_M + P_M h^M (z^M - \hat{\mu}_M), \tag{5.4.36}$$

with

$$P_M = P_{M-1} - \frac{P_{M-1} h^M (h^M)^* P_{M-1}}{(h^M)^* P_{M-1} h^M + \hat{W}_M^{-1}}. \tag{5.4.37}$$

If we want to solve the system (5.4.35) by the iterative method (5.4.20), we get the sequence

$$\hat{x}^{k+1} = \hat{x}^k + (\sum_{j=1}^{M} W^j(\hat{x}^k) h^j (h^j)^*)^{-1} \sum_{j=1}^{M} (z^j - \mu^j(\hat{x}^k)) h^j. \tag{5.4.38}$$

5.4.5.2 The Poisson Distribution

We consider

$$d=1,\ b(\theta)=\exp\theta,\ c(z)=-\log z!,\ \Sigma=1,\ g(\mu)=\log\mu, \qquad (5.4.39)$$
$$Hx+h=h^*x,$$

and z is an integer, so

$$f(z,\theta)=\frac{\theta^z}{z!}\exp -\theta.$$

Therefore,

$$\mu(x)=\exp h^*x,\ \varphi(\eta)=\eta,\ \theta(x)=h^*x,$$
$$V(x)=\exp h^*x,\ W(x)=\exp h^*x.$$

Then the system (5.4.19) becomes

$$\sum_{j=1}^{M}(z^j-\exp((h^j)^*\hat{x}_M))h^j=0. \qquad (5.4.40)$$

The algorithm (5.4.20) reduces to

$$\hat{x}^{k+1}=\hat{x}^k+(\sum_{j=1}^{M}\mu^j(\hat{x}^k)h^j(h^j)^*)^{-1}\sum_{j=1}^{M}(z^j-\mu^j(\hat{x}^k))h^j, \qquad (5.4.41)$$

and the recursive algorithm (5.4.33) yields

$$\hat{x}_{M+1}=\hat{x}_M+P_Mh^M(z^M-\hat{\mu}_M), \qquad (5.4.42)$$

with

$$P_M=P_{M-1}-\frac{P_{M-1}h^M(h^M)^*P_{M-1}}{(h^M)^*P_{M-1}h^M+\hat{\mu}_M^{-1}}. \qquad (5.4.43)$$

5.4.5.3 The Gamma Distribution

The case of the gamma distribution is a little bit more complex. We take again $d=1$ and (ν is a positive constant)

$$\Sigma = \frac{1}{\nu},\ b(\theta) = -\nu \log(-\theta),\ \theta \in R^-$$

$$g(\mu) = \frac{1}{\mu},\ \mu > 0,\ c(z) = (\nu - 1)\log z - \log \Gamma(\nu) + \nu \log \nu,\ z > 0.$$

We have

$$f(z, \theta) = \frac{(-\theta\nu)^\nu}{\Gamma(\nu)} z^{\nu-1} \exp \nu\theta z,$$

where we recall that $\theta < 0$. Then $Hx + h = h^* x$. We can easily check that

$$\varphi(\eta) = -\eta,\ \theta(x) = -h^* x.$$

We note that x must satisfy the constraint

$$h^* x \geq 0. \tag{5.4.44}$$

Then

$$\mu(x) = \frac{1}{h^* x},\ V(x) = \frac{1}{\nu (h^* x)^2},\ W(x) = \frac{\nu}{(h^* x)^2}, \tag{5.4.45}$$

so we can write

$$D_x L^j(z, x) = -h^j \nu (z - \frac{1}{(h^j)^* x}). \tag{5.4.46}$$

Because of the constraint (5.4.44), we cannot write directly (5.4.19). The necessary condition of optimality of $\hat{x}_M$ can be written as follows (Kuhn–Tucker condition):

$$\sum_{j=1}^{M} (D_x L^j(z, \hat{x}_M))^* x \leq 0,\ \forall x, \text{such that},\ (h^j)^* x \geq 0,\ \forall j \tag{5.4.47}$$

$$\sum_{j=1}^{M} (D_x L^j(z, \hat{x}_M))^* \hat{x}_M = 0,\ (h^j)^* \hat{x}_M \geq 0,\ \forall j, \tag{5.4.48}$$

and using formulas (5.4.45), we obtain

$$\sum_{j=1}^{M} (h^j)^* x (z^j - \frac{1}{(h^j)^* \hat{x}_M}) \geq 0,\ \forall x, \text{such that},\ (h^j)^* x \geq 0,\ \forall j \tag{5.4.49}$$

$$\sum_{j=1}^{M} (h^j)^* \hat{x}_M (z^j - \frac{1}{(h^j)^* \hat{x}_M}) = 0,\ (h^j)^* \hat{x}_M \geq 0,\ \forall j. \tag{5.4.50}$$

5.5 Dynamic Models

5.5.1 General Bayesian Approach

5.5.1.1 Preliminaries

In all the preceding sections, we have been considering a fixed parameter x. The problem is thus an estimation problem, and the maximum likelihood is an appropriate method to achieve this estimation. When the parameter itself evolves with time, it can be considered the state of a dynamical system. In general, there is an evolution law for this state, but uncertainties affect this evolution, in contrast to the situation of a fixed parameter. The problem is to estimate the current state, what we call a tracking problem. An adequate approach is the Bayesian approach. We are going to describe it in general, then apply it for dynamic generalized models.

Instead of a fixed parameter, we shall speak of the state of the system at time j, denoted by $x^j \in R^n$. The observation is still denoted by $z^j \in R^d$. The pair x^j, z^j evolves as a Markov chain, with the particularity that the transition probability depends only on x. In other words, we consider a sequence of functions $\varpi^j(\eta, \zeta, x)$, where $\eta, x \in R^n$ and $\zeta \in R^d$. Defining the σ-algebra generated by $z^1, \cdots, z^j$,

$$\mathcal{Z}^j = \sigma(z^1, \cdots, z^j),$$

and similarly

$$\mathcal{F}^j = \sigma(x^0, x^1, z^1, \cdots, x^j, z^j),$$

we have, for a continuous bounded function $\varphi(x, z)$ on $R^n \times R^d$,

$$E(\varphi(x^{j+1}, z^{j+1})|\mathcal{F}^j) = \int\int \varphi(\eta, z)\varpi^j(\eta, z, x^j)d\eta dz. \tag{5.5.1}$$

To complete the description of the evolution, we need an initial probability density for x^0, denoted by $\mu(\eta)$.

5.5.1.2 Recurrence Formulas

We begin by considering the joint probability density of the variables $z^1, \cdots, z^M, x^M$ given by

$$\pi^M(\zeta^1, \cdots, \zeta^M, \eta^M) = \int \cdots \int \mu(\eta^0)\varpi^0(\eta^1, \zeta^1, \eta^0) \cdots \varpi^{M-1}(\eta^M, \zeta^M, \eta^{M-1})$$
$$d\eta^0 \cdots d\eta^{M-1}, \tag{5.5.2}$$

and we see immediately that it satisfies a recurrence equation

$$\pi^{M+1}(\zeta^1, \cdots, \zeta^{M+1}, \eta^{M+1}) = \int \pi^M(\zeta^1, \cdots, \zeta^M, \eta)\varpi^M(\eta^{M+1}, \zeta^{M+1}, \eta)d\eta. \tag{5.5.3}$$

We then derive the conditional probability density of x^M, given the σ-algebra $\mathcal{Z}^M$. For a given bounded continuous function φ on R^n, we consider $E[\varphi(x^M)|\mathcal{Z}^M]$. It is a random variable that is $\mathcal{Z}^M$-measurable. It is standard to check that it is obtained through a conditional probability density by the formula

$$E[\varphi(x^M)|\mathcal{Z}^M] = \int p^M(z^1, \cdots, z^M, \eta)\varphi(\eta)d\eta, \tag{5.5.4}$$

and the function $p^M(\zeta^1, \cdots, \zeta^M, \eta)$ is given by the formula

$$p^M(\zeta^1, \cdots, \zeta^M, \eta) = \frac{\pi^M(\zeta^1, \cdots, \zeta^M, \eta)}{\int \pi^M(\zeta^1, \cdots, \zeta^M, \eta')d\eta'}. \tag{5.5.5}$$

From the recurrence (5.5.3), we obtain easily a recurrence for the function p^M. We have

$$p^{M+1}(\zeta^1, \cdots, \zeta^{M+1}, \eta) = \frac{\int p^M(\zeta^1, \cdots, \zeta^M, \eta')\varpi^M(\eta, \zeta^{M+1}, \eta')d\eta'}{\int\int p^M(\zeta^1, \cdots, \zeta^M, \eta')\varpi^M(\eta'', \zeta^{M+1}, \eta')d\eta' d\eta''}. \tag{5.5.6}$$

We start this recurrence with

$$p^0(\eta) = \mu(\eta).$$

5.5.2 Dynamic GLM

5.5.2.1 Conditional Probability

At time j, the observation z^{j+1} has a conditional probability density when $x^j = x$ given by

$$f^j(z, x) = \exp(\theta^j(x)^*(R^j)^{-1}z - b^j(\theta^j(x)) + c^j(z)). \tag{5.5.7}$$

The dynamical system x^j evolves according to the model

$$x^{j+1} = F^j x^j + f^j + G^j w^j, \tag{5.5.8}$$

$$x^0 = N(\bar{\xi}, P_0),$$

in which the w^j are independent random variables, normal with mean 0 and covariance matrix Q^j. The variables w^j take values in R^m. They are independent of x^0. Also, for given x^j, the variables x^{j+1} and z^{j+1} are independent. Therefore, the pair x^j, z^j is a Markov chain, as described in Section 5.5.1.1. The function $\varpi^j(\eta, \zeta, x)$ defined in (5.5.1) is given by

$$\varpi^j(\eta,\zeta,x) = \frac{\begin{array}{c}\exp[-\frac{1}{2}(\eta - F^j x - f^j)^*(G^j Q^j (G^j)^*)^{-1}(\eta - F^j x - f^j) + \\ \theta^j(x)^*(R^j)^{-1}\zeta - b^j(\theta^j(x)) + c^j(\zeta)]\end{array}}{(2\pi)^{\frac{n}{2}}|G^j Q^j (G^j)^*|^{\frac{1}{2}}}. \tag{5.5.9}$$

We introduce the conditional probability of x^M given the filtration $\mathcal{Z}^M$, denoted by $p^M(\eta)$, in which we omit the dependence with respect to the arguments $\zeta^1, \cdots \zeta^M$. To simplify notation, we define

$$\gamma^j(\eta, x) = \frac{\exp[-\frac{1}{2}(\eta - F^j x - f^j)^*(G^j Q^j (G^j)^*)^{-1}(\eta - F^j x - f^j)]}{(2\pi)^{\frac{n}{2}}|G^j Q^j (G^j)^*|^{\frac{1}{2}}}$$

and

$$g^j(\zeta, x) = \exp(\theta^j(x)^*(R^j)^{-1}\zeta - b^j(\theta^j(x))).$$

Then formula (5.5.6) leads to

$$p^{M+1}(\eta) = \frac{\int p^M(\eta')\gamma^M(\eta, \eta')g^M(\zeta^{M+1}, \eta')d\eta'}{\int p^M(\eta')g^M(\zeta^{M+1}, \eta')d\eta'}, \tag{5.5.10}$$

and we obtain the following result.

PROPOSITION 5.3. *For the model (5.5.7), (5.5.8), the conditional probability density of x^M given the σ-algebra $\mathcal{Z}^M$, denoted $p^M(\eta) = p^M(\zeta^1, \cdots, \zeta^M, \eta)$, is defined recursively by formula (5.5.10) with $p^0(\eta) = \mu(\eta) = N(\bar{\xi}, P_0)$.*

5.5.2.2 The First Two Moments

The best estimate of x^M, denoted by $\hat{x}_M$, is defined from the conditional probability density $p^M(\eta)$ simply by the formula

$$\hat{x}_M = \int \eta p^M(\eta) d\eta. \tag{5.5.11}$$

Unfortunately, there is no recursive formula for $\hat{x}_M$. Noting that

$$\int \eta \gamma^M(\eta, \eta') = F^M \eta' + f^M,$$

we can write

$$\hat{x}_{M+1} = F^M \hat{y}_M + f^M, \tag{5.5.12}$$

with

$$\hat{y}_M = \frac{\int \eta\, p^M(\eta) g^M(\zeta^{M+1}, \eta) d\eta}{\int p^M(\eta) g^M(\zeta^{M+1}, \eta) d\eta}. \tag{5.5.13}$$

It is possible to obtain a recursive formula for $\hat{x}_M$ by considering approximations. The idea is to introduce the covariance operator

$$P_M = \int \eta\eta^* p^M(\eta) d\eta - \hat{x}_M (\hat{x}_M)^* \tag{5.5.14}$$

and to approximate $p^M(\eta)$ by a Gaussian

$$p^M(\eta) = \frac{\exp[-\frac{1}{2}(\eta - \hat{x}_M)^* P_M^{-1}(\eta - \hat{x}_M)]}{(2\pi)^{\frac{n}{2}} |P_M|^{\frac{1}{2}}}.$$

Therefore,

$$\hat{y}_M = \frac{\int \eta\, g^M(\zeta^{M+1}, \eta) \exp[-\frac{1}{2}\eta^* P_M^{-1}\eta + (\hat{x}_M)^* P_M^{-1}\eta] d\eta}{\int g^M(\zeta^{M+1}, \eta) \exp[-\frac{1}{2}\eta^* P_M^{-1}\eta + (\hat{x}_M)^* P_M^{-1}\eta] d\eta}. \tag{5.5.15}$$

Recalling (5.5.12), we see that $\hat{x}_{M+1}$ can be obtained from the knowledge of $\hat{x}_M$ and P_M.

We next have to define P_{M+1}. We use

$$P_{M+1} = \int \eta\eta^* p^{M+1}(\eta) d\eta - \hat{x}_{M+1}(\hat{x}_{M+1})^*,$$

with $p^{M+1}(\eta)$ given by (5.5.10) and $\hat{x}_{M+1}$ given by (5.5.12). We then introduce

$$\begin{aligned} \Gamma_M &= \frac{\int \eta\eta^*\, g^M(\zeta^{M+1}, \eta) \exp[-\frac{1}{2}\eta^* P_M^{-1}\eta + (\hat{x}_M)^* P_M^{-1}\eta] d\eta}{\int g^M(\zeta^{M+1}, \eta) \exp[-\frac{1}{2}\eta^* P_M^{-1}\eta + (\hat{x}_M)^* P_M^{-1}\eta] d\eta} - \hat{y}_M \hat{y}_M^* \\ &= \frac{\int (\eta - \hat{y}_M)(\eta - \hat{y}_M)^*\, g^M(\zeta^{M+1}, \eta) \exp[-\frac{1}{2}\eta^* P_M^{-1}\eta + (\hat{x}_M)^* P_M^{-1}\eta] d\eta}{\int g^M(\zeta^{M+1}, \eta) \exp[-\frac{1}{2}\eta^* P_M^{-1}\eta + (\hat{x}_M)^* P_M^{-1}\eta] d\eta}. \end{aligned} \tag{5.5.16}$$

We check easily the formula

$$P_{M+1} = F^M \Gamma_M (F^M)^* + G^M Q^M (G^M)^*, \tag{5.5.17}$$

and so we propose the recursive algorithm for $\hat{x}_M$ and P_M defined by formulas (5.5.12) and (5.5.17), in which the random quantities $\hat{y}_M$ and Γ_M are given by formulas (5.5.15) and (5.5.16).

5.5.3 *Applications*

5.5.3.1 Kalman Filter

We consider the situation

$$\theta^j(x) = H^j x + h^j,\ b^j(\theta) = \frac{1}{2}\theta^*(R^j)^{-1}\theta,$$

so that

$$g^j(\zeta, x) = \exp((H^j x + h^j)^*(R^j)^{-1}\zeta - \frac{1}{2}(H^j x + h^j)^*(R^j)^{-1}(Hx + h^j)).$$

We will check that $p^M(\eta)$ is indeed a Gaussian. Assuming that such is the case for $p^M(\eta)$, we prove it for $p^{M+1}(\eta)$ by computing the characteristic function

$$\mathcal{L}^{M+1}(\lambda) = \int \exp i\lambda^*\eta\, p^{M+1}(\eta)d\eta.$$

Since

$$\int \exp i\lambda^*\eta\, \gamma^M(\eta, \eta')d\eta = \exp[i\lambda^*(F^M\eta' + f^M) - \frac{1}{2}\lambda^* G^M Q^M (G^M)^*\lambda],$$

we get

$$\mathcal{L}^{M+1}(\lambda) = \exp[i\lambda^* f^M - \frac{1}{2}\lambda^* G^M Q^M (G^M)^*\lambda]\frac{N(\lambda)}{N(0)},$$

with

$$N(\lambda) =$$

$$\int \exp\Big[-\frac{1}{2}(\eta^*(P_M^{-1} + (H^M)^*(R^M)^{-1}H^M)\eta) + (i(F^M)^*\lambda + P_M^{-1}\hat{x}_M + (H^M)^*$$
$$(R^M)^{-1}(\zeta^{M+1} - h^M))^*\eta\Big]d\eta.$$

We then check easily that

$$\frac{N(\lambda)}{N(0)} = \exp\left[-\frac{1}{2}\lambda^*(P_M^{-1} + (H^M)^*(R^M)^{-1}H^M)^{-1}\lambda\right] \times$$
$$\exp i\lambda^* \left(F^M(P_M^{-1} + (H^M)^*(R^M)^{-1}H^M)^{-1}(P_M^{-1}\hat{x}_M + (H^M)^*(R^M)^{-1}\right.$$
$$\left.(\zeta^{M+1} - h^M))\right).$$

Collecting results, we see that $\mathcal{L}^{M+1}(\lambda)$ is the exponential of a quadratic form in λ. Therefore, $p^{M+1}(\eta)$ is a Gaussian with mean

$$\hat{x}_{M+1} = f^M + F^M(P_M^{-1} + (H^M)^*(R^M)^{-1}H^M)^{-1}(P_M^{-1}\hat{x}_M + (H^M)^*(R^M)^{-1}$$
$$(\zeta^{M+1} - h^M))$$

and covariance matrix

$$P_{M+1} = G^M Q^M (G^M)^* + F^M(P_M^{-1} + (H^M)^*(R^M)^{-1}H^M)^{-1}(F^M)^*.$$

We can rewrite these expressions as follows:

$$\hat{x}_{M+1} = F^M\hat{x}_M + f^M + F^M(P_M^{-1} + (H^M)^*(R^M)^{-1}H^M)^{-1}(H^M)^*(R^M)^{-1}$$
$$(\zeta^{M+1} - H^M\hat{x}_M - h^M). \tag{5.5.18}$$

Now we check easily that

$$(P_M^{-1} + (H^M)^*(R^M)^{-1}H^M)^{-1} = P^M - P^M(H^M)^*(R^M + H^M P^M (H^M)^*)^{-1}H^M P^M.$$

So

$$P_{M+1} = F^M P^M (F^M)^* - F^M P^M (H^M)^*(R^M + H^M P^M (H^M)^*)^{-1}$$
$$H^M P^M (F^M)^* + G^M Q^M (G^M)^*. \tag{5.5.19}$$

We use also

$$(P_M^{-1} + (H^M)^*(R^M)^{-1}H^M)^{-1}(H^M)^*(R^M)^{-1} = P^M(H^M)^*(R^M + H^M P^M (H^M)^*)^{-1}$$

to write

$$\hat{x}_{M+1} = F^M\hat{x}_M + f^M + F^M P^M (H^M)^*(R^M + H^M P^M (H^M)^*)^{-1}(\zeta^{M+1} - H^M\hat{x}_M - h^M), \tag{5.5.20}$$

and we obtain the Kalman filter.

Considering formulas (5.5.15) and (5.5.16), we check easily that

$$\hat{y}_M = (P_M^{-1} + (H^M)^*(R^M)^{-1}H^M)^{-1}(P_M^{-1}\hat{x}_M + (H^M)^*(R^M)^{-1}(\zeta^{M+1} - h^M)), \tag{5.5.21}$$

$$\Gamma_M = (P_M^{-1} + (H^M)^*(R^M)^{-1}H^M)^{-1}, \tag{5.5.22}$$

and thus these formulas are no longer approximations.

5.5.3.2 The Poisson Distribution

Consider the situation of Section 5.4.5.2. Then

$$\theta^j(x) = (h^j)^*x, \; b^j(\theta) = \exp\theta,$$

$$g^j(\zeta, x) = \exp(\zeta(h^j)^*x - \exp(h^j)^*x).$$

Therefore, formulas (5.5.15) and (5.5.16) yield

$$\hat{y}_M = \frac{\int \eta \exp[-\frac{1}{2}\eta^* P_M^{-1}\eta - \exp(h^j)^*\eta + ((\hat{x}_M)^* P_M^{-1} + \zeta^{M+1}(h^j)^*)\eta]d\eta}{\int \exp[-\frac{1}{2}\eta^* P_M^{-1}\eta - \exp(h^j)^*\eta + ((\hat{x}_M)^* P_M^{-1} + \zeta^{M+1}(h^j)^*)\eta]d\eta}, \tag{5.5.23}$$

$$\Gamma_M = \frac{\begin{array}{c}\int(\eta - \hat{y}_M)(\eta - \hat{y}_M)^* \exp[-\frac{1}{2}\eta^* P_M^{-1}\eta - \exp(h^j)^*\eta + ((\hat{x}_M)^* P_M^{-1} \\ +\zeta^{M+1}(h^j)^*)\eta]d\eta\end{array}}{\int \exp[-\frac{1}{2}\eta^* P_M^{-1}\eta - \exp(h^j)^*\eta + ((\hat{x}_M)^* P_M^{-1} + \zeta^{M+1}(h^j)^*)\eta]d\eta}, \tag{5.5.24}$$

and $\hat{x}_{M+1}$, P_{M+1} are given by (5.5.12) and (5.5.17).

5.5.3.3 The Kalman Filter Revisited

We consider formulas (5.5.21), (5.5.22). Note that the Kalman filter $\hat{x}_{M+1}$ and the covariance error P_{M+1} are given by formulas (5.5.12) and (5.5.17). We rewrite them as

$$\begin{aligned}\hat{y}_M &= \hat{x}_M + (P_M^{-1} + (H^M)^*(R^M)^{-1}H^M)^{-1}(H^M)^* \\ &\quad (R^M)^{-1}(z^{M+1} - (H^M\hat{x}_M + h^M)) \qquad (5.5.25) \\ \Gamma_M &= (P_M^{-1} + (H^M)^*(R^M)^{-1}H^M)^{-1}. \qquad (5.5.26)\end{aligned}$$

We have reinstated z^{M+1} in formula (5.5.25) in lieu of ζ^{M+1} to consider $\hat{y}_M$ a random variable that is $\mathcal{Z}^{M+1}$-measurable. In fact, in this section, we will use the same notation for random variables and arguments, to save notation. We want to prove the following result.

PROPOSITION 5.4. *We have*

$$\hat{y}_M = E[x_M|\mathcal{Z}^{M+1}], \tag{5.5.27}$$

$$\Gamma_M = E[(x_M - \hat{y}_M)(x_M - \hat{y}_M)^*|\mathcal{Z}^{M+1}]. \tag{5.5.28}$$

Note that this interpretation fits perfectly with $\hat{x}_{M+1} = E[x^{M+1}|\mathcal{Z}^{M+1}]$ and

$$x^{M+1} = F^M x^M + f^M + G^M w^M,$$

where we note that w^M is independent of $\mathcal{Z}^{M+1}$.

PROOF. In proving Proposition 5.4, we shall use an approach inspired by [50]. This approach focuses on the canonical parameter and the GLM form of the observation probability density. So we introduce the sequence of random variables

$$\theta^M = H^M x^M + h^M.$$

The probability density of z^{M+1} given θ^M is a Gaussian with mean θ^M and covariance matrix R^M. We shall use the notation

$$\text{Prob}(z^{M+1}|\theta^M) = N(\theta^M|R^M).$$

We recall that we use the same notation for random variables and arguments representing their values. We write this probability in GLM format as

$$\text{Prob}(z^{M+1}|\theta^M) = B(z^{M+1}, (R^M)^{-1}) \exp[(z^{M+1})^*(R^M)^{-1}\theta^M - b(\theta^M, (R^M)^{-1})], \tag{5.5.29}$$

with, of course,

$$B(z^{M+1}, (R^M)^{-1}) = \frac{\exp -\frac{1}{2}(z^{M+1})^*(R^M)^{-1}z^{M+1}}{(2\pi)^{\frac{d}{2}}|R^M|^{\frac{1}{2}}},$$

and the function $b(\theta, \Sigma)$ defined on $R^d \times \mathcal{L}(R^d, R^d)$ is given by

$$b(\theta, \Sigma) = \frac{1}{2}\theta^*\Sigma\theta.$$

We recognize the function b depending on the canonical argument. We have introduced an additional dependence, with respect to a symmetric matrix. This dependence is linear. This will play a key role in the following. Since we are employing a recursive argument, we know that

$$\text{Prob}(\theta^M|\mathcal{Z}^M) = N(H^M\hat{x}_M + h^M, H^M P_M(H^M)^*).$$

The key idea is to write it as follows:

$$\text{Prob}(\theta^M|\mathcal{Z}^M) = c(\alpha_M, \beta_M)\exp[\alpha_M^*\theta^M - b(\theta^M, \beta_M)], \tag{5.5.30}$$

where α_M is a vector in R^d and $\beta_M \in \mathcal{L}(R^d, R^d)$. They are given explicitly by

$$\alpha_M = (H^M P_M(H^M)^*)^{-1}(H^M\hat{x}_M + h^M),$$
$$\beta_M = (H^M P_M(H^M)^*)^{-1},$$

and

$$c(\alpha, \beta) = \frac{|\beta|^{\frac{1}{2}}\exp -\frac{1}{2}\alpha^*\beta^{-1}\alpha}{(2\pi)^{\frac{d}{2}}}.$$

We note the analogy between formulas (5.5.29) and (5.5.30), in terms of using the function b, but with different arguments.

From the probabilities (5.5.29) and (5.5.30) we deduce, using the linearity of b with respect to the second argument,

$$\text{Prob}(z^{M+1}|\mathcal{Z}^M) = c(\alpha_M, \beta_M)B(z^{M+1}, (R^M)^{-1})\int \exp[(\alpha_M + (R^M)^{-1}z^{M+1})^*$$
$$\theta - b(\theta, (R^M)^{-1} + \beta_M)]d\theta.$$

Hence clearly,

$$\text{Prob}(z^{M+1}|\mathcal{Z}^M) = \frac{c(\alpha_M, \beta_M)B(z^{M+1}, (R^M)^{-1})}{c(\alpha_M + (R^M)^{-1}z^{M+1}, (R^M)^{-1} + \beta_M)}. \tag{5.5.31}$$

We can then compute $\text{Prob}(\theta^M|\mathcal{Z}^{M+1})$. Indeed,

$$\text{Prob}(\theta^M|\mathcal{Z}^{M+1}) = \frac{\text{Prob}(\theta^M, z^{M+1}|\mathcal{Z}^M)}{\text{Prob}(z^{M+1}|\mathcal{Z}^M)}$$
$$= \frac{\text{Prob}(z^{M+1}|\theta^M, \mathcal{Z}^M)\text{Prob}(\theta^M|\mathcal{Z}^M)}{\text{Prob}(z^{M+1}|\mathcal{Z}^M)}$$
$$= \frac{\text{Prob}(z^{M+1}|\theta^M)\text{Prob}(\theta^M|\mathcal{Z}^M)}{\text{Prob}(z^{M+1}|\mathcal{Z}^M)},$$

and thus

$$\begin{aligned}\mathrm{Prob}(\theta^M|\mathcal{Z}^{M+1}) &= c(\alpha_M + (R^M)^{-1}z^{M+1}, (R^M)^{-1} + \beta_M) \\ &\quad \times \exp[(\alpha_M + (R^M)^{-1}z^{M+1})^*\theta^M - b(\theta^M, (R^M)^{-1} + \beta_M)].\end{aligned} \tag{5.5.32}$$

Using the value of the α_M, β_M and the definition of the functions c and b, we obtain, after a lengthy calculation, that

$$\mathrm{Prob}(\theta^M|\mathcal{Z}^{M+1}) = N(g_M, \Pi_M), \tag{5.5.33}$$

with

$$g_M = H^M\hat{x}_M + h^M + \Pi_M(R^M)^{-1}(z^{M+1} - H^M\hat{x}_M - h^M), \tag{5.5.34}$$

$$\Pi_M = \left((R^M)^{-1} + (H^M P_M (H^M)^*)^{-1}\right)^{-1}. \tag{5.5.35}$$

In order to compute the conditional probability $\mathrm{Prob}(x^M|\mathcal{Z}^{M+1})$, we compute the joint conditional probability $\mathrm{Prob}(x^M, \theta^M|\mathcal{Z}^{M+1})$. This leads to

$$\begin{aligned}\mathrm{Prob}(x^M, \theta^M|\mathcal{Z}^{M+1}) &= \mathrm{Prob}(x^M, \theta^M|z^{M+1}, \mathcal{Z}^M) \\ &= \frac{\mathrm{Prob}(x^M, \theta^M, z^{M+1}|\mathcal{Z}^M)}{\mathrm{Prob}(z^{M+1}|\mathcal{Z}^M)} \\ &= \frac{\mathrm{Prob}(z^{M+1}|x^M, \theta^M, \mathcal{Z}^M)\mathrm{Prob}(x^M, \theta^M|\mathcal{Z}^M)}{\mathrm{Prob}(z^{M+1}|\mathcal{Z}^M)} \\ &= \frac{\mathrm{Prob}(z^{M+1}|\theta^M)\mathrm{Prob}(x^M, \theta^M|\mathcal{Z}^M)}{\mathrm{Prob}(z^{M+1}|\mathcal{Z}^M)} \\ &= \frac{\mathrm{Prob}(z^{M+1}|\theta^M)\mathrm{Prob}(\theta^M|\mathcal{Z}^M)\mathrm{Prob}(x^M|\theta^M, \mathcal{Z}^M)}{\mathrm{Prob}(z^{M+1}|\mathcal{Z}^M)} \\ &= \frac{\mathrm{Prob}(z^{M+1}, \theta^M|\mathcal{Z}^M)\mathrm{Prob}(x^M|\theta^M, \mathcal{Z}^M)}{\mathrm{Prob}(z^{M+1}|\mathcal{Z}^M)} \\ &= \mathrm{Prob}(\theta^M|\mathcal{Z}^{M+1})\mathrm{Prob}(x^M|\theta^M, \mathcal{Z}^M).\end{aligned}$$

We know $\mathrm{Prob}(\theta^M|\mathcal{Z}^{M+1})$. We can define $\mathrm{Prob}(x^M|\theta^M, \mathcal{Z}^M)$ from the knowledge of the joint conditional probability of the pair x^M, θ^M given the σ-algebra $\mathcal{Z}^M$. However, this joint probability has no density, since θ^M is linked to x^M by a deterministic relation. However, since the pair is Gaussian, it is well known that the conditional probability is also Gaussian with mean $E[x^M|\theta^M, \mathcal{Z}^M]$ and covariance

$$E\left((x^M - E[x^M|\theta^M, \mathcal{Z}^M])(x^M - E[x^M|\theta^M, \mathcal{Z}^M])^*|\theta^M, \mathcal{Z}^M\right) = \Delta_M.$$

Classical linear estimation theory for Gaussian variables tells us that

$$E[x^M|\theta^M, \mathcal{Z}^M] = \hat{x}_M + \Theta_M(\theta^M - H^M\hat{x}_M - h^M), \tag{5.5.36}$$

with

$$\Theta_M = P_M(H^M)^*(H^M P_M(H^M)^*)^{-1}$$

and

$$\Delta_M = P_M - Q_M H^M P_M(H^M)^* Q_M^*. \tag{5.5.37}$$

Hence we have

$$\text{Prob}(x^M, \theta^M|\mathcal{Z}^{M+1}) = N(\theta^M; g_M, \Pi_M)N(x^M; \hat{x}_M + \Theta_M(\theta^M - H^M\hat{x}_M - h^M), \Delta_M). \tag{5.5.38}$$

From this formula, the conditional probability $\text{Prob}(x^M|\mathcal{Z}^{M+1})$ is obtained by integrating in θ^M. It is Gaussian with mean

$$\begin{aligned} E(x^M|\mathcal{Z}^{M+1}) &= \hat{x}_M + \Theta_M(g_M - H^M\hat{x}_M - h^M) \\ &= \hat{x}_M + \Theta_M\Pi_M(R^M)^{-1}(z^{M+1} - H^M\hat{x}_M - h^M). \end{aligned}$$

It remains to show that

$$\Gamma_M(H^M)^* = \Theta_M\Pi_M,$$

which is left to the reader. We have proven the result (5.5.27). Formula (5.5.36) shows also that

$$E\left((x^M - E(x^M|\mathcal{Z}^{M+1}))(x^M - E(x^M|\mathcal{Z}^{M+1}))^*|\mathcal{Z}^{M+1}\right) =$$

$$E\left((x^M - E(x^M|\theta^M, \mathcal{Z}^M))(x^M - E(x^M|\theta^M, \mathcal{Z}^M))^*|\mathcal{Z}^{M+1}\right) +$$

$$+E\left((E(x^M|\theta^M,\mathcal{Z}^M) - E(x^M|\mathcal{Z}^{M+1}))(E(x^M|\theta^M,\mathcal{Z}^M) - E(x^M|\mathcal{Z}^{M+1}))^*|\mathcal{Z}^{M+1}\right)$$

$$= \Delta_M + \Theta_M\Pi_M\Theta_M^*,$$

and it remains to show that

$$\Delta_M + \Theta_M\Pi_M\Theta_M^* = \Gamma_M,$$

which completes the proof. □

5.5.4 First Two Moments Revisited

5.5.4.1 General Ideas

In Section 5.5.2.2, we formulated an approximate recurrence for $\hat{x}_M$ and P_M, namely formulas (5.5.12), (5.5.15) and (5.5.16), (5.5.17). It is obtained in two steps, first defining quantities $\hat{y}_M$ and Γ_M in terms of $\hat{x}_M$ and P_M, then obtaining $\hat{x}_{M+1}$ and P_{M+1}. The major approximation stems from considering that the conditional probability of x^M given $\mathcal{Z}^M$ is a Gaussian with mean $\hat{x}_M$ and covariance matrix P_M. In the case of the Kalman filter, we have interpreted $\hat{y}_M$ as $E[x^M|\mathcal{Z}^{M+1}]$ and Γ_M as Cov $(x^M|\mathcal{Z}^{M+1})$. Also, in the case of the Kalman filter, the Gaussian property is not an approximation. We have revisited the Kalman filter, focusing on the canonical parameter θ^M instead of the state x^M. We have considered the canonical parameter a random variable, linked to x^M by a deterministic relation. Thanks to linearity, we could remain in the Gaussian framework and recover all formulas.

In this section, we will follow the same idea for the dynamic GLM and focus on the canonical parameter. It is still linked to the state, but this time through a nonlinear deterministic relation. To get a recurrence, an approximation will be needed, but of a different type. This method was introduced in [50] for the case in which the canonical parameter is a scalar.

5.5.4.2 Model and Approximation

We follow the notation of Section 5.5.3.3; see Proposition 5.4. We have first to make more precise the probability density (5.5.2), which defines the dependence of the observation on the canonical parameter. We write

$$f^j(z,\theta) = \exp(\theta^*(R^j)^{-1}z - b(\theta,(R^j)^{-1}))B(z,(R^j)^{-1}). \tag{5.5.39}$$

So the function $b^j(\theta)$ entering into the definition of (5.5.2) is clarified by formula (5.5.39). The function b is linear in the second argument. The relation between the canonical parameter and the state is defined as

$$\theta^j(x) = \varphi^j(H^j x + h^j)$$

or by the inverse

$$H^j x + h^j = \gamma^j(\theta^j). \tag{5.5.40}$$

We next consider the recurrence from M to $M+1$. We suppose we know

$$\hat{x}_M = E[x^M|\mathcal{Z}^M],\ P_M = \mathrm{Cov}(x^M|\mathcal{Z}^M). \tag{5.5.41}$$

Note that P_M is not necessarily deterministic. We define the random variable θ^M by

$$H^M x^M + h^M = \gamma^M(\theta^M)$$

and recall that

$$x^{M+1} = F^M x^M + f^M + G^M w^M.$$

Therefore,

$$E(\gamma^M(\theta^M)|\mathcal{Z}^M) = H^M \hat{x}^M + h^M, \ \mathrm{Cov}(\gamma^M(\theta^M)|\mathcal{Z}^M) = H^M P_M (H^M)^* \tag{5.5.42}$$

and

$$\hat{x}^{M+1} = F^M \, E(x|\mathcal{Z}^{M+1}) + f^M, \tag{5.5.43}$$

$$P_{M+1} = F^M \, \mathrm{Cov}(x^M|\mathcal{Z}^{M+1})(F^M)^* + G^M Q^M (G^M)^*. \tag{5.5.44}$$

In contrast to the Gaussian case, we do not know the conditional probability of θ^M, given $\mathcal{Z}^M$, except for two relations that must be satisfied, namely (5.5.42). We then postulate that it has the form (5.5.30),

$$\mathrm{Prob}(\theta^M|\mathcal{Z}^M) = c(\alpha_M, \beta_M) \exp[\alpha_M^* \theta^M - b(\theta^M, \beta_M)], \tag{5.5.45}$$

where α_M, β_M are parameters, which we can define by writing conditions (5.5.42). So we write

$$H^M \hat{x}^M + h^M = c(\alpha_M, \beta_M) \int \gamma^M(\theta) \exp[\alpha_M^* \theta - b(\theta, \beta_M)] \, d\theta, \tag{5.5.46}$$

$$H^M P_M (H^M)^* + (H^M \hat{x}^M + h^M)(H^M \hat{x}^M + h^M)^* = c(\alpha_M, \beta_M) \int \gamma^M(\theta)(\gamma^M(\theta))^* \exp[\alpha_M^* \theta - b(\theta, \beta_M)] \, d\theta. \tag{5.5.47}$$

These two relations allow us in principle to compute α_M, β_M. We can then proceed as in Proposition 5.4 to show that

$$\begin{aligned} \mathrm{Prob}(\theta^M|\mathcal{Z}^{M+1}) &= c(\alpha_M + (R^M)^{-1} z^{M+1}, (R^M)^{-1} + \beta_M) \\ &\quad \times \exp[(\alpha_M + (R^M)^{-1} z^{M+1})^* \theta^M - b(\theta^M, (R^M)^{-1} + \beta_M)] \end{aligned} \tag{5.5.48}$$

and

$$\mathrm{Prob}(x^M, \theta^M|\mathcal{Z}^{M+1}) = \mathrm{Prob}(\theta^M|\mathcal{Z}^{M+1}) \mathrm{Prob}(x^M|\theta^M, \mathcal{Z}^M).$$

Therefore, again we have

$$E(x^M|\mathcal{Z}^{M+1}) = E\left(E(x^M|\theta^M, \mathcal{Z}^M)|\mathcal{Z}^{M+1}\right), \tag{5.5.49}$$

$$\mathrm{Cov}\,(x^M|\mathcal{Z}^{M+1}) = E\left(\mathrm{Cov}\,(x^M|\theta^M, \mathcal{Z}^M)|\mathcal{Z}^{M+1}\right)+\mathrm{Cov}\left(E[x^M|\theta^M, \mathcal{Z}^M]|\mathcal{Z}^{M+1}\right). \tag{5.5.50}$$

However, in contrast to the Kalman filter case, we do not know the conditional probability Prob$(x^M|\theta^M, \mathcal{Z}^M)$. Fortunately, to compute the quantities on the left-hand side of (5.5.49), (5.5.50), we do not need the full conditional probability. We only need the quantities

$$E(x^M|\theta^M, \mathcal{Z}^M),\ \mathrm{Cov}\,(x^M|\theta^M, \mathcal{Z}^M).$$

The first term is the best estimate of x^M given $\theta^M, \mathcal{Z}^M$, and the second one is the covariance of the estimation error. Since knowing θ^M is equivalent to knowing $\gamma^M(\theta^M)$, we need to compute

$$E(x^M|\gamma^M(\theta^M), \mathcal{Z}^M),\ \mathrm{Cov}\,(x^M|\gamma^M(\theta^M), \mathcal{Z}^M).$$

The first quantity is the best estimate of x^M given $\gamma^M(\theta^M), \mathcal{Z}^M$, and the second is the covariance of the residual error. We cannot compute these quantities, because we do not know the conditional probability of x^M given $\gamma^M(\theta^M), \mathcal{Z}^M$. However, we can compute the best linear estimate, because we know

$$E\begin{pmatrix} x^M \\ \gamma^M(\theta^M) \end{pmatrix}|\mathcal{Z}^M\Big) = \begin{pmatrix} \hat{x}^M \\ H^M\hat{x}^M + h^M \end{pmatrix}$$

and

$$\mathrm{Cov}\begin{pmatrix} x^M \\ \gamma^M(\theta^M) \end{pmatrix}|\mathcal{Z}^M\Big) = \begin{pmatrix} P_M & H^M P_M \\ P_M(H^M)^* & H^M P_M(H^M)^* \end{pmatrix}.$$

This best linear estimate has been obtained in the case of the Kalman filter. We thus take the approximation

$$E(x^M|\theta^M, \mathcal{Z}^M) \sim \hat{x}_M + \Theta_M(\gamma^M(\theta^M) - H^M\hat{x}_M - h^M), \tag{5.5.51}$$

with

$$\Theta_M = P_M(H^M)^*(H^M P_M(H^M)^*)^{-1} \tag{5.5.52}$$

and

$$\mathrm{Cov}\,(x^M|\theta^M, \mathcal{Z}^M) \sim \Delta_M = P_M - \Theta_M H^M P_M(H^M)^*\Theta_M^*. \tag{5.5.53}$$

Next, using (5.5.48), we have

$$\hat{y}_M = E(x^M|\mathcal{Z}^{M+1}) \sim \hat{x}_M + c(\alpha_M + (R^M)^{-1}z^{M+1}, (R^M)^{-1} + \beta_M)\Theta_M \tag{5.5.54}$$

$$\times \left(\int \gamma^M(\theta) \exp[(\alpha_M + (R^M)^{-1}z^{M+1})^*\theta - b(\theta, (R^M)^{-1} + \beta_M)]d\theta \right)$$

$$- \Theta_M(H^M\hat{x}_M + h^M),$$

and from (5.5.50), (5.5.53), (5.5.51), we obtain

$$\text{Cov}\,(x^M|\mathcal{Z}^{M+1}) \sim \Delta_M + \Theta_M\Pi_M\Theta_M^*, \tag{5.5.55}$$

with

$$\Pi_M = c(\alpha_M + (R^M)^{-1}z^{M+1}, (R^M)^{-1} + \beta_M) \tag{5.5.56}$$

$$\times \left(\int \gamma^M(\theta)(\gamma^M(\theta))^* \exp[(\alpha_M+(R^M)^{-1}z^{M+1})^*\theta - b(\theta, (R^M)^{-1}+\beta_M)]d\theta \right) -$$

$$-(c(\alpha_M + (R^M)^{-1}z^{M+1}, (R^M)^{-1} + \beta_M))^2$$

$$\times \left(\int \gamma^M(\theta) \exp[(\alpha_M + (R^M)^{-1}z^{M+1})^*\theta - b(\theta, (R^M)^{-1} + \beta_M)]d\theta \right)^2.$$

So summarizing, we have that $\hat{x}_{M+1}$ and P_{M+1} are obtained by formulas (5.5.43), (5.5.44), with $E(x|\mathcal{Z}^{M+1})$ given by formula (5.5.54), $\text{Cov}\,(x^M|\mathcal{Z}^{M+1})$ given by formulas (5.5.55), (5.5.56), in which α_M and β_M are computed from relations (5.5.46), (5.5.47).

5.5.4.3 Further Approximation

We define

$$\hat{\theta}_M = E(\theta^M|\mathcal{Z}^M), \ \text{Cov}\,(\theta^M|\mathcal{Z}^M) = E(\theta^M(\theta^M)^*|\mathcal{Z}^M) - \hat{\theta}_M(\hat{\theta}_M)^*.$$

Then we consider the following approximation:

$$\gamma_i^M(\theta^M) = \gamma_i^M(\hat{\theta}_M) + D\gamma_i^M(\hat{\theta}_M)(\theta^M - \hat{\theta}_M) + \frac{1}{2}\text{tr}\left(D^2\gamma_i^M(\hat{\theta}_M)(\theta^M - \hat{\theta}_M)(\theta^M - \hat{\theta}_M)^*\right). \tag{5.5.57}$$

We can write

$$E(\gamma_i^M(\theta^M)|\mathcal{Z}^M) = \gamma_i^M(\hat{\theta}_M) + \frac{1}{2}\text{tr}\left(D^2\gamma_i^M(\hat{\theta}_M)\text{Cov}(\theta^M|\mathcal{Z}^M)\right), \tag{5.5.58}$$

the second term being small with respect to the first one. Similarly, we can check that

$$\mathrm{Cov}_{ij}(\gamma^M(\theta^M)|\mathcal{Z}^M) = \frac{1}{2}\mathrm{tr}\left(((D\gamma_i^M(\hat{\theta}_M)(D\gamma_j^M(\hat{\theta}_M))^* + D\gamma_i^M(\hat{\theta}_M)(D\gamma_j^M(\hat{\theta}_M))^*)\right.$$
$$\left.\mathrm{Cov}(\theta^M|\mathcal{Z}^M)\right).$$

Consider the family of matrices

$$K_{ij}(\hat{\theta}_M) = \frac{1}{2}\left(D\gamma_i^M(\hat{\theta}_M)(D\gamma_j^M(\hat{\theta}_M))^* + D\gamma_i^M(\hat{\theta}_M)(D\gamma_j^M(\hat{\theta}_M))^*\right). \tag{5.5.59}$$

Then we write

$$\mathrm{Cov}_{ij}(\gamma^M(\theta^M)|\mathcal{Z}^M) = \mathrm{tr}(K_{ij}(\hat{\theta}_M)\mathrm{Cov}(\theta^M|\mathcal{Z}^M)). \tag{5.5.60}$$

We can rewrite (5.5.46), (5.5.47) as

$$H^M\hat{x}^M + h^M = \gamma^M(\hat{\theta}_M), \tag{5.5.61}$$

$$(H^M P_M(H^M)^*)_{ij} = \mathrm{tr}(K_{ij}(\hat{\theta}_M)\mathrm{Cov}(\theta^M|\mathcal{Z}^M)). \tag{5.5.62}$$

Now $\hat{\theta}_M$ and $\mathrm{Cov}(\theta^M|\mathcal{Z}^M)$ can be expressed as functions of α_M, β_M. We have, from (5.5.30),

$$\hat{\theta}_M = -D_{\alpha_M}\log c(\alpha_M, \beta_M), \tag{5.5.63}$$

$$\mathrm{Cov}(\theta^M|\mathcal{Z}^M) = -\frac{D^2_{\alpha_M}c}{c}(\alpha_M, \beta_M) + D_{\alpha_M}\log c(\alpha_M, \beta_M)(D_{\alpha_M}\log c(\alpha_M, \beta_M))^*, \tag{5.5.64}$$

and thus (5.5.61), (5.5.62) is a nonlinear system of algebraic equations, but it does not involve integrals. Similarly, we define

$$\hat{\theta}_M^{M+1} = E(\theta^M|\mathcal{Z}^{M+1}),\ \mathrm{Cov}\ (\theta^M|\mathcal{Z}^{M+1}) = E(\theta^M(\theta^M)^*|\mathcal{Z}^{M+1}) - \hat{\theta}_M^{M+1}(\hat{\theta}_M^{M+1})^*. \tag{5.5.65}$$

These quantities can be computed from the probability (5.5.48), but this time we cannot avoid computing the integrals

$$\hat{\theta}_M^{M+1} = c(\alpha_M + (R^M)^{-1}z^{M+1}, (R^M)^{-1} + \beta_M) \tag{5.5.66}$$
$$\times \int \theta \exp[(\alpha_M + (R^M)^{-1}z^{M+1})^*\theta - b(\theta^M, (R^M)^{-1} + \beta_M)]d\theta,$$

$$E(\theta^M(\theta^M)^*|\mathcal{Z}^{M+1}) = c(\alpha_M + (R^M)^{-1}z^{M+1}, (R^M)^{-1} + \beta_M) \tag{5.5.67}$$
$$\times \int \theta\theta^* \exp[(\alpha_M + (R^M)^{-1}z^{M+1})^*\theta - b(\theta^M, (R^M)^{-1} + \beta_M)]d\theta.$$

Knowing these quantities, we can approximate formulas (5.5.54) and (5.5.56). We write

$$\hat{y}_M = E(x^M|\mathcal{Z}^{M+1}) = \hat{x}_M + \Theta_M(\gamma^M(\hat{\theta}_M^{M+1}) - (H^M\hat{x}_M + h^M)) \tag{5.5.68}$$

and

$$(\Pi_M)_{ij} = \text{tr}(K_{ij}(\hat{\theta}_M^{M+1})\text{Cov}(\theta^M|\mathcal{Z}^{M+1})). \tag{5.5.69}$$

5.5.5 Example of a Beta Model

We report here a simplified example discussed in [48], with state $x \in R^n$ and observation $z \in (0, 1)$. The canonical parameter is $\theta \in (0, 1)$. The probability density for a value θ of the canonical parameter is given by

$$f(z, \theta) = \frac{z^{\theta-1}(1-z)^{-\theta}}{\Gamma(\theta)\Gamma(1-\theta)}. \tag{5.5.70}$$

This is not in the GLM form, but the methodology will be easily adapted. At each experiment (each time) j, the canonical parameter will depend on the state x^j by the relation

$$\theta^j(x^j) = \frac{\exp(h^j)^*x^j}{1+\exp(h^j)^*x^j}, \tag{5.5.71}$$

and the evolution of the state is given by

$$x^{j+1} = F^jx^j + f^j + G^jw^j. \tag{5.5.72}$$

We describe next the procedure between M and $M+1$. We know that

$$\hat{x}_M = E(x^M|\mathcal{Z}^M), \ P_M = \text{Cov}(x^M|\mathcal{Z}^M).$$

Considering the function

$$\gamma(\theta) = \log\frac{\theta}{1-\theta}, \ 0 < \theta < 1,$$

we define the variable θ^M by the relation

$$\gamma(\theta^M) = (h^M)^* x^M,$$

and we choose as Prob$(\theta^M|\mathcal{Z}^M)$ the probability density, depending on parameters α_M,β_M,

$$\text{Prob}(\theta^M|\mathcal{Z}^M) = \frac{\Gamma(\alpha_M+\beta_M)}{\Gamma(\alpha_M)\Gamma(\beta_M)}(\theta^M)^{\alpha_M-1}(1-\theta^M)^{\beta_M-1}. \tag{5.5.73}$$

Setting

$$p^M(\theta) = \frac{\Gamma(\alpha_M+\beta_M)}{\Gamma(\alpha_M)\Gamma(\beta_M)}(\theta)^{\alpha_M-1}(1-\theta)^{\beta_M-1}$$

$$= \exp[(\alpha_M-1)\log\theta + (\beta_M-1)\log(1-\theta) - B(\alpha_M,\beta_M)],$$

with

$$B(\alpha_M,\beta_M) = -\log\Gamma(\alpha_M+\beta_M) + \log\Gamma(\alpha_M) + \log\Gamma(\beta_M),$$

we check easily the formulas

$$\frac{\partial B}{\partial \alpha_M} = \int_0^1 \log\theta \; p^M(\theta)d\theta, \tag{5.5.74}$$

$$\frac{\partial B}{\partial \beta_M} = \int_0^1 \log(1-\theta) \; p^M(\theta)d\theta, \tag{5.5.75}$$

and thus

$$\frac{\partial B}{\partial \alpha_M} - \frac{\partial B}{\partial \beta_M} = \int_0^1 \gamma(\theta)p^M(\theta)d\theta.$$

Therefore,

$$(h^M)^*\hat{x}_M = \Psi(\alpha_M) - \Psi(\beta_M), \tag{5.5.76}$$

with $\Psi(x) = \frac{d}{dx}\log\Gamma(x)$.

Equation (5.5.76) provides a first relation to compute the pair α_M,β_M. To get a second one, we note that

$$\text{Var}(\gamma(\theta)|\mathcal{Z}^M) = \text{Var}\,(\log\theta^M|\mathcal{Z}^M) + \text{Var}\,(\log(1\text{-}\theta^M)|\mathcal{Z}^M)$$
$$- 2\text{Cov}\,(\log\theta^M, \log(1-\theta^M)|\mathcal{Z}^M).$$

We then check

$$\text{Var}\,(\log\theta^M|\mathcal{Z}^M) = \frac{\partial^2 B}{\partial\alpha_M^2} = \Psi'(\alpha_M) - \Psi'(\alpha_M+\beta_M),$$

$$\text{Var}\,(\log(1\text{-}\theta^M)|\mathcal{Z}^M) = \frac{\partial^2 B}{\partial\beta_M^2} = \Psi'(\beta_M) - \Psi'(\alpha_M+\beta_M),$$

$$\text{Cov}\,(\log\theta^M, \log(1-\theta^M)|\mathcal{Z}^M) = \frac{\partial^2 B}{\partial\alpha_M\partial\beta_M} = -\Psi'(\alpha_M+\beta_M).$$

Collecting results, we obtain

$$\text{Var}(\gamma(\theta)|\mathcal{Z}^M) = \Psi'(\alpha_M) + \Psi'(\beta_M),$$

which leads to the second relation

$$(h^M)^* P_M h^M = \Psi'(\alpha_M) + \Psi'(\beta_M), \tag{5.5.77}$$

and relations (5.5.76), (5.5.77) allow us to obtain α_M, β_M. If we accept the approximation $\Psi(x) \sim \log x$, we obtain

$$\alpha_M = \frac{1+\exp(h^M)^*\hat{x}_M}{(h^M)^* P_M h^M}, \tag{5.5.78}$$

$$\beta_M = \frac{1-\exp(h^M)^*\hat{x}_M}{(h^M)^* P_M h^M}.$$

We can next formulate $\text{Prob}(\theta^M|\mathcal{Z}^{M+1})$. From (5.5.70), we have

$$\text{Prob}(z^{M+1}|\theta^M, \mathcal{Z}^M) = \frac{(z^{M+1})^{\theta^M-1}(1-z^{M+1})^{-\theta^M}}{\Gamma(\theta^M)\Gamma(1-\theta^M)},$$

and using (5.5.73), we can write

$$\text{Prob}(\theta^M|\mathcal{Z}^{M+1}) = \frac{(z^{M+1})^{\theta^M-1}(1-z^{M+1})^{-\theta^M}(\theta^M)^{\alpha_M-1}(1-\theta^M)^{\beta_M-1}}{\Gamma(\theta^M)\Gamma(1-\theta^M)D(z^{M+1})}, \tag{5.5.79}$$

in which

$$D(z^{M+1}) = \int_0^1 \frac{(z^{M+1})^{\theta-1}(1-z^{M+1})^{-\theta}(\theta)^{\alpha_M-1}(1-\theta)^{\beta_M-1}}{\Gamma(\theta)\Gamma(1-\theta)}d\theta. \tag{5.5.80}$$

We can then compute

$$\hat{\theta}_M^{M+1} = E(\theta^M|\mathcal{Z}^{M+1})$$
$$= \int_0^1 \frac{(z^{M+1})^{\theta-1}(1-z^{M+1})^{-\theta}\theta^{\alpha_M}(1-\theta)^{\beta_M-1}}{\Gamma(\theta)\Gamma(1-\theta)D(z^{M+1})}d\theta,$$

$$V_M^{M+1} = \mathrm{Var}(\theta^M|\mathcal{Z}^{M+1})$$

$$= \int_0^1 \frac{(z^{M+1})^{\theta-1}(1-z^{M+1})^{-\theta}\theta^{\alpha_M+1}(1-\theta)^{\beta_M-1}}{\Gamma(\theta)\Gamma(1-\theta)D(z^{M+1})}d\theta \; - (\hat{\theta}_M^{M+1})^2.$$

We set next

$$\Theta_M = \frac{P_M h^M}{(h^M)^* P_M h^M},$$

and we obtain

$$\hat{y}_M = E(x^M|\mathcal{Z}^{M+1}) = \hat{x}_M + \Theta_M(\log \frac{\hat{\theta}_M^{M+1}}{1-\hat{\theta}_M^{M+1}} - (h^M)^*\hat{x}_M). \tag{5.5.81}$$

We next set

$$\Delta_M = P_M - \frac{P_M h^M (P_M h^M)^*}{(h^M)^* P_M h^M}$$

and

$$K(\theta) = (\gamma'(\theta))^2 - \frac{1}{\theta^2(1-\theta)^2},$$

$$\mathrm{Cov}(x^M|\mathcal{Z}^{M+1}) = P_M + \frac{P_M h^M (P_M h^M)^*}{((h^M)^* P_M h^M)^2}(K(\hat{\theta}_M^{M+1})V_M^{M+1} - (h^M)^* P_M h^M). \tag{5.5.82}$$

Finally, we have

$$\hat{x}_{M+1} = F^M \hat{y}_M + f^M, \tag{5.5.83}$$

$$P_{M+1} = F^M \mathrm{Cov}(x^M|\mathcal{Z}^{M+1})(F^M)^* + G^M Q^M (G^M)^*. \tag{5.5.84}$$

5.6 Seasonal Factors

5.6.1 *Setting of the Problem*

We write the discrete time t as a period divided into subperiods. For instance, it might be one year divided into quarters, months, or weeks. If k represents the number of subperiods in one period, we can write

$$t = (n - 1)k + j, \tag{5.6.1}$$

where n is the index of the period and $j = 1, \cdots, k$. So in fact, $t = (n, j)$. The observation at time t is given by

$$z_t = z_{n,j} = (an + b)s_j + \epsilon_t. \tag{5.6.2}$$

Here $an + b$ is called the level; a is called the trend. The parameter s_j represents the seasonal factors. We have the relation

$$\sum_{j=1}^{k} s_j = k. \tag{5.6.3}$$

Indeed, when there are no seasons, $k = 1$ and $s_1 = 1$. So in fact, the model (5.6.2) contains $k + 1$ parameters. The quantities ϵ_t are observation errors, modeled as independent random variables. The problem is to estimate the $k + 1$ parameters, based on the observations.

5.6.2 *Moving Averages*

A simple method to estimate the seasonal factors is to use the technique of moving averages. If we omit the errors, we can write the relations

$$z_t = z_{n,j} = (an + b)s_j.$$

The definition of the moving average depends on whether k is odd or even. Suppose k is odd. We define the averages

$$\tilde{z}_t = \frac{\sum_{h=1}^{\frac{k-1}{2}} z_{t-h} + \sum_{h=0}^{\frac{k-1}{2}} z_{t+h}}{k}.$$

This definition is valid when $t - \frac{k-1}{2} \geq 1$, i.e., $t \geq \frac{k+1}{2}$. If $k = 1$, we have $\tilde{z}_t = z_t$. If k is even, the formulas are slightly more complex. We begin by defining

$$\tilde{z}_{t-\frac{1}{2}} = \frac{\sum_{h=1}^{\frac{k}{2}} z_{t-h} + \sum_{h=0}^{\frac{k}{2}-1} z_{t+h}}{k},$$

$$\tilde{z}_{t+\frac{1}{2}} = \frac{\sum_{h=0}^{\frac{k}{2}-1} z_{t-h} + \sum_{h=1}^{\frac{k}{2}} z_{t+h}}{k},$$

and we set

$$\tilde{z}_t = \frac{\tilde{z}_{t-\frac{1}{2}} + \tilde{z}_{t+\frac{1}{2}}}{2}.$$

Again, this definition is valid whenever $t - \frac{k}{2} \geq 1$, i.e., $t \geq \frac{k}{2} + 1$. We can combine the formulas to obtain

$$\tilde{z}_t = \frac{1}{2k}\left(z(t - \frac{k}{2}) + z(t + \frac{k}{2}) + 2[\sum_{h=0}^{\frac{k}{2}-1'} z(t+h) + \sum_{h=1}^{\frac{k}{2}-1'} z(t-h) \right).$$

In the sequel, we shall assume k odd to simplify calculations. We recall that $t = (n-1)k + j$, and we first assume that $1 \leq j \leq \frac{k-1}{2}$. This term is irrelevant when $k = 1$. We have

$$\tilde{z}_t = \frac{1}{k}\left(\sum_{h=1}^{\frac{k-1}{2}} z_{(n-1)k+j-h} + \sum_{h=0}^{\frac{k-1}{2}} z_{(n-1)k+j+h} \right)$$

$$= \frac{1}{k}\left(\sum_{h=1}^{j-1} z_{(n-1)k+j-h} + \sum_{h=j}^{\frac{k-1}{2}} z_{(n-1)k+j-h} + \sum_{h=0}^{\frac{k-1}{2}} z_{(n-1)k+j+h} \right).$$

But

$$\sum_{h=j}^{\frac{k-1}{2}} z_{(n-1)k+j-h} = \sum_{h=j}^{\frac{k-1}{2}} z_{(n-2)k+j+k-h} = \sum_{h=j+\frac{k+1}{2}}^{k} z_{(n-2)k+h}$$

and

$$\sum_{h=1}^{j-1} z_{(n-1)k+j-h} + \sum_{h=0}^{\frac{k-1}{2}} z_{(n-1)k+j+h} = \sum_{h=1}^{j+\frac{k-1}{2}} z_{(n-1)k+h}.$$

Therefore,

$$\tilde{z}_t = \frac{1}{k}\left(\sum_{h=j+\frac{k+1}{2}}^{k} z_{(n-2)k+h} + \sum_{h=1}^{j+\frac{k-1}{2}} z_{(n-1)k+h}\right), \text{ if } 1 \leq j \leq \frac{k-1}{2}. \tag{5.6.4}$$

Similar calculations lead to

$$\tilde{z}_t = \frac{1}{k}\sum_{h-1}^{k} z_{(n-1)k+h}, \text{ if } j = \frac{k+1}{2}, \tag{5.6.5}$$

$$\tilde{z}_t = \frac{1}{k}\left(\sum_{h=j-\frac{k-1}{2}}^{k} z_{(n-1)k+h} + \sum_{h=1}^{j-\frac{k+1}{2}} z_{nk+h}\right), \text{ if } \frac{k+1}{2}+1 \leq j \leq k. \tag{5.6.6}$$

Replacing $z_{(n-2)k+h}$ by $(a(n-1)+b)s_h$ and $z_{(n-1)k+h}$ by $(an+b)s_h$ in formula (5.6.4) and rearranging, we obtain easily

$$\tilde{z}_t = an + b - a\frac{\sum_{h=j+\frac{k+1}{2}}^{k} s_h}{k}, \text{ if } 1 \leq j \leq \frac{k-1}{2}, \tag{5.6.7}$$

and similarly

$$\tilde{z}_t = an + b, \text{ if } j = \frac{k+1}{2}, \tag{5.6.8}$$

$$\tilde{z}_t = an + b + a\frac{\sum_{h=1}^{j-\frac{k+1}{2}} s_h}{k}, \text{ if } \frac{k+1}{2}+1 \leq j \leq k, \tag{5.6.9}$$

and thus

$$\frac{z_t}{\tilde{z}_t} = \frac{s_j}{1 - \frac{a\rho_j}{an+b}}, \tag{5.6.10}$$

with

$$\rho_j = \left| \begin{array}{ll} \frac{\sum_{h=j+\frac{k+1}{2}}^{k} s_h}{k} & \text{if } 1 \leq j \leq \frac{k-1}{2}, \\ 0 & \text{if } j = \frac{k+1}{2}, \\ -\frac{\sum_{h=1}^{j-\frac{k+1}{2}} s_h}{k} & \text{if } \frac{k+1}{2}+1 \leq j \leq k. \end{array} \right. \tag{5.6.11}$$

Consider

$$\frac{1}{N}\sum_{n=1}^{N}\frac{z_t}{\tilde{z}_t}=\frac{1}{N}\sum_{n=1}^{N}\frac{z_{n,j}}{\tilde{z}_{n,j}}=s_j+s_j\rho_j\frac{1}{N}\sum_{n=1}^{N}\frac{1}{n+\frac{b}{a}},$$

assuming of course that $n+\frac{b}{a}$ is not 0. Otherwise, we delete this integer from the summation. Since

$$\frac{1}{N}\sum_{n=1}^{N}\frac{1}{n+\frac{b}{a}}\to 0,\ \text{as } N\to +\infty,$$

we obtain

$$\frac{1}{N}\sum_{n=1}^{N}\frac{z_{n,j}}{\tilde{z}_{n,j}}\to s_j,\ \text{as } N\to +\infty. \tag{5.6.12}$$

This provides an estimate of the seasonal factor s_j. Note that when $j=\frac{k+1}{2}$, we have $\frac{z_{n,j}}{\tilde{z}_{n,j}}=s_j, \forall n$.

5.6.3 *Exponential Smoothing*

We consider first the simple case in which there are no seasonal factors and no trend. So we have the very simple model

$$z_n=b+\epsilon_n, \tag{5.6.13}$$

and we assume that errors ϵ_n are independent variables that are normal with mean 0 and variance R_n. This is the problem considered in Chapter 4, Section 4.7.1, (4.7.1), with $x=b$ and matrices $H_n=1$. Based on the observations $z_1,\cdots,z_n$, the best estimate of x, denoted by $\hat{x}_n$, is given by formula (4.7.3), with $P_0^{-1}=0$, which leads to

$$\hat{x}_n=\frac{\sum_{j=1}^{n}z_jR_j^{-1}}{\sum_{j=1}^{n}R_j^{-1}}. \tag{5.6.14}$$

We can define recursive formulas—see Section 4.7.2—namely

$$\hat{x}_{n+1}=\hat{x}_n+K_{n+1}(z_{n+1}-\hat{x}_n),\ \hat{x}_1=z_1, \tag{5.6.15}$$

with

$$K_{n+1}=P_n(R_{n+1}+P_n)^{-1}. \tag{5.6.16}$$

The sequence P_n is given by induction:

$$P_{n+1} = P_n - P_n(R_{n+1} + P_n)^{-1}P_n, \;\; P_1 = R_1.$$

This is equivalent to

$$P_{n+1}^{-1} = P_n^{-1} + R_{n+1}^{-1}$$

or

$$P_n^{-1} = \sum_{j=1}^{n} R_j^{-1}.$$

We check easily that

$$K_{n+1} = P_{n+1}^{-1} R_{n+1}^{-1}. \tag{5.6.17}$$

We know that

$$P_n = E(\hat{x}_n - x)^2$$

and

$$P_n = \frac{1}{\sum_{j=1}^{n} R_j^{-1}}.$$

Within reasonable assumptions on the sequence R_n, such as $R_n \leq R$, we can write $P_n \leq \frac{R}{n}$. Therefore, the error tends to 0 in mean square as $n \to +\infty$.

The method of *exponential smoothing* consists in using an estimate like (5.6.13) and replacing K_{n+1} by a fixed constant $K \in (0, 1)$. Note that K_{n+1} is indeed between 0 and 1. Therefore, we can write

$$\hat{x}_{n+1} = \hat{x}_n + K(z_{n+1} - \hat{x}_n), \;\; \hat{x}_1 = z_1. \tag{5.6.18}$$

This leads to the explicit form

$$\hat{x}_{n+1} = (1 - K)^n z_1 + K \sum_{j=0}^{n-1} (1 - K)^j z_{n+1-j}. \tag{5.6.19}$$

The estimation error $\tilde{x}_n = \hat{x}_n - x$ satisfies the relation

$$\tilde{x}_{n+1} = \tilde{x}_n + K(\epsilon_{n+1} - \tilde{x}_n), \;\; \tilde{x}_1 = \epsilon_1,$$

and hence

$$\tilde{x}_{n+1} = (1-K)^n \epsilon_1 + K \sum_{j=0}^{n-1} (1-K)^j \epsilon_{n+1-j},$$

and thus

$$E\tilde{x}_{n+1}^2 = (1-K)^{2n} R_1 + K^2 \sum_{j=0}^{n-1} (1-K)^{2j} R_{n+1-j}.$$

This expression does not go to 0. For instance, for $R_n = R$, it remains bounded. The exponential smoothing method is simple, but not as efficient as the Kalman filter.

5.6.4 Estimation of the Trend

We still neglect seasonal factors, but consider a trend, so the model becomes

$$z_n = an + b + \epsilon_n. \tag{5.6.20}$$

We define

$$x = \begin{pmatrix} a \\ b \end{pmatrix}.$$

Then (5.6.20) can be written as

$$z_n = H_n x + \epsilon_n, \tag{5.6.21}$$

with $H_n = (n \quad 1)$. The measurement errors ϵ_n are independent random variables, which are Gaussian with mean 0 and variance R_n. We are in the situation of (4.7.1), Section 4.7.1. Defining the best estimate $\hat{x}_n$ of x, based on the observations $z_1, \cdots, z_n$, we have the formulas (4.7.6), (4.7.7), namely

$$\hat{x}_{n+1} = \hat{x}_n + K_{n+1}(z_{n+1} - H_{n+1}\hat{x}_n), \tag{5.6.22}$$

with

$$K_{n+1} = P_n H_{n+1}^* (R_{n+1} + H_{n+1} P_n H_{n+1}^*)^{-1} \tag{5.6.23}$$

and

$$P_{n+1} = P_n - P_n H_{n+1}^* (R_{n+1} + H_{n+1} P_n H_{n+1}^*)^{-1} H_{n+1} P_n. \tag{5.6.24}$$

This is equivalent to

$$P_{n+1} = (P_n^{-1} + H_{n+1}^* R_{n+1}^{-1} H_{n+1})^{-1}, \tag{5.6.25}$$

and also

$$P_n = (\sum_{j=1}^{n} H_j^* R_j^{-1} H_j)^{-1}, \tag{5.6.26}$$

$$K_{n+1} = P_{n+1} H_{n+1}^* R_{n+1}^{-1}. \tag{5.6.27}$$

However, the expression (5.6.26) is valid only when $n \geq 2$, since the matrix

$$P_1^{-1} = H_1^* R_1^{-1} H_1 = \frac{1}{R_1} \begin{pmatrix} 1 & 1 \\ 1 & 1 \end{pmatrix}$$

is not invertible. Beginning with $n = 2$, the sequence (5.6.24) is valid, starting with $P_2 = (\sum_{j=1}^{2} H_j^* R_j^{-1} H_j)^{-1}$. Using (5.6.27), formula (5.6.22) can be used, starting with $n = 1$, provided we define $\hat{x}_1$. We cannot define $\hat{x}_1$ uniquely. We can only claim that

$$(1 \quad 1)\hat{x}_1 = z_1.$$

So it is natural to take

$$\hat{x}_1 = \begin{pmatrix} 0 \\ z_1 \end{pmatrix},$$

and all subsequent estimates $\hat{x}_n, n \geq 2$ are well defined by the formulas above, (5.6.22), (5.6.23), (5.6.24), where the last one is initiated with $n = 2$, and P_2 is defined as indicated.

Recall that the level at time n is $S_n = an + b$. From the best estimate $\hat{x}_n$ of the vector x, we deduce the best estimates of the pair a and S_n, denoted by $\hat{a}_n$ and $\hat{S}_n$, as follows:

$$\begin{aligned} \hat{a}_n &= (1 \quad 0)\hat{x}_n, \\ \hat{S}_n &= H_n \hat{x}_n. \end{aligned} \tag{5.6.28}$$

From the induction (5.6.22) we obtain easily (using $H_{n+1}\hat{x}_n = \hat{S}_n + \hat{a}_n$) the relations

$$\hat{S}_{n+1} = \hat{S}_n + \hat{a}_n + H_{n+1} P_{n+1} H_{n+1}^* R_{n+1}^{-1} (z_{n+1} - \hat{S}_n - \hat{a}_n), \tag{5.6.29}$$

$$\hat{a}_{n+1} = \hat{a}_n + \frac{(1 \quad 0) P_{n+1} H_{n+1}^*}{H_{n+1} P_{n+1} H_{n+1}^*} (\hat{S}_{n+1} - \hat{S}_n - \hat{a}_n), \tag{5.6.30}$$

and we start this induction with $\hat{a}_1 = 0$ and $\hat{S}_1 = z_1$. Defining the gains

$$\begin{aligned}\alpha_n &= H_n P_n H_n^* R_n^{-1},\\ \beta_n &= \frac{(1 \quad 0) P_n H_n^*}{H_n P_n H_n^*}, \quad n \geq 2,\end{aligned} \tag{5.6.31}$$

we can write formulas (5.6.29), (5.6.30) as follows:

$$\begin{aligned}\hat{S}_{n+1} &= \hat{S}_n + \hat{a}_n + \alpha_{n+1}(z_{n+1} - \hat{S}_n - \hat{a}_n),\\ \hat{a}_{n+1} &= \hat{a}_n + \beta_{n+1}(\hat{S}_{n+1} - \hat{S}_n - \hat{a}_n).\end{aligned} \tag{5.6.32}$$

We note that $0 < \alpha_n < 1$ and $0 < \beta_n < 1$. Indeed, we first have

$$P_n^{-1} \geq \frac{1}{R_n} H_n^* H_n,$$

and we have for all $\xi \in R^2$,

$$\xi^* P_n^{-1} \xi \geq \frac{1}{R_n} \xi^* H_n^* H_n \xi = \frac{1}{R_n} (H_n \xi)^2.$$

Taking $\xi = P_n H_n^*$, it follows that

$$H_n P_n H_n^* \geq \frac{1}{R_n} (H_n P_n H_n^*)^2,$$

which implies $H_n P_n H_n^* \leq R_n$, i.e., $0 < \alpha_n < 1$. In fact, we can give explicit expressions for $\alpha_n, \beta_n, n \geq 2$. First,

$$\sum_{j=1}^{n} H_j^* R_j^{-1} H_j = \begin{pmatrix} \sum_{j=1}^n \frac{j^2}{R_j} & \sum_{j=1}^n \frac{j}{R_j} \\ \sum_{j=1}^n \frac{j}{R_j} & \sum_{j=1}^n \frac{1}{R_j} \end{pmatrix},$$

and thus

$$P_n = \frac{1}{\Delta_n} \begin{pmatrix} \sum_{j=1}^n \frac{1}{R_j} & -\sum_{j=1}^n \frac{j}{R_j} \\ -\sum_{j=1}^n \frac{j}{R_j} & \sum_{j=1}^n \frac{j^2}{R_j} \end{pmatrix},$$

with

$$\Delta_n = \sum_{j=1}^{n} \frac{j^2}{R_j} \sum_{j=1}^{n} \frac{1}{R_j} - (\sum_{j=1}^{n} \frac{j}{R_j})^2.$$

Therefore,

$$H_n P_n H_n^* = \frac{\sum_{j=1}^{n} \frac{(n-j)^2}{R_j}}{\Delta_n}.$$

We can check that

$$H_n P_n H_n^* = \frac{\sum_{j=1}^{n-1} \frac{(n-j)^2}{R_j}}{\Delta_{n-1} + \frac{1}{R_n} \sum_{j=1}^{n-1} \frac{(n-j)^2}{R_j}} \leq R_n.$$

Also

$$\beta_n = \frac{\sum_{j=1}^{n-1} \frac{n-j}{R_j}}{\sum_{j=1}^{n-1} \frac{(n-j)^2}{R_j}},$$

which implies $0 \leq \beta_n \leq 1$.

Following the exponential smoothing approach, we introduce the double exponential smoothing approach, which consists in taking two constants $\alpha, \beta \in (0, 1)$ and setting the relations

$$\begin{aligned} \hat{S}_{n+1} &= \hat{S}_n + \hat{a}_n + \alpha(z_{n+1} - \hat{S}_n - \hat{a}_n), \\ \hat{a}_{n+1} &= \hat{a}_n + \beta(\hat{S}_{n+1} - \hat{S}_n - \hat{a}_n). \end{aligned} \tag{5.6.33}$$

This method is due to Holt and Winters [23, 52]. The parameter β is called the *trend smoothing factor*, and the parameter α is the *data smoothing factor*. The data applies to the sequence z_n.

5.6.5 Holt–Winters Formulas with Seasonality

We return to the model (5.6.2). We introduce the vector $z_n \in R^k$, with components $z_{n,j}$. Similarly, we set $s \in R^k$, with components s_j, the seasonal factors. Therefore, we write

$$z_n = (an + b)s + \epsilon_n, \tag{5.6.34}$$

where ϵ_n is a sequence of independent random variables in R^k, with covariance matrix R_n, assumed invertible. We recall that $\sum_{j=1}^{k} s_j = k$, so in fact, there are only $k + 1$ degrees of freedom. A convenient way to proceed is to set $\theta = bs,\ \lambda = \frac{a}{b}$, so that the model becomes

$$z_n = (\lambda n + 1)\theta + \epsilon_n, \tag{5.6.35}$$

in which θ is a free vector in R^k, and λ is a scalar. Unfortunately, the model is nonlinear in the parameters λ and θ. The notable exception is the case $k = 1$, in which the pair λ, θ is equivalent to the pair $\lambda\theta, \theta$ in which the model becomes linear. We of course recover a, b, s from λ, θ, by the formulas

$$b = \frac{1}{k}\sum_{j=1}^{k}\theta_j, \ s = \frac{\theta}{b}, \ a = b\lambda.$$

Because of the nonlinearity, the best estimate of the parameters is not a trivial concept. So it is natural to consider the least squares estimator LSE. We introduce the least squares functional

$$J_n(\lambda, \theta) = \sum_{i=1}^{n}(z_i - (\lambda i + 1)\theta)^* R_i^{-1}(z_i - (\lambda i + 1)\theta). \tag{5.6.36}$$

Since there are no constraints on the parameters, we compute the gradient in λ, θ and obtain conditions by writing that the gradient is 0. We get

$$\sum_{i=1}^{n}(\hat{\lambda}_n i + 1)R_i^{-1}(z_i - (\hat{\lambda}_n i + 1)\hat{\theta}_n) = 0,$$

$$\sum_{i=1}^{n} i\hat{\theta}_n^* R_i^{-1}(z_i - (\hat{\lambda}_n i + 1)\hat{\theta}_n) = 0, \tag{5.6.37}$$

from which we get also the relation

$$\sum_{i=1}^{n}\hat{\theta}_n^* R_i^{-1}(z_i - (\hat{\lambda}_n i + 1)\hat{\theta}_n) = 0.$$

From these relations we obtain easily the system

$$\hat{\theta}_n = \left(\sum_{i=1}^{n}(\hat{\lambda}_n i + 1)^2 R_i^{-1}\right)^{-1}\sum_{i=1}^{n}(\hat{\lambda}_n i + 1)R_i^{-1}z_i, \tag{5.6.38}$$

$$\hat{\lambda}_n = \frac{\hat{\theta}_n^* \sum_{i=1}^{n} R_i^{-1}(z_i - \hat{\theta}_n)}{\hat{\theta}_n^* \sum_{i=1}^{n} i R_i^{-1}\hat{\theta}_n}. \tag{5.6.39}$$

We deduce

$$\hat{b}_n = \frac{1}{k}\sum_{j=1}^{k}\hat{\theta}_{n,j}, \ \hat{s}_n = \frac{\hat{\theta}_n}{\hat{b}_n}, \ \hat{a}_n = \hat{b}_n\hat{\lambda}_n.$$

It follows also that

$$\hat{s}_n = \left(\sum_{i=1}^{n}(\hat{a}_n i + \hat{b}_n)^2 R_i^{-1}\right)^{-1} \sum_{i=1}^{n}(\hat{a}_n i + \hat{b}_n) R_i^{-1} z_i, \tag{5.6.40}$$

$$\hat{s}_n^* \sum_{i=1}^{n} R_i^{-1} z_i = \hat{s}_n^* \sum_{i=1}^{n} R_i^{-1}(\hat{a}_n i + \hat{b}_n)\hat{s}_n. \tag{5.6.41}$$

From (5.6.37) we get also

$$\sum_{i=1}^{n} i \hat{s}_n^* R_i^{-1}(z_i - (\hat{a}_n i + \hat{b}_n)\hat{s}_{\cdot n}) = 0 \tag{5.6.42}$$

For consistency with the case without seasonal factors, (5.6.4), we shall set

$$x = \begin{pmatrix} a \\ b \end{pmatrix}, \ \hat{x}_n = \begin{pmatrix} \hat{a}_n \\ \hat{b}_n \end{pmatrix}$$

and recall $H_n = (n \quad 1)$. We can then write the pair (5.6.41), (5.6.42) in matrix form as follows:

$$\sum_{i=1}^{n} H_i^* \hat{s}_n^* R_i^{-1} z_i = \sum_{i=1}^{n} H_i^* \hat{s}_n^* R_i^{-1} \hat{s}_n H_i \hat{x}_n. \tag{5.6.43}$$

Hence

$$\hat{x}_n = (\sum_{i=1}^{n} H_i^* \hat{s}_n^* R_i^{-1} \hat{s}_n H_i)^{-1} \sum_{i=1}^{n} H_i^* \hat{s}_n^* R_i^{-1} z_i, \tag{5.6.44}$$

and (5.6.40) yields

$$\hat{s}_n = \left(\sum_{i=1}^{n}(H_i \hat{x}_n)^2 R_i^{-1}\right)^{-1} \sum_{i=1}^{n} H_i \hat{x}_n R_i^{-1} z_i. \tag{5.6.45}$$

The set of equations (5.6.44), (5.6.45) defines the LSE as a fixed point, whose solution is the pair $\hat{x}_n, \hat{s}_n$. For $n = 1$, we have only

$$H_1 \hat{x}_1 = \frac{\sum_{j=1}^{k} z_{1,j}}{k}, \ \hat{s}_1 = \frac{z_1}{H_1 \hat{x}_1}, \tag{5.6.46}$$

and we take, as in the case $k = 1$,

$$\hat{x}_1 = \begin{pmatrix} \hat{a}_1 \\ \hat{b}_1 \end{pmatrix}, \ \hat{a}_1 = 0, \ \hat{b}_1 = \frac{\sum_{j=1}^{k} z_{1,j}}{k}. \tag{5.6.47}$$

To define recursive formulas from (5.6.44), (5.6.45), we have to proceed with approximations. We begin by replacing (5.6.44), (5.6.45) with

$$\hat{x}_n = (\sum_{i=1}^{n} H_i^* \hat{s}_i^* R_i^{-1} \hat{s}_i H_i)^{-1} \sum_{i=1}^{n} H_i^* \hat{s}_i^* R_i^{-1} z_i, \tag{5.6.48}$$

$$\hat{s}_n = \left(\sum_{i=1}^{n} (H_i \hat{x}_i)^2 R_i^{-1} \right)^{-1} \sum_{i=1}^{n} H_i \hat{x}_i R_i^{-1} z_i. \tag{5.6.49}$$

For $n \geq 2$, we define P_n by

$$P_n = (\sum_{i=1}^{n} H_i^* \hat{s}_i^* R_i^{-1} \hat{s}_i H_i)^{-1},$$

$$\Gamma_n = \left(\sum_{i=1}^{n} (H_i \hat{x}_i)^2 R_i^{-1} \right)^{-1}.$$

Then

$$\begin{aligned} P_{n+1}^{-1} &= P_n^{-1} + H_{n+1}^* \hat{s}_{n+1}^* R_{n+1}^{-1} \hat{s}_{n+1} H_{n+1}, \ n \geq 1, \\ \Gamma_{n+1}^{-1} &= \Gamma_n^{-1} + (H_{n+1} \hat{x}_{n+1})^2 R_{n+1}^{-1}. \end{aligned} \tag{5.6.50}$$

We interpret $P_1^{-1} = H_1^* \hat{s}_1^* R_1^{-1} \hat{s}_1 H_1$, although this is not an invertible matrix. Similarly, $\Gamma_1^{-1} = (H_1 \hat{x}_1)^2 R_1^{-1}$.

Next (5.6.48) becomes

$$\hat{x}_{n+1} = \hat{x}_n + P_{n+1} H_{n+1}^* \hat{s}_{n+1}^* R_{n+1}^{-1} (z_{n+1} - \hat{s}_{n+1} H_{n+1} \hat{x}_n), \tag{5.6.51}$$

and (5.6.49) becomes

$$\hat{s}_{n+1} = \hat{s}_n + \Gamma_{n+1} H_{n+1} \hat{x}_{n+1} R_{n+1}^{-1} (z_{n+1} - H_{n+1} \hat{x}_{n+1} \hat{s}_n). \tag{5.6.52}$$

The system (5.6.50), (5.6.51), (5.6.52) can be considered a system in which $P_n^{-1}, \Gamma_n^{-1}, \hat{x}_n, \hat{s}_n$ are known, and we compute $P_{n+1}^{-1}, \Gamma_{n+1}^{-1}, \hat{x}_{n+1}, \hat{s}_{n+1}$. So it is recursive, but we have a complex system to solve. So we proceed with a further step of approximation, by writing

$$\Gamma_{n+1}^{-1} = \Gamma_n^{-1} + (H_{n+1}\hat{x}_n)^2 R_{n+1}^{-1}, \tag{5.6.53}$$

$$\hat{s}_{n+1} = \hat{s}_n + \Gamma_{n+1} H_{n+1}\hat{x}_n R_{n+1}^{-1}(z_{n+1} - H_{n+1}\hat{x}_n\hat{s}_n), \tag{5.6.54}$$

$$P_{n+1}^{-1} = P_n^{-1} + H_{n+1}^* \hat{s}_{n+1}^* R_{n+1}^{-1}\hat{s}_{n+1}H_{n+1}, \tag{5.6.55}$$

$$\hat{x}_{n+1} = \hat{x}_n + P_{n+1}H_{n+1}^*\hat{s}_{n+1}^* R_{n+1}^{-1}(z_{n+1} - \hat{s}_{n+1}H_{n+1}\hat{x}_n). \tag{5.6.56}$$

Recalling that

$$\hat{x}_n = \begin{pmatrix} \hat{a}_n \\ \hat{b}_n \end{pmatrix}, \; \hat{S}_n = H_n\hat{x}_n = n\hat{a}_n + \hat{b}_n$$

and thus $H_{n+1}\hat{x}_n = \hat{S}_n + \hat{a}_n$, we get

$$\Gamma_{n+1}^{-1} = \Gamma_n^{-1} + (\hat{S}_n + \hat{a}_n)^2 R_{n+1}^{-1}, \tag{5.6.57}$$

$$\hat{s}_{n+1} = \hat{s}_n + \Gamma_{n+1}(\hat{S}_n + \hat{a}_n)R_{n+1}^{-1}(z_{n+1} - (\hat{S}_n + \hat{a}_n)\hat{s}_n), \tag{5.6.58}$$

$$P_{n+1}^{-1} = P_n^{-1} + H_{n+1}^*\hat{s}_{n+1}^* R_{n+1}^{-1}\hat{s}_{n+1}H_{n+1}, \tag{5.6.59}$$

$$\hat{S}_{n+1} = \hat{S}_n + \hat{a}_n + H_{n+1}P_{n+1}H_{n+1}^*\hat{s}_{n+1}^* R_{n+1}^{-1}(z_{n+1} - \hat{s}_{n+1}(\hat{S}_n + \hat{a}_n)), \tag{5.6.60}$$

$$\hat{a}_{n+1} = \hat{a}_n + \frac{(1 \quad 0)P_{n+1}H_{n+1}^*(\hat{S}_{n+1} - \hat{S}_n + \hat{a}_n)}{H_{n+1}P_{n+1}H_{n+1}^*}. \tag{5.6.61}$$

We recall that $z_t{=}z_{n,j}$, with $t = (n-1)k + j$. In a way, similar to the case without seasonal factors, see (5.6.33), it is possible to modify the Holt–Winters formulas to introduce seasonality. One writes

$$\hat{S}_{t+1} = \hat{S}_t + \hat{a}_t + \alpha\left(\frac{z_{t+1}}{s_{t+1-k}} - \hat{S}_t - \hat{a}_t\right) \tag{5.6.62}$$

$$\hat{a}_{t+1} = \hat{a}_t + \beta(\hat{S}_{t+1} - \hat{S}_t - \hat{a}_t), \tag{5.6.63}$$

$$\hat{s}_{t+1} = \hat{s}_{t+1-k} + \gamma\left(\frac{z_{t+1}}{S_{t+1}} - \hat{s}_{t+1-k}\right), \tag{5.6.64}$$

where α, β, γ are adequate constants in the interval (0, 1). Note that formulas (5.6.58), (5.6.60) are not particular cases of (5.6.64), (5.6.62), because z_n, $\hat{s}_n$ are vectors, whereas z_t, $\hat{s}_t$ are scalars.

Chapter 6
Complements on Probability Theory

6.1 Probability Concepts

6.1.1 *Review of Basic Probability Concepts*

A probability setup begins with a triple $\Omega, \mathcal{A}, P$, called a probability space, where Ω is a set, and $\mathcal{A}$ is a σ-algebra on Ω, which is a set of subsets of Ω that contains Ω and is stable with respect to taking the complement and taking countable unions and intersections. The elements of $\mathcal{A}$ are called *events*. Therefore, Ω is an event, $\emptyset$ is an event, and a countable union or intersection of events is an event. The elements of Ω are called *elementary events*. So events are also composed of elementary events. However, since events are defined by a specific σ-algebra on Ω, elementary events are not necessarily events.

If $\mathcal{A}$ and $\mathcal{B}$ are two σ-algebras on Ω, we say that $\mathcal{B}$ is a sub-σ-algebra of $\mathcal{A}$ if all events of $\mathcal{B}$ are also events of $\mathcal{A}$, $\mathcal{B} \subset \mathcal{A}$. Note that $\{\Omega, \emptyset\}$ is a σ-algebra and thus included in every σ-algebra on Ω.

The set of all subsets of Ω is a σ-algebra on Ω. It is clearly the largest σ-algebra on Ω. However, it is too big to be used unless Ω is finite or countable. This is why an operational σ-algebra has to be defined together with the set Ω of elementary events. When Ω has only a countable (in particular finite) number of elementary events, one can choose the set of all subsets of Ω as the operational σ-algebra. In that case, it is omitted and not introduced explicitly. Also, in this case elementary events are also events. This explains why in elementary probability theory, the concept of σ-algebra is not necessary.

However, even in the finite or countable case, the concept of σ-algebra has an important information content. One can think of Ω as the set of outcomes of an experiment. The outcome will be a specific $\omega \in \Omega$. So the result of an experiment is an elementary event. The operational σ-algebra $\mathcal{A}$ associated with Ω describes all

A. Bensoussan, *Estimation and Control of Dynamical Systems*, Interdisciplinary Applied Mathematics 48, https://doi.org/10.1007/978-3-319-75456-7_6

events that we consider in the framework of the experiment. Some are observable, which means that after the experiment is performed, we know whether they have occurred. In general, not all events are observable, and we say that we have partial information. The events that are observable are also structured as in the σ-algebra $\mathcal{B}$, which will be a sub-σ-algebra of $\mathcal{A}$. So by specifying $\mathcal{B}$, we describe the information that is available about the outcomes of the experiment. Full information means that we have access to all the events in $\mathcal{A}$. Note that if elementary events do not belong to $\mathcal{A}$, it is impossible, even with full information, to have access to them.

If Ω is finite or countable, even though $\mathcal{A}$ is the set of all subsets of Ω and thus omitted, we still need to describe the information by a sub-σ-algebra $\mathcal{B}$.

If $\mathcal{B} = \{\Omega, \emptyset\}$, then we have no information. Knowing the event Ω means simply that the experiment will occur and lead to some outcome belonging to Ω. This event is called the "certain event" and belongs to every sub-σ-algebra $\mathcal{B}$. The event $\emptyset$ is the empty event. This is why $\emptyset$ is called the "impossible event" and belongs also to every sub-σ-algebra $\mathcal{B}$.

The last element of the triple is P. This is a map from $\mathcal{A}$ to $[0, 1]$. We require $P(\Omega) = 1$, $P(\emptyset) = 0$ and

$$P(A \cap B) + P(A \cup B) = P(A) + P(B),$$

whence clearly

$$P(\bar{A}) = 1 - P(A),$$

where $\bar{A}$ denotes the complement of A. Note that if $P(A) = 1$, then A is not necessarily Ω, but if it is not Ω, it differs from Ω by a set of probability 0. It is useful to complete the σ-algebra $\mathcal{A}$ by adding all subsets of events of probability 0. If $P(A) = 1$, we shall say that A is almost sure (or certain). Note that if A_n is a sequence of events that are disjoint (there is no elementary event that belongs to the intersection of two events of the sequence), then one has

$$P(\cup_n A_n) = \sum_n P(A_n).$$

Such an equality, which would be impossible to write for an uncountable sequence, explains why we cannot take as the operational σ-algebra the set of all subsets of Ω.

It is useful to consider the pair $\Omega, \mathcal{A}$ as such. It is not a probability space, since P is not defined. Nevertheless, it is called a **measurable space.** This means that the pair is ready to be completed as a probability space by defining the probability measure P.

The set of real numbers R will have an important specific σ-algebra. It is the one that contains the open intervals and is the smallest that contains them. It is obtained by taking complements of open intervals as well as countable unions and intersections of open intervals and performing the same operations on the new subsets of R constructed in this manner. We say that it is generated by the open intervals. It is also generated by the closed intervals, or half-closed intervals. We call this σ-algebra the Borel σ-algebra on R, denoted by $\mathcal{R}$. Note that

$$\{a\} = (-\infty, a] \cap [a, +\infty).$$

Therefore, singletons are Borelian (elements of $\mathcal{R}$). Of course, we can extend the definition to R^n by taking open cubes instead of intervals.

A random variable is a map X from Ω to R such that the reciprocal of a Borelian is an event. In particular, if we consider any interval (a, b), then

$$\{\omega | X(\omega) \in (a, b)\} \subset \mathcal{A}.$$

We say that X is a measurable map. So a random variable is simply a measurable map from Ω to R. It follows that for every interval (a, b), we can define

$$P(a < X < b) = P(\{\omega | X(\omega) \in (a, b)\}),$$

which is called the probability that X is in the interval (a, b). We set

$$F(x) = P(X \leq x), \tag{6.1.1}$$

which defines a function, called the cumulative distribution function (CDF). This is a monotonically increasing function. It is continuous to the right, but not necessarily to the left. We have

$$P(\{x\}) = F(x) - F(x - 0).$$

Therefore, if $F(x)$ is continuous, the probability of singletons is 0. If the CDF is not only continuous but differentiable, then

$$f(x) = F'(x)$$

is called the probability density of the random variable (R.V.) X.

In the elementary theory of probability, without the concepts of σ-algebra and measurability, the concept of random variable reduces to that of CDF, and (6.1.1) is simply a notation that gives a probability measure to the interval $(-\infty, x]$ and more generally

$$P(a < X \leq b) = F(b) - F(a).$$

One can do a lot with CDF. We shall further comment on this issue below; see Remark 6.1.

The moments of X are expressed by

$$EX^p = \int x^p dF(x), \; p \geq 1.$$

We also mention the concepts of convergence. Let X_k be a sequence of random variables. We have pointwise convergence of X_k to X if

$$X_k(\omega) \to X(\omega), \quad \text{a.s.},$$

where a.s. (almost surely) means outside a subset of Ω of probability 0. This concept is the one naturally derived from the case of convergence of deterministic sequences. But in the probability setup, one has additional important concepts of convergence.

We say that X_k converges to X in probability whenever $\forall \varepsilon > 0$,

$$P(|X_k - X| > \epsilon) \to 0, \text{ as } k \to +\infty,$$

and we say that X_k converges to X in L^2 whenever

$$E|X_k - X|^2 \to 0, \text{ as } k \to +\infty.$$

REMARK 6.1. It is important to have in mind that the concept of random variable carries two aspects. First is the concept of a measurable map X from a probability space $\Omega, \mathcal{A}, P$ to $R, \mathcal{R}$. This concept does not depend on P. The second concept is that of CDF, cumulative distribution of X,

$$F(x) = P(\{-\infty < X \leq x\}),$$

which of course depends on the probability P. If we change P, we change $F(x)$, in spite of the fact that $X(\omega)$ does not change. When we define the function $X(\omega)$, we say that we define a random variable in a strong sense. If we consider purely a CDF, we say that it defines a random variable X in a weak sense. Indeed, it is possible to take $\Omega = R, \ \mathcal{A} = \mathcal{R}$, and P the probability distribution induced by $F(x)$ on $R, \mathcal{R}$. One then defines X as the identity. So there is always one random variable in a strong sense for which the CDF is indeed $F(x)$, but there may be several. As long as we manipulate only the CDF, it does not matter. In that sense the concept is weaker. This degree of freedom, which may look somewhat abstract, turns out to be extremely useful in many applications of stochastic calculus. We shall illustrate this idea in stochastic control in most of the subsequent chapters. In elementary probability theory, one works with random variables in a weak sense, without mentioning it. Only CDF are considered.

6.1.2 Conditional Expectation

Conditional expectation is a very important concept linked to information. Let X be a real R.V. and $\mathcal{B}$ a sub-σ-algebra of $\mathcal{A}$ representing the status of information. We say that X is $\mathcal{B}$ measurable if

$$\{a < X < b\} \subset \mathcal{B}.$$

Since a and b are arbitrary, this implies that we know the value of X once we can observe the events of $\mathcal{B}$. If X is not $\mathcal{B}$-measurable, then we do not know the value of X if the information is limited to $\mathcal{B}$. Nevertheless, how can we use the information provided by $\mathcal{B}$ to estimate the best value of X? If X is in L^2, the best estimate will be the $\mathcal{B}$-measurable variable $\hat{X}$ that minimizes the distance $E|X - Y|^2$ among all $\mathcal{B}$-measurable variables that are in L^2. This is the MMSE, in the sense of Chapter 4, Section 4.4. Since the space of $\mathcal{B}$-measurable variables that are in L^2 is a closed subspace of L^2, $\hat{X}$ is the projection of X on this subspace. It is uniquely defined by the necessary and sufficient condition

$$E[(\hat{X} - X)Y] = 0, \forall Y\ \mathcal{B}\,\text{measurable}. \tag{6.1.2}$$

This concept is easily extended when X is simply integrable, i.e., $X \in L^1$. First, we claim that there exists a unique $\hat{X}$ in L^1 such that

$$E[(\hat{X} - X)Y] = 0, \forall Y\ \mathcal{B}\,\text{measurable, bounded}. \tag{6.1.3}$$

To prove the existence, we note that we can assume $X \geq 0$. Indeed, It is clear that

$$\hat{X} = \hat{X}^+ - \hat{X}^-.$$

If now $X \geq 0$, we can define $X_n = \min(X, n)$, which is bounded, and thus $\hat{X}_n$ is well defined. It is easy to check that $\hat{X}_n$ is increasing, and thus since $\hat{X}_n \leq X$, the sequence $\hat{X}_n$ converges pointwise to an R.V. $\hat{X}$ that satisfies the necessary and sufficient condition (6.1.3).

The sequence $\hat{X}$ is called the conditional expectation of X given $\mathcal{B}$, also denoted by

$$\hat{X} = E[X|\mathcal{B}] = E^{\mathcal{B}}X.$$

The conditional expectation is the MMSE in the L^2 case.

EXERCISE 6.1. Show that the sequence $\hat{X}_n$ is monotonically increasing.

We note the two rules of conditional expectations

$$\begin{aligned} &E(E^{\mathcal{B}}X) = EX, \\ &E^{\mathcal{B}}(XY) = YE^{\mathcal{B}}X, \ \forall Y\ \mathcal{B}\,\text{measurable, bounded}. \end{aligned} \tag{6.1.4}$$

EXERCISE 6.2. Show that the two rules (6.1.4) define a unique element that coincides with the conditional expectation.

6.2 Stochastic Processes

6.2.1 General Concepts

Let $(\Omega, \mathcal{A}, P)$ be a probability space. A stochastic process in continuous time is a function $X(t;\omega)$ that is measurable in both arguments. The σ-algebra is the product of the Borel σ-algebra on R^+ and $\mathcal{A}$. A trajectory is the function

$$t \to X(t;\omega),\ \forall\omega\text{fixed.}$$

A process is said to be continuous (right-continuous, left-continuous) if its trajectories satisfy this property a.s.

A filtration $\mathcal{F}^t$ is a family of σ-algebras, indexed by time, such that $\mathcal{F}^t \subset \mathcal{F}^{t+1}$. If we think of $\mathcal{F}^t$ as the σ-algebra of events that are observable at time t, then this inclusion reflects the fact that information increases with time. We say that $X(t)$ is adapted to $\mathcal{F}^t$ if

$$\forall t,\ X(t) \text{ is } \mathcal{F}^t \text{ measurable.}$$

An R.V. $T \geq 0$ is a stopping time with respect to the filtration $\mathcal{F}^t$ if

$$\forall t,\ \{T \leq t\} \in \mathcal{F}^t.$$

We next define the σ-algebra $\mathcal{F}^T$ by

$$A \in \mathcal{F}^T \Rightarrow A \cap \{T \leq t\} \in \mathcal{F}^t, \forall t.$$

Consider a process $X(t)$ that is adapted to $\mathcal{F}^t$. We say that it is a P, $\mathcal{F}^t$ martingale if $E|X(t)| < +\infty, \forall t$ and

$$E(X(t)|\mathcal{F}^s) = X(s),\ \forall t \geq s.$$

It is a submartingale if

$$E(X(t)|\mathcal{F}^s) \geq X(s),\ \forall t \geq s,$$

and a supermartingale if $-X(t)$ is a submartingale.

6.2.2 Wiener Process

We shall use extensively the Wiener process, denoted by $w(t)$, which has the following properties:

$$\forall t_1, \cdots, t_n \quad w(t_1), \cdots, w(t_n) \text{ is a Gaussian vector,} \tag{6.2.1}$$
$$Ew(t)w(s) = \min(t,s)\ \forall t, s.$$

It is easy to check that $w(t)$ has independent increments, i.e., considering $t_1 < t_2 < t_3 < t_4$, the variables $w(t_2) - w(t_1)$ and $w(t_4) - w(t_3)$ are independent. The Wiener process is continuous at every point but is nowhere differentiable. The definition (6.2.1) does not involve any σ-algebra and was known prior to the introduction of the concept of σ-algebra in probability theory. This definition is now superseded by a new one as follows.

The definition relies on the following important result, called Lévy's theorem. Consider a continuous process $w(t)$ that is a $P, \mathcal{F}^t$ martingale such that $w(0) = 0$. It we have the property

$$E[(w(t) - w(s))^2|\mathcal{F}^s] = t - s, \ \forall t \geq s,$$

then $w(t)$ is a Wiener process. This result might look surprising, since we do not make any assumption about Gaussian properties. It has become now the most common definition of a Wiener process. We use the terminology $P, \mathcal{F}^t$ Wiener process to emphasize the filtration that is involved in the definition. We add the word **standard** to indicate that the mean of $w(t)$ is 0 and the variance t.

The concept extends easily to an n-dimensional $P, \mathcal{F}^t$ standard Wiener process by writing

$$Ew(t)w(s)^* = I \min(t, s),$$

$$E[(w(t) - w(s))(w(t) - w(s))^*|\mathcal{F}^s] = I(t - s),$$

where I is the identity matrix.

The Wiener process is at the basis of a stochastic integral and differential calculus, due to Itô, which has become a major tool in stochastic control and filtering theory.

REMARK 6.2. Whether a process is a Wiener process depends on the probability P. If we change the probability, a given process $w(t; \omega)$ may cease to be a Wiener process.

We shall use extensively this degree of freedom in stochastic control.

6.3 Stochastic Calculus

6.3.1 *Stochastic Integrals*

We begin with the concept of stochastic integral. Consider a filtration $\mathcal{F}^t$ and a $P, \mathcal{F}^t$ standard Wiener process (scalar). We consider a process $\varphi(t)$ that is adapted to $\mathcal{F}^t$ and satisfies

$$E \int_0^T |\varphi(t)|^2 dt < +\infty. \tag{6.3.1}$$

We call $L^2_{\mathcal{F}}(0,T)$ the space of processes satisfying the preceding properties. It can be structured as a Hilbert space, with norm defined by (6.3.1). Note that $L^2_{\mathcal{F}}(0,T)$ is a closed subspace of $L^2((0,T)\times\Omega; dt\otimes dP; R)$ for the norm (6.3.1). The objective is to define the concept

$$I(\varphi)=\int_0^T \varphi(t)dw(t), \tag{6.3.2}$$

called the stochastic integral of $\varphi(.)$ with respect to the Wiener process $w(.)$. The definition is easy when the integrand $\varphi(.)$ is piecewise constant. This means that there exists a sequence of deterministic times

$$t_0=0<t_1<t_2<\cdots t_N=T$$

such that

$$\varphi(t)=\varphi_n, \forall t\in[t_n,t_{n+1})$$

and φ_n is an R.V. that is $\mathcal{F}^{t_n}$-measurable and in L^2. In that case, the notation (6.3.2) implies

$$I(\varphi)=\sum_{n=0}^{N-1}\varphi_n(w(t_{n+1})-w(t_n)). \tag{6.3.3}$$

EXERCISE 6.3. Show that in the case of a piecewise constant integrand, the following properties hold:

$$\begin{aligned} &EI(\varphi)=0,\\ &E(I(\varphi)I(\psi))=E\int_0^T \varphi(t)\psi(t)dt. \end{aligned} \tag{6.3.4}$$

If we consider now any element in $L^2_{\mathcal{F}}(0,T)$, we can approximate it by a sequence of piecewise constant processes $\varphi^k(.)$ that converges to $\varphi(.)$ as $k\to 0$. We choose $k=\dfrac{T}{N}$ and $N\to+\infty$. For $1\le n\le N-1$, we write $t_n=nk$ and

$$\varphi_n=\frac{1}{k}\int_{t_{n-1}}^{t_n}\varphi(t)dt,$$

and we set $\varphi_0=0$. The function $\varphi^k(.)$ is defined by

$$\varphi^k(t)=\varphi_n,\ t_n\le t<t_{n+1}.$$

We thus can consider $I(\varphi^k)$ and let $k \to +\infty$. Since $I(\varphi^k)$ is a random variable, we can consider several concepts of convergence. There is no almost sure convergence, since the Wiener process is not differentiable, nor can extensions like Stieltjes integrals be applied. However, there is convergence in L^2. Indeed, we have

$$E\int_0^T |\varphi^k(t) - \varphi(t)|^2 dt \to 0, \text{ as } k \to +\infty.$$

Therefore,

$$E\int_0^T |\varphi^k(t) - \varphi^{k'}(t)|^2 dt \to 0, \text{ as } k, k' \to +\infty.$$

Since

$$\begin{aligned} E|I(\varphi^k) - I(\varphi^{k'})|^2 &= E|I(\varphi^k - \varphi^{k'})|^2 \\ &= E\int_0^T |\varphi^k(t) - \varphi^{k'}(t)|^2 dt, \end{aligned}$$

it follows that $I(\varphi^k)$ is a Cauchy sequence in the Hilbert space L^2. There exists, consequently, a unique random variable, denoted by $I(\varphi)$, such that

$$I(\varphi^k) \to I(\varphi) \text{ in } L^2.$$

This is the definition of the stochastic integral (6.3.2). The properties (6.3.4) are valid for the stochastic integral.

EXERCISE 6.4. Check that if S, T are two stopping times with $S \leq T$, then

$$E[\int_S^T \varphi(t)dw(t)|\mathcal{F}^S] = 0. \tag{6.3.5}$$

6.3.2 *Stochastic Differential*

For $t \in (0, T)$, we consider $I(\mathbb{1}_{(0,t)}\varphi)$, which is a modification of $\int_0^t \varphi(s)dw(s)$ in the sense

$$I(\mathbb{1}_{(0,t)}\varphi) = \int_0^t \varphi(s)dw(s), \text{ a.s., } \forall t.$$

Since we are interested in the dependence in t, it is convenient to set

$$I(t) = \int_0^t \varphi(s)dw(s)$$

and to consider it a stochastic process. We have

$$EI(t)^2 = E\int_0^t |\varphi(s)|^2 ds.$$

But more can be obtained (**martingale inequality**):

$$E \sup_{0\leq s\leq t} E|I(s)|^2 \leq CEI(t)^2. \tag{6.3.6}$$

EXERCISE 6.5. Check that thanks to (6.3.5), the stochastic process $I(t)$ is a $P, \mathcal{F}^t$ martingale. It can be shown that it is a continuous process, but like the Wiener process, it is not differentiable.

A very important result is the following (Kunita–Watanabe [31]): Suppose that $\mathcal{F}^t$ is generated by the Wiener process $w(.)$, i.e.,

$$\mathcal{F}^t = \sigma(w(s),\ s \leq t).$$

Then every $P, \mathcal{F}^t$ square integrable martingale $M(t)$ can be written as

$$M(t) = M(0) + \int_0^t \varphi(s)dw(s),\ \varphi(.) \in L^2_{\mathcal{F}}(0,T) < +\infty, \forall T. \tag{6.3.7}$$

This result is called the representation of martingales theorem. Moreover, $M(0)$ is a constant.

We now give the name semimartingale to a process defined as follows:

$$\xi(t) = \xi(0) + \int_0^t a(s)ds + \int_0^t b(s)dw(s), \tag{6.3.8}$$

where a and b are in $L^2_{\mathcal{F}}(0,T)$. We assume that $\xi(0)$ is $\mathcal{F}^0$-measurable. A process of the form (6.3.8) is said to have an Itô differential, written as follows:

$$d\xi(t) = a(t)dt + b(t)dw(t) \tag{6.3.9}$$

A process that has an Itô differential is called an **Itô process**.

If $b(t) = 0$, the concept reduces to that of ordinary differential, and $a(t)$ is the derivative of $\xi(t)$. Note that because of the ordinary integral, $\xi(t)$ is not a martingale.

The martingale inequality extends to semimartingales:

$$E \sup_{0\leq s\leq t} |\xi(s)|^2 \leq CE|\xi(t)|^2 \leq C(E|\xi(0)|^2 + E\int_0^T |a(t)|^2 dt + E\int_0^T |b(t)|^2 dt). \tag{6.3.10}$$

The reason we need to consider semimartingales and not just martingales is that if we consider a deterministic bounded function $\Psi(x,t)$ that is C^2 in x and C^1 in t, then the process $\Psi(\xi(t),t)$ is also a semimartingale. Its Itô differential is given by Itô's formula

$$d\Psi(\xi(t),t) = (\frac{\partial\Psi}{\partial t}(\xi(t),t) + \frac{1}{2}\frac{\partial^2\Psi}{\partial x^2}(\xi(t),t)b^2(t))dt + \frac{\partial\Psi}{\partial x}(\xi(t),t))d\xi(t). \tag{6.3.11}$$

Comparing with ordinary differential calculus, one can note the additional term involving the second derivative with respect to x.

This formula can be generalized to an n-dimensional semimartingale. We consider $w(t)$ an m-dimensional standard Wiener process, and $b(t)$ becomes an $n \times m$ matrix-valued process. Formula (6.3.11) becomes

$$d\Psi(\xi(t),t) = [\frac{\partial\Psi}{\partial t}(\xi(t),t) + \frac{1}{2}b^*(t)\frac{\partial^2\Psi}{\partial x^2}(\xi(t),t)b(t)]dt + \frac{\partial\Psi}{\partial x}(\xi(t),t).d\xi(t). \tag{6.3.12}$$

6.4 Stochastic Differential Equations

In a way very similar to that of ordinary differential equations, it is possible to solve stochastic differential equations as follows. Consider an m-dimensional standard $P, \mathcal{F}^t$ Wiener process $w(t)$ and functions $g(x,t) \in R^n$, $\sigma(x,t) \in \mathcal{L}(R^m; R^n)$. We look for a process $y(t)$ that is a semimartingale whose Itô differential satisfies the equation

$$dy = g(y(t),t)dt + \sigma(y(t),t)dw(t). \tag{6.4.1}$$

The value $y(0)$ is a given $\mathcal{F}^0$-measurable random variable y_0. We also say that $y(t)$ is a **diffusion process**. The following result generalizes the deterministic case.

THEOREM 6.1. *Assume that*

$$|g(x,t) - g(x',t)| + |\sigma(x,t) - \sigma(x',t)| \le K|x - x'|,$$
$$|g(x,t)|^2 + |\sigma(x,t)|^2 \le K_0^2(1 + |x|^2),$$
$$E|y_0|^2 < +\infty.$$

Then there exists a unique semimartingale $y(t)$ solution of (6.4.1) such that $y(0) = y_0$ and

$$E \sup_{0 \le t \le T} |y(t)|^2 \le C_T(1 + E|y_0|^2).$$

EXERCISE 6.6. Consider the linear case ($n = 1$):

$$g(x,t) = x\alpha(t), \quad \sigma(x,t) = x\sigma(t).$$

Give explicitly the solution of (6.4.1). Compare with the deterministic case.

REMARK 6.3. If $\mathcal{O}$ is a smooth domain of R^n, and $y(.)$ is a diffusion process, as above, we shall often consider **exit times** of the process $y(.)$ out of the domain, defined as follows:

$$\tau = \inf\{t | y(t) \notin \mathcal{O}\}. \tag{6.4.2}$$

The exit time τ is a stopping time with respect to the filtration $\mathcal{F}^t$.

6.5 Girsanov's Theorem

Consider a $P, \mathcal{F}^t$ standard Wiener process $w(.)$. Let $b(.)$ be adapted and bounded. Define

$$\eta(t) = \exp[\int_0^t b(s)dw(s) - \frac{1}{2}\int_0^t |b(s)|^2 ds],$$

$$\hat{w}(t) = w(t) - \int_0^t b(s)ds.$$

Then $\eta(t)$ satisfies

$$d\eta = \eta b dw$$

and $E\eta(t) = 1$. We define on each measurable space $\Omega, \mathcal{F}^t$ a probability P^t by setting

$$P^t(A) = E\mathbb{1}_A \eta(t), \; \forall A \in \mathcal{F}^t. \tag{6.5.1}$$

If the σ-algebra $\mathcal{A}$ on Ω is

$$\mathcal{A} = \cup_{t \geq 0} \mathcal{F}^t,$$

then by a famous theorem of Kolmogorov, the family P^t derives from a common probability measure $\hat{P}$ on $\Omega, \mathcal{A}$, and one has the compatibility property

$$P^t(A) = \hat{P}(A), \; \forall A \in \mathcal{F}^t.$$

We write

$$\frac{d\hat{P}}{dP}|_{\mathcal{F}^t} = \eta(t),$$

and $\eta(t)$ is called the Radon–Nikodym derivative of $\hat{P}$ with respect to P. This terminology is no more than (6.5.1).

Girsanov's theorem states that $\hat{w}(t)$ is a $\hat{P},\mathcal{F}^t$ standard Wiener process. Of course, $\hat{w}(t)$ is not a $P, \mathcal{F}^t$ standard Wiener process, and $w(t)$ is not a $\hat{P},\mathcal{F}^t$ standard Wiener process.

EXERCISE 6.7. Prove Girsanov's theorem by showing that $\hat{w}(t)$ satisfies the conditions of Lévy's theorem.

An important topic has been to extend Girsanov's theorem to cases in which $b(t)$ is not bounded. Note that in general,

$$E\eta(t) \leq 1. \tag{6.5.2}$$

Indeed, if we define the sequence of stopping times

$$\tau_N = \inf\{t | \, |b(t)| \geq N\},$$

then we have

$$E\eta(t \wedge \tau_N) = 1$$

and $\eta(t \wedge \tau_N) \uparrow \eta(t)$ as $N \uparrow +\infty$. By Fatou's lemma, (6.5.2) follows. Therefore, P^t is always a positive bounded measure on $\Omega, \mathcal{F}^t$ but may not be a probability.

Chapter 7
Filtering Theory in Continuous Time

7.1 Kalman Filters in Continuous Time

7.1.1 *Statement of the Problem*

We consider the analogue of (4.7.14) in continuous time. The model is described as follows:

$$dx = (F(t)x(t) + f(t))dt + dw(t), \tag{7.1.1}$$
$$x(0) = \xi,$$

$$dz = (H(t)x(t) + h(t))dt + d\epsilon(t), \tag{7.1.2}$$
$$z(0) = 0,$$

where $x(t) \in R^n$, $z(t) \in R^d$. We assume that $w(.)$ and $\epsilon(.)$ are independent Wiener processes in R^n and R^d respectively. For the sake of generality they are not standard, although they have mean 0, since the mean can always be included in the deterministic terms $f(.), h(.)$, and covariance matrices $Q(.)$, $R(.)$ respectively. By that we mean

$$Ew(t)w(s)^* = \int_0^{\min(t,s)} Q(\tau)d\tau,$$

$$E\epsilon(t)\epsilon(s)^* = \int_0^{\min(t,s)} R(\tau)d\tau.$$

It is important to allow this degree of freedom, since the covariance measures the intensity of uncertainty. The matrices $F(.)$, $H(.)$ and the vectors $f(.), h(.)$

A. Bensoussan, *Estimation and Control of Dynamical Systems*, Interdisciplinary Applied Mathematics 48, https://doi.org/10.1007/978-3-319-75456-7_7

are deterministic locally bounded functions. The random variable ξ is Gaussian, with mean $\bar{\xi}$ and covariance matrix P_0. It is independent of the Wiener processes $w(.), \epsilon(.)$. We set

$$\mathcal{F}^t = \sigma(\xi, w(s), \epsilon(s), 0 \leq s \leq t),$$

and $w(.), \epsilon(.)$ are $P, \mathcal{F}^t$ martingales. The process $x(.)$ represents the state of a dynamical system, and $z(.)$ the output or observation process. We define

$$\mathcal{Z}^t = \sigma(z(s), 0 \leq s \leq t).$$

The problem is to compute the MMSE, which is the conditional expectation

$$\hat{x}(t) = E(x(t)|\mathcal{Z}^t).$$

We shall have a result that is the analogue of the discrete-time case, Theorem 4.2. We consider the estimation error

$$\tilde{x}(t) = x(t) - \hat{x}(t)$$

and its covariance

$$P(t) = E\tilde{x}(t)\tilde{x}(t)^*. \tag{7.1.3}$$

Then $P(t)$ is the solution of the Riccati equation

$$\begin{aligned} \frac{dP}{dt} &= F(t)P(t) + P(t)F(t)^* - P(t)H^*(t)R^{-1}(t)H(t)P(t) + Q(t), \\ P(0) &= P_0. \end{aligned} \tag{7.1.4}$$

We then have the following theorem.

THEOREM 7.1. *The process $\hat{x}(t)$ is the solution of the stochastic differential equation*

$$\begin{aligned} d\hat{x} &= (F(t)\hat{x}(t) + f(t))dt + \\ &\quad + P(t)H^*(t)R^{-1}(t)(dz - (H(t)\hat{x}(t) + h(t))dt), \qquad (7.1.5) \\ \hat{x}(0) &= \bar{\xi}. \end{aligned}$$

Equation (7.1.5), which defines a linear filter on the observation process, is called a Kalman filter. Note that the interpretation (7.1.3) shows indirectly that the Riccati equation (7.1.4) has a solution for every value of t.

7.1.2 The Innovation Process

There are several approaches to obtaining a Kalman filter. We shall develop first that of the innovation process, which is the analogue of that introduced in the discrete-time case, Theorem 4.2. The innovation process is defined by

$$\nu(t) = z(t) - \int_0^t (H(s)\hat{x}(s) + h(s))ds. \tag{7.1.6}$$

We have the following lemma.

LEMMA 7.1. *The process* $\nu(t)$ *is a* $P, \mathcal{Z}^t$ *Wiener process, with covariance matrix* $R(.)$.

PROOF. Clearly, $\nu(t)$ is adapted to the filtration $\mathcal{Z}^t$. It is a continuous process, and $\nu(0) = 0$. We use Lévy's theorem. It is sufficient to show that it is a P, $\mathcal{Z}^t$ martingale and that

$$E[(\nu(t) - \nu(s))(\nu(t) - \nu(s))^*|\mathcal{Z}^s] = \int_s^t R(\tau)d\tau. \tag{7.1.7}$$

But

$$\nu(t) - \nu(s) = \int_s^t H(\tau)\tilde{x}(\tau)d\tau + \epsilon(t) - \epsilon(s),$$

whence

$$E(\nu(t) - \nu(s))|\mathcal{Z}^s) = \int_s^t H(\tau)E(\tilde{x}(\tau)|\mathcal{Z}^s)d\tau = 0,$$

since $\tilde{x}(\tau)$ is independent of $\mathcal{Z}^\tau$. Hence the martingale property.

Next, for $\tau > s$, we have

$$\begin{aligned} d(\nu(\tau) - \nu(s)) \ \ (\nu(\tau) - \nu(s))^* &= (H(\tau)\tilde{x}(\tau)d\tau + d\epsilon(\tau))(\nu(\tau) - \nu(s))^* + \\ &(\nu(\tau) - \nu(s))(\tilde{x}(\tau)^* H(\tau)^* dt + d\epsilon(\tau)^*) + R(\tau)d\tau. \end{aligned}$$

We can take the conditional expectation with respect to $\mathcal{Z}^s$ and obtain easily

$$d\, E[(\nu(\tau) - \nu(s))(\nu(\tau) - \nu(s))^*|\mathcal{Z}^s] = R(\tau)d\tau;$$

and integrating between s and t, we obtain (7.1.7). This concludes the proof. □

7.1.3 Proof of Theorem 7.1

Consider the process

$$M(t) = \hat{x}(t) - \bar{\xi} - \int_0^t (F(s)\hat{x}(s) + f(s))ds.$$

We first check that it is a P, $\mathcal{Z}^t$ martingale. Indeed,

$$\begin{aligned}
E[M(t) - M(s)|\mathcal{Z}^s] &= E[\hat{x}(t) - \hat{x}(s) - \int_s^t (F(\tau)\hat{x}(\tau) + f(\tau))d\tau|\mathcal{Z}^s] \\
&= E[x(t) - x(s) - \int_s^t (F(\tau)x(\tau) + f(\tau))d\tau|\mathcal{Z}^s] \\
&= E[w(t) - w(s)|\mathcal{Z}^s] \\
&= 0.
\end{aligned}$$

Let $\mathcal{N}^t = \sigma(\nu(s), s \leq t)$ be the σ-algebra generated by the innovation process. Clearly, $\mathcal{N}^t \subset \mathcal{Z}^t$. It will turn out that the two are identical. But at this stage, we cannot claim this. We have

$$E[M(t) - M(s)|\mathcal{N}^s] = E[(M(t) - M(s)|\mathcal{Z}^s)|\mathcal{N}^s] = 0,$$

but we cannot assert that $M(t)$ is adapted to the filtration $\mathcal{N}^t$. Hence we cannot claim that $M(t)$ is a $P, \mathcal{N}^t$ martingale and use the representation of the martingale theorem. However, thanks to the special relation between $\nu(t)$ and $z(t)$, see (7.1.6), the representation theorem extends to P, $\mathcal{Z}^t$ square integrable martingales. This is due to Liptser and Shiryaev [34]. Therefore, we can write

$$M(t) = \int_0^t K(s)R^{-1}(s)d\nu(s),$$

where $K(t)$ is in $\mathcal{L}(R^d; R^n)$ and is adapted to the filtration $\mathcal{Z}^t$. Moreover,

$$E\int_0^T ||K(t)||^2 dt < +\infty, \forall T.$$

We have put $R^{-1}(s)$ in the formula for convenience. We shall identify completely the process $K(t)$ (in fact, it will be deterministic). It follows that $\hat{x}(t)$ is the solution of

$$\begin{aligned}
d\hat{x} &= (F(t)\hat{x}(t) + f(t))dt + K(t)R^{-1}(t)(dz - (H(t)\hat{x}(t) + h(t))dt), \qquad (7.1.8) \\
\hat{x}(0) &= \bar{\xi}.
\end{aligned}$$

We identify $K(t)$ as follows. Consider a process

$$\eta(t) = \int_0^t \lambda^*(s)R^{-1}(s)d\nu(s)$$

with $\lambda(t)$ adapted to $\mathcal{Z}^t$ and square integrable, with values in R^d. We can compute $EM(t)\nu(t)$ in two ways. Using first the fact that $\nu(t)$ is a P, $\mathcal{Z}^t$ Wiener process with covariance matrix $R(t)$, we have

$$EM(t)\eta(t) = E\int_0^t K(s)R^{-1}(s)\lambda(s)ds. \qquad (7.1.9)$$

On the other hand, we have

$$EM(t)\eta(t) = E(\hat{x}(t) - \bar{\xi})\eta(t) - E\int_0^t (F(s)\hat{x}(s) + f(s))\eta(t)ds,$$

and using the fact that $\eta(t)$ is a P, $\mathcal{Z}^t$ martingale, we obtain

$$\begin{aligned} EM(t)\eta(t) &= E\hat{x}(t)\eta(t) - E\int_0^t (F(s)\hat{x}(s) + f(s))\eta(s)ds \\ &= Ex(t)\eta(t) - E\int_0^t (F(s)x(s) + f(s))\eta(s)ds. \end{aligned}$$

We now note that

$$\begin{aligned} d\eta(t) &= \lambda^*(t)R^{-1}(t)(H(t)\tilde{x}(t)dt + d\epsilon(t)), \\ dx(t) &= (F(t)x(t) + f(t))dt + dw(t). \end{aligned}$$

Therefore,

$$\begin{aligned} dx(t)\eta(t) = {} & x(t)\tilde{x}^*(t)H^*(t)R^{-1}(t)\lambda(t)dt + x(t)\lambda^*(t)R^{-1}(t)d\epsilon(t) \\ & +(F(t)x(t) + f(t))\eta(t)dt + \eta(t)dw(t), \end{aligned}$$

whence

$$Ex(t)\eta(t) = E\int_0^t x(s)\tilde{x}^*(s)H^*(s)R^{-1}(s)\lambda(s)ds + E\int_0^t (F(s)x(s) + f(s))\eta(s)ds$$

and

$$\begin{aligned} EM(t)\eta(t) &= E\int_0^t x(s)\tilde{x}^*(s)H^*(s)R^{-1}(s)\lambda(s)ds \\ &= E\int_0^t \tilde{x}(s)\tilde{x}^*(s)H^*(s)R^{-1}(s)\lambda(s)ds \\ &= E\int_0^t P(s)H^*(s)R^{-1}(s)\lambda(s)ds, \end{aligned}$$

where $P(s) = E\tilde{x}(s)\tilde{x}^*(s)$. We have used the fact that $\tilde{x}(s)$ is independent of $\mathcal{Z}^s$. Comparing with (7.1.9) and using the fact that $\lambda(s)$ is arbitrary, we obtain

$$K(s) = P(s)H^*(s).$$

Therefore, $\hat{x}(t)$ is the solution of (7.1.5), with $P(t)$ given by (7.1.3).

It remains to show that $P(t)$ is the solution of (7.1.4). We first state the equation for $\tilde{x}(t)$. From (7.1.1), (7.1.5), (7.1.2), we get

$$\begin{aligned} d\tilde{x} &= (F(t) - P(t)H^*(t)R^{-1}(t)H(t))\tilde{x}(t)dt \\ &\quad + dw(t) - P(t)H^*(t)R^{-1}(t)d\epsilon(t), \qquad (7.1.10) \\ \tilde{x}(0) &= \xi - \bar{\xi}. \end{aligned}$$

We compute $d(\tilde{x}(t)\tilde{x}^*(t))$ by Itô's formula. We obtain easily, by taking the mathematical expectation,

$$\begin{aligned} \frac{dP}{dt} &= (F(t)-P(t)H^*(t)R^{-1}(t)H(t))P(t)+P(t)(F(t)-P(t)H^*(t)R^{-1}(t)H(t))^* \\ &\quad + Q(t) + P(t)H^*(t)R^{-1}(t)H(t)P(t), \\ P(0) &= P_0, \end{aligned}$$

and we obtain immediately that $P(t)$ is the solution of the Riccati equation (7.1.4). The proof is complete. ■

7.2 Complements

Looking at the equation of the Kalman filter, written as

$$\begin{aligned} d\hat{x} &= (F(t)\hat{x}(t) + f(t))dt + P(t)H^*(t)R^{-1}(t)d\nu(t), \qquad (7.2.1) \\ \hat{x}(0) &= \bar{\xi}, \end{aligned}$$

it is easy to convince oneself that $\hat{x}(t)$ is adapted to $\mathcal{N}^t$, the filtration generated by $\nu(t)$. But then, since

$$z(t) = \nu(t) + \int_0^t (H(s)\hat{x}(s) + h(s))ds,$$

we see that $z(t)$ is adapted to $\mathcal{N}^t$. Therefore, $\mathcal{Z}^t \subset \mathcal{N}^t$, and in fact, $\mathcal{Z}^t = \mathcal{N}^t$. However, this equivalence will be proved afterward, so it cannot be used in the proof of Theorem 7.1.

Nevertheless, it is possible to prove directly that $\hat{x}(t)$ is the solution of (7.1.5), with $P(t)$, the solution of (7.1.4), the conditional expectation of $x(t)$ given $\mathcal{Z}^t$. Indeed, define

$$\tilde{x}(t) = x(t) - \hat{x}(t).$$

Then $\tilde{x}(t)$ is the solution of (7.1.10). We do not know that $P(t) = E(\tilde{x}(t)\tilde{x}^*(t))$, but this can be checked easily. Indeed, if we set temporarily

$$\Pi(t) = E(\tilde{x}(t)\tilde{x}^*(t)),$$

then from (7.1.10), we obtain easily

$$\begin{aligned}\frac{d\Pi}{dt} &= (F(t)-P(t)H^*(t)R^{-1}(t)H(t))\Pi(t)+\Pi(t)(F(t)-P(t)H^*(t)R^{-1}(t)H(t))^* \\ &\quad + Q(t) + P(t)H^*(t)R^{-1}(t)H(t)P(t), \\ \Pi(0) &= P_0,\end{aligned}$$

and this equation defines a unique solution $\Pi(t)$ for every given $P(t)$. Since $P(t)$ is obviously a solution, we have indeed $\Pi(t) = P(t)$.

Let us next show that $\tilde{x}(t)$ is not correlated to $z(s)$, $\forall s \leq t$. Looking at the equation for $\tilde{x}(t)$, see (7.1.10), we can set

$$\tilde{x}(t) = \Phi_P(t,s)\tilde{x}(s) + \int_s^t \Phi_P(t,\tau)(dw(\tau) - P(\tau)H^*(\tau)R^{-1}(\tau)d\epsilon(\tau)),$$

where $\Phi_P(t,s)$ is the fundamental matrix corresponding to $F(t) - P(t)H^*(t)R^{-1}(t)H(t)$. It follows that

$$E\tilde{x}(t)z(s)^* = \Phi_P(t,s)E\tilde{x}(s)z(s)^*.$$

It is thus sufficient to prove that $\tilde{x}(t)$ and $z(t)$ are not correlated. But

$$\begin{aligned}d\,\tilde{x}(t)z(t)^* &= (F(t) - P(t)H^*(t)R^{-1}(t)H(t))\tilde{x}(t)z(t)^*dt + \\ &\quad + (dw(t) - P(t)H^*(t)R^{-1}(t)dc(t))z(t)^* + \\ &\quad + \tilde{x}(t)(\tilde{x}^*(t)H^*(t)dt + d\varepsilon^*(t)) - P(t)H^*(t)dt,\end{aligned}$$

and therefore,

$$\begin{aligned}\frac{d}{dt}E\tilde{x}(t)z(t)^* &= (F(t) - P(t)H^*(t)R^{-1}(t)H(t))\,E\tilde{x}(t)z(t)^*, \\ E\tilde{x}(0)z(0)^* &= 0.\end{aligned}$$

This implies that $E\tilde{x}(t)z(t)^* = 0, \forall t$, and $E\tilde{x}(t)z(s)^* = 0, \forall t \geq s$. Therefore, $\tilde{x}(t)$ is not correlated to $z(s)$, $\forall s \leq t$. Since they are both Gaussian variables, they are in fact independent. Consequently, $\tilde{x}(t)$ is independent of $\mathcal{Z}^t$. Therefore,

$$E(\tilde{x}(t)|\mathcal{Z}^t) = E\tilde{x}(t) = 0.$$

Since $\hat{x}(t)$ is $\mathcal{Z}^t$-measurable, this implies

$$\hat{x}(t) = E(x(t)|\mathcal{Z}^t),$$

which proves the property.

7.3 Control Problems Related to Filtering Theory

7.3.1 *Minimum-Variance Linear Estimator*

We restrict the class of estimators to those that are linear functionals of the observation process. In fact, in view of the the form of the Kalman filter, we know that the conditional expectation is a linear functional of the observation process. But we start from scratch in the present context. We shall consider unbiased linear estimators. We set $\bar{x}(t) = Ex(t)$, which is the solution of

$$\begin{aligned} \frac{d\bar{x}}{dt} &= F(t)\bar{x}(t) + f(t), \\ \bar{x}(0) &= \bar{\xi}. \end{aligned} \tag{7.3.1}$$

Similarly, we set

$$\bar{z}(t) = \int_0^t (H(s)\bar{x}(s) + h(s))ds = Ez(t).$$

We set $\zeta(t) = z(t) - \bar{z}(t)$. To fix the ideas, we are interested in the minimum-variance linear estimator of $x(T)$, given the observation $z(t), t \leq T$.

A linear operator is defined by a linear map $\Lambda(t) \in \mathcal{L}(R^d; R^n)$, possibly depending on T, and the corresponding estimate is defined by

$$x^{\Lambda}(T) = \bar{x}(T) + \int_0^T \Lambda(t)d\zeta(t).$$

We note that

$$d\zeta(t) = H(t)(x(t) - \bar{x}(t))dt + d\epsilon(t).$$

Let

$$\begin{aligned} \tilde{x}^{\Lambda}(T) &= x(T) - x^{\Lambda}(T), \\ P^{\Lambda}(T) &= E\tilde{x}^{\Lambda}(T)(\tilde{x}^{\Lambda}(T))^*. \end{aligned}$$

The objective is to find $\Lambda(.)$ to minimize $P^{\Lambda}(T)$ (possibly in the sense of positive matrices). We shall associate a control problem to this optimization. Let $\psi \in R^n$. We compute

$$\begin{aligned}\psi^* x^{\Lambda}(T) &= \psi^* \bar{x}(T) + \psi^* \int_0^T \Lambda(t) d\zeta(t) \\ &= \psi^* \bar{x}(T) + \psi^* \int_0^T \Lambda(t) H(t)(x(t) - \bar{x}(t)) dt + \psi^* \int_0^T \Lambda(t) d\epsilon(t).\end{aligned}$$

We define now $q(t)$ by solving the backward differential equation

$$\begin{aligned}-\frac{dq}{dt} &= F^*(t) q(t) - H(t)^* \Lambda(t)^* \psi, \\ q(T) &= \psi.\end{aligned} \tag{7.3.2}$$

Combining this with the expression of $\psi^* x^{\Lambda}(T)$, we finally get

$$\psi^* x^{\Lambda}(T) = q(0)(\xi - \bar{\xi}) + \int_0^T q^*(t) dw(t) - \int_0^T \psi^* \Lambda(t) d\epsilon(t),$$

which leads to the expression

$$\psi^* P^{\Lambda}(T) \psi = q(0)^* P_0 q(0) + \int_0^T q^*(t) Q(t) q(t) dt + \int_0^T \psi^* \Lambda(t) R(t) \Lambda^*(t) \psi dt. \tag{7.3.3}$$

Setting

$$\mu(t) = \Lambda^*(t) \psi, \tag{7.3.4}$$

we get the control problem

$$\begin{aligned}-\frac{dq}{dt} &= F^*(t) q(t) - H(t)^* \mu(t), \\ q(T) &= \psi;\end{aligned} \tag{7.3.5}$$

minimize

$$J_{\psi}(\mu(.)) = q^*(0) P_0 q(0) + \int_0^T q^*(t) Q(t) q(t) dt + \int_0^T \mu^* R(t) \mu(t) dt. \tag{7.3.6}$$

We solve this control problem and show that it also solves the problem of finding the optimal Λ. To solve the problem (7.3.5), (7.3.6), we use the technique described in Chapter 3, Theorem 3.1, consisting in changing the control. More precisely, we set

$$\mu(t) = R^{-1}(t) H(t) \Pi(t) q(t) + \tilde{\mu}(t).$$

The function $\Pi(t)$ has to be determined. Then $\tilde{\mu}(t)$ is the new control. The state equation (7.3.5) becomes

$$\begin{aligned}-\frac{dq}{dt} &= (F^*(t) - H(t)^* R^{-1}(t) H(t)\Pi(t))q(t) - H(t)^*\tilde{\mu}(t),\\ q(T) &= \psi.\end{aligned} \tag{7.3.7}$$

We transform the objective functional $J_\psi(\mu(.))$ into

$$J_\psi(\mu(.)) = q^*(0)P_0 q(0) + \int_0^T q^*(t)(Q(t) + \Pi(t)H(t)^* R^{-1}(t)H(t)\Pi(t))q(t)dt +$$
$$+ \int_0^T \tilde{\mu}^*(t)R(t)\tilde{\mu}(t)dt + 2\int_0^T \tilde{\mu}^*(t)H(t)\Pi(t)q(t)dt.$$

In the last integral, we replace $\tilde{\mu}^*(t)H(t)$ using (7.3.7). We take $\Pi(t)$ to be symmetric and differentiable. We note that

$$\frac{d}{dt}q^*(t)\Pi(t)q(t) = 2\frac{dq}{dt}^*(t)\Pi(t)q(t) + q^*(t)\frac{d}{dt}\Pi(t)q(t).$$

Collecting results, we obtain

$$\begin{aligned}J_\psi(\mu(.)) = {} & q^*(0)P_0 q(0) - q^*(0)\Pi(0)q(0) + \\ & + \int_0^T q^*(t)(Q(t) - \Pi(t)H(t)^* R^{-1}(t)H(t)\Pi(t) \\ & - \frac{d}{dt}\Pi(t) + F(t)\Pi(t) + \Pi(t)F^*(t))q(t)dt \\ & + \psi^*\Pi(T)\psi + \int_0^T \tilde{\mu}^*(t)R(t)\tilde{\mu}(t)dt.\end{aligned}$$

If we choose $\Pi(t) = P(t)$, the solution of the Riccati equation (7.1.4), we get immediately

$$J_\psi(\mu(.)) = \psi^* P(T)\psi + \int_0^T \tilde{\mu}^*(t)R(t)\tilde{\mu}(t)dt. \tag{7.3.8}$$

The optimal $\hat{\mu}(.)$ corresponds clearly to $\tilde{\mu}(.) = 0$. This means that

$$\hat{\mu}(t) = R^{-1}(t)H(t)\Pi(t)\hat{q}(t), \tag{7.3.9}$$

where $\hat{q}(t)$ is the solution of

$$\begin{aligned}-\frac{d\hat{q}}{dt} &= (F^*(t) - H(t)^* R^{-1}(t)H(t)P(t))\hat{q}(t),\\ \hat{q}(T) &= \psi.\end{aligned} \tag{7.3.10}$$

But equation (7.3.10) defines a linear map $\hat{\Lambda}(t)$ by the formula

$$R^{-1}(t)H(t)P(t)\hat{q}(t) = \hat{\Lambda}^*(t)\psi \tag{7.3.11}$$

and

$$J_\psi(\hat{\mu}(.)) = \psi^* P^{\hat{\Lambda}}(T)\psi = \psi^* P(T)\psi.$$

Since ψ is arbitrary, we obtain $P^{\hat{\Lambda}}(T) = P(T)$. Moreover, for every map $\Lambda(.)$, we have

$$J_\psi(\mu(.)) = \psi^* P^{\Lambda}(T)\psi \geq \psi^* P(T)\psi,$$

and thus $P^{\Lambda}(T) \geq P^{\hat{\Lambda}}(T)$ in the sense of positive matrices. Therefore, $\hat{\Lambda}$ is optimal.

We then check that

$$\begin{aligned} x^{\hat{\Lambda}}(T) &= \bar{x}(T) + \int_0^T \hat{\Lambda}(t)(dz(t) - (H(t)\bar{x}(t) + h(t))dt) \\ &= \hat{x}(T), \end{aligned} \tag{7.3.12}$$

where $\hat{x}(T)$ is the value of the Kalman filter at time T; see (7.1.5). We have

$$\psi^* x^{\hat{\Lambda}}(T) = \psi^* \bar{x}(T) + \int_0^T q^*(t)P(t)H^*(t)R^{-1}(t)(dz(t) - (H(t)\bar{x}(t) + h(t))dt).$$

But from the equation of the Kalman filter, we get easily

$$\begin{aligned} d(\hat{x}(t) - \bar{x}(t)) = (F(t) - P(t)H^*(t)R^{-1}(t)H(t))(\hat{x}(t) - \bar{x}(t))dt + \\ + P(t)H^*(t)R^{-1}(t)(dz(t) - (H(t)\bar{x}(t) + h(t))dt), \end{aligned}$$

and from the equation of $q(t)$, we obtain

$$d\, q^*(t)(\hat{x}(t) - \bar{x}(t)) = q^*(t)P(t)H^*(t)R^{-1}(t)(dz(t) - (H(t)\bar{x}(t) + h(t))dt),$$

and thus, on integrating between 0 and T,

$$\psi^*(\hat{x}(T) - \bar{x}(T)) = \int_0^T q^*(t)P(t)H^*(t)R^{-1}(t)(dz(t) - (H(t)\bar{x}(t) + h(t))dt),$$

which proves $\psi^* x^{\hat{\Lambda}}(T) = \psi^* \hat{x}(T)$. And since ψ is arbitrary, we obtain (7.3.12).

7.3.2 Least Squares Estimator

Because we have continuous random processes in the model, the least squares estimator is not justified as easily as in the discrete-time case. So we shall proceed only formally, and define a control problem, the dual of (7.3.5), (7.3.6), whose solution will also lead to the Kalman filter.

The dynamical system is

$$\frac{dx}{dt} = F(t)x(t) + f(t) + w(t), \tag{7.3.13}$$
$$x(0) = \xi.$$

This a deterministic system, and $w(.)$ and ξ are control variables to be defined in order to minimize

$$J_{z(.)}(\xi, w(.)) = (\xi - \bar{\xi})^* P_0^{-1}(\xi - \bar{\xi}) + \int_0^T w^*(t)Q(t)^{-1}w(t)dt +$$
$$+ \int_0^T (z(t) - H(t)x(t) - h(t))^* R^{-1}(t)(z(t) - H(t)x(t) - h(t))dt. \tag{7.3.14}$$

The notation is reminiscent of the probabilistic origin. The function $z(t)$ is a given $L^2(0, T; R^d)$ function. It is reminiscent of the observation process, in fact rather the derivative of the observation process (which, as we know, does not exist). Similarly, $w(t)$ is reminiscent of the noise that perturbs the system (again its derivative), and ξ is the value of the initial condition, which we do not know. The cost functional (7.3.14) contains weights related to the covariance matrices that were part of the initial probabilistic model. One can compare this functional to that introduced in Chapter 4; see (4.5.6).

In the present context, we simply want to solve this control problem and find the optimal $\hat{x}(T)$, which we will compare to the Kalman filter. We solve the problem by the method of change of control function. However, this concerns only the control $w(.)$. We leave ξ as it is. Note that the problem makes sense only when P_0 and $Q(t)$ are invertible.

So we make the transformation

$$w(t) = Q(t)\Pi(t)(x(t) - r(t)) + \tilde{w}(t).$$

In this transformation, $\Pi(t)$ and $r(t)$ are to be defined. Equation (7.3.13) becomes

$$\frac{dx}{dt} = (F(t) + Q(t)\Pi(t))(x(t) - r(t)) + F(t)r(t) + f(t) + \tilde{w}(t), \tag{7.3.15}$$
$$x(0) = \xi.$$

We have

$$
\begin{aligned}
J_{z(.)}(\xi, w(.)) &= (\xi - \bar{\xi})^* P_0^{-1}(\xi - \bar{\xi}) + \int_0^T \tilde{w}^*(t)Q(t)^{-1}\widetilde{w}(t)dt + \\
&+ \int_0^T [(x(t) - r(t))^*\Pi(t)Q(t)\Pi(t)(x(t) - r(t)) + \qquad (7.3.16)\\
&+ (z(t) - H(t)x(t) - h(t))^* R^{-1}(t)(z(t) - H(t)x(t) - h(t))]dt \\
&+ 2\int_0^T (x(t) - r(t))^*\Pi(t)\widetilde{w}(t)dt.
\end{aligned}
$$

After rearrangements, we obtain

$$
\begin{aligned}
J_{z(.)}(\xi, w(.)) &= (\xi - \bar{\xi})^* P_0^{-1}(\xi - \bar{\xi}) + \int_0^T \tilde{w}^*(t)Q(t)^{-1}\widetilde{w}(t)dt \\
&+ \int_0^T [(x(t) - r(t))^*(-\frac{d\Pi}{dt} - \Pi(t)F(t) - F(t)^*\Pi(t) - \Pi(t)Q(t)\Pi(t) \\
&+ H^*(t)R^{-1}(t)H(t)\ )(x(t) - r(t))]dt + \\
&+ \int_0^T (z(t) - H(t)r(t) - h(t))^* R^{-1}(t)(z(t) - H(t)r(t) - h(t))dt + \\
&+ 2\int_0^T (x(t) - r(t))^*[\Pi(t)(\frac{dr}{dt} - F(t)r(t) - f(t)) \\
&- H^*(t)R^{-1}(t)(z(t) - H(t)r(t) - h(t))]dt \\
&+ (x(T) - r(T))^*\Pi(T)(x(T) - r(T)) - (\xi - r(0))^*\Pi(0)(\xi - r(0)).
\end{aligned}
$$

We then choose $\Pi(t)$ and $r(t)$ as follows:

$$
\frac{d\Pi}{dt} + \Pi(t)F(t) + F(t)^*\Pi(t) + \Pi(t)Q(t)\Pi(t) - H^*(t)R^{-1}(t)H(t) = 0,
$$

$$
\Pi(0) = P_0^{-1}, \quad (7.3.17)
$$

$$
\frac{dr}{dt} - F(t)r(t) - f(t)) - H^*(t)R^{-1}(t)(z(t) - H(t)r(t) - h(t)) = 0,
$$

$$
r(0) = \bar{\xi}. \qquad (7.3.18)
$$

We then get immediately

$$
\begin{aligned}
J_{z(.)}(\xi, w(.)) &= \int_0^T (z(t) - H(t)r(t) - h(t))^* R^{-1}(t)(z(t) - H(t)r(t) - h(t))dt \\
&+ \int_0^T \tilde{w}^*(t)Q(t)^{-1}\widetilde{w}(t)dt + (x(T) - r(T))^*\Pi(T)(x(T) - r(T)),
\end{aligned}
$$

and thus

$$J_{z(.)}(\xi, w(.)) \geq \int_0^T (z(t) - H(t)r(t) - h(t))^* R^{-1}(t)(z(t) - H(t)r(t) - h(t))dt.$$

On the other hand, there exists a pair $\xi^*, w^*(.)$ for which there is equality. It corresponds to the choice $\widetilde{w}(t) = 0$ and a value of the initial condition such that $x(T) = r(T)$. The optimal state $x^*(.)$ is then the solution of

$$\begin{aligned} \frac{dx^*}{dt} &= F(t)x^*(t) + f(t) + Q(t)\Pi(t)(x^*(t) - r(t)), \\ x^*(T) &= r(T), \end{aligned} \tag{7.3.19}$$

and we have

$$\xi^* = x^*(0), \quad w^*(t) = Q(t)\Pi(t)(x^*(t) - r(t)). \tag{7.3.20}$$

EXERCISE 7.1. Prove directly that

$$J_{z(.)}(\xi^*, w^*(.)) \geq \int_0^T (z(t) - H(t)r(t) - h(t))^* R^{-1}(t)(z(t) - H(t)r(t) - h(t))dt.$$

We first note immediately that

$$\Pi(t) = P^{-1}(t),$$

and $r(t)$ is the solution of

$$\frac{dr}{dt} = F(t)r(t) + f(t) + P(t)H^*(t)R^{-1}(t)(z(t) - H(t)r(t) - h(t)), \tag{7.3.21}$$
$$r(0) = \bar{\xi}.$$

This equation is very similar to the Kalman filter if we think of $z(t)$ as the derivative of the observation process. Note that in the least squares problem we have considered the variable of interest to be $x^*(T)$, since we want to estimate the state $x(T)$ with the observation available at time T. Since $x^*(T) = r(T)$, in fact the variable of interest is $r(T)$, for which we get by (7.3.21) a recursive expression.

Recall the minimum-variance linear estimator; see Section 7.3.1. It was defined by a linear map $\hat{\Lambda}(t)$ such that

$$R^{-1}(t)H(t)P(t)\hat{q}(t) = \hat{\Lambda}^*(t)\psi,$$

with $\hat{q}(t)$ defined by (7.3.10).

EXERCISE 7.2. Introduce the function

$$\zeta(t) = z(t) - H(t)\bar{x}(t) - h(t).$$

Show that

$$r(T) - \bar{x}(T) = \int_0^T \hat{\Lambda}(t)\zeta(t)dt. \tag{7.3.22}$$

Hence the least squares estimate coincides with the minimum-variance linear estimator, as expected.

Chapter 8
Stochastic Control of Linear Dynamical Systems with Full Information

8.1 The Basic Problem

We consider the dynamical system of Chapter 3, (3.1.1). The system is now perturbed by a noise. So the evolution looks as follows (a stochastic differential equation replaces a deterministic differential equation):

$$dx = (F(t)x(t) + G(t)v(t) + f(t))dt + dw, \tag{8.1.1}$$
$$x(0) = x_0.$$

We have to make precise the probability setup. We consider a probability space $(\Omega, \mathcal{A}, P)$ and a filtration $\mathcal{F}^t$. The process $w(.)$ is a $P, \mathcal{F}^t$ Wiener process with covariance matrix $Q(.)$, with values in R^n. The variable x_0 is deterministic.

We have perfect information, which means that we can observe all the events of $\mathcal{F}^t$ at time t. The control $v(.)$ is a square integrable process adapted to the filtration $\mathcal{F}^t$, so it belongs to $L^2_{\mathcal{F}}(0, T; R^m)$. We want to minimize the payoff functional

$$J(v(.)) = E[\int_0^T (x(t)^*M(t)x(t)+v(t)^*N(t)v(t)+2m(t)^*x(t)+2n(t)^*v(t))dt+$$
$$+x(T)^*M_Tx(T) + 2m_T^*x(T)]. \tag{8.1.2}$$

The assumptions on $M(t)$, $N(t)$, M_T, $m(t)$, $n(t)$, M_T, m_T are identical to the deterministic case; see Chapter 3, Section 3.1.

The solution will be obtained by the same feedback rule as in the deterministic case, namely

$$\hat{v}(x, t) = -N^{-1}(t)[n(t) + G^*(t)(\pi(t)x + r(t))], \tag{8.1.3}$$

A. Bensoussan, *Estimation and Control of Dynamical Systems*, Interdisciplinary Applied Mathematics 48, https://doi.org/10.1007/978-3-319-75456-7_8

in which $\pi(t)$ and $r(t)$ are deterministic functions that are solutions of

$$\frac{d\pi}{dt}+\pi(t)F(t)+F(t)^{*}\pi(t)-\pi(t)G(t)N^{-1}(t)G^{*}(t)\pi(t)+M(t)=0,$$
$$\pi(T)=M_T, \qquad (8.1.4)$$

$$\frac{dr}{dt}+(F(t)^{*}-\pi(t)G(t)N^{-1}(t)G^{*}(t))r(t)+\pi(t)(f(t)-G(t)N^{-1}(t)n(t))+m(t)=0,$$
$$r(T)=m_T. \qquad (8.1.5)$$

In fact, if we use the feedback rule (8.1.3) in the state equation (8.1.1), we obtain the process $y(t)$, the solution of

$$\begin{aligned} dy &= ((F(t)-G(t)N^{-1}(t)G^{*}(t)\pi(t))y(t)-G(t)N^{-1}(t)(n(t)\\ &\quad +G^{*}(t)r(t))+f(t))dt+dw,\\ y(0)&=x_0. \end{aligned}$$

It belongs to $L^2_{\mathcal{F}}(0,T;R^n)$. Therefore,

$$u(t)=\hat{v}(y(t),t) \qquad (8.1.6)$$

belongs to $L^2_{\mathcal{F}}(0,T;R^m)$. It is very important to notice that although the feedback does not change from the deterministic case to the stochastic case, the state evolution is different. It remains random in the stochastic case. We want to prove the following theorem.

THEOREM 8.1. *The process $u(.)$ given by (8.1.6) minimizes the functional $J(v(.))$ defined by (8.1.2).*

PROOF. We proceed with the method of change of control, as in the deterministic case; see Chapter 3, Section 3.1.2. For a control $v(.)$ in $L^2_{\mathcal{F}}(0,T;R^m)$, we define $\mu(.)\in L^2_{\mathcal{F}}(0,T;R^m)$ by the formula

$$v(t)=\Lambda(t)x(t)+\lambda(t)+\mu(t),$$

in which $\Lambda(t)$, $\lambda(t)$ will be defined below. So $x(t)$ is the solution of

$$dx=((F(t)+G(t)\Lambda(t))x(t)+f(t)+G(t)\lambda(t)+G(t)\mu(t))dt+dw, \qquad (8.1.7)$$
$$x(0)=x_0,$$

and we consider the process corresponding to $\mu(t)=0$, namely $y(t)$, the solution of

$$dy=((F(t)+G(t)\Lambda(t))y(t)+f(t)+G(t)\lambda(t))dt+dw, \qquad (8.1.8)$$
$$y(0)=x_0.$$

We define next $\tilde{x}(t)$ by

$$x(t) = y(t) + \tilde{x}(t),$$

and $\tilde{x}(t)$ is the solution of

$$\frac{d\tilde{x}}{dt} = (F(t) + G(t)\Lambda(t))\tilde{x}(t) + G(t)\mu(t),$$
$$\tilde{x}(0) = 0,$$

since the Wiener process disappears when we subtract the two equations. Note that

$$v(t) = u(t) + \Lambda(t)\tilde{x}(t) + \mu(t),$$

with

$$u(t) = \Lambda(t)y(t) + \lambda(t).$$

We can then compute, as in the deterministic case,

$$\begin{aligned} J(v(.)) = J(u(.)) &+ \\ &+ E\int_0^T [\tilde{x}^* M\, \tilde{x} + (\Lambda\tilde{x} + \mu)^* N(\Lambda\tilde{x} + \mu)]dt + E\tilde{x}(T)^* M_T \tilde{x}(T) \\ &+ 2E\int_0^T [\tilde{x}^*(M\, y + m) + (\Lambda\tilde{x} + \mu)^*(Nu + n)]dt \\ &+ 2E\, \tilde{x}(T)^*(M_T y(T) + m_T). \end{aligned}$$

We set

$$X = E\int_0^T [\tilde{x}^*(M\, y + m) + (\Lambda\tilde{x} + \mu)^*(Nu + n)]dt + E\tilde{x}(T)^*(M_T y(T) + m_T),$$

and we introduce functions $\pi(t), r(t)$, where $\pi(t)$ is a symmetric matrix, with the only conditions

$$\pi(T) = M_T,\ r(T) = m_T.$$

We assume differentiability of the functions $\pi(t), r(t)$. We set next

$$p(t) = \pi(t)y(t) + r(t).$$

We then compute

$$\begin{aligned} d(\tilde{x}(t)^* p(t)) = \{&[\tilde{x}^*(F^* + \Lambda^* G^*) + \mu^* G^*]p + \\ &+ \tilde{x}^*[\frac{d\pi}{dt} y + \pi((F + G\Lambda)y + f + G\lambda) \\ &+ \frac{dr}{dt}]\}dt + \tilde{x}(t)^*\pi(t)dw(t). \end{aligned}$$

Note that

$$\tilde{x}(T)^*(M_T y(T) + m_T) = \tilde{x}(T)^* p(T)$$
$$= \int_0^T d(\tilde{x}(t)^* p(t)).$$

We can then integrate and take the mathematical expectation; the stochastic integral vanishes, and we are led to exactly the same algebraic conditions as in the deterministic case. We can then conclude the proof. □

REMARK 8.1. The result of Theorem 8.1 justifies the use of the feedback even in the deterministic case. Indeed, in the deterministic case we may use at any time $u(t)$ the optimal control, or the feedback on the current state. In both cases the trajectory is the same $y(t)$, the optimal trajectory, called in engineering the nominal one. In the stochastic case they differ. Indeed, the feedback is applied to the real state, not to the nominal one, so the effective value is random. Since every real system will be perturbed by noise, it is much more robust to use the feedback value instead of the nominal value. To clarify this difference, we call the nominal $u(t)$ the optimal **open loop control**, whereas the feedback $\hat{v}(x,t)$ is called the optimal **closed loop control**. An open loop control is a deterministic function, while a closed loop control (or feedback) leads to a random process. The fact that the feedback is the same is the consequence of minimizing an average stochastic cost in the random case. The real payoff is random, its average will never occur, but it is the most common payoff that we can think of. The solution would be completely different if the payoff to be optimized were a combination of the mean and the variance of the random payoff. This makes a lot of sense when one wants to take account of risk aspects, a very important consideration in applications. What we do in considering only the mean of the random payoff is called the "certainty equivalence." We do not study payoffs with risk aspects in this book. These problems belong to mean field type control.

EXERCISE 8.1. We want to compute $J(u(.))$. First show as in Chapter 3, Exercise 3.1, that

$$J(u(.)) = E\int_0^T [y(t)^*(M(t) + \pi(t)G(t)N^{-1}(t)G^*(t)\pi(t))y(t) +$$
$$+ 2y(t)^*(\pi(t)G(t)N^{-1}(t)G^*(t)r(t) + m(t))$$
$$+ r^*(t)G(t)N^{-1}(t)G^*(t)r(t) - n^*(t)N^{-1}(t)n(t)]dt$$
$$+ E(y(T)^* M_T y(T) + 2m_T^* y(T)).$$

Then define

$$\Phi(t) = E[y(t)^*\pi(t)y(t) + 2r(t)^*y(t) + \rho_1(t)]$$

with

$$\rho_1(t) = \int_t^T (2r^*(s)f(s) - (G^*(s)r(s) + n(s))^* N^{-1}(s)(G^*(s)r(s) + n(s)) \\ + \operatorname{tr} \pi(s)Q(s))ds,$$

then

$$J(u(.)) = \Phi(0) = x_0^* \pi(0) x_0 + 2r(0)^* x_0 + \rho_1(0).$$

If we compare this with the optimal value in the deterministic case, the difference comes only from the third term $\rho_1(0) \neq \rho(0)$. We have

$$\rho_1(0) - \rho(0) = \int_0^T \operatorname{tr} \pi(t)Q(t))dt.$$

8.2 A More Elaborate Model with Control on the Diffusion Term

We now consider the following model:

$$dx = (Fx + Gv + f)dt + \sum_{i=1}^{I} (B_i x + D_i v + g_i)dw_i + dw,$$
$$x(0) = x_0. \tag{8.2.1}$$

All the functions depend on t, but to simplify notation we omit writing it explicitly. As in the previous section, we assume that $w(.)$ is a $P, \mathcal{F}^t$ Wiener process with covariance matrix $Q(.)$, with values in R^n. The processes $w_i(.)$ are scalar standard $P, \mathcal{F}^t$ Wiener processes. To simplify, we assume that all these Wiener processes are mutually independent. The cost functional is still defined by (8.1.2). This problem has been considered by Wonham [53] and Bismut [11].

The optimal solution will still be defined by a linear feedback. We first state the solution. The Riccati equation is changed into

$$\frac{d}{dt}\Pi + F^*\Pi + \Pi F + \sum_{i=1}^{I} B_i^* \Pi B_i + M - \tag{8.2.2}$$

$$-(\Pi G + \sum_{i=1}^{I} B_i^* \Pi D_i)(N + \sum_{i=1}^{I} D_i^* \Pi D_i)^{-1} (G^*\Pi + \sum_{i=1}^{I} D_i^* \Pi B_i) = 0,$$
$$\Pi(T) = M_T,$$

and the function $r(t)$ is the solution of

$$\frac{dr}{dt}+(F^*-(\Pi G+\sum_{i=1}^{I}B_i^*\Pi D_i)(N+\sum_{i=1}^{I}D_i^*\Pi D_i)^{-1}G^*)r+m+\Pi f+ \tag{8.2.3}$$

$$+\sum_{i=1}^{I}B_i^*\Pi g_i-(\Pi G+\sum_{i=1}^{I}B_i^*\Pi D_i)(N+\sum_{i=1}^{I}D_i^*\Pi D_i)^{-1}(n+\sum_{i=1}^{I}D_i^*\Pi g_i)=0,$$

$$r(T)=m_T.$$

We next define the feedback

$$\hat{v}(x,t)=-(N+\sum_{i=1}^{I}D_i^*\Pi D_i)^{-1}((G^*\Pi+\sum_{i=1}^{I}D_i^*\Pi B_i)x+G^*r+n+\sum_{i=1}^{I}D_i^*\Pi g_i), \tag{8.2.4}$$

and we can state the following theorem.

THEOREM 8.2. *The feedback given by (8.2.4) defines the optimal control of the problem (8.2.1), (8.1.2).*

PROOF. We proceed as in the basic case. We make the change of control

$$v(t)=\Lambda(t)x(t)+\lambda(t)+\mu(t),$$

and the state equation becomes

$$\begin{aligned} dx &= ((F+G\Lambda)x+G\lambda+f+G\mu)dt+ \\ &\quad+\sum_{i=1}^{I}((B_i+D_i\Lambda)x+D_i\lambda+g_i+D_i\mu)dw_i+dw, \qquad (8.2.5)\\ x(0)&=x_0. \end{aligned}$$

We denote by $y(t)$ the state corresponding to $\mu(t)=0$, the solution of

$$\begin{aligned} dy &= ((F+G\Lambda)y+G\lambda+f)dt+ \\ &\quad+\sum_{i=1}^{I}((B_i+D_i\Lambda)y+D_i\lambda+g_i)dw_i+dw, \qquad (8.2.6)\\ x(0)&=x_0, \end{aligned}$$

and the corresponding control is

$$u(t)=\Lambda(t)y(t)+\lambda(t). \tag{8.2.7}$$

We consider

$$\tilde{x}(t)=x(t)-y(t),$$

and hence $\tilde{x}(t)$ is the solution of

$$d\tilde{x} = ((F + G\Lambda)\tilde{x} + G\mu)dt + \sum_{i=1}^{I}((B_i + D_i\Lambda)\tilde{x} + D_i\mu)dw_i,$$

$$\tilde{x}(0) = 0,$$

$$v(t) = u(t) + \Lambda(t)\tilde{x}(t) + \mu(t).$$

We can then make the same computation as in the proof of Theorem 8.1, and we have to achieve $X = 0$, where

$$X = E\int_0^T [\tilde{x}^*(M\,y + m) + (\Lambda\tilde{x} + \mu)^*(Nu + n)]dt + E\tilde{x}(T)^*(M_T y(T) + m_T).$$

We define

$$p(t) = \Pi(t)y(t) + r(t),$$

so we have four functions to define: $\Lambda(t), \Pi(t), \lambda(t), r(t)$. We set

$$\Pi(T) = M_T, \quad r(T) = m_T$$

and compute

$$\begin{aligned}
d\,(\tilde{x}^* p) &= (\tilde{x}^*(F^* + \Lambda^* G^*)p + \mu^* G^*)p)dt + \\
&+ \sum_{i=1}^{I}(\tilde{x}^*(B_i^* + \Lambda^* D_i^*)p + \mu^* D_i^* p)dw_i \\
&+ \tilde{x}^*(\frac{d}{dt}\Pi\,y + \frac{dr}{dt}) + \tilde{x}^*\Pi((F + G\Lambda)y + G\lambda + f)dt \\
&+ \sum_{i=1}^{I}\tilde{x}^*\Pi((B_i + D_i\Lambda)y + D_i\lambda + g_i)dw_i + \tilde{x}^*\Pi dw \\
&+ \sum_{i=1}^{I}(\tilde{x}^*(B_i^* + \Lambda^* D_i^*) + \mu^* D_i^*)\Pi((B_i + D_i\Lambda)y + D_i\lambda + g_i).
\end{aligned}$$

Collecting results, we get the expression

$$\begin{aligned}
X = E\int_0^T &\{\tilde{x}^*(M\,y + m) + (\Lambda\tilde{x} + \mu)^*(Nu + n) + \\
&+ (\tilde{x}^*(F^* + \Lambda^* G^*) + \mu^* G^*)p + \\
&+ \tilde{x}^*(\frac{d}{dt}\Pi\,y + \frac{dr}{dt}) + \tilde{x}^*\Pi((F + G\Lambda)y + G\lambda + f) + \\
&+ \sum_{i=1}^{I}(\tilde{x}^*(B_i^* + \Lambda^* D_i^*) + \mu^* D_i^*)\Pi((B_i + D_i\Lambda)y + D_i\lambda + g_i)\}dt.
\end{aligned}$$

To achieve $X = 0$, we equate to 0 the coefficients of $\tilde{x}^*$ and μ^*, thus obtaining

$$M\,y + m + \Lambda^*(Nu+n) + (F^* + \Lambda^* G^*)p + \frac{d}{dt}\Pi\, y + \frac{dr}{dt}$$

$$+\Pi((F+G\Lambda)y+G\lambda+f)+\sum_{i=1}^{I}(B_i^*+\Lambda^* D_i^*)\Pi((B_i+D_i\Lambda)y+D_i\lambda+g_i)=0,$$

$$Nu+n+G^*p+\sum_{i=1}^{I}D_i^*\Pi((B_i+D_i\Lambda)y+D_i\lambda+g_i)=0. \tag{8.2.8}$$

We replace in these equations $u(t)$ by $\Lambda(t)y(t)+\lambda(t)$ and $p(t)$ by $\Pi(t)y(t)+r(t)$. We then equate to 0 the coefficient of $y(t)$ and the term independent of $y(t)$. We obtain four relations:

$$M + \Lambda^* N\Lambda + (F^* + \Lambda^* G^*)\Pi + \frac{d}{dt}\Pi + \Pi(F+G\Lambda) +$$

$$+\sum_{i=1}^{I}(B_i^* + \Lambda^* D_i^*)\Pi(B_i + D_i\Lambda) = 0,$$

$$m + \Lambda^* N\lambda + \Lambda^* n + (F^* + \Lambda^* G^*)r + \frac{dr}{dt} + \Pi(G\lambda + f) +$$

$$+\sum_{i=1}^{I}(B_i^* + \Lambda^* D_i^*)\Pi(D_i\lambda + g_i) = 0,$$

$$N\Lambda + G^*\Pi + \sum_{i=1}^{I}D_i^*\Pi(B_i + D_i\Lambda) = 0,$$

$$N\lambda + n + G^* r + \sum_{i=1}^{I}D_i^*\Pi(D_i\lambda + g_i) = 0.$$

The two last relations allow us to obtain $\Lambda(t)$ and $\lambda(t)$. We get the formulas

$$\Lambda = -(N + \sum_{i=1}^{I}D_i^*\Pi D_i)^{-1}(G^*\Pi + \sum_{i=1}^{I}D_i^*\Pi B_i),$$

$$\lambda = -(N + \sum_{i=1}^{I}D_i^*\Pi D_i)^{-1}(n + G^* r + \sum_{i=1}^{I}D_i^*\Pi g_i).$$

Inserting these expressions into the first two relations, we obtain the equations of $\Pi(t)$ and $r(t)$. □

EXERCISE 8.2. We can compute $J(u(.))$. We define

$$\Phi(t) = E[y(t)^*\pi(t)y(t) + 2r(t)^*y(t) + \rho_2(t)],$$

with

$$\rho_2(t) = \int_t^T [2r^* f + \sum_i g_i^*\pi g_i + \operatorname{tr} \pi Q -$$
$$- (G^*r+n+\sum_i D_i^*\pi g_i)^*(N+\sum_i D_i^*\Pi D_i)^{-1}(G^*r+n+\sum_i D_i^*\pi g_i)]dt.$$

Show that

$$\begin{aligned} J(u(.)) &= \Phi(0) \\ &= x_0^*\pi(0)x_0 + 2r(0)^*x_0 + \rho_2(0). \end{aligned}$$

8.3 Exponential-of-Integral Payoff

8.3.1 *Setting of the Problem*

We consider the state equation

$$\begin{aligned} dx &= (F(t)x(t) + G(t)v(t) + f(t))dt + dw, \\ x(0) &= x_0, \end{aligned} \tag{8.3.1}$$

and the probabilistic setup is the same as in Section 8.1. We shall, however, need to restrict the space of controls as a subset of $L^2_{\mathcal{F}}(0, T; R^m)$. The payoff we want to minimize is now

$$\begin{aligned} J(v(.)) = E\{\theta \exp\theta[\int_0^T (x^*Mx + v^*Nv + 2m^*x + 2n^*v)dt + \\ + x(T)^*M_Tx(T) + 2m_T^*x(T)]\}, \end{aligned} \tag{8.3.2}$$

and θ is a given real number (possibly negative). When θ is positive, we do not need to multiply by θ, but it does not hurt. When θ is negative, it is necessary to do so. In this way, the multiplication by θ will allow both cases to be combined in a single functional. It is convenient to introduce the random variable

$$\begin{aligned} Z(v(.)) = \int_0^T (x^*Mx + v^*Nv + 2m^*x + 2n^*v)dt \\ + x(T)^*M_Tx(T) + 2m_T^*x(T), \end{aligned}$$

which we call the random cost. In Section 8.1, we were interested in minimizing its expected value $EZ(v(.))$. The functional $J(v(.))$ clearly involves all moments of the random cost, so it incorporates risk aspects. Note that

$$J(v(.)) = \theta + \theta^2 EZ + \theta^3 E[Z^2 \int_0^1 \int_0^1 \lambda \exp(\lambda\mu\theta Z) d\lambda d\mu].$$

We see that if $\theta > 0$, the second moment is an additional cost. We can say that the decision-maker is **risk-averse**, since variability is a penalty. On the other hand, if $\theta < 0$, the second moment reduces the cost, so variability is a benefit. We say that the decision-maker is **risk-tolerant**. Moreover, when θ is small, the important term to minimize is EZ. For $\theta = 0$, the payoff is interpreted as EZ, and we say that the decision-maker is **risk-neutral**. This model tries to capture the risk-sensitivity of the decision-maker. See P. Whittle [51] for more details.

The only problem we may have in the definition of $J(v(.))$ is that it can take the value $+\infty$ if $\theta > 0$. Since we want to minimize, the corresponding controls will certainly not be optimal.

8.3.2 The Formal Method

We cannot proceed as for the basic problem. However, we shall use a technique that is more general and can be applied to the basic problem as well. It is reminiscent of the approach used to compute $J(u(.))$; see Exercise 8.1. For a control in $L^2_{\mathcal{F}}(0, T; R^m)$, we consider the process

$$\varphi_t = \int_0^t (x^* M x + v^* N v + 2m^* x + 2n^* v) ds + x_t^* \pi_t x_t + 2r_t^* x_t + \rho_t, \tag{8.3.3}$$

in which π_t, r_t, ρ_t have to be found, and π_t is symmetric. We save notation by using indices instead of arguments. We impose, however,

$$\pi_T = M_T,\ r_T = m_T,\ \rho_T = 0. \tag{8.3.4}$$

Thus $\varphi_T = Z(v(.))$, the random cost. We can write, omitting indices, when there is no confusion,

$$\begin{aligned} d\varphi_t = {} & [x^*(\dot{\pi} + F^*\pi + \pi F + M)x + v^* N v + 2x^* \pi G v + \\ & + 2x^*(\dot{r} + F^* r + m + \pi f) + 2v^*(G^* r + n) + 2r^* f + \operatorname{tr} \pi Q + \dot{\rho}]dt + \\ & + 2(x^* \pi + r^*) dw. \end{aligned}$$

From Itô's formula, we deduce

$$\begin{aligned}
d\,\theta\exp\theta\varphi_t &= \theta\exp\theta\varphi_t(\theta d\varphi_t + 2\theta^2(x^*\pi + r^*)Q(\pi x + r)dt)\\
&= \theta^2\exp\theta\varphi_t\{\,[x^*(\dot{\pi} + F^*\pi + \pi F + 2\theta\pi Q\pi + M)x +\\
&\quad + v^*Nv + 2x^*\pi Gv + 2x^*(\dot{r} + (F^* + 2\theta\pi Q)r + m + \pi f) +\\
&\quad + 2v^*(G^*r + n) + 2r^*f + \operatorname{tr}\pi Q + 2\theta r^*Qr + \dot{\rho}]dt +\\
&\quad + 2(x^*\pi + r^*)dw\}.
\end{aligned}$$

Completing the square yields

$$\begin{aligned}
d\,\theta\exp\theta\varphi_t &= \theta^2\exp\theta\varphi_t\{\,[x^*(\dot{\pi}+F^*\pi+\pi F-\pi GN^{-1}G^*\pi+2\theta\pi Q\pi + M)x+\\
&\quad + (v + N^{-1}(G^*(\pi x + r) + n))^*N(v + N^{-1}(G^*(\pi x + r) + n))\\
&\quad + 2x^*(\dot{r} + (F^*+2\theta\pi Q - \pi GN^{-1}G^*)r + m+\pi f - \pi GN^{-1}n)+\\
&\quad - (G^*r + n)^*N^{-1}(G^*r + n) + 2r^*f + \operatorname{tr}\pi Q + 2\theta r^*Qr + \dot{\rho}]dt\\
&\quad + 2(x^*\pi + r^*)dw\}.
\end{aligned}$$

We now define π_t, r_t, ρ_t as follows:

$$\dot{\pi} + F^*\pi + \pi F - \pi GN^{-1}G^*\pi + 2\theta\pi Q\pi + M = 0, \tag{8.3.5}$$

$$\dot{r} + (F^* + 2\theta\pi Q - \pi GN^{-1}G^*)r + m + \pi f - \pi GN^{-1}n = 0, \tag{8.3.6}$$

$$\dot{\rho} - (G^*r + n)^*N^{-1}(G^*r + n) + 2r^*f + \operatorname{tr}\pi Q + 2\theta r^*Qr = 0. \tag{8.3.7}$$

We assume that the Riccati equation has a solution. With the final conditions (8.3.4), the differential equations define completely π_t, r_t, ρ_t. With this choice we get

$$\begin{aligned}
d\,\theta\exp\theta\varphi_t &= \theta^2\exp\theta\varphi_t\{(v + N^{-1}(G^*(\pi x + r) + n))^*\\
&\qquad N(v + N^{-1}(G^*(\pi x + r) + n)) + 2(x^*\pi + r^*)dw\}.
\end{aligned} \tag{8.3.8}$$

8.3.3 Solution

We try to achieve

$$v_t + N_t^{-1}(G_t^*(\pi_t x_t + r_t) + n_t) = 0.$$

Replacing in the state equation (8.3.1), we obtain

$$\begin{aligned}
dy &= [(F_t - G_tN_t^{-1}G_t^*\pi_t)y_t - G_tN_t^{-1}(G_t^*r_t + n_t) + f_t]dt + dw,\\
y(0) &= x_0,
\end{aligned} \tag{8.3.9}$$

and the control

$$u_t = -N_t^{-1}(G_t^*(\pi_t y_t + r_t) + n_t). \tag{8.3.10}$$

We consider the class of controls $v(.) \in L^2_{\mathcal{F}}(0, T; R^m)$ such that

$$E \exp[\int_0^T 2\theta(x^*\pi + r^*)dw - 2\theta^2 \int_0^T (x^*\pi + r^*)Q(\pi x + r)dt] = 1. \tag{8.3.11}$$

It is a technical result to check that $u(.)$ defined by (8.3.10) belongs to this class; see [5] for details. We can then state the following result.

THEOREM 8.3. *We assume that the Riccati equation (8.3.5) with $\pi(T) = M_T$ has a solution that is positive symmetric. Then the control $u(.)$ defined by (8.3.10) is optimal for the problem (8.3.1), (8.3.3), in the class of controls that satisfy (8.3.11).*

PROOF. If we take $v(.) = u(.)$, then from (8.3.8) we have

$$d\,\theta \exp\theta\varphi_t = \theta^2 \exp\theta\varphi_t 2(y^*\pi + r^*)dw.$$

Therefore, on solving this stochastic differential equation in $\exp\theta\varphi_t$, we have

$$\theta \exp\theta\varphi_t = \theta \exp\theta\varphi_0 \exp[\int_0^t 2\theta(y^*\pi + r^*)dw - 2\theta^2 \int_0^t (y^*\pi + r^*)Q(\pi y + r)ds].$$

Applying this at time T and taking the expectation, we get

$$J(u(.)) = \theta \exp\theta\varphi_0. \tag{8.3.12}$$

On the other hand, for every admissible control we have from (8.3.8) an inequality instead of an equality,

$$\theta \exp\theta\varphi_T \geq \theta \exp\theta\varphi_0 \exp[\int_0^T 2\theta(y^*\pi + r^*)dw - 2\theta^2 \int_0^T (y^*\pi + r^*)Q(\pi y + r)ds],$$

and taking the mathematical expectation, we get

$$J(v(.)) \geq \theta \exp\theta\varphi_0,$$

which proves the optimality of $u(.)$. □

Chapter 9
Stochastic Control of Linear Dynamical Systems with Partial Information

9.1 General Discussion

The state of the system is described by

$$dx = (Fx + Gv + f)dt + dw, \tag{9.1.1}$$
$$x(0) = \xi.$$

We shall clarify later how we choose the control process. Unlike what we did in the preceding chapter, here we do not observe the state of the system, so we cannot use a feedback on the state. The observation process is defined by

$$dz = (Hx + h)dt + d\epsilon, \tag{9.1.2}$$
$$z(0) = 0.$$

We consider a probability space $(\Omega, \mathcal{A}, P)$. The processes $w(.), \epsilon(.)$ are independent Wiener processes with covariance matrices $Q(.)$ and $R(.)$ respectively. The state $x(t)$ takes values in R^n, and $z(.)$ takes values in R^d. The initial value ξ is a random variable that is independent of the Wiener processes. It is Gaussian with mean $\bar{\xi}$ and covariance matrix P_0. To fix ideas, we define

$$\mathcal{F}^t = \sigma(\xi, w(s), \epsilon(s), s \leq t).$$

We want to minimize the payoff functional

$$J(v(.)) = E[\int_0^T (x^*M + v^*Nv + 2m^*x + 2n^*v)dt + \tag{9.1.3}$$
$$+ x(T)^*M_Tx(T) + 2m_T^*x(T)].$$

A. Bensoussan, *Estimation and Control of Dynamical Systems*, Interdisciplinary Applied Mathematics 48, https://doi.org/10.1007/978-3-319-75456-7_9

The basic difficulty is that in contrast to the full information case, we cannot take $v(.) \in L^2_{\mathcal{F}}(0,T;R^m)$. What is accessible is the observation process. So if we define $\mathcal{Z}^t = \sigma(z(s), s \leq t)$, we should have $v(.)$ adapted to the filtration $\mathcal{Z}^t$. However, this filtration cannot be defined independently of the control. So we face a chicken-and-egg problem. How to begin?

Thanks to the linearity, it is possible to separate the uncertainties from the control and obtain a starting point. Indeed, we first define the processes $\alpha(.)$, $\beta(.)$ corresponding to the situation in which the control is 0, namely,

$$\begin{aligned} d\alpha &= (F\alpha + f)dt + dw, \\ \alpha(0) &= \xi, \end{aligned} \tag{9.1.4}$$

$$\begin{aligned} d\beta &= (H\alpha + h)dt + d\epsilon, \\ \beta(0) &= 0. \end{aligned} \tag{9.1.5}$$

We next consider a process $v(t)$ that is square integrable in the two arguments t, ω, with no adaptation requirements. We can define the processes x_1, z_1 by the formulas

$$\begin{aligned} \frac{dx_1}{dt} &= Fx_1 + Gv, \\ x_1(0) &= 0, \end{aligned} \tag{9.1.6}$$

$$\begin{aligned} \frac{dz_1}{dt} &= Hx_1, \\ z_1(0) &= 0. \end{aligned} \tag{9.1.7}$$

So we can set

$$\begin{aligned} x(t) &= \alpha(t) + x_1(t), \\ z(t) &= \beta(t) + z_1(t), \end{aligned} \tag{9.1.8}$$

and we call $x(t)$ the state of the system at time t. Similarly, we call $z(t)$ the observation on the system at time t. Note that $x(t)$ is not an Itô process, since the control is not adapted. We cannot exploit the definition very much, and obviously, this definition is possible only thanks to the linearity. We next define

$$\mathcal{Z}_v^t = \sigma(z(s), s \leq t).$$

We have emphasized in the notation the fact that $\mathcal{Z}_v^t$ depends on the control. So far, this construction shows that for every square integrable control, we can define the state and the observation, and we can clearly also define the payoff $J(v(.))$.

It remains to define what the admissible controls are. They must be adapted to the observation. So a natural definition of admissibility is

$$v(.) \text{ is admissible } \Leftrightarrow v(.) \text{ is adapted to } \mathcal{Z}_v^t. \tag{9.1.9}$$

This is a constraint that is not easy to handle. Anyway, we can define the problem of minimizing $J(v(.))$ on the set of admissible controls. This problem is difficult but can be completely solved. See [5] for a full solution. The optimal solution will have an additional structure property, which we introduce here beforehand. This is because proving that it is optimal against all controls that benefit from the same property will be much simpler. This is the approach we will follow in this presentation.

Let us introduce $\mathcal{B}^t = \sigma(\beta(s), s \leq t)$. For an admissible control, the second equation (9.1.8) implies immediately that

$$v(.) \text{ is admissible } \Longrightarrow \mathcal{B}^t \subset \mathcal{Z}_v^t. \tag{9.1.10}$$

The limitation we want to introduce is the following:

$$v(.) \text{ is admissible } \Leftrightarrow v(.) \text{ is adapted to } \mathcal{Z}_v^t \text{ and } \mathcal{B}^t. \tag{9.1.11}$$

With this limitation, it is immediate from the second relation (9.1.8) that if $v(.)$ is admissible in this sense, then $\mathcal{Z}_v^t = \mathcal{B}^t$, and the filtration of the observation becomes fixed. This simplifies considerably the treatment. In the next section, we show that the candidate optimal control satisfies the property. Besides, it is guessed by a very intuitive argument, very appealing to engineers, called the **separation principle**, which we will comment on in Section 9.3.

9.2 A Class of Admissible Controls

We first note that an immediate consequence of the new definition of admissibility is that both the state and the observation are Itô processes, and (9.1.1), (9.1.2) are well defined. Moreover, the Kalman filter

$$\hat{x}(t) = E[x(t)|\mathcal{Z}_v^t] = E[x(t)|\mathcal{B}^t]$$

is easily obtained. We have the following.

PROPOSITION 9.1. *If $v(.)$ is admissible in the sense (9.1.11), then the Kalman filter is the solution of*

$$\begin{aligned} d\hat{x} &= (F\hat{x}(t) + f + Gv)dt + PH^*R^{-1}(dz - (H\hat{x} + h)dt), \\ \hat{x}(0) &= \bar{\xi}, \end{aligned} \tag{9.2.1}$$

with $P(t)$ the solution of the Riccati equation

$$\begin{aligned} \frac{dP}{dt} &= FP + PF^* - PH^*R^{-1}HP + Q, \\ P(0) &= P_0. \end{aligned} \tag{9.2.2}$$

PROOF. We have

$$\hat{x}(t) = x_1(t) + E[\alpha(t)|\mathcal{B}^t]$$
$$= x_1(t) + \hat{\alpha}(t),$$

and we know that $\hat{\alpha}(t)$ satisfies

$$d\hat{\alpha} = (F\hat{\alpha}(t) + f)dt + PH^*R^{-1}(d\beta - (H\hat{\alpha} + h)dt), \tag{9.2.3}$$
$$\hat{\alpha}(0) = \bar{\xi},$$

where $P(t)$ is the solution of (9.2.2). Then clearly

$$d\hat{x} = (F\hat{x}(t) + f + Gv)dt + PH^*R^{-1}(d\beta - (H\hat{\alpha} + h)dt), \tag{9.2.4}$$
$$\hat{x}(0) = \bar{\xi}.$$

But

$$d\beta - (H\hat{\alpha} + h)dt = dz - Hx_1dt - (H\hat{\alpha} + h)dt$$
$$= dz - (H\hat{x} + h)dt,$$

whence the result. ■

We now construct a class of admissible controls, defined by a linear feedback. We will prove that the optimal control for the class (9.1.11) is in this subclass. □

If we introduce the innovation process

$$\nu(t) = \beta(t) - \int_0^t (H\hat{\alpha} + h)ds,$$

then we know from Chapter 7, Section 7.1.2, that $\nu(t)$ is a $P, \mathcal{B}^t$ Wiener process with values in R^m. We now consider a feedback rule $\Lambda(t)x + \lambda(t)$, where $\Lambda(t) \in \mathcal{L}(R^n; R^m)$ and $\lambda(t) \in R^m$, bounded in time. We solve the stochastic differential equation

$$d\eta = ((F + G\Lambda)\eta + f + G\lambda)dt + PH^*R^{-1}d\nu, \tag{9.2.5}$$
$$\eta(0) = \bar{\xi}.$$

We next define the control associated with the feedback by the formula

$$v(t) = \Lambda(t)\eta(t) + \lambda(t),$$

which is clearly adapted to the filtration $\mathcal{B}^t$. To this control we associate the state $x(t)$ and the observation $z(t)$ defined by formulas (9.1.8). It is easy to check that

$$\eta(t) = \hat{\alpha}(t) + x_1(t).$$

Next, the observation $z(t)$ satisfies

$$dz = d\beta + Hx_1 dt,$$

and therefore,

$$\begin{aligned} d\nu &= d\beta - (H\hat{\alpha} + h)dt \\ &= dz - (H\eta + h)dt, \end{aligned}$$

so equation (9.2.5) can be written

$$\begin{aligned} d\eta &= ((F + G\Lambda)\eta + f + G\lambda)dt + PH^* R^{-1}(dz - (H\eta + h)dt), \quad (9.2.6) \\ \eta(0) &= \bar{\xi}, \end{aligned}$$

and $z(t)$ appears to be the only source of noise in equation (9.2.6). It is easy to show that $\eta(t)$ is adapted to $\mathcal{Z}_v^t = \sigma(z(s), s \le t)$. Therefore, $v(t)$ is also adapted to $\mathcal{Z}_v^t$. Hence $v(t)$ is admissible in the sense of (9.1.11). The Kalman filter $\hat{x}(t)$ is equal to $\eta(t)$, and hence the Kalman filter is the solution of

$$\begin{aligned} d\hat{x} &= ((F + G\Lambda)\hat{x} + f + G\lambda)dt + PH^* R^{-1}(dz - (H\hat{x} + h)dt), \quad (9.2.7) \\ \hat{x}(0) &= \bar{\xi}, \end{aligned}$$

and the control associated with the feedback is

$$v(t) = \Lambda(t)\hat{x}(t) + \lambda(t). \quad (9.2.8)$$

9.3 The Separation Principle

We can naturally think of a specific linear feedback, the one obtained in solving the problem of optimal control with full information; see (8.1.3). We recall

$$\begin{aligned} \Lambda(t) &= -N^{-1}(t)G^*(t)\pi(t), \quad (9.3.1) \\ \lambda(t) &= -N^{-1}(t)(G^*(t)r(t) + n(t)). \end{aligned}$$

We want to prove that it is a good candidate. We have the following theorem.

THEOREM 9.1. *For the problem (9.1.1), (9.1.2), (9.1.3), the control obtained from the linear feedback (9.3.1) is optimal in the class of admissible controls defined in (9.1.11).*

This result is very appealing to practitioners. Indeed, one knows the optimal feedback to be applied on the state, when the state is known. If the state is not known, one cannot apply the feedback. The result says that it is optimal to estimate the state and then apply the rule to the estimate, instead of the state itself. So one proceeds in two steps, an estimation phase first, and then one operates as if the state were known. This remarkable property is called the **separation principle**. One separates the operations of estimation and control. In general, for non-linear–quadratic cases, this is impossible. Estimation and control are fully coupled.

PROOF. We shall use again the technique of change of controls. Consider $\Lambda(t)$, $\lambda(t)$ to be defined. We associate to this pair an admissible control, as defined above. Since it will be optimal for a convenient choice of $\Lambda(t)$, $\lambda(t)$, we call it $u(t)$. The corresponding state is called $y(t)$, the corresponding observation is called $\zeta(t)$, and the Kalman filter is called $\hat{y}(t)$. So we have the sequence of relations

$$dy = (Fy + f + Gu)dt + dw, \quad y(0) = \xi, \tag{9.3.2}$$

$$d\zeta = (Hy + h)dt + d\epsilon, \quad \zeta(0) = 0, \tag{9.3.3}$$

$$d\hat{y} = (F\hat{y} + f + Gu)dt + PH^*R^{-1}(d\zeta - (H\hat{y} + h)dt), \quad \hat{y}(0) = \bar{\xi}, \tag{9.3.4}$$

$$u(t) = \Lambda(t)\hat{y}(t) + \lambda(t). \tag{9.3.5}$$

Now consider any admissible control $v(t)$, the corresponding state $x(t)$, the corresponding observation $z(t)$, and the Kalman filter $\hat{x}(t)$. We define $\mu(t)$ by

$$v(t) = \Lambda(t)\hat{x}(t) + \lambda(t) + \mu(t). \tag{9.3.6}$$

The process $\mu(t)$ is also adapted to $\mathcal{Z}_v^t$ and $\mathcal{B}^t$. We have the relations

$$dx = (Fx + f + Gv)dt + dw, \quad y(0) = \xi, \tag{9.3.7}$$

$$dz = (Hx + h)dt + d\epsilon, \quad z(0) = 0, \tag{9.3.8}$$

$$d\hat{x} = (F\hat{x} + f + Gv)dt + PH^*R^{-1}(dz - (H\hat{x} + h)dt), \quad \hat{x}(0) = \bar{\xi}. \tag{9.3.9}$$

We note that

$$\begin{aligned} dz - (H\hat{x} + h)dt &= d\zeta - (H\hat{y} + h)dt \\ &= d\beta - (H\hat{\alpha} + h)dt. \end{aligned} \tag{9.3.10}$$

We define

$$\tilde{x}(t) = \hat{x}(t) - \hat{y}(t),$$

and because of (9.3.10), we see immediately that

$$\frac{d\tilde{x}}{dt} = (F + G\Lambda)\tilde{x} + G\mu, \tag{9.3.11}$$
$$\tilde{x}(0) = 0.$$

But then

$$\tilde{x} = x_1 - y_1 = x - y,$$

and therefore,

$$\begin{aligned} J(v(.)) = J(u(.)) + \\ &+ E\int_0^T [\tilde{x}^* M\,\tilde{x} + (\Lambda\tilde{x} + \mu)^* N(\Lambda\tilde{x} + \mu)]dt + E\tilde{x}(T)^* M_T\tilde{x}(T) \\ &+ 2E\int_0^T [\tilde{x}^*(M\,y + m) + (\Lambda\tilde{x} + \mu)^*(Nu + n)]dt \\ &+ 2E\,\tilde{x}(T)^*(M_T y(T) + m_T). \end{aligned}$$

We set

$$X = E\int_0^T [\tilde{x}^*(M\,y + m) + (\Lambda\tilde{x} + \mu)^*(Nu + n)]dt + E\tilde{x}(T)^*(M_T y(T) + m_T)$$

and note that

$$\begin{aligned} E\tilde{x}^*(t)(M(t)\,y(t) + m(t)) &= E\tilde{x}^*(t)(M(t)\,E(y(t)|\mathcal{B}^t) + m(t)) \\ &= E\tilde{x}^*(t)(M(t)\,\hat{y}(t) + m(t)). \end{aligned}$$

In view of the equations of $\hat{x}$ and $\hat{y}$, the expression of X becomes identical to that of the basic problem of Chapter 8, Section 8.1. Therefore, we can choose $\Lambda$and λ as in in the proof of Theorem 8.1. We obtain (9.3.1), which completes the proof. □

REMARK 9.1. We have discussed the key difficulty of the two σ-algebras $\mathcal{B}^t$ and $\mathcal{Z}_v^t$. It is possible to overcome it by defining the observation in a weak sense and relying on Girsanov's theorem. We refer to [5] for details. In fact, we shall use it to consider the situation of an exponential-of-integral payoff and partial information. This will be developed in the next section.

9.4 Exponential-of-Integral Payoff and Partial Information

9.4.1 Setting of the Problem

We want to consider the system

$$dx = (Fx + Gv + f)dt + dw, \tag{9.4.1}$$
$$x(0) = \xi.$$

The observation process is defined by

$$dz = (Hx + h)dt + d\epsilon, \tag{9.4.2}$$
$$z(0) = 0.$$

However, we now want to minimize

$$J(v(.)) = E\{\theta \exp \theta[\int_0^T (x^*Mx + v^*Nv + 2m^*x + 2n^*v)dt + $$
$$+ x(T)^*M_T x(T) + 2m_T^* x(T)]\}. \tag{9.4.3}$$

We must first define the problem correctly. The method of change of controls developed for the case $\theta = 0$, see preceding sections, does not carry over in this case. This is why we are going to use Girsanov's approach.

We consider a probability space $\Omega, \mathcal{A}, P$ and a filtration $\mathcal{F}^t$. We have next

$$\xi = \text{normal random variable}, \tag{9.4.4}$$
$$E\xi = \bar{\xi}, \ \text{cov}\xi = P_0,$$
$$\xi \text{ is } \mathcal{F}^0\text{-measurable}$$

$$w(.),\ z(.) \text{ independent } P, \mathcal{F}^t \text{ Wiener processes} \tag{9.4.5}$$
$$\text{with covariance } Q(.),\ R(.) \text{ respectively.}$$

We assume $R(.)$ invertible, with bounded inverse.

To simplify, we shall assume that

$$\mathcal{F}^t = \sigma(\xi, w(s), z(s), s \leq t) \tag{9.4.6}$$

and set

$$\mathcal{Z}^t = \sigma(z(s), s \leq t). \tag{9.4.7}$$

The notation $z(.)$ indicates that it is the observation process. However, the important aspect is that it is defined a priori, not after the control is introduced. In fact, an admissible control $v(.)$ is any process in $L^2_{\mathcal{Z}}(0, T; R^m)$. The horizon T is given, since the problem is to optimize a payoff on $(0, T)$.

Now, for an admissible control $v(.)$, we define the state $x(.)$ as the solution of (9.4.1). The process $x(.)$ belongs to $L^2_{\mathcal{F}}(0, T; R^n)$. It depends on $v(.)$. In fact, we may write

$$x(t) = \alpha(t) + x_1(t), \tag{9.4.8}$$

where $\alpha(.)$, $x_1(.)$ are defined by (9.1.4), (9.1.6).

To recover (9.4.2) in a weak sense, we rely on Girsanov's theorem. We introduce the process η^v depending on the control $v(.)$ by the formula

$$\eta^v(t) = \exp[\int_0^t (x^* H^* + h^*) R^{-1} dz - \frac{1}{2}\int_0^t (x^* H^* + h^*) R^{-1}(Hx + h) ds], \tag{9.4.9}$$

which is also the solution of

$$\begin{aligned} d\eta^v &= \eta^v (x^* H^* + h^*) R^{-1} dz, \\ \eta^v(0) &= 1. \end{aligned} \tag{9.4.10}$$

We know only that $E\eta^v(t) \leq 1$. If we had $E\eta^v(t) = 1$, then we could define a new probability on $\Omega, \mathcal{F}^T$ (we do not need to consider times larger than T), called P^v, by the Radon–Nikodym derivative

$$\frac{dP^v}{dP} = \eta^v(T). \tag{9.4.11}$$

Defining then $\epsilon^v(t)$ by the formula

$$\epsilon^v(t) = z(t) - \int_0^t (Hx(s) + h(s)) ds, \tag{9.4.12}$$

by Girsanov's theorem, the process $\epsilon^v(.)$ becomes a $P^v, \mathcal{F}^t$ Wiener process, with covariance matrix $R(.)$. Therefore, when we equip $\Omega, \mathcal{F}^T$ with P^v, the observation process $z(.)$ looks as if it has the Itô differential

$$dz = (Hx + h)dt + d\epsilon^v$$

exactly as in (9.4.2). Therefore, we can define the payoff as

$$\begin{aligned} J(v(.)) = E^v\{\theta \exp\theta[\int_0^T (x^* M x + v^* N v + 2m^* x + 2n^* v) dt + \\ + x(T)^* M_T x(T) + 2m_T^* x(T)]\}, \end{aligned}$$

and by the definition of P^v, we have also

$$\begin{aligned} J(v(.)) = E\{\theta\eta^v(T) \exp\theta[\int_0^T (x^* M x + v^* N v + 2m^* x + 2n^* v) dt + \\ + x(T)^* M_T x(T) + 2m_T^* x(T)]\}. \end{aligned} \tag{9.4.13}$$

If we now look at this formula, it is defined even when $E\eta^v(T) \neq 1$. Since it is the expectation of a positive random variable, it always makes sense, provided we accept the value $+\infty$.

It is now convenient to define our problem as minimizing $J(v(.))$ defined by (9.4.13) with $v(.) \in L^2_{\mathcal{Z}}(0, T; R^m)$ and $x(.)$ defined by (9.4.1).

9.4.2 Statement of the Solution

Define

$$X^v = \theta\eta^v(T)\exp\theta[\int_0^T (x^*Mx + v^*Nv + 2m^*x + 2n^*v)dt + \\ + x(T)^*M_Tx(T) + 2m_T^*x(T)]. \quad (9.4.14)$$

Then the payoff functional is

$$J(v(.)) = EX^v.$$

We are going to show that it is possible to reduce this problem to a situation with full information. We shall introduce a system whose evolution depends only on the random process $z(t)$, so it is fully observable, with a corresponding functional, and we shall obtain a solution as we did in Chapter 8. The state of the system is of course not $x(.)$, which is not observable. Nor is it the Kalman filter, as in Section 9.3. The separation principle will not apply in this case. The evolution of the system will incorporate elements of the payoff and will depend on θ. Nevertheless, it is remarkable that it can be obtained. It will represent a **sufficient statistic**. By that we mean a state with the same dimension as the original state that is fully observable and that can perfectly replace the original state as far as the optimal stochastic control problem is concerned.

In order to state the result, we introduce the Riccati equation

$$\dot{P} - FP - PF^* + P(H^*R^{-1}H - 2\theta M)P - Q = 0, \quad (9.4.15)$$
$$P(0) = P_0.$$

We shall assume that this equation has a well-defined symmetric positive definite solution on the interval $(0, T)$. We then introduce the process

$$dr = [Fr + Gv + f + 2\theta P(Mr + m)]dt \quad (9.4.16)$$
$$+ PH^*R^{-1}(dz - (Hr + h)dt),$$
$$r(0) = \bar{\xi}.$$

Equations (9.4.15) and (9.4.16) depend on θ. For $\theta = 0$, we recover the Kalman filter; see Proposition 9.1. The important result is the following.

PROPOSITION 9.2. *We assume that the Riccati equation (9.4.15) has a positive symmetric invertible solution on* $(0, T)$. *Assume also that*

$$I - 2\theta M_T P(T) \text{ is invertible.} \tag{9.4.17}$$

Then

$$\begin{aligned} E[X^v|\mathcal{Z}^T] = C_\theta \theta \exp\Big\{&\theta r(T)^*(I - 2\theta M_T P(T))^{-1}(M_T r(T) + 2m_T) \\ &+ \int_0^T (r^* H^* + h^*)R^{-1} dz - \frac{1}{2}\int_0^T (r^* H^* + h^*)R^{-1}(Hr + h)dt \\ &+ \theta \int_0^T (r^* M r + v^* N v + 2m^* r + 2n^* v)dt\Big\}. \end{aligned} \tag{9.4.18}$$

The constant C_θ *is given by the formula*

$$C_\theta = \frac{\exp(2\theta^2 m_T^* P(T)(I - 2\theta M_T P(T))^{-1} m_T + \theta \int_0^T \operatorname{tr} PM\, dt)}{|I - 2\theta M_T P(T)|^{\frac{1}{2}}}. \tag{9.4.19}$$

We postpone the proof of this result to the next section. Thanks to this result, we have reduced the partial information case to the full information case. The system is $r(.)$, whose evolution is described by (9.4.16). It is observable, and the payoff is, omitting the positive constant C_θ,

$$\begin{aligned} J(v(.)) = \theta E \exp\Big\{&\theta r(T)^*(I - 2\theta M_T P(T))^{-1}(M_T r(T) + 2m_T) \\ &+ \int_0^T (r^* H^* + h^*)R^{-1} dz - \frac{1}{2}\int_0^T (r^* H^* + h^*)R^{-1}(Hr + h)dt \\ &+ \theta \int_0^T (r^* M r + v^* N v + 2m^* r + 2n^* v)dt\Big\}. \end{aligned} \tag{9.4.20}$$

To describe the solution, we introduce the deterministic functions $S(t), s(t), \rho(t)$ defined by

$$\begin{aligned} &\dot{S} + S(F + 2\theta PM) + (F^* + 2\theta MP)S \\ &-S(GN^{-1}G^* - 2\theta PH^*R^{-1}HP)S + M = 0, \end{aligned} \tag{9.4.21}$$

$$S(T) = \frac{1}{2}[(I - 2\theta M_T P(T))^{-1} M_T + M_T(I - 2\theta P(T) M_T)^{-1}],$$

$$\dot{s} + (F^* + 2\theta SPH^*R^{-1}HP + 2\theta MP - SGN^{-1}G^*)s \qquad (9.4.22)$$
$$+m + S(f + 2\theta Pm - GN^{-1}n) = 0,$$

$$s(T) = (I - 2\theta M_T P(T))^{-1} m_T,$$

$$\dot{\rho} + 2\theta s^* PH^*R^{-1}HPs - (s^*G + n^*)N^{-1}(G^*s + n) \qquad (9.4.23)$$
$$+(f^* + 2\theta m^* P)s + \text{tr}\, PSPH^*R^{-1}H = 0,$$

$$\rho(T) = 0.$$

We shall assume that the Riccati equation (9.4.21) has a global solution on $(0, T)$. As in Chapter 8, (8.3.11), we define the set of admissible controls as $v(.) \in L^2_{\mathcal{Z}}(0, T; R^m)$ such that

$$E \exp\left(\int_0^T (Hr + h + 2\theta P(Sr + s))^* R^{-1} dz \right. \qquad (9.4.24)$$
$$\left. -\frac{1}{2}\int_0^T (Hr + h + 2\theta P(Sr + s))^* R^{-1}(Hr + h + 2\theta P(Sr + s))dt\right) = 1.$$

We then state the following theorem.

THEOREM 9.2. *We make the assumptions of Proposition 9.2, and assume that the Riccati equation (9.4.21) has a global solution on* $(0, T)$. *Define the feedback on* r:

$$\hat{v}(r, t) = -N_t^{-1}G_t^*(S_t r + s_t) - N_t^{-1}n_t. \qquad (9.4.25)$$

Then used in the state equation (9.4.16), it defines an optimal control within the class of admissible controls.

PROOF. The proof is similar to that of Theorem 8.3. For a control $v(.)$, we define the process

$$\varphi_t = \int_0^t (r^*H^* + h^*)R^{-1}dz - \frac{1}{2}\int_0^t (r^*H^* + h^*)R^{-1}(Hr + h)d\tau$$
$$+\theta \int_0^t (r^*Mr + v^*Nv + 2m^*r + 2n^*v)d\tau$$
$$+\theta r_t^* S_t r_t + 2\theta s_t^* r_t + \rho_t,$$

with the choice of $S(T), s(T), \rho(T)$ given above. Therefore,

$$J(v(.)) = \theta E \exp \varphi_T.$$

Now we compute the Itô differential of $\exp\varphi_t$. With the choice of S_t, s_t, ρ_t, we get the formula

$$\begin{aligned}d\exp\varphi_t &= \exp\varphi_t\,(Hr+h+2\theta P(Sr+s))^* R^{-1}dz\\ &\quad+\exp\varphi_t[v_t+N_t^{-1}G_t^*(S_tr+s_t)+N_t^{-1}n_t]^* N_t[v_t+N_t^{-1}G_t^*(S_tr+s_t)\\ &\quad+N_t^{-1}n_t]dt,\end{aligned}$$

and thus for every admissible control, we have

$$J(v(.)) \geq \theta\exp\varphi_0.$$

We omit proving that the control defined by the feedback (9.4.25) is admissible. It achieves the lower bound, and thus clearly it is optimal. □

9.4.3 Proof of Proposition 9.2

The result (9.4.18) is interesting in itself. It is due to Bensoussan and van Schuppen [9]. The proof we are going to give needs some technical assumptions, which do not appear in the final result. It would be interesting to find a direct proof that does not rely on these technical steps. We assume first the differentiability properties

$$R^{-1}(t)H(t),\;R^{-1}(t)h(t) \text{ are differentiable.} \tag{9.4.26}$$

With this assumption we can express the stochastic integral

$$\begin{aligned}\int_0^T (x^*H^*+h^*)R^{-1}dz &= (x(T)^*H(T)^*+h(T)^*)R^{-1}(T)z(T)\\ &\quad-\int_0^T (x^*F^*+f^*+v^*G^*)H^*R^{-1}zdt-\int_0^T z^*R^{-1}Hdw\\ &\quad-\int_0^T x^*(H^*R^{-1})'zdt-\int_0^T (h^*R^{-1})'zdt.\end{aligned}$$

We introduce a quadratic form

$$x^*\Sigma(t)x-2x^*\sigma(t)+\beta(t),$$

where the function $\Sigma(t)$ is deterministic and the functions $\sigma(t), \beta(t)$ will depend only on $z(.)$ but are not adapted. We assume

$$\begin{aligned}\Sigma(T) &= -2\theta M_T,\\ \sigma(T) &= 2\theta m_T+H(T)^*R^{-1}(T)z(T),\\ \beta(T) &= -2h(T)^*R^{-1}(T)z(T).\end{aligned} \tag{9.4.27}$$

We then have

$$
\begin{aligned}
X^v = \theta \exp\Bigg\{ & -\int_0^T (x^* F^* + f^* + v^* G^*) H^* R^{-1} z dt - \int_0^T z^* R^{-1} H dw \\
& - \int_0^T x^* (H^* R^{-1})' z dt - \int_0^T (h^* R^{-1})' z dt - \frac{1}{2} \int_0^T (x^* H^* + h^*) R^{-1} (Hx + h) dt \\
& + \theta \int_0^T (x^* M x + v^* N v + 2m^* x + 2n^* v) dt \\
& - \frac{1}{2} (x(T)^* \Sigma(T) x(T) - 2x(T)^* \sigma(T) + \beta(T)) \Bigg\}.
\end{aligned}
$$

We compute the Itô differential of

$$
x(t)^* \Sigma(t) x(t) - 2x(t)^* \sigma(t) + \beta(t).
$$

However, although $\sigma(t), \beta(t)$ are random, they will be differentiable, so Itô's rule applies only to the first term. After tedious but easy calculations we obtain

$$
\begin{aligned}
X^v = \theta \exp\Bigg\{ & -\int_0^T (x^* \Sigma + z^* R^{-1} H - \sigma^*) dw \\
& - \frac{1}{2} \int_0^T (x^* \Sigma + z^* R^{-1} H - \sigma^*) Q (\Sigma x + H^* R^{-1} z - \sigma) dt \\
& + \frac{1}{2} \int_0^T (x^* \Sigma + z^* R^{-1} H - \sigma^*) Q (\Sigma x + H^* R^{-1} z - \sigma) dt \\
& - \int_0^T (x^* F^* + f^* + v^* G^*) H^* R^{-1} z dt - \int_0^T x^* (H^* R^{-1})' z dt \\
& - \int_0^T (h^* R^{-1})' z dt - \frac{1}{2} \int_0^T (x^* H^* + h^*) R^{-1} (Hx + h) dt \\
& + \theta \int_0^T (x^* M x + v^* N v + 2m^* x + 2n^* v) dt \\
& - \int_0^T x^* \Sigma (Fx + f + Gv) dt - \frac{1}{2} \int_0^T x^* \dot{\Sigma} x dt - \frac{1}{2} \int_0^T \mathrm{tr}\, \Sigma Q \, dt \\
& + \int_0^T \sigma^* (Fx + f + Gv) dt + \int_0^T x^* \dot{\sigma} dt - \frac{1}{2} \int_0^T \dot{\beta} dt \\
& - \frac{1}{2} (\xi^* \Sigma(0) \xi - 2\xi^* \sigma(0) + \beta(0)) \Bigg\}.
\end{aligned}
$$

We next choose $\Sigma(.), \sigma(.), \beta(.)$ as follows:

$$\dot{\Sigma} + \Sigma F + F^*\Sigma - \Sigma Q \Sigma + (H^* R^{-1} H - 2\theta M) = 0, \qquad (9.4.28)$$
$$\Sigma(T) = -2\theta M_T,$$

$$\dot{\sigma} + (F^* - \Sigma Q)\sigma - ((F^* - \Sigma Q)H^* R^{-1} + (H^* R^{-1})')z$$
$$-H^* R^{-1} h + 2\theta m - \Sigma(f + Gv) = 0, \quad (9.4.29)$$
$$\sigma(T) = 2\theta m_T + H(T)^* R^{-1}(T) z(T),$$

$$\dot{\beta} = (z^* R^{-1} H - \sigma^*) Q (H^* R^{-1} z - \sigma) - 2z^* R^{-1} H(f + Gv) \quad (9.4.30)$$
$$- 2z^*(R^{-1}h)' - h^* R^{-1} h + 2\theta(v^* N v + 2n^* v)$$
$$- \mathrm{tr}\Sigma Q + 2\sigma^*(f + Gv),$$
$$\beta(T) = -2h(T)^* R^{-1}(T) z(T).$$

We need to assume that the Riccati equation (9.4.28) has a global solution.

With these choices, the expression of X^v simplifies considerably. We have indeed

$$X^v = \theta \exp\left\{ - \int_0^T (x^*\Sigma + z^* R^{-1} H - \sigma^*) dw \right.$$
$$- \frac{1}{2} \int_0^T (x^*\Sigma + z^* R^{-1} H - \sigma^*) Q (\Sigma x + H^* R^{-1} z - \sigma) dt$$
$$\left. - \frac{1}{2}(\xi^* \Sigma(0) \xi - 2\xi^* \sigma(0) + \beta(0)) \right\}.$$

We take the conditional expectation with respect to the σ-algebra generated by $\mathcal{Z}^t$ and ξ. Thanks to the mutual independence of ξ, $w(.)$, $z(.)$, we can consider the terms not equal to x as frozen. Therefore,

$$E[\exp\left\{ - \int_0^T (x^*\Sigma + z^* R^{-1} H - \sigma^*) dw \right.$$
$$\left. -\frac{1}{2} \int_0^T (x^*\Sigma + z^* R^{-1} H - \sigma^*) Q (\Sigma x + H^* R^{-1} z - \sigma) dt \right\} | \mathcal{Z}^T, \xi] = 1,$$

and hence

$$E[X^v | \mathcal{Z}^T, \xi] = \theta \exp[-\frac{1}{2}(\xi^* \Sigma(0) \xi - 2\xi^* \sigma(0) + \beta(0))],$$

and also

$$E[X^v | \mathcal{Z}^T] = \theta \int p(x, 0) \pi(x) dx, \qquad (9.4.31)$$

where we have set

$$p(x,t) = \exp[-\frac{1}{2}(x^*\Sigma(t)x - 2x^*\sigma(t) + \beta(t))], \tag{9.4.32}$$

and $\pi(x)$ is the probability density of a Gaussian random variable of mean $\bar{\xi}$ and variance P_0. Namely,

$$\begin{aligned}\pi(x) &= \frac{1}{((2\pi)^n|P_0|)^{\frac{1}{2}}}\exp[-\frac{1}{2}(x-\bar{\xi})^* P_0^{-1}(x-\bar{\xi})] \\ &= \exp -\frac{1}{2}[x^* P_0^{-1} x - 2x^* P_0^{-1}\bar{\xi} + \bar{\xi}^* P_0^{-1}\bar{\xi} + \log((2\pi)^n|P_0|).\end{aligned} \tag{9.4.33}$$

We then give another expression for $E[X^v|\mathcal{Z}^T]$ as follows:

$$E[X^v|\mathcal{Z}^T] = \theta \int p(x,T)q(x,T)dx, \tag{9.4.34}$$

where $q(x,t)$ is to be defined with

$$q(x,0) = \pi(x). \tag{9.4.35}$$

In fact, we are going to define $q(x,t)$ by requiring

$$\int p(x,t)q(x,t)dx = \int p(x,0)\pi(x)dx, \forall t,$$

and hence

$$\frac{d}{dt}\int p(x,t)q(x,t)dx = 0.$$

We shall use the fact that $p(x,t)$ is the solution of a partial differential equation (PDE), namely

$$\begin{aligned}-\frac{\partial p}{\partial t} &= Dp^*(Fx + Gv + f - QH^*R^{-1}z) + \frac{1}{2}\text{tr}\ D^2p\ Q \\ &+ \frac{p}{2}\{-x^*(H^*R^{-1}H - 2\theta M)x - 2x^*H^*R^{-1}h - h^*R^{-1}h \\ &- 2z^*[R^{-1}H(Fx + Gv + f) + (R^{-1}H)'x + (R^{-1}h)'] \\ &+ 4\theta m^*x + 2\theta(v^*Nv + 2n^*v) + z^*R^{-1}HQH^*R^{-1}z\}.\end{aligned} \tag{9.4.36}$$

It is clear that we must choose $q(x,t)$ as the solution of the dual PDE

$$\begin{aligned}\frac{\partial q}{\partial t} &= -Dq^*(Fx+Gv+f-QH^*R^{-1}z)+\frac{1}{2}\text{tr } D^2q\, Q\\ &+\frac{q}{2}\{-2\text{tr } F-x^*(H^*R^{-1}H-2\theta M)x-2x^*H^*R^{-1}h-h^*R^{-1}h\\ &-2z^*[R^{-1}H(Fx+Gv+f)+(R^{-1}H)'x+(R^{-1}h)'] \qquad (9.4.37)\\ &+4\theta m^*x+2\theta(v^*Nv+2n^*v)+z^*R^{-1}HQH^*R^{-1}z\}.\end{aligned}$$

The solution of (9.4.37) is also the exponential of a quadratic form, which we write, in view of the form of $q(x,0)$, see (9.4.35), as

$$q(x,t)=\exp -\frac{1}{2}[x^*P_t^{-1}x-2x^*(P_t^{-1}r_t-H^*R^{-1}z)+\gamma_t], \qquad (9.4.38)$$

and we try to identify the functions $P(t), r(t), \gamma(t)$, in order to satisfy the PDE (9.4.37).

After lengthy calculations, we can see that $P(t)$ and $r(t)$ satisfy (9.4.15) and (9.4.16). Moreover, we have

$$\begin{aligned}\frac{d\gamma}{dt} &= \text{tr } (P^{-1}Q+2F)-r^*P^{-1}QP^{-1}r+2r^*P^{-1}(Gv+f)\\ &+2z^*(R^{-1}h)'+h^*R^{-1}h-2\theta(v^*Nv+2n^*v),\end{aligned}$$

with

$$\gamma(0)=\bar{\xi}^*P_0^{-1}\bar{\xi}+\log((2\pi)^n|P_0|).$$

We can check that

$$\begin{aligned}\gamma_t &= r_t^*P_t^{-1}r_t+\log((2\pi)^n|P_t|)+2h^*R^{-1}z \qquad (9.4.39)\\ &-2\int_0^t(r^*H^*+h^*)R^{-1}dz+\int_0^t(r^*H^*+h^*)R^{-1}(Hr+h)d\tau\\ &-2\theta\int_0^t(r^*Mr+v^*Nv+2m^*r+2n^*v)d\tau-2\theta\int_0^t\text{tr}PM\,d\tau.\end{aligned}$$

This can be seen by taking the Itô differential of the previous expression, while noting the following relation:

$$\begin{aligned}\frac{d}{dt}\log|P_t| &= \text{tr } \dot{P}P^{-1} \qquad (9.4.40)\\ &= \text{tr}(2F-H^*R^{-1}H+2\theta PM+QP^{-1}).\end{aligned}$$

We can now compute the right-hand side of (9.4.34). We see that

$$E[X^v|\mathcal{Z}^T] = \theta \int \exp -\frac{1}{2}[x^*(P_T^{-1} - 2\theta M_T)x - 2x^*(P_T^{-1}r_T + 2\theta m_T) \\ + \gamma_T - 2h^*(T)R^{-1}(T)z(T)]dx.$$

We can compute this integral. After some easy but tedious calculations, we finally obtain the formula (9.4.18), which completes the proof of the proposition. ■

Chapter 10
Deterministic Optimal Control

10.1 Pontryagin's Maximum Principle

10.1.1 Setting of the Problem

We turn now to nonlinear systems with nonquadratic payoffs. We cannot hope for closed-form (explicit) solutions. However, since optimal control is an optimization problem, it is natural to look for necessary conditions of optimality for an optimal control. We explore this question in this section. Since it is a dynamic optimization, the role of time in expressing the necessary condition is important. Pontryagin's maximum principle, see [45], expresses the necessary condition as a sequence of static optimization problems taking place at each time, which taken together constitute a necessary condition of optimality for the global problem. Note that this is a major improvement with respect to a calculus of variations approach, consisting in writing a global necessary condition. We shall explore this path in Section 10.1.3.

We begin by stating the problem. The state $x(t)$ is in R^n, and the control $v(t)$ is in R^m. The evolution is described by the differential equation

$$\frac{dx}{dt} = g(x(t), v(t)), \tag{10.1.1}$$
$$x(0) = x_0, ,$$

and we assume that

$$g(x, v) : \ R^n \times R^m \rightarrow R^n, \text{ is continuously differentiable} \tag{10.1.2}$$
$$g_x, g_v \text{ are bounded,}$$
$$|g(x, v)| \ \leq \ \bar{g}(1 + |x| + |v|).$$

A. Bensoussan, *Estimation and Control of Dynamical Systems*, Interdisciplinary Applied Mathematics 48, https://doi.org/10.1007/978-3-319-75456-7_10

We define now what an admissible control is. Consider a subset of R^m,

$$\mathcal{U}_{\text{ad}} = \text{nonempty closed subset of } R^m. \tag{10.1.3}$$

An admissible control is a function $v(.) \in L^2(0, T; R^m)$ such that

$$v(t) \in \mathcal{U}_{\text{ad}}\ , \text{a.e. } t \tag{10.1.4}$$

With these assumptions, the solution of (10.1.1) is well defined by standard results on differential equations, and we have

$$x(.) \in H^1(0, T; R^n) = \{x(.) \in L^2(0, T; R^n) | \frac{dx}{dt} \in L^2(0, T; R^n)\}. \tag{10.1.5}$$

We now define the payoff function. It will be written as

$$J(v(.)) = \int_0^T l(x(t), v(t))dt + h(x(T)), \tag{10.1.6}$$

and we make the following assumptions:

$$\begin{aligned} l(x, v) : R^n \times R^m &\to R, \text{ is continuously differentiable,} \\ h(x) : R^n &\to R, \text{ is continuously differentiable,} \end{aligned} \tag{10.1.7}$$

$$\begin{aligned} &|l(x, v)| \leq \bar{l}(1 + |x|^2 + |v|^2),\ |h(x)| \leq \bar{h}(1 + |x|^2), \\ &|l_x|, |l_v| \leq \bar{l}(1 + |x| + |v|),\ |h_x| \leq \bar{h}(1 + |x|). \end{aligned} \tag{10.1.8}$$

REMARK 10.1. All the functions can depend on time. We avoid indicating this to save notation. The quantities $\bar{g}, \bar{l}, \bar{h}$ are constants. Note that the linear–quadratic case is a particular case of what we are considering here.

With these assumptions, the functional $J(v(.))$ is well defined. An optimal control, if it exists, is an admissible control that minimizes the functional $J(v(.))$. We shall denote an optimal control by $u(.)$ and the corresponding trajectory by $y(.)$. We are interested in obtaining a necessary condition of optimality for an optimal control.

10.1.2 Necessary Condition of Optimality

We define the **Hamiltonian** by

$$H(x, v, q) = l(x, v) + q.g(x, v), \tag{10.1.9}$$

and the **adjoint state**$p(.)$ is a solution of a backward differential equation

$$-\frac{dp}{dt} = H_x(y(t), u(t), p(t)),$$
$$p(T) = h_x(y(T)). \tag{10.1.10}$$

Note that

$$H_x(y(t), u(t), p(t)) = g_x^*(y(t), u(t))p(t) + l_x(y(t), u(t)), \tag{10.1.11}$$

where $g_x^*(y(t), u(t))$ is the transpose of the matrix $g_x(y(t), u(t)) \in \mathcal{L}(R^n; R^n)$.

In the adjoint equation, the state $y(t)$ and the control $u(t)$ are given, and they represent respectively the optimal state and the optimal control. So they are related by the equation

$$\frac{dy}{dt} = g(y(t), u(t)), \tag{10.1.12}$$
$$y(0) = x_0.$$

Then the necessary condition is provided by the following theorem.

THEOREM 10.1. *We assume (10.1.2), (10.1.3), (10.1.7), (10.1.8). Then the optimal control $u(.)$ satisfies*

$$H(y(t), v, p(t)) \geq H(y(t), u(t), p(t)), \ \forall v \in \mathcal{U}_{ad}, a.e.\ t \in (0, T). \tag{10.1.13}$$

REMARK 10.2. To determine the optimal control, one should solve (10.1.10), (10.1.12), (10.1.13). We see that the global optimization problem min $J(v(.))$ as been changed into a family of local optimization problems, occurring at each time t. Theorem 10.1 expresses Pontryagin's maximum principle (in our formulation, it is a minimum principle, since we minimize the payoff).

PROOF. Let $t_0 \in (0, T)$. We can find $\theta > 0$ such that $t_0 + \theta < T$. We modify the optimal control $u(.)$ on the interval $(t_0, t_0 + \theta)$ as follows:

$$u_\theta(t) = \begin{cases} u(t), & t \in (0, t_0), \\ v, & t \in (t_0, t_0 + \theta), \\ u(t), & t \in (t_0 + \theta, T), \end{cases}$$

where v is an arbitrary element of $\mathcal{U}_{\text{ad}}$. The control $u_\theta(.)$ is admissible, so

$$J(u_\theta(.)) \geq J(u(.)). \tag{10.1.14}$$

We are going to check that

$$\frac{1}{\theta}(J(u_\theta(.)) - J(u(.))) = H(y(t_0), v, p(t_0)) - \frac{1}{\theta}\int_{t_0}^{t_0+\theta} H(y(t), u(t), p(t))dt + X_\theta(t_0), \tag{10.1.15}$$

with the property

$$X_\theta(t_0) \to 0, \text{ as } \theta \to 0, \ \forall t_0.$$

Note that $H(y(t), u(t), p(t))$ is an $L^1(0, T)$ function. So we can claim that

$$\frac{1}{\theta}\int_{t_0}^{t_0+\theta} H(y(t), u(t), p(t))dt \to H(y(t_0), u(t_0), p(t_0)), \text{ a.e. } t_0.$$

The set of points t_0 for which the preceding limit property holds is called the set of Lebesgue points of the function $H(y(t), u(t), p(t))$.

At a Lebesgue point, we can pass to the limit in (10.1.15) to obtain

$$H(y(t_0), v, p(t_0)) - H(y(t_0), u(t_0), p(t_0)) \geq 0, \ \forall v \in \mathcal{U}_{\text{ad}}, \text{a.e.} t_0,$$

and this represents the desired result. So there remains to prove (10.1.15). We first denote by $y_\theta(.)$ the state trajectory corresponding to $u_\theta(.)$.

We introduce the function $z_\theta(t)$, defined for $t \geq t_0 + \theta$, a solution of

$$\frac{dz_\theta}{dt} = g_x(y(t), u(t))z_\theta(t) \; t > t_0 + \theta$$

$$z_\theta(t_0 + \theta) = g(y(t_0), v) - \frac{1}{\theta}(y(t_0 + \theta) - y(t_0)), \tag{10.1.16}$$

and set, also for $t \geq t_0 + \theta$,

$$\tilde{y}_\theta(t) = \frac{1}{\theta}(y_\theta(t) - y(t)) - z_\theta(t). \tag{10.1.17}$$

Then we have

$$\tilde{y}_\theta(t_0 + \theta) = \frac{1}{\theta}\int_{t_0}^{t_0+\theta} (g(y_\theta(t), v) - g(y(t_0), v))dt \tag{10.1.18}$$

and

$$\frac{d\tilde{y}_\theta}{dt} = \frac{1}{\theta}(g(y_\theta, u) - g(y, u)) - g_x(y, u)z_\theta \; t > t_0 + \theta$$

$$= \int_0^1 g_x(y + \lambda(y_\theta - y), u)d\lambda \, \tilde{y}_\theta + \int_0^1 (g_x(y + \lambda(y_\theta - y), u) - g_x(y, u))d\lambda z_\theta. \tag{10.1.19}$$

From this differential equation in $\tilde{y}_\theta$ we first deduce that

$$\sup_{t_0+\theta \leq t \leq T} |\tilde{y}_\theta(t)| \leq C,$$

where C depends only on the constants of the problem and not on θ.

But then from (10.1.17), we get

$$\sup_{t_0+\theta\leq t\leq T} |y_\theta(t) - y(t)| \to 0, \text{ as } \theta \to 0.$$

Using this property in the differential equation (10.1.19) again, we obtain easily the property

$$\sup_{t_0+\theta\leq t\leq T} |\tilde{y}_\theta(t)| \to 0, \text{ as } \theta \to 0. \tag{10.1.20}$$

We then compute

$$\begin{aligned}
J(u_\theta(.)) - J(u(.)) = &\int_{t_0}^{t_0+\theta} (l(y_\theta, v) - l(y,u))dt \\
&+ \int_{t_0+\theta}^{T} (l(y+\theta(z_\theta+\tilde{y}_\theta), u) - l(y,u))dt \\
&+ h(y(T) + \theta(z_\theta+\tilde{y}_\theta)(T)) - h(y(T)),
\end{aligned}$$

then

$$\begin{aligned}
\frac{1}{\theta}(J(u_\theta(.)) - J(u(.))) = &\, l(y(t_0), v) - \frac{1}{\theta}\int_{t_0}^{t_0+\theta} l(y,u)dt \\
&+ \int_{t_0+\theta}^{T} l_x(y,u)z_\theta dt + h_x(y(T))z_\theta(T) + \\
&+ \tilde{X}_\theta(t_0),
\end{aligned}$$

with

$$\begin{aligned}
\tilde{X}_0(t_0) = &\frac{1}{\theta}\int_{t_0}^{t_0+\theta} (l(y_\theta, v) - l(y(t_0), v))dt \\
&+ \frac{1}{\theta}\int_{t_0+\theta}^{T} (l(y+\theta(z_\theta+\tilde{y}_\theta), u) - l(y,u) - \theta l_x(y,u)z_\theta)dt + \\
&+ \frac{1}{\theta}[h(y(T)+\theta(z_\theta+\tilde{y}_\theta)(T)) - h(y(T)) - \theta h_x(y(T))z_\theta(T)],
\end{aligned}$$

or

$$\begin{aligned}
\tilde{X}_\theta(t_0) = &\frac{1}{\theta}\int_{t_0}^{t_0+\theta} (l(y_\theta, v) - l(y(t_0), v))dt \\
&+ \int_0^1\int_{t_0+\theta}^{T} (l_x(y+\lambda\theta(z_\theta+\tilde{y}_\theta), u) - l_x(y,u))z_\theta dt d\lambda + \\
&+ \int_0^1\int_{t_0+\theta}^{T} l_x(y+\lambda\theta(z_\theta+\tilde{y}_\theta), u)\tilde{y}_\theta dt d\lambda +
\end{aligned}$$

$$+\int_0^1 (h_x(y(T)+\lambda\theta(z_\theta+\tilde{y}_\theta)(T)) - h_x(y(T)))z_\theta(T)d\lambda$$

$$+\int_0^1 h_x(y(T)+\lambda\theta(z_\theta+\tilde{y}_\theta)(T))\tilde{y}_\theta(T)d\lambda,$$

and from the smoothness assumptions of functions l, l_x, h_x, we obtain

$$\tilde{X}_\theta(t_0) \to 0, \text{ as } \theta \to 0.$$

Next, from the definition of p,

$$\begin{aligned}\int_{t_0+\theta}^T l_x(y,u)z_\theta dt + h_x(y(T))z_\theta(T) &= p(t_0+\theta)z_\theta(t_0+\theta)\\ &= p(t_0)g(y(t_0),v) + (p(t_0+\theta)-p(t_0))g(y(t_0),v)\\ &\quad -\frac{1}{\theta}\int_{t_0}^{t_0+\theta} pg(y,u)dt - \frac{1}{\theta}\int_{t_0}^{t_0+\theta}(p(t_0+\theta)-p(t))g(y,u)dt.\end{aligned}$$

Defining finally

$$\begin{aligned}X_\theta(t_0) &= \tilde{X}_\theta(t_0) + (p(t_0+\theta)-p(t_0))g(y(t_0),v)\\ &\quad -\frac{1}{\theta}\int_{t_0}^{t_0+\theta}(p(t_0+\theta)-p(t))g(y,u)dt,\end{aligned}$$

we obtain

$$\begin{aligned}\frac{1}{\theta}(J(u_\theta(.)) - J(u(.))) &= H(y(t_0),v,p(t_0))\\ &\quad -\frac{1}{\theta}\int_{t_0}^{t_0+\theta} H(y(t),u(t),p(t))dt + X_\theta(t_0)\end{aligned}$$

and

$$X_\theta(t_0) \to 0, \text{ as } \theta \to 0, \ \forall t_0.$$

Therefore, (10.1.15) has been obtained. This completes the proof. □

EXERCISE 10.1. Apply the maximum principle to the linear quadratic case (3.1.1), (3.1.2). Obtain the optimal control.

10.1.3 Gâteaux Differential

We have seen in the proof of Pontryagin's maximum principle the role of a "spike modification" of the optimal control (modification only on a small interval). Let us envisage now a calculus of variations approach. We note that $J(v(.))$ is a functional on $L^2(0, T; R^m)$. If we want to write a necessary condition of optimality, it is useful to compute its Gâteaux differential. We say that $J(v(.))$ has a Gâteaux differential at $u(.)$, given by $K(.) \in L^2(0, T; R^m)$, if one has the property

$$\frac{d}{d\theta} J(u(.) + \theta v(.))|_{\theta=0} = \int_0^T K(t).v(t)dt, \ \forall v(.) \in L^2(0, T; R^m). \tag{10.1.21}$$

We can relate the Gâteaux differential to the Hamiltonian. We have indeed the following result.

PROPOSITION 10.1. *The functional $J(v(.))$ has a Gâteaux differential at $u(.)$ equal to $H_v(y(t), u(t), p(t))$.*

PROOF. We set $u_\theta(.) = u(.) + \theta v(.)$ and let $y_\theta(.)$ be the trajectory corresponding to $u_\theta(.)$. Define next $z(.)$ by

$$\begin{aligned} \frac{dz}{dt} &= g_x(y, u)z + g_v(y, u)v, \\ z(0) &= 0, \end{aligned} \tag{10.1.22}$$

and set

$$\tilde{y}_\theta(t) = \frac{1}{\theta}(y_\theta(t) - y(t)) - z(t).$$

Then $\tilde{y}_\theta(.)$ is the solution of the differential equation

$$\begin{aligned} \frac{d\tilde{y}_\theta}{dt} &= \frac{1}{\theta}(g(y_\theta, u_\theta) - g(y, u)) - g_x(y, u)z - g_v(y, u)v \\ &= \int_0^1 g_x(y + \lambda(y_\theta - y), u + \lambda\theta v)d\lambda \ \tilde{y}_\theta + \\ &+ \int_0^1 (g_x(y + \lambda(y_\theta - y), u + \lambda\theta v) - g_x(y, u))d\lambda z \\ &+ \int_0^1 (g_v(y + \lambda(y_\theta - y), u + \lambda\theta v) - g_v(y, u))d\lambda v, \end{aligned} \tag{10.1.23}$$

and $\tilde{y}_\theta(0) = 0$. We can check, from the continuity of the gradients g_x, g_v, that

$$\sup_{t\in[0,T]} |\tilde{y}_\theta(t)| \to 0 \text{ as } \theta \to 0.$$

This is done by checking that

$$\tilde{y}_\theta(.) \to 0 \in L^2(0, T; R^n), \ \frac{d}{dt}\tilde{y}_\theta \to 0 \in L^2(0, T; R^n).$$

We can next write

$$\frac{1}{\theta}(J(u_\theta(.)) - J(u(.))) = \int_0^T [l_x(y,u)z + l_v(y,u)v]dt + h_x(y(T))z(T) + X_\theta,$$

where

$$\begin{aligned} X_\theta = &\int_0^1 \int_0^T l_x(y + \lambda\theta(z + \tilde{y}_\theta), u + \lambda\theta v)\tilde{y}_\theta dt d\lambda \\ &+ \int_0^1 \int_0^T [l_x(y + \lambda\theta(z + \tilde{y}_\theta), u + \lambda\theta v) - l_x(y,u)]z dt d\lambda \\ &+ \int_0^1 \int_0^T [l_v(y + \lambda\theta(z + \tilde{y}_\theta), u + \lambda\theta v) - l_v(y,u)]v dt d\lambda \\ &+ \int_0^1 h_x(y(T) + \lambda\theta(z + \tilde{y}_\theta)(T))\tilde{y}_\theta(T) d\lambda \\ &+ \int_0^1 [h_x(y(T) + \lambda\theta(z + \tilde{y}_\theta)(T)) - h_x(y(T))]z(T) d\lambda \end{aligned}$$

and

$$X_\theta \to 0, \text{ as } \theta \to 0,$$

so we get the formula

$$\frac{d}{d\theta}J(u(.) + \theta v(.))|_{\theta=0} = \int_0^T [l_x(y,u)z + l_v(y,u)v]dt + h_x(y(T))z(T).$$

Using now the adjoint equation (10.1.10), (10.1.11), we obtain easily

$$\begin{aligned} \frac{d}{d\theta}J(u(.) + \theta v(.))|_{\theta=0} &= \int_0^T [l_v(y,u)v + pg_v(y,u)v]dt \\ &= \int_0^T H_v(y(t), u(t), p(t))v(t)dt, \end{aligned}$$

which completes the proof. □

Thanks to the formula for the Gâteaux differential, it is easy to derive a necessary condition of optimality, at least in the case

$$\mathcal{U}_{\text{ad}} \text{ is convex.} \tag{10.1.24}$$

Indeed, if $v(.)$ is an admissible control, then $u(.) + \theta(v(.) - u(.))$ is also admissible. Since

$$J(u(.) + \theta(v(.) - u(.))) \geq J(u(.)),$$

we obtain

$$\frac{d}{d\theta}J(u(.)+\theta(v(.)-u(.))|_{\theta=0}\geq 0,$$

and therefore,

$$\int_0^T H_v(y(t),u(t),p(t))(v(t)-u(t))dt\geq 0$$

for every admissible control $v(.)$.

Let now $v\in\mathcal{U}_{\mathrm{ad}}$, and for all t_0 with $t_0+\theta<T$,

$$u_\theta(t)=\begin{cases}u(t), & t\in(0,t_0),\\ v, & t\in(t_0,t_0+\theta),\\ u(t), & t\in(t_0+\theta,T).\end{cases}$$

Taking $v(.)=u_\theta(.)$, we obtain

$$\int_{t_0}^{t_0+\theta} H_v(y(t),u(t),p(t))(v-u(t))dt\geq 0.$$

Dividing by θ and letting $\theta\to 0$, we obtain convergence for almost all t_0. Therefore, we have proven

$$H_v(y(t),u(t),p(t))(v-u(t))\geq 0,\ \text{a.e.}t,\ \forall v\in\mathcal{U}_{\mathrm{ad}}. \tag{10.1.25}$$

This condition is itself a consequence of the maximum principle, which expresses that $u(t)$ achieves the minimum in $\mathcal{U}_{\mathrm{ad}}$ of $H(y(t),v,p(t))$ for a.e. t. So (10.1.25) is a weaker result than the maximum principle, and besides, it requires the convexity of $\mathcal{U}_{\mathrm{ad}}$. However, this necessary condition of optimality is commonly used to identify an optimal control $u(t)$.

REMARK 10.3. The maximum principle is a necessary condition of optimality. One can also derive from it a sufficient condition, provided additional conditions are assumed. An obvious one is that the functional $J(v(.))$ be convex. But there are other possibilities, which we shall not develop. In fact, when one looks for sufficient conditions, it is more convenient to turn to dynamic programming, which is the objective of the next section.

10.1.4 Example

We discuss an application in which we use the maximum principle to obtain the optimal control. This example is inspired by cybersecurity; see [8]. The system is a diagnostic device, whose state is its accuracy $x(t)$ at time t. This accuracy is a percentage $0\leq x(t)\leq 1$. The evolution is given by the differential equation

$$\frac{dx}{dt} = c_0 v(t)(1 - x(t)) - c_1 x(t), \tag{10.1.26}$$
$$x(0) = x_0,$$

in which $v(t)$ is the control, $0 \leq v(t) \leq 1$. This control is a maintenance effort, expressed as a percentage. The impact of the maintenance effort depends on the accuracy level, and it decreases with $x(t)$. In particular, it vanishes when $x(t) = 1$. The term $c_1 x(t)$ represents a level of attacks. It is external and proportional to the accuracy $x(t)$. The functional to be minimized is a cost functional given by

$$J(v(.)) = \int_0^T (f(x(t)) + kv(t))dt. \tag{10.1.27}$$

The term $f(x(t))$ is an operational cost depending on the accuracy. The function $f(x)$ is decreasing with x. The higher the accuracy, the lower the cost, which in particular contains the cost of false alarms. The second term $kv(t)$ represents the maintenance effort cost. The problem is to find an optimal control.

EXERCISE 10.2. If $u(t)$ represents the optimal control and $y(t)$ the corresponding optimal state, write down the maximum principle. Consider the function

$$h(x) = c_0(1 - x)^2 f'(x) + c_1 k. \tag{10.1.28}$$

Assume $h(x)$ monotonically increasing (which is true if $f(x)$ is convex) and $h(0) = c_0 f'(0) + c_1 k < 0$. Then there is a single value $\hat{x}$ such that $h(\hat{x}) = 0$. Define $\hat{u} = \dfrac{c_1 \hat{x}}{c_0(1 - \hat{x})}$. Show that when $x_0 < \hat{x}$ and for T sufficiently large, there exist two times $0 < \tau < \theta < T$ such that an optimal control is described as follows:

$$u(t) = \begin{vmatrix} 1, \text{ if } 0 < t < \tau, \\ \hat{u}, \text{ if } \tau < t < \theta, \\ 0, \text{ if } \theta < t < T. \end{vmatrix} \tag{10.1.29}$$

What is the optimal control when $x_0 > \hat{x}$?

10.2 Dynamic Programming

10.2.1 Invariant Embedding and Optimality Principle

We begin by associating to the optimal control problem (10.1.1), (10.1.3), (10.1.4) a family of optimal control problems, indexed by the initial conditions. So we take a time $t \in [0, T]$ and a vector $x \in R^n$, which is the value of the state of the system at time t, representing the initial time.

The state $x(s)$, $s \in (t, T)$, is the solution of

$$\frac{dx}{ds} = g(x(s), v(s)) \; t < s < T, \tag{10.2.1}$$
$$x(t) = x.$$

The admissible controls satisfy

$$v(s) \in \mathcal{U}_{\text{ad}}, \tag{10.2.2}$$

and the payoff becomes

$$J_{x,t}(v(.)) = \int_t^T l(x(s), v(s))ds + h(x(T)), \tag{10.2.3}$$

in which we have emphasized the dependence on the initial conditions. We make the same assumptions as in Section 10.1.1.

We call the function

$$\Phi(x, t) = \inf_{v(.)} J_{x,t}(v(.)) \tag{10.2.4}$$

the **value function**. The initial problem corresponds to $t = 0$, $x = x_0$. It has been embedded in a two-parameter family. This procedure is called **invariant embedding**.

REMARK 10.4. The term "value function" is coined from its interpretation in an economic context. In such a case, the state represents in general capital or wealth (assume $n = 1$). The initial value of the state can be interpreted as an initial endowment, which is given. With this initial capital, the decision-maker executes the best possible management in minimizing the payoff. Therefore, $\Phi(x, t)$ can be interpreted as the value of capital x at time t. Its derivative $\Phi_x(x, t)$ is interpreted as the price of capital, since it is the increase of value that is expected from an additional dollar. When x is a vector, the argument is valid for each component. Each component of the gradient is interpreted as the price of a resource.

Optimality Principle The optimality principle (also called dynamic programming principle) is the relation, called in the sequel the Bellman equation,

$$\Phi(x, t) = \inf_{v(.)}[\int_t^{t+\epsilon} l(x(s), v(s))ds + \Phi(x(t+\epsilon), t+\epsilon)], \; \forall \epsilon \; t+\epsilon < T,$$
$$\Phi(x, T) = h(x). \tag{10.2.5}$$

To understand this principle, we can assume that there exists an optimal control for every pair of initial values. Let $\hat{v}_{x,t}(.)$ be the optimal control for the problem with initial conditions x, t. Therefore,

$$\Phi(x, t) = J_{x,t}(\hat{v}_{x,t}(.)).$$

Let $\hat{x}_{x,t}(.)$ be the optimal trajectory, corresponding to the control $\hat{v}_{x,t}(.)$. Then

$$J_{x,t}(\hat{v}_{x,t}(.)) = \int_t^{t+\epsilon} l(\hat{x}_{x,t}(s), \hat{v}_{x,t}(s))ds + J_{\hat{x}_{x,t}(t+\epsilon),t+\epsilon}(\hat{v}_{x,t}(.))$$

$$\geq \int_t^{t+\epsilon} l(\hat{x}_{x,t}(s), \hat{v}_{x,t}(s))ds + \Phi(\hat{x}_{x,t}(t+\epsilon), t+\epsilon)$$

$$\geq \inf_{v(.)}[\int_t^{t+\epsilon} l(x(s), v(s))ds + \Phi(x(t+\epsilon), t+\epsilon)].$$

Therefore, we have

$$\Phi(x,t) \geq \inf_{v(.)}[\int_t^{t+\epsilon} l(x(s), v(s))ds + \Phi(x(t+\epsilon), t+\epsilon)],\ \forall \epsilon\ t+\epsilon < T.$$

Take now any control $v(.)$. At time $t+\epsilon$, the state is $x(t+\epsilon)$. Consider the optimal control for the problem with initial conditions $x(t+\epsilon)$, $t+\epsilon$. With our notation, it is $\hat{v}_{x(t+\epsilon),t+\epsilon}(.)$. We can then consider a new control consisting of $v(s)$, $t < s < t+\epsilon$, and $\hat{v}_{x(t+\epsilon),t+\epsilon}(s)$ for $t+\epsilon < s < T$. It is an admissible control, denoted by $\tilde{v}(.)$, and clearly,

$$J_{x,t}(\tilde{v}(.)) = \int_t^{t+\epsilon} l(x(s), v(s))ds + \Phi(x(t+\epsilon), t+\epsilon).$$

But then

$$\Phi(x,t) \leq \int_t^{t+\epsilon} l(x(s), v(s))ds + \Phi(x(t+\epsilon), t+\epsilon),$$

whence also, since $v(.)$ is arbitrary,

$$\Phi(x,t) \leq \inf_{v(.)}[\int_t^{t+\epsilon} l(x(s), v(s))ds + \Phi(x(t+\epsilon), t+\epsilon)],$$

and therefore (10.2.5) holds.

EXERCISE 10.3. Extend the proof of the optimality principle, without assuming the existence of optimal controls for any initial conditions. Use controls that achieve the infimum up to ϵ arbitrarily small.

Note also the relation

$$\Phi(x,t) = \int_t^{t+\epsilon} l(\hat{x}_{x,t}(s), \hat{v}_{x,t}(s))ds + \Phi(\hat{x}_{x,t}(t+\epsilon), t+\epsilon) \tag{10.2.6}$$

and

$$\Phi(\hat{x}_{x,t}(t+\epsilon), t+\epsilon) = J_{\hat{x}_{x,t}(t+\epsilon),t+\epsilon}(\hat{v}_{x,t}(.)), \tag{10.2.7}$$

which are consequences of the above proof. We interpret this second relation as saying that considering an optimal control for the problem starting at t, with initial condition x, the portion of the control between $t+\epsilon$ and T remains optimal for the problem starting at $t+\epsilon$, provided the initial condition is $\hat{x}_{x,t}(t+\epsilon)$, the value attained at $t+\epsilon$ after using $\hat{v}_{x,t}(.)$ on the interval $(t, t+\epsilon)$. This statement is also called the **optimality principle**.

It means also that after $t+\epsilon$, the initial condition x_0 intervenes only through $\hat{x}_{x,t}(t+\epsilon)$, the state at time $t+\epsilon$. This important property is called **time consistency**. This property, which may look natural, is in fact rather rare. It is related to the specific form of the payoff, an integral along the path of the state trajectory, plus a final cost. As soon as we modify this structure, for instance by taking a functional of the integral, the time consistency property is lost.

10.2.2 HJB Equation

We now use a notation that unfortunately differs from that of the maximum principle. What was called the Hamiltonian, see (10.1.9), namely

$$L(x, v, q) = l(x, v) + q.g(x, v), \tag{10.2.8}$$

we call the Lagrangian, and we call the function

$$H(x, q) = \inf_{v \in \mathcal{U}_{\text{ad}}} L(x, v, q) \tag{10.2.9}$$

the Hamiltonian. This discrepancy is unfortunately now a tradition that, it seems, cannot be changed.

We make additional assumptions to those of Section 10.1.1, namely

$$\begin{aligned} l(x, v) &\geq l_0 |v|^2 - c_0, \\ h(x) &\geq -c_0. \end{aligned} \tag{10.2.10}$$

In taking the infimum in (10.2.9), it is sufficient to restrict v to satisfy

$$L(x, v, q) \leq L(x, v_0, q), \ \ v_0 \in \mathcal{U}_{\text{ad}}, \text{ fixed.}$$

Using (10.2.10) and (10.1.2), we see that we can restrict v to belong to a bounded set (depending on x and q). Since $L(x, v, q)$ is continuous in v, the infimum is attained. It is a classical result on optimization theory with parameter dependency that one can define a measurable map $\bar{v}(x, q)$ such that

$$\begin{aligned} L(x, \bar{v}(x, q), q) &= \inf_{v \in \mathcal{U}_{\text{ad}}} L(x, v, q), \text{ a.e. } x, q \\ &= H(x, q). \end{aligned} \tag{10.2.11}$$

EXERCISE 10.4. Prove that if the minimum is unique, then the minimum $\bar{v}(x, q)$ is a continuous function of x, q.

Consider now the value function. We first notice that

$$\Phi(x, T) = h(x). \tag{10.2.12}$$

Suppose now that the value function is locally Lipschitz continuous, and more precisely satisfies

$$\begin{aligned} |\Phi| &\le c(1+|x|^2), \\ |\frac{\partial \Phi}{\partial t}| &\le c(1+|x|^2), \\ |D\Phi| &\le c(1+|x|). \end{aligned} \tag{10.2.13}$$

We then have the following.

PROPOSITION 10.2. *We make the assumptions (10.1.2), (10.1.3), (10.1.7), (10.1.8), (10.2.10). Assume also that the value function satisfies (10.2.13). Then it satisfies the Hamilton–Jacobi–Bellman (HJB) equation*

$$\frac{\partial \Phi}{\partial t} + H(x, D\Phi) = 0, \;\; a.e. \tag{10.2.14}$$

PROOF. The value function satisfies the optimality principle (10.2.5) and by assumption the regularity properties (10.2.13). In the optimality principle, we need to consider square integrable controls defined on $(t, t+\epsilon)$. We are going to show that one can consider controls $v(.)$ such that

$$\int_t^{t+\epsilon} |v(s)|^2 ds \le C\epsilon(1+|x|^2). \tag{10.2.15}$$

Suppose this is done. Then for such controls, the state $x(.)$ satisfies

$$\sup_{t \le s \le t+\epsilon} |x(s) - x| \le C\epsilon(1+|x|). \tag{10.2.16}$$

Now, from the regularity assumptions, Φ is differentiable a.e. So if x, t is a point where Φ is differentiable, we have

$$|\Phi(x(t+\epsilon), t+\epsilon) - \Phi(x,t) - \epsilon\frac{\partial \Phi}{\partial t} - D\Phi.(x(t+\epsilon)-x)| \le \epsilon\, 0(\epsilon), \tag{10.2.17}$$

where $0(\epsilon)$ tends to 0 as $\epsilon \to 0$ and depends only on the point x, t, thanks to (10.2.16).

From the optimality principle, we can write

$$\begin{aligned} 0 = \inf_{v(.)}[&\int_t^{t+\epsilon} l(x(s), v(s))ds + D\Phi \int_t^{t+\epsilon} g(x(s), v(s))ds + \epsilon\frac{\partial \Phi}{\partial t} \\ &+ \Phi(x(t+\epsilon), t+\epsilon) - \Phi(x,t) - \epsilon\frac{\partial \Phi}{\partial t} - D\Phi \int_t^{t+\epsilon} g(x(s), v(s))ds] \\ \ge \inf_{v(.)}[&\int_t^{t+\epsilon} l(x(s), v(s))ds + D\Phi \int_t^{t+\epsilon} g(x(s), v(s))ds] + \epsilon\frac{\partial \Phi}{\partial t} - \epsilon 0(\epsilon). \end{aligned}$$

In view of (10.2.16), we also get easily

$$\inf_{v(.)}[\int_t^{t+\epsilon} l(x, v(s))ds + D\Phi \int_t^{t+\epsilon} g(x, v(s))ds] + \epsilon \frac{\partial \Phi}{\partial t} \leq \epsilon 0(\epsilon),$$

which implies

$$\frac{\partial \Phi}{\partial t} + H(x, D\Phi) \leq 0(\epsilon),$$

and letting $h \to 0$, we obtain

$$\frac{\partial \Phi}{\partial t} + H(x, D\Phi) \leq 0.$$

On the other hand, for fixed v, we can write, from the optimality principle,

$$\Phi(x, t) \leq \int_t^{t+\epsilon} l(x(s), v)ds + \Phi(x(t+\epsilon), t+\epsilon),$$

and at a point of differentiability, we get

$$\Phi(x, t) \leq \epsilon l(x, v) + \Phi(x, t) + \epsilon \frac{\partial \Phi}{\partial t} + D\Phi.g(x, v)\epsilon + \epsilon 0(\epsilon).$$

Therefore,

$$\frac{\partial \Phi}{\partial t} + l(x, v) + D\Phi.g(x, v) \geq 0,$$

whence

$$\frac{\partial \Phi}{\partial t} + H(x, D\Phi) \geq 0,$$

and thus we obtain (10.2.14). It remains to check (10.2.15). Consider a fixed control v_0 and let $x_0(.)$ be the corresponding trajectory. We can restrict $v(.)$ to satisfy

$$\int_t^{t+\epsilon} l(x(s), v(s))ds + \Phi(x(t+\epsilon), t+\epsilon) \leq \int_t^{t+\epsilon} l(x_0(s), v_0)ds \\ + \Phi(x_0(t+\epsilon), t+\epsilon).$$

We use the regularity assumptions (10.2.13) to write

$$|\Phi(x(t+\epsilon), t+\epsilon) - \Phi(x, t)| \leq C|x(t+\epsilon) - x|(1 + |x| + |x(t+\epsilon)|) \\ + C\epsilon(1 + |x|^2 + |x(t+\epsilon)|^2),$$

$$|\Phi(x_0(t+\epsilon), t+\epsilon) - \Phi(x, t)| \leq C\epsilon(1 + |x|^2).$$

In the second relation we use the fact that v_0 is constant. Using the first assumption (10.2.10) and collecting results, we obtain

$$l_0 \int_t^{t+\epsilon} |v|^2 ds \le C\epsilon(1+|x|^2) + C\epsilon|x(t+\epsilon)|^2 \\ + C|x(t+\epsilon)-x|(1+|x|+|x(t+\epsilon)|). \tag{10.2.18}$$

Now we have, from the state equation,

$$\sup_{t \le s \le t+\epsilon} |x(s)| \le C(|x| + \sqrt{\epsilon} + \sqrt{\int_t^{t+\epsilon} |v|^2 ds}), \tag{10.2.19}$$

$$|x(t+\epsilon)-x| \le C\epsilon(1+|x|) + C\epsilon\sqrt{\int_t^{t+\epsilon} |v|^2 ds}) \\ + C\int_t^{t+\epsilon} |v(s)|ds, \tag{10.2.20}$$

and from (10.2.18), we obtain

$$l_0 \int_t^{t+\epsilon} |v|^2 ds \le C\epsilon(1+|x|^2) + C\sqrt{\epsilon}\int_t^{t+\epsilon} |v|^2 ds \\ + C(1+|x|)\int_t^{t+\epsilon} |v(s)|ds, \tag{10.2.21}$$

and if we choose ϵ sufficiently small, the property (10.2.15) follows easily. This completes the proof. □

EXERCISE 10.5. Prove the estimates (10.2.19), (10.2.20).

10.2.3 Verification Principle; Regularity

We check here that the value function satisfies the regularity properties and is the maximum solution of the HJB equation.

THEOREM 10.2. *We make the assumptions (10.1.2), (10.1.3), (10.1.7), (10.1.8), (10.2.10). Then the value function satisfies the regularity properties (10.2.13) and is the maximum solution of the HJB equation satisfying the initial condition (10.2.12) and the regularity properties (10.2.13).*

PROOF. Let $v_0 \in \mathcal{U}_{ad}$ and $x_0(.)$ be the state corresponding to the constant control v_0. We have

$$\Phi(x,t) \le J_{x,t}(v_0).$$

We check easily that

$$|x_0(s)| \leq C(1+|x|),$$

where the constant depends naturally of the choice of v_0 and the constants of the problem. Therefore,

$$J_{x,t}(v_0) \leq C(1+|x|^2),$$

and thus

$$\Phi(x,t) \leq C(1+|x|^2).$$

On the other hand, from the assumption (10.2.10) we obtain immediately

$$J_{x,t}(v(.)) \geq -C_1, \ \forall v(.),$$

and hence $\Phi(x,t) \geq -C_1$. This completes the proof of the first property (10.2.13). Thanks to this property and assumption (10.2.10), from arguments already used we can restrict the controls to satisfy

$$\int_t^T |v(s)|^2 ds \leq C_1(1+|x|^2). \tag{10.2.22}$$

We now prove the third inequality (10.2.13). We take $x, \tilde{x} \in R^n$ and controls satisfying (in view of (10.2.22))

$$\int_t^T |v(s)|^2 ds \leq C_1(1+|x|^2+|\tilde{x}|^2). \tag{10.2.23}$$

We consider the states $x(.)$ and $\tilde{x}(.)$ corresponding to the initial conditions $x, \tilde{x}$ and the same control satisfying the bound (10.2.23). It is then easy to check that

$$|x(s)-\tilde{x}(s)| \leq C|x-\tilde{x}|,$$

and consequently,

$$|J_{x,t}(v(.)) - J_{\tilde{x},t}(v(.))| \leq C(1+|x|+|\tilde{x}|)|x-\tilde{x}|, \tag{10.2.24}$$

which implies

$$|\Phi(x,t)-\Phi(\tilde{x},t)| \leq C(1+|x|+|\tilde{x}|)|x-\tilde{x}|. \tag{10.2.25}$$

This proves that $\Phi(x,t)$ is locally Lipschitz continuous in x. It is not continuously differentiable, but from Rademacher's theorem it is a.e. differentiable, and the third property (10.2.13) holds.

For the second property (10.2.13) we use the optimality principle. We first take a fixed control v_0 and the corresponding state $x_0(.)$. We have

$$\Phi(x,t) \leq \int_t^{t+\epsilon} l(x_0(s), v_0)ds + \Phi(x_0(t+\epsilon), t+\epsilon).$$

We can check easily that

$$|x_0(t+\epsilon) - x| \leq C\epsilon(1+|x|).$$

In view of the third inequality (10.2.13), we have

$$|\Phi(x_0(t+\epsilon), t+\epsilon) - \Phi(x, t+\epsilon)| \leq C\epsilon(1+|x|^2).$$

Therefore, we have obtained

$$\Phi(x,t) \leq \Phi(x, t+\epsilon) + C\epsilon(1+|x|^2). \tag{10.2.26}$$

On the other hand, for every control satisfying the bound (10.2.22) we have, from the third inequality (10.2.13),

$$|\Phi(x(t+\epsilon), t+\epsilon) - \Phi(x, t+\epsilon)| \leq C|x(t+\epsilon) - x|(1+|x|).$$

From the optimality principle, we now have

$$\begin{aligned}\Phi(x,t) - \Phi(x,t+\epsilon) &= \inf_{v(.)}[\int_t^{t+\epsilon} l(x,v)ds + \Phi(x(t+\epsilon), t+\epsilon) - \Phi(x, t+\epsilon)] \\ &\geq \inf_{v(.)}[\int_t^{t+\epsilon} l(x,v)ds - C|x(t+\epsilon) - x|(1+|x|)].\end{aligned} \tag{10.2.27}$$

For $k \geq \bar{g}$, we can now check the estimate

$$\frac{1}{2}\frac{d}{ds}(\exp(-2ks)\,|x(s)-x|^2) \leq \bar{g}\exp(-2ks)\,|x(s)-x|)(1+|x|+|v(s)|).$$

Integrating between t and $\tau < t+\epsilon$, we obtain

$$(\exp -2k\tau)\,|x(\tau)-x|^2 \leq \bar{g} \sup_{t\leq s\leq t+\epsilon} [(\exp -ks)\,|x(s)-x|](\epsilon(1+|x|) + \int_t^{t+\epsilon} |v(s)|ds).$$

Therefore,

$$\sup_{t\leq s\leq t+\epsilon} [(\exp -ks)|x(s)-x|] \leq \bar{g}(\epsilon(1+|x|) + \int_t^{t+\epsilon} |v(s)|),$$

which implies in particular

$$|x(t+\epsilon)-x| \leq C(\epsilon(1+|x|)+\int_t^{t+\epsilon} |v(s)|). \tag{10.2.28}$$

Using this inequality in (10.2.27), we obtain

$$\Phi(x,t)-\Phi(x,t+\epsilon) \geq \inf_{v(.)}[\int_t^{t+\epsilon} l(x,v)ds - C(1+|x|)(\epsilon(1+|x|)+\int_t^{t+\epsilon} |v(s)|)],$$

and from the assumption (10.2.10), we get

$$\begin{aligned}\Phi(x,t)-\Phi(x,t+\epsilon) &\geq \inf_{v(.)}[\int_t^{t+\epsilon} l_0|v|^2 ds - C(1+|x|)(\epsilon(1+|x|)+\int_t^{t+\epsilon} |v(s)|)]\\ &\geq -C\epsilon(1+|x|^2).\end{aligned}$$

On combining this with (10.2.26), we obtain

$$|\Phi(x,t+\epsilon)-\Phi(x,t)| \leq C\epsilon(1+|x|^2). \tag{10.2.29}$$

So $\Phi(x,t)$ is Lipschitz continuous in t, and we have the second property (10.2.13). So the proof of (10.2.13) is complete. □

Suppose now that $\Psi(x,t)$ satisfies (10.2.13) and (10.2.12), (10.2.14). Then one has

$$\Psi(x,t) \leq \Phi(x,t). \tag{10.2.30}$$

Indeed, consider an admissible control $v(.)$ and the corresponding state $x(.)$. Then one has

$$\begin{aligned}\frac{d}{dt}\Psi(x(s),s) &= \frac{\partial \Psi}{\partial t}(x(s),s) + D\Psi(x(s),s).g(x(s),v(s))\\ &\geq -l(x(s),v(s)).\end{aligned}$$

Integrating between t and T, we get

$$\Psi(x,t) \leq J_{x,t}(v(.)),$$

and since $v(.)$ is arbitrary, (10.2.30) follows. So the value function is the maximum solution of the HJB equation satisfying the regularity property (10.2.13) and the final condition (10.2.12). The proof is complete. ■

EXERCISE 10.6. Prove the properties (10.2.24), (10.2.25).

With additional regularity, the HJB equation will provide an optimal control with a meaningful interpretation. Consider a solution Ψ of the HJB equation satisfying (10.2.12), (10.2.13), and (10.2.14). Recall the measurable map $\bar{v}(x,q)$ defined in (10.2.11). Moreover, one has the inequality

$$|\bar{v}(x,q)| \leq C(1+|x|+|q|). \tag{10.2.31}$$

We can next define a measurable map

$$\hat{v}(x,t) = \bar{v}(x, D\Psi(x,t)) \tag{10.2.32}$$

that satisfies the bound

$$|\hat{v}(x,t)| \leq C(1+|x|), \tag{10.2.33}$$

and we have the property

$$\frac{\partial \Psi}{\partial t} + l(x, \hat{v}(x,t)) + D\Psi.g(x, \hat{v}(x,t)) = 0, \text{ a.e. } x, t. \tag{10.2.34}$$

The map $\hat{v}(x,t)$ is called a **feedback**. The idea is to insert this feedback rule into the state equation and to look for the solution of

$$\begin{aligned} \frac{dy}{ds} &= g(y, \hat{v}(y,s)), \\ y(t) &= x. \end{aligned} \tag{10.2.35}$$

Unfortunately, the function $g(x, \hat{v}(x,s))$ is only measurable, so the differential equation is not solvable. Suppose, however, that we can solve this equation and obtain a solution $y(.) \in L^2(t,T;R^n)$ (and consequently in $H^1(t,T;R^n)$). Then on setting

$$u(s) = \hat{v}(y(s), s), \tag{10.2.36}$$

we see that this is the optimal control. Indeed, writing (10.2.34) with $x = y(s)$ and $t = s$, we have

$$\begin{aligned} \frac{\partial \Psi(y(s),s)}{\partial s} + l(y(s), u(s)) + D\Psi(y(s),s).g(y(s),u(s)) &= 0, \\ \Psi(y(T),T) &= h(y(T)), \end{aligned}$$

and hence

$$\frac{d}{ds}\Psi(y(s),s) + l(y(s),u(s)) = 0.$$

Integrating between t and T, we obtain

$$\Psi(x,t) = J_{x,t}(u(.)).$$

Therefore, $\Psi(x,t) \geq \Phi(x,t)$, the value function. However, $\Phi(x,t)$ is the maximum function satisfying (10.2.12), (10.2.13), and (10.2.14). Hence $\Psi(x,t) = \Phi(x,t)$, and $u(.)$ is the optimal control. We have proved the following result.

PROPOSITION 10.3. *The HJB equation (10.2.12), (10.2.14) has a unique solution satisfying the regularity assumptions (10.2.13). It is also the solution for which the state equation (10.2.35) has a square integrable solution. This unique solution is the value function, and there exists an optimal control obtained from a feedback rule.*

REMARK 10.5. We do not claim uniqueness of the optimal control. The fact that the optimal control is obtained from a feedback rule proves also the time consistency property. To compute the optimal control at any time, we need to know the current state. We do not need to remember the initial state.

Proposition 10.3 is often referred to as the **verification principle**. We prove the existence of an optimal control by verifying that the optimal feedback rule allows one to construct an optimal control. This is also why we call dynamic programming a theory of sufficient conditions, in contrast to the maximum principle, which is a theory of necessary conditions. As soon as the HJB equation has a sufficiently smooth solution, an optimal control exists and can be constructed. Note that in our framework of regularity, (10.1.2), (10.1.3), (10.1.7), (10.1.8), (10.2.10), the value function is necessarily a solution of the HJB equation (and the maximum solution), but we cannot guarantee the existence of an optimal control unless additional smoothness is satisfied.

We may wonder how we can guarantee that (10.2.35) has a solution. As soon as $\hat{v}(x,t)$ is continuous in x, this is true. We say that $\hat{v}$ is Carathéodory.

Indeed, set

$$\hat{g}(x,t) = g(x, \hat{v}(x,t)).$$

Then $\hat{g}$ is Carathéodory and satisfies

$$|\hat{g}(x,t)| \leq C(1+|x|),$$

and then the differential equation

$$\begin{aligned} \frac{dy}{dt} &= \hat{g}(y(t),t), \\ y(0) &= x_0, \end{aligned} \tag{10.2.37}$$

has a solution in $H^1(0,T;R^n)$.

EXERCISE 10.7. Find a theorem of existence of solutions of differential equations that guarantees that (10.2.37) has a solution.

To obtain the Carathéodory property for $\hat{v}(x,t)$, one may assume the uniqueness of the minimum $\bar{v}(x,q)$, in which case it is a continuous function, and we obtain the existence of a solution of the HJB equation that is C^1 in x.

REMARK 10.6. The optimal control $u(t)$ is obtained through a feedback function $\hat{v}(x,t)$; see (10.2.36). A function of time only is called an *open-loop control.* A control defined by a feedback is called a *closed-loop control.* We see that the optimal control in the class of open-loop controls and the optimal control in the class of closed-loop controls coincide. This property has been noted already in the linear–quadratic case. In fact, the two classes are equivalent. Indeed, an open-loop control is obviously a particular closed-loop control. Similarly, to a closed-loop control we can associate an open-loop control, as we have done in (10.2.35), (10.2.36). Of course, this makes sense only for the feedbacks for which we can solve the equivalent to (10.2.35).

10.3 Links Between the Maximum Principle and Dynamic Programming

We consider the problem $J_{x_0,0}(v(.))$. If $u(.)$ is an optimal control and $y(.)$ is the corresponding state, then we have introduced in the statement of the maximum principle the adjoint state $p(.)$, the solution of

$$-\frac{dp}{dt} = g_x^*(y,u)p + l_x(y,u), \tag{10.3.1}$$
$$p(T) = h_x(y(T)).$$

Consider, on the other hand, the HJB equation

$$\frac{\partial \Phi}{\partial t} + \inf_v(l(x,v) + D\Phi.g(x,v)) = 0, \tag{10.3.2}$$
$$\Phi(x,T) = h(x).$$

We assume (this can be relaxed)

$$\mathcal{U}_{\text{ad}} = R^m, \tag{10.3.3}$$

in which case the optimal feedback $\hat{v}(x,t)$ satisfies

$$l_v(x,\hat{v}) + g_v^*(x,\hat{v})D\Phi(x,t) = 0. \tag{10.3.4}$$

In view of the difference of notation between the maximum principle and dynamic programming, we have not used Lagrangians and Hamiltonians but rather explicit expressions.

In the sequel we proceed formally, assuming sufficient smoothness and that the derivations are legitimate. We claim that

$$p(t) = D\Phi(y(t), t). \tag{10.3.5}$$

The relation is clearly satisfied at time T. Now if we define $p(t)$ by formula (10.3.5), we shall check that (10.3.1) holds. Since it is a linear equation in p, it has a unique solution given by formula (10.3.5).

Thanks to the smoothness, we can differentiate (10.3.5) to obtain

$$\frac{dp}{dt} = D^2\Phi(y(t), t)g(y(t), u(t)) + D\frac{\partial \Phi}{\partial t}(y(t), t). \tag{10.3.6}$$

However, taking the gradient in x of the HJB equation, we obtain, using (10.3.4),

$$D\frac{\partial \Phi}{\partial t} + l_x(x, \hat{v}) + D^2\Phi.g(x, \hat{v}) + g_x^*(x, \hat{v})D\Phi = 0.$$

Taking $x = y(t)$, we have $\hat{v}(y(t), t) = u(t)$. Comparing with (10.3.6), we obtain immediately (10.3.1).

Recalling Remark 10.4, we know that $D\Phi(x_0, 0)$ is the price of capital x_0 at time 0. This is also $p(0)$, the initial value of the adjoint variable.

10.4 No Smoothness Case and Viscosity Approach

10.4.1 Characterization That Does Not Require Derivatives

When the value function is not smooth, how can we characterize it? What replaces the HJB equation? We first turn to the optimality principle and get a characterization that does not require derivatives. To fix the ideas, we assume only

$$\begin{aligned} g(x, v) : R^n \times R^m &\to R^n, \text{ continuous}, \\ g_x &\quad \text{bounded}, \\ |g(x, v)| &\leq \bar{g}(1 + |x| + |v|), \end{aligned} \tag{10.4.1}$$

$$\begin{aligned} l(x, v) : R^n \times R^m &\to R, \text{ continuous}, \\ h(x) : R^n &\to R, \text{ continuous}, \end{aligned} \tag{10.4.2}$$

$$|l(x, v)| \leq \bar{l}(1 + |x|^2 + |v|^2), \; |h(x)| \leq \bar{h}(1 + |x|^2), \tag{10.4.3}$$

$$\mathcal{U}_{\text{ad}} = \text{nonempty closed convex subset of } R^m. \tag{10.4.4}$$

The problem $J_{x,t}(v(.))$ is well defined, and we can define the value function $\Phi(x,t)$. It satisfies the optimality principle. But the HJB equation is not valid. Nevertheless, it is possible to proceed as follows. Consider the set of functions $\Psi(x,t)$ that satisfy

$$\Psi(x,t) \le \int_t^{t+\epsilon} l(x(s),v)ds + \Psi(x(t+\epsilon),t+\epsilon),\ \forall\epsilon,\ t+\epsilon \le T, \forall v \in \mathcal{U}_{ad}, \tag{10.4.5}$$

$$\Psi(x,T) \le h(x).$$

We call this set the set of **subsolutions** of Bellman's equation (10.2.5).

We look for functions that are continuous and satisfy a growth condition

$$|\Psi(x,t)| \le C(1+|x|^2). \tag{10.4.6}$$

In (10.4.5), $x(s)$ is the trajectory corresponding to a fixed control v and initial conditions x,t. It is clear from the optimality principle that the value function belongs to this set. We call this set $\mathcal{S}$.

We then have the following.

THEOREM 10.3. *We make the assumptions (10.4.1), (10.4.2), (10.4.3), (10.4.4). The value function is the maximum element of the set $\mathcal{S}$.*

PROOF. It is sufficient to prove that for every element of the set, one has $\Psi(x_0,0) \le \Phi(x_0,0)$. The proof is similar for an arbitrary point x,t. Set $k=\frac{T}{N}$, with $N \to +\infty$. Consider an admissible control $v(.)$. We approximate it by a sequence $v^k(.)$ such that

$$v^k(t) = v_j^k = \frac{1}{k}\int_{(j-1)k}^{jk} v(s)ds,\ j=1,\cdots,N,$$
$$\forall\ t \in ((j-1)k, jk),$$

which is admissible by the convexity assumption. We have

$$v^k(.) \to v(.) \text{ in } L^2(0,T;R^m).$$

Let $x^k(.)$ be the state corresponding to $v^k(.)$, and $x(.)$ the state corresponding to $v(.)$. Then also

$$x^k(.) \to x(.) \text{ in } L^2(0,T;R^n)$$

and

$$J_{x_0,0}(v^k(.)) \to J_{x_0,0}(v(.)).$$

We can write

$$\Psi(x^k((j-1)k), (j-1)k) \leq \int_{(j-1)k}^{jk} l(x^k(s), v_j^k)ds + \Psi(x^k(jk), jk)$$
$$\forall\ j = 1, \cdots N,$$

and on summing, we get immediately

$$\Psi(x_0, 0) \leq J_{x_0,0}(v^k(.)).$$

Letting $N \to +\infty$, we get

$$\Psi(x_0, 0) \leq J_{x_0,0}(v(.)),$$

and since $v(.)$ is arbitrary, we have obtained $\Psi(x_0, 0) \leq \Phi(x_0, 0)$. The proof is complete. □

REMARK 10.7. The function $\Phi(x, t)$ is characterized as a sequence of linear programs with infinitely many constraints. Indeed, (10.3.1) defines a set of linear constraints on the function $\Psi(x, t)$, and we want to find the maximum one. This approach leads to an algorithm to solve the optimal control problem numerically.

We can define also the set $\bar{S}$ of **supersolutions** of Bellman equation (10.2.5). The set $\bar{S}$ is defined by the set of functions satisfying (10.4.6) and

$$\Psi(x, t) \geq \inf_{v(.)}[\int_t^{t+\epsilon} l(x(s), v(s))ds + \Psi(x(t+\epsilon), t+\epsilon)],\ \forall\epsilon,\ t+\epsilon \leq T, \tag{10.4.7}$$

$$\Psi(x, T) \geq h(x).$$

It is clear that the value function is the minimal element of the set $\bar{S}$.

10.4.2 Viscosity Solutions

10.4.2.1 Differentials

Viscosity solutions were introduced by M.G. Crandall and P.L. Lions; see [32] for a complete description. We follow the presentation of W. Fleming and M. Soner [22]. We consider functions that are in $C^0(R^n \times [0, T])$. Let $\Phi(x, t)$ belong to $C^0(R^n \times [0, T])$. For $x, t \in R^n \times (0, T)$ we define the set of **superdifferentials** of Φ at x, t by

$$D^+\Phi(x,t) = \{q \in R^n, \rho \in R | \limsup_{y\to x, s\to t} \frac{\Phi(y,s) - \Phi(x,t) - q.(y-x) - \rho(s-t)}{|y-x| + |s-t|} \leq 0\}, \tag{10.4.8}$$

and the set of **subdifferentials** of Φ at x, t by

$$D^-\Phi(x,t) = \{q \in R^n, \rho \in R | \liminf_{y\to x, s\to t} \frac{\Phi(y,s) - \Phi(x,t) - q.(y-x) - \rho(s-t)}{|y-x| + |s-t|} \geq 0\}. \tag{10.4.9}$$

Clearly, $D^+(-\Phi)(x,t) = -D^-\Phi(x,t)$. These sets are convex sets. They may be empty.

LEMMA 10.1. . *If $D^+\Phi(x,t)$ and $D^-\Phi(x,t)$ are both nonempty, then the function Φ is differentiable at x, t, and*

$$D^+\Phi(x,t) = D^-\Phi(x,t) = \{D\Phi(x,t), \frac{\partial \Phi}{\partial t}(x,t)\}. \tag{10.4.10}$$

If $\Phi(x,t)$ is convex, then the subdifferential is not empty and coincides with the subdifferential in the sense of convex analysis ($x, t \in R^n \times (0,T)$):

$$D^-\Phi(x,t) = \partial\Phi(x,t) = \tag{10.4.11}$$

$$= \{q \in R^n, \rho \in R | \Phi(y,s) - \Phi(x,t) - q.(y-x) - \rho(s-t) \geq 0, \forall y \in R^n, s \in (0,T)\}.$$

Therefore, $D^+\Phi(x,t)$ is empty, unless the function is differentiable.

If $\Phi(x,t)$ is Lipschitz continuous on $R^n \times [0,T]$, i.e.,

$$|\Phi(x,t) - \Phi(y,s)| \leq M|x-y| + M_1|s-t|, \tag{10.4.12}$$

then

$$|q| \leq M,\ |\rho| \leq M_1, \forall q, \rho \in D^+\Phi(x,t) \cup D^-\Phi(x,t).$$

PROOF. Suppose that $(q,\rho) \in D^+\Phi(x,t)$ and $(q',\rho') \in D^-\Phi(x,t)$. Then

$$\limsup_{y\to x, s\to t} \frac{\Phi(y,s) - \Phi(x,t) - q.(y-x) - \rho(s-t)}{|y-x| + |s-t|} \leq 0,$$

$$\liminf_{y\to x, s\to t} \frac{\Phi(y,s) - \Phi(x,t) - q'.(y-x) - \rho'(s-t)}{|y-x| + |s-t|} \geq 0.$$

Hence

$$\limsup_{y\to x, s\to t} \frac{(q'-q).(y-x) + (\rho'-\rho)(s-t)}{|y-x| + |s-t|} \leq 0.$$

But we can take $s - t = \theta(\rho' - \rho)$, $y - x = \theta(q' - q)$, with $\theta \to 0$. It follows immediately that $q' = q$, $\rho' = \rho$. But then

$$\lim_{y\to x, s\to t} \frac{\Phi(y,s) - \Phi(x,t) - q.(y-x) - \rho(s-t)}{|y-x| + |s-t|} = 0,$$

whence the result (10.4.10). Suppose now that $\Phi(x,t)$ is a convex function. It is clear that $\partial\Phi(x,t) \subset D^-\Phi(x,t)$. Suppose the two sets do not coincide. Then there exists a pair $(q,\rho) \in D^-\Phi(x,t)$ such that one can find $(\bar{y},\bar{s})$ satisfying

$$\Phi(\bar{y},\bar{s}) - \Phi(x,t) - q.(\bar{y}-x) - \rho(\bar{s}-t) = -\delta < 0.$$

We define the sequence

$$y_n = x + \frac{1}{2^n}(\bar{y} - x),\ s_n = t + \frac{1}{2^n}(\bar{s} - t).$$

By convexity, one checks easily that

$$\Phi(y_n, s_n) \le \Phi(x,t) + q.(y_n - x) + \rho(s_n - t) - \delta\frac{|y_n - x| + |s_n - t|}{|\bar{y} - x| + |\bar{s} - t|},$$

and thus

$$\liminf_{n\to\infty} \frac{\Phi(y_n,s_n) - \Phi(x,t) - q.(y_n - x) - \rho(s_n - t)}{|y_n - x| + |s_n - t|} \le \frac{-\delta}{|\bar{y} - x| + |\bar{s} - t|} < 0,$$

which contradicts the fact that $(q,\lambda) \in D^-\Phi(x,t)$. So we have proven (10.4.11). Finally, let us assume the Lipschitz property (10.4.12). Then

$$\frac{\Phi(y,t) - \Phi(x,t) - q.(y-x)}{|y-x|} \ge -M - \frac{q.(y-x)}{|y-x|}.$$

We consider the sequence $y - x = -\frac{q}{n}$. It follows that

$$n\frac{\Phi(x - \frac{q}{n}, t) - \Phi(x,t) + \frac{|q|^2}{n}}{|q|} \ge -M + |q|.$$

Since the lim inf of the left-hand side is negative, we obtain $|q| \le M$. The remaining properties are proven in a similar manner. ■

EXERCISE 10.8. Let $\Phi(x,t) = \min(1 - |x|, 1 - t)$, with $x \in [-1,+1]$, $t \in [0,1]$. Show that for $x \in (-1,+1)$, $t \in (0,1)$, $t \neq |x|$, the function is differentiable. For $t = |x| \in (0,1)$, we have

$$D^+\Phi(x,t) = \{-\theta\frac{x}{|x|}, -1+\theta\}, \theta \in (0,1),\ D^-\Phi(x,t) = \emptyset\,.$$

We then give an important characterization of the superdifferential and of the subdifferential.

PROPOSITION 10.4. *If* $(q,\rho) \in D^+\Phi(x,t)$, *one can construct a function* $\varphi_{x,t}(y,s) \in C^1(R^n \times [0,T)) \cap C^0(R^n \times [0,T])$ *such that* $q = D\varphi_{x,t}(x,t)$, $\rho = \frac{\partial \varphi_{x,t}}{\partial s}(x,t)$ *and*

$$\Phi(y,s) - \varphi_{x,t}(y,s) \leq \Phi(x,t) - \varphi_{x,t}(x,t), \forall y, s. \tag{10.4.13}$$

Alternatively, if a function $\varphi_{x,t}(y,s)$ *satisfies (10.4.13), then* $q = D\varphi_{x,t}(x,t)$, $\rho = \frac{\partial \varphi_{x,t}}{\partial s}(x,t)$ *belongs to* $D^+\Phi(x,t)$. *A symmetric characterization holds for the subdifferential* $D^-\Phi(x,t)$.

PROOF. If we find a function $\varphi_{x,t}(y,s)$ such that (10.4.13) holds, then taking $q = D\varphi_{x,t}(x,t)$, $\rho = \frac{\partial \varphi_{x,t}}{\partial s}(x,t)$, we have, for $x \in R^n$ and $t \in (0,T)$,

$$\limsup_{y\to x, s\to t} \frac{\Phi(y,s) - \Phi(x,t) - q.(y-x) - \lambda(s-t)}{|y-x|+|s-t|}$$
$$\leq \limsup_{y\to x, s\to t} \frac{\varphi_{x,t}(y,s) - \varphi_{x,t}(x,t) - q.(y-x) - \rho(s-t)}{|y-x|+|s-t|} = 0,$$

since $\varphi_{x,t}(y,s)$ is continuously differentiable. So $(q,\rho) \in D^+\Phi(x,t)$. We have thereby obtained the second part. Let us check the first part. We set

$$d_{x,t}(y,s) = \sqrt{(|y-x|^2 + |s-t|^2)},$$

and for $\tau > 0$ we define

$$h_{xt}(\tau) = \sup_{\{y,s|0<d_{x,t}(y,s)\leq\tau\}} \frac{(\Phi(y,s) - \Phi(x,t) - q.(y-x) - \rho(s-t))^+}{d_{xt}(y,s)}.$$

The function $h_{xt}(\tau)$ is monotonically increasing, and since $(q,\rho) \in D^+\Phi(x,t)$, we have $h_{xt}(\tau) \to 0$ as $\tau \to 0$. We set

$$\varphi_{x,t}(y,s) = \Phi(x,t) + q.(y-x) + \rho(s-t) + \int_{d_{x,t}(y,s)}^{2d_{x,t}(y,s)} h(\tau)d\tau.$$

Then $\varphi_{x,t}(y,s)$ is $C^1(R^n \times [0,T)) \cap C^0(R^n \times [0,T])$, and the gradient in (x,t) is (q,ρ). Moreover, we have

$$\int_{d_{x,t}(y,s)}^{2d_{x,t}(y,s)} h(\tau)d\tau \geq h(d_{x,t}(y,s))d_{x,t}(y,s) \geq (\Phi(y,s) - \Phi(x,t)$$
$$-q.(y-x) - \rho(s-t))^+ \geq \Phi(y,s) - \Phi(x,t) - q.(y-x) - \rho(s-t),$$

and hence $\Phi(y,s) \leq \varphi_{x,t}(y,s)$ and $\varphi_{x,t}(x,t) = \Phi(x,t)$. The proof is complete. □

10.4.2.2 Viscosity Solution

In (10.4.5) and (10.4.7) we have defined the concept of subsolution and supersolution for the Bellman equation. We now define the concept of viscosity subsolution and viscosity supersolution for the HJB equation (10.2.12), (10.2.14). A **viscosity subsolution** $\Psi(x,t)$ of the HJB equation (10.2.12), (10.2.14) satisfies

$$\Psi(x,T) \leq h(x), \tag{10.4.14}$$
$$\rho + H(x,t,q) \geq 0, \ \forall q, \rho \in D^+\Psi(x,t),$$

and similarly, a viscosity supersolution of the HJB equation (10.2.12), (10.2.14) satisfies

$$\Psi(x,T) \leq h(x), \tag{10.4.15}$$
$$\rho + H(x,t,q) \leq 0, \ \forall q, \rho \in D^-\Psi(x,t).$$

A function $\Phi(x,t)$ that is at the same time a viscosity subsolution and a viscosity supersolution is called a viscosity solution of the HJB equation.

PROPOSITION 10.5. *A subsolution of the Bellman equation is a viscosity subsolution of the HJB equation. A supersolution of the Bellman equation is a viscosity supersolution of the HJB equation. Therefore, the value function is a viscosity solution of the HJB equation.*

PROOF. Consider a subsolution of the Bellman equation. We have

$$\Psi(x,t) \leq \int_t^{t+\epsilon} l(x(s),v)ds + \Psi(x(t+\epsilon),t+\epsilon), \ \forall \epsilon, \ t+\epsilon \leq T, \forall v \in \mathcal{U}_{ad}.$$

Let $q,\rho \in D^+\Psi(x,t)$. We can construct the function $\varphi_{x,t}(y,s)$ that is $C^1(R^n \times (0,T)) \cap C^0(R^n \times [0,T])$ such that its gradient at x,t is q,ρ and satisfies (10.4.13). Without loss of generality, we can assume $\varphi_{x,t}(x,t) = \Psi(x,t)$ and thus $\varphi_{x,t}(y,s) \geq \Psi(y,s)$. We then have

$$\varphi_{x,t}(x,t) \leq \int_t^{t+\epsilon} l(x(s),v)ds + \varphi_{x,t}(x(t+\epsilon),t+\epsilon).$$

But now the function $\varphi_{x,t}(y,s)$ is differentiable. So it is easy to check that

$$0 \leq \frac{\partial \varphi_{xt}}{\partial t}(x,t) + l(x,v) + D\varphi_{xt}(x,t).g(x,v),$$

which means that

$$0 \leq \rho + l(x,v) + q.g(x,v),$$

and since v is arbitrary, we obtain (10.4.14). If $\Psi(x,t)$ is a supersolution of the Bellman equation, we have

$$\Psi(x,t) \geq \inf_{v(.)}[\int_t^{t+\epsilon} l(x(s),v(s))ds + \Psi(x(t+\epsilon),t+\epsilon)],\ \forall\epsilon,\ t+\epsilon \leq T.$$

Let $(q,\rho) \in D^-\Psi(x,t)$. We can find a function $\varphi_{xt}(y,s)$ that is $C^1(R^n \times (0,T)) \cap C^0(R^n \times [0,T])$, with gradient at x,t equal to (q,ρ), such that $\varphi_{xt}(x,t) = \Psi(x,t)$ and $\varphi_{x,t}(y,s) \leq \Psi(y,s)$. Therefore,

$$\varphi_{xt}(x,t) \geq \inf_{v(.)}[\int_t^{t+\epsilon} l(x(s),v(s))ds + \varphi_{xt}(x(t+\epsilon),t+\epsilon)],$$

$$0 \geq \inf_{v(.)}[\int_t^{t+\epsilon} (l(x,v(s)) + q.g(x,v(s))ds + \rho\epsilon] + o(\epsilon),$$

$$0 \geq \int_t^{t+\epsilon} H(x,s,q)ds + \rho\epsilon + o(\epsilon).$$

Dividing by ϵ and letting $\epsilon \to 0$, we obtain (10.4.15), which completes the proof. □

10.4.2.3 Maximum Principle

We want to prove the following property (called the maximum principle, not to be confused with Pontryagin's maximum principle).

PROPOSITION 10.6. *Viscosity subsolutions are smaller than viscosity supersolutions.*

As a consequence, we obtain the following characterization. We do not detail the assumptions.

THEOREM 10.4. *The value function is the unique viscosity solution of the HJB equation.*

Proof of Proposition 10.6

PROOF. The statement of the proposition will be a consequence of the maximum principle. Let $\Psi(x,t)$ be a viscosity subsolution and $\chi(x,t)$ a viscosity supersolution. So they satisfy

$$\begin{aligned}\Psi(x,T) &\leq h(x), \\ \rho + H(x,t,q) &\geq 0,\ \forall (q,\rho) \in D^+\Psi(x,t),\end{aligned} \tag{10.4.16}$$

$$\begin{aligned}\chi(x,T) &\geq h(x), \\ \rho + H(x,t,q) &\leq 0,\ \forall (q,\rho) \in D^-\Psi(x,t).\end{aligned} \tag{10.4.17}$$

We claim that

$$\Psi(x,t) \leq \chi(x,t),\ \forall x,t. \tag{10.4.18}$$

We first consider the case in which the functions $\Psi(x,t)$ and $\chi(x,t)$ are differentiable. So they satisfy

$$\begin{aligned}\frac{\partial \Psi}{\partial t} + H(x,t,D\Psi) &\geq 0, \\ \frac{\partial \chi}{\partial t} + H(x,t,D\chi) &\leq 0.\end{aligned} \tag{10.4.19}$$

First define $\tilde{\chi}(x,t) = \chi(x,t) + \beta(T-t),\ \beta > 0$. Therefore,

$$\frac{\partial \tilde{\chi}}{\partial t} + H(x,t,D\tilde{\chi}) + \beta \leq 0. \tag{10.4.20}$$

Suppose there is a point x_0, t_0 with $0 < t_0 < T$ such that $\Psi(x_0,t_0) - \tilde{\chi}(x_0,t_0) > 0$. Under **reasonable conditions** (not detailed), we can claim that there will be a positive local maximum of the function $\Psi(x,t) - \tilde{\chi}(x,t)$ in $R^n \times (0,T)$. Let $\bar{x}, \bar{t}$ be such a point. Necessarily, we have

$$D\Psi(\bar{x},\bar{t}) = D\tilde{\chi}(\bar{x},\bar{t}),\ \frac{\partial \Psi}{\partial t}(\bar{x},\bar{t}) = \frac{\partial \tilde{\chi}}{\partial t}(\bar{x},\bar{t}).$$

Writing the first inequality (10.4.19) and the inequality (10.4.20) and then subtracting, we get $-\beta \geq 0$, which is a contradiction. So necessarily $t_0 = T$. But then $\Psi(x_0,T) - \tilde{\chi}(x_0,T) = \Psi(x_0,T) - \chi(x_0,T) \leq 0$. Again a contradiction. So there cannot exist such a point x_0, t_0, and then $\Psi(x,t) \leq \tilde{\chi}(x,t) = \chi(x,t) + \beta(T-t)$, $\forall x,t$. But β is arbitrary, so (10.4.18) follows.

In the nondifferentiable case, let us introduce the function

$$\psi_{\epsilon\beta}(x,t;y,s) = \Psi(x,t) - \chi(y,s) - \beta(T-s) - \frac{1}{\epsilon}(|x-y|^2 + |s-t|^2).$$

The parameter β has the same meaning as in the differentiable case. This time, we need to take different points (x,t) and (y,s), but there is a penalty term if they are far away. If we assume also that the functions $\Psi(x,t)$ and $\chi(y,s)$ are bounded or do not grow to infinity too fast as $|x|, |y| \to +\infty$, then for ϵ sufficiently small, the function $\psi_{\epsilon\beta}(x,t;y,s)$ will have a global maximum, denoted by $\bar{x}, \bar{t}, \bar{y}, \bar{s}$, of course depending on the parameters ϵ, β. It follows that the function $\psi_{\epsilon\beta}(x,t;\bar{y},\bar{s})$ has a maximum at $\bar{x}, \bar{t}$. Also, the function $\Psi(x,t) - \frac{1}{\epsilon}(|x-\bar{y}|^2 + |t-\bar{s}|^2)$ has a maximum at $\bar{x}, \bar{t}$. Define then

$$\varphi_{\bar{x},\bar{t}}(x,t) = \Psi(\bar{x},\bar{t}) + \frac{1}{\epsilon}(|x-\bar{y}|^2 + |t-\bar{s}|^2) - \frac{1}{\epsilon}(|\bar{x}-\bar{y}|^2 + |\bar{t}-\bar{s}|^2).$$

It follows that $\Psi(x,t) - \varphi_{\bar{x},\bar{t}}(x,t) \leq 0,\ \Psi(\bar{x},\bar{t}) - \varphi_{\bar{x},\bar{t}}(\bar{x},\bar{t}) = 0$. This implies that $q = D\varphi_{\bar{x},\bar{t}}(\bar{x},\bar{t}),\ \rho = \dfrac{\partial \varphi_{\bar{x},\bar{t}}}{\partial t}(\bar{x},\bar{t})$ belongs to $D^+\Psi(\bar{x},\bar{t})$. Hence $q = \dfrac{1}{\epsilon}(\bar{x}-\bar{y}),\ \rho = \dfrac{1}{\epsilon}(\bar{t}-\bar{s})$ belongs to $D^+\Psi(\bar{x},\bar{t})$. It follows that

$$\frac{1}{\epsilon}(\bar{t}-\bar{s}) + H(\bar{x},\bar{t},\frac{1}{\epsilon}(\bar{x}-\bar{y})) \geq 0. \tag{10.4.21}$$

Similarly, the function $\psi_{\epsilon\beta}(\bar{x},\bar{t};y,s)$ of the pair (y,s) has a maximum at $(\bar{y},\bar{s})$. Therefore, $\chi(y,s) + \beta(T-s) + \frac{1}{\epsilon}(|y-\bar{x}|^2 + |s-\bar{t}|^2)$ has a minimum at $(\bar{y},\bar{s})$. Define

$$\zeta_{\bar{y},\bar{s}}(y,s) = \chi(\bar{y},\bar{s}) + \beta(T-\bar{s}) + \frac{1}{\epsilon}(|\bar{y}-\bar{x}|^2 + |\bar{s}-\bar{t}|^2) -$$
$$-(\beta(T-s) + \frac{1}{\epsilon}(|y-\bar{x}|^2 + |s-\bar{t}|^2)).$$

Then $\chi(y,s) - \zeta_{\bar{y},\bar{s}}(y,s) \geq 0$ and $\chi(\bar{y},\bar{s}) - \zeta_{\bar{y},\bar{s}}(\bar{y},\bar{s}) = 0$. This implies that $q = D\zeta_{\bar{y},\bar{s}}(\bar{y},\bar{s}),\ \rho = \dfrac{\partial \zeta_{\bar{y},\bar{s}}}{\partial s}(\bar{y},\bar{s})$ belongs to $D^-\chi(\bar{y},\bar{s})$. We have $q = \dfrac{1}{\epsilon}(\bar{x}-\bar{y}),\ \rho = \beta + \dfrac{1}{\epsilon}(\bar{t}-\bar{s})$. Therefore, we can write

$$\beta + \frac{1}{\epsilon}(\bar{t}-\bar{s}) + H(\bar{y},\bar{s},\frac{1}{\epsilon}(\bar{x}-\bar{y})) \leq 0. \tag{10.4.22}$$

Combining (10.4.21), (10.4.22), we can assert that

$$-\beta + H(\bar{x},\bar{t},\frac{1}{\epsilon}(\bar{x}-\bar{y})) - H(\bar{y},\bar{s},\frac{1}{\epsilon}(\bar{x}-\bar{y})) \geq 0. \tag{10.4.23}$$

To establish a contradiction at this stage, we must guarantee that for fixed β, the quantity $H(\bar{x}, \bar{t}, \frac{1}{\epsilon}(\bar{x}-\bar{y})) - H(\bar{y}, \bar{s}, \frac{1}{\epsilon}(\bar{x}-\bar{y}))$ tends to 0 as $\epsilon \to 0$. This is trivial if $H(x, t, q)$ depends only on q and easy if it is of the form $H_0(x, t) + H_1(q)$. **Assuming that we can guarantee this**, we get a contradiction, and thus one of the numbers $\bar{t}, \bar{s}$ or both is equal to T. Suppose, for instance, $\bar{t} = T$. Then

$$\begin{aligned}\psi_{\epsilon\beta}(\bar{x}, \bar{t}; \bar{y}, \bar{s}) &= \Psi(\bar{x}, T) - \chi(\bar{y}, \bar{s}) - \beta(T-\bar{s}) - \frac{1}{\epsilon}(|\bar{x}-\bar{y}|^2 + |\bar{s}-\bar{t}|^2)\\ &\leq \chi(\bar{x}, \bar{t}) - \chi(\bar{y}, \bar{s}),\end{aligned}$$

and if χ is uniformly continuous, we can conclude that

$$\limsup_{\epsilon\to 0} \psi_{\epsilon\beta}(\bar{x}, \bar{t}; \bar{y}, \bar{s}) \leq 0. \tag{10.4.24}$$

Noting that

$$\begin{aligned}\Psi(x, t) - \chi(x, t) &= \psi_{\epsilon\beta}(x, t; x, t) + \beta(T-t)\\ &\leq \psi_{\epsilon\beta}(\bar{x}, \bar{t}; \bar{y}, \bar{s}) + \beta(T-t),\end{aligned}$$

we conclude that

$$\begin{aligned}\Psi(x, t) - \chi(x, t) &\leq \limsup_{\epsilon\to 0} \psi_{\epsilon\beta}(\bar{x}, \bar{t}; \bar{y}, \bar{s}) + \beta(T-t)\\ &\leq \beta(T-t),\end{aligned}$$

which implies (10.4.18) and concludes the proof. □

Chapter 11
Stochastic Optimal Control

11.1 Stochastic Maximum Principle

11.1.1 Setting of the Problem

We consider

$$
\begin{aligned}
&g(x, v) : R^n \times R^m \to R^n, \\
&\sigma(x, v) : R^n \times R^m \to \mathcal{L}(R^k; R^n), \\
&g, \sigma \text{ are continuously differentiable and have} \\
&\qquad\qquad \text{bounded derivatives,} \\
&|g(x, v)| \leq \bar{g}(1 + |x| + |v|), \\
&|\sigma(x, v)| \leq \bar{\sigma}(1 + |x| + |v|).
\end{aligned}
\tag{11.1.1}
$$

We consider a probability space $\Omega, \mathcal{A}, P$ equipped with a filtration $\mathcal{F}^t$ and a standard $P, \mathcal{F}^t$ Wiener process with values in R^k. Let

$$\mathcal{U}_{\text{ad}} = \text{nonempty closed convex subset of } R^m. \tag{11.1.2}$$

An admissible control is a process $v(.) \in L^2_{\mathcal{F}}(0, T; R^m)$ such that

$$v(t) \in \mathcal{U}_{\text{ad}}, \text{ a.e a.s.}$$

For each admissible control, the state of the system is the solution of the stochastic differential equation (SDE)

A. Bensoussan, *Estimation and Control of Dynamical Systems*, Interdisciplinary Applied Mathematics 48, https://doi.org/10.1007/978-3-319-75456-7_11

$$dx = g(x(t), v(t))dt + \sigma(x(t), v(t))dw, \tag{11.1.3}$$
$$x(0) = \xi.$$

The initial state ξ is $\mathcal{F}^0$-measurable. We also assume that ξ is independent of the Wiener process. We have the estimate

$$E \sup_{0 \le t \le T} |x(t)|^2 \le C(1 + E|\xi|^2 + E \int_0^T |v(t)|^2 dt). \tag{11.1.4}$$

We now define the payoff functional. We consider

$$\begin{aligned} l(x, v) : R^n \times R^m &\to R, \\ h(x) : R^n &\to R, \\ l, h \text{ are continuously differentiable}, \\ |l(x, v)| &\le \bar{l}(1 + |x|^2 + |v|^2), \\ |h(x)| &\le \bar{h}(1 + |x|^2), \\ |l_x|,\ |l_v| &\le \bar{l}(1 + |x| + |v|), \\ |h_x| &\le \bar{h}(1 + |x|), \end{aligned} \tag{11.1.5}$$

and we set

$$J(v(.)) = E[\int_0^T l(x(t), v(t))dt + h(x(T))], \tag{11.1.6}$$

which is well defined by virtue of the assumptions.

11.1.2 Gâteaux Differential

Let $u(.)$ be an optimal control and $y(.)$ the corresponding optimal state. We shall follow the direction of Chapter 10, Proposition 10.1, and compute the Gâteaux differential

$$\frac{d}{d\theta} J(u(.) + \theta v(.))|_{\theta=0},$$

where $v(.) \in L^2_{\mathcal{F}}(0, T; R^m)$.

We introduce the process $z(.) \in L^2_{\mathcal{F}}(0, T; R^m)$, the solution of the linear equation

$$dz = (g_x(y,u)z + g_v(y,u)v)dt + \sum_{j=1}^{k}(\sigma_x^j(y,u)z + \sigma_v^j(y,u)v)dw_j, \quad (11.1.7)$$

$$z(0) = 0.$$

Then we state the following result

LEMMA 11.1. *Assume (11.1.1), (11.1.5). Then we have the formula*

$$\frac{d}{d\theta}J(u(.) + \theta v(.))|_{\theta=0} = E\int_0^T [l_x(y,u)z + l_v(y,u)v]dt + Eh_x(y(T))z(T). \quad (11.1.8)$$

PROOF. Let $y_\theta(.)$ be the state associated with $u(.) + \theta v(.)$. Define

$$\tilde{y}_\theta = \frac{y_\theta - y}{\theta} - z. \quad (11.1.9)$$

The first task is to show that

$$E \sup_{0\le t\le T} |\tilde{y}_\theta(t)|^2 \to 0, \text{ as } \theta \to 0. \quad (11.1.10)$$

If this is true, then we can write

$$\begin{aligned}
J(u(.) + \theta v(.)) &- J(u(.)) - \theta E[\int_0^T (l_x(y,u)z + l_v(y,u)v)dt + h_x(y(T))z(T)] \\
&= \theta\left\{E\int_0^T\int_0^1 l_x(y + \lambda(y_\theta - y), u + \lambda\theta v)\tilde{y}_\theta dtd\lambda + \right. \\
&+ E\int_0^T\int_0^1 (l_x(y + \lambda(y_\theta - y), u + \lambda\theta v) - l_x(y,u))zdtd\lambda \\
&+ E\int_0^T\int_0^1 (l_v(y + \lambda(y_\theta - y), u + \lambda\theta v) - l_v(y,u))vdtd\lambda \\
&+ E\int_0^1 h_x(y(T) + \lambda(y_\theta - y)(T))\tilde{y}_\theta(T)d\lambda \\
&\left. + E\int_0^1 (h_x(y(T) + \lambda(y_\theta - y)(T)) - h_x(y(T))z(T)d\lambda\right\} \\
&= \theta 0(\theta),
\end{aligned}$$

with $0(\theta) \to 0$, as $\theta \to 0$. Dividing by θ and letting $\theta \to 0$, we obtain the property (11.1.8). It remains to prove (11.1.10). We can see that $\tilde{y}_\theta(.)$ satisfies

$$
\begin{aligned}
d\tilde{y}_\theta &= \int_0^1 g_x(y+\lambda(y_\theta - y), u+\lambda\theta v)\tilde{y}_\theta d\lambda dt + \qquad (11.1.11)\\
&+ \sum_{j=1}^k \int_0^1 \sigma_x^j(y+\lambda(y_\theta - y), u+\lambda\theta v)\tilde{y}_\theta d\lambda dw_j + \\
&+ \int_0^1 (g_x(y+\lambda(y_\theta - y), u+\lambda\theta v) - g_x(y,u))z d\lambda dt + \\
&+ \int_0^1 (g_v(y+\lambda(y_\theta - y), u+\lambda\theta v) - g_v(y,u))v d\lambda dt + \\
&+ \sum_{j=1}^k \int_0^1 (\sigma_x^j(y+\lambda(y_\theta - y), u+\lambda\theta v) - \sigma_x^j(y,u))z d\lambda dw_j + \\
&+ \sum_{j=1}^k \int_0^1 (\sigma_v^j(y+\lambda(y_\theta - y), u+\lambda\theta v) - \sigma_v^j(y,u))v d\lambda dw_j, \\
\tilde{y}_\theta(0) &= 0.
\end{aligned}
$$

Hence

$$
\begin{aligned}
\frac{1}{2}\frac{d}{dt}E|\tilde{y}_\theta|^2 &= E\tilde{y}_\theta \int_0^1 g_x(y+\lambda(y_\theta - y), u+\lambda\theta v)\tilde{y}_\theta d\lambda + \\
&+ E\tilde{y}_\theta[\int_0^1 (g_x(y+\lambda(y_\theta - y), u+\lambda\theta v) - g_x(y,u))z d\lambda] + \\
&+ E\tilde{y}_\theta[\int_0^1 (g_v(y+\lambda(y_\theta - y), u+\lambda\theta v) - g_v(y,u))v d\lambda] + \\
&+ \sum_{j=1}^k E|\int_0^1 [(\sigma_x^j(y+\lambda(y_\theta - y), u+\lambda\theta v) - \sigma_x^j(y,u))z + \\
&+ (\sigma_v^j(y+\lambda(y_\theta - y), u+\lambda\theta v) - \sigma_v^j(y,u))v]d\lambda|^2.
\end{aligned}
$$

Also

$$
\begin{aligned}
\frac{1}{2}\frac{d}{dt}E|\tilde{y}_\theta|^2 &\le CE|\tilde{y}_\theta|^2 + \\
&+ CE|\int_0^1 [(g_x(y+\lambda(y_\theta - y), u+\lambda\theta v) - g_x(y,u))z + \\
&+ (g_v(y+\lambda(y_\theta - y), u+\lambda\theta v) - g_v(y,u))v]d\lambda|^2
\end{aligned}
$$

$$+\sum_{j=1}^{k}E|\int_0^1[(\sigma_x^j(y+\lambda(y_\theta-y),u+\lambda\theta v)-\sigma_x^j(y,u))z+$$

$$+(\sigma_v^j(y+\lambda(y_\theta-y),u+\lambda\theta v)-\sigma_v^j(y,u))v]d\lambda|^2.$$

It easily follows that

$$\sup_{0\leq t\leq T}E|\tilde{y}_\theta(t)|^2\leq C.$$

But applying the semimartingale inequality, see Chapter 6, (6.3.10), we can also state that

$$E\sup_{0\leq t\leq T}|\tilde{y}_\theta(t)|^2\leq C. \tag{11.1.12}$$

But then from (11.1.9), one gets

$$E\sup_{0\leq t\leq T}|y_\theta(t)-y(t)|^2\leq C\theta^2. \tag{11.1.13}$$

It follows that for a subsequence, we have

$$\sup_{0\leq t\leq T}|y_\theta(t)-y(t)|^2\to 0,\text{ a.s., as }\theta\to 0.$$

Using this information in the SDE of $\tilde{y}_\theta(.)$, see (11.1.11), and using the continuity of $g_x, g_v, \sigma_x, \sigma_v$ and the fact that they are bounded, we obtain by standard techniques the property (11.1.10). The proof is then complete. □

Of course the drawback of formula (11.1.8) is that z remains on the right-hand side. So it is not explicit in $v(.)$. We thus need to introduce adjoint states. For that purpose, we consider $\varphi, \psi_j,\ j=1,\cdots,k,\ \in R^n$. We define the process $\zeta(.)$, the solution of

$$d\zeta=(g_x(y,u)\zeta+\varphi)dt+\sum_{j=1}^{k}(\sigma_x^j(y,u)\zeta+\psi^j)dw_j,$$

$$\zeta(0)=0. \tag{11.1.14}$$

We next consider the map

$$\varphi,\psi^j\to E[\int_0^T l_x(y,u).\zeta dt\ +h_x(y(T)).\zeta(T)],$$

which is linear and continuous on $(L^2_{\mathcal{F}}(0, T; R^n))^{k+1}$. Therefore, there exist $q, r^j \in L^2_{\mathcal{F}}(0, T; R^n)$ uniquely defined such that

$$E[\int_0^T l_x(y, u).\zeta dt \ + h_x(y(T)).\zeta(T)] = E[\int_0^T \varphi.p \qquad (11.1.15)$$
$$+ \sum_{j=1}^{k} \psi^j r^j]dt \forall \varphi, \psi^j.$$

Comparing this with the Gâteaux differential (11.1.8), we obtain

$$\frac{d}{d\theta} J(u(.) + \theta v(.))|_{\theta=0} = E \int_0^T [l_v(y, u)v + p.g_v(y, u)v + \sum_{j=1}^{k} r^j.\sigma_v^j(y, u)v]dt.$$

We can then introduce the Hamiltonian

$$H(x, v, q, r) = l(x, v) + q.g(x, v) + \sum_{j=1}^{k} r^j.\sigma^j(x, v), \qquad (11.1.16)$$

where r stands for the matrix whose jth column is r^j.

We obtain the Gâteaux differential formula

$$\frac{d}{d\theta} J(u(.) + \theta v(.))|_{\theta=0} = E \int_0^T H_v(y(t), u(t), p(t), r(t))v(t)dt. \qquad (11.1.17)$$

PROPOSITION 11.1. *We assume (11.1.1), (11.1.2), (11.1.5). If $u(.)$ is an optimal control for (11.1.3), (11.1.6), then one has*

$$H_v(y(t), u(t), p(t), r(t))(v - u(t)) \geq 0, \ \text{a.s., a.e. (product measure)}$$
$$\forall \ v \in \mathcal{U}_{ad}. \qquad (11.1.18)$$

PROOF. From the Gâteaux differential formula (11.1.17), it follows, where we recall that $\mathcal{U}_{\text{ad}}$ is convex, that for every admissible control $v(.)$, one has

$$E \int_0^T H_v(y(t), u(t), p(t), r(t))(v(t) - u(t))dt \geq 0. \qquad (11.1.19)$$

We derive (11.1.18) by a localization argument. For $v \in \mathcal{U}_{\text{ad}}$ fixed, we define

$$\lambda(t; \omega) = H_v(y(t), u(t), p(t), r(t))(v - u(t)),$$

and we consider

$$A^t = \{\omega | \lambda(t; \omega) < 0\} \in \mathcal{F}^t.$$

We can then set

$$v(t) = \begin{vmatrix} v & \text{if } \omega \in A^t, \\ u(t) & \text{if } \omega \notin A^t. \end{vmatrix}$$

We get an admissible control, and

$$E \int_0^T H_v(y(t), u(t), p(t), r(t))(v(t) - u(t))dt = \int_0^T E(\lambda(t; \omega)\mathbb{1}_{A^t})dt < 0,$$

which contradicts (11.1.19). □

11.1.3 Equations for p and r

We first notice that since the uncertainties in the state evolution are limited to ξ and $w(.)$, we may as well consider that

$$\mathcal{F}^t = \sigma(\xi, w(s)\ s \leq t). \tag{11.1.20}$$

We then can state the following theorem.

THEOREM 11.1. *We make the assumptions of Proposition 11.1, as well as (11.1.20). Then the pair p, r is the unique solution of the backward stochastic differential equation*

$$- dp = (g_x^*(y, u)p + l_x(y, u) + \sum_{j=1}^{k} \sigma_x^{j*}(y, u)r^j)dt - \sum_{j=1}^{k} r^j dw_j, \tag{11.1.21}$$
$$p(T) = h_x(y(T)).$$

PROOF. We first show that if $p, r^j \in L^2_{\mathcal{F}}(0, T; R^n)$ is a solution of (11.1.21), then it is unique. It suffices to check that p, r^j satisfy the relation (11.1.15) for all φ, ψ^j. Considering ζ to be the solution of (11.1.14), we can compute the Itô differential of $p(t).\zeta(t)$ and obtain (11.1.15) after easy calculations. It remains to show that (11.1.21) has a solution.

We consider the fundamental matrix $\Phi(t)$, a solution of

$$d\Phi = g_x(y, u)\Phi dt + \sum_{j=1}^{k} \sigma_x^j(y, u)\Phi dw_j, \tag{11.1.22}$$
$$\Phi(0) = I.$$

Let us show that it has an inverse $\Psi(t)$, the solution of

$$
\begin{aligned}
-d\Psi &= (\Psi g_x(y,u) - \Psi \sum_{j=1}^{k} \sigma_x^j \sigma_x^j (y,u))dt + \Psi \sum_{j=1}^{k} \sigma_x^j (y,u) dw_j, \quad (11.1.23)\\
\Psi(0) &= I.
\end{aligned}
$$

These are two linear equations, so the solution is well defined. Considering $\Phi\Psi$, we can check that □

$$
\begin{aligned}
d\Phi\Psi &= (g_x \Phi\Psi - \Phi\Psi g_x + \Phi\Psi \sum_{j=1}^{k} \sigma_x^j \sigma_x^j - \sum_{j=1}^{k} \sigma_x^j \Phi\Psi \sigma_x^j)dt\\
&\quad + \sum_{j=1}^{k} (\sigma_x^j \Phi\Psi - \Phi\Psi \sigma_x^j) dw_j,\\
\Phi\Psi(0) &= I.
\end{aligned}
$$

This is a linear equation in $\Phi\Psi$, and the identity is a solution, hence the unique solution, so $\Phi\Psi(t) = I$. Considering next $\Psi\Phi$, we check easily that $d\Psi\Phi = 0$, whence also $\Psi\Phi(t) = I$.

We now identify $p(t)$. Consider (11.1.14) with $\psi^j = 0$. Hence

$$
\begin{aligned}
d\zeta &= (g_x(y,u)\zeta + \varphi)dt + \sum_{j=1}^{k} \sigma_x^j (y,u) \zeta dw_j,\\
\zeta(0) &= 0. \quad (11.1.24)
\end{aligned}
$$

We check easily that

$$
d(\Psi(t)\zeta(t)) = \Psi(t)\varphi(t)dt,
$$

and hence

$$
\zeta(t) = \Phi(t) \int_0^t \Psi(s)\varphi(s)ds. \quad (11.1.25)
$$

From (11.1.15) we get

$$
\begin{aligned}
E \int_0^T p(t).\varphi(t)dt &= E \int_0^T l_x(y(t), u(t)).\Phi(t)(\int_0^t \Psi(s)\varphi(s)ds)dt\\
&\quad + Eh_x(y(T)).\Phi(T) \int_0^T \Psi(s)\varphi(s)ds.
\end{aligned}
$$

Therefore,

$$E\int_0^T p(s).\varphi(s)ds = E\int_0^T [\Psi^*(s)(\Phi^*(T)h_x(y(T)) + \int_s^T \Phi^*(t)l_x(y(t), u(t))dt)].\varphi(s)ds$$
$$= E\int_0^T [\Psi^*(s)(X - \int_0^s \Phi^*(t)l_x(y(t), u(t))dt)], \varphi(s)ds$$

with

$$X = \Phi^*(T)h_x(y(T)) + \int_0^T \Phi^*(t)l_x(y(t), u(t))dt.$$

Since φ is arbitrary in $L^2_{\mathcal{F}}(0, T; R^n)$, we have

$$p(t) = -\Psi^*(t)\int_0^t \Phi^*(s)l_x(y(s), u(s))ds + \Psi^*(t)E[X|\mathcal{F}^t]. \tag{11.1.26}$$

This defines $p(t)$. We use next the representation theorem of martingales [31] to write

$$E[X|\mathcal{F}^t] = \sum_{j=1}^k \int_0^t G^j dw_j + X_0,$$

with

$$X_0 = E[X|\mathcal{F}^0]$$

and $G^j \in L^2_{\mathcal{F}}(0, T; R^n)$. We note that

$$p(t) = \Psi^*(t)[X_0 - \int_0^t \Phi^*(s)l_x(y(s), u(s))ds + \sum_{j=1}^k \int_0^t G^j dw_j]$$
$$= \Psi^*(t)\eta(t).$$

It is then easy to compute the Itô differential of $p(t)$ and to check that it satisfies the first relation (11.1.21) with

$$r^j(t) = \Psi^*(t)G^j(t) - (\sigma_x^j)^*(y(t), u(t))p(t). \tag{11.1.27}$$

Moreover, one checks easily the final condition in (11.1.21). This concludes the proof. ■

We can summarize the set of conditions for $u(.)$ to be an optimal control:

$$dy = g(y,u)dt + \sigma(y,u)dw, \tag{11.1.28}$$
$$y(0) = \xi,$$
$$-dp = (g_x^*(y,u)p + l_x(y,u) + \sum_{j=1}^{k} \sigma_x^{j*}(y,u)r^j)dt - \sum_{j=1}^{k} r^j dw_j, \tag{11.1.29}$$
$$p(T) = h_x(y(T)),$$

$$H_v(y(t), u(t), p(t), r(t))(v - u(t)) \geq 0, \text{ a.s., a.e. (product measure)}$$
$$\forall\ v \in \mathcal{U}_{\text{ad}}. \tag{11.1.30}$$

This set of relations forms the **stochastic maximum principle**.

EXERCISE 11.1. Apply the stochastic maximum principle to the problem of Chapter 8, Section 8.2.

REMARK 11.1. The second equation (11.1.29) is a backward stochastic differential equation (BSDE). Indeed, the initial condition is not given at time 0, but at time T, in spite of the fact that the process $p(t)$, which is the solution is adapted. We shall consider these equations in greater depth in the sequel. Note that a BSDE has two unknowns, namely $p(t)$ and $r(t)$, in the case of (11.1.29), which are defined jointly. The unknown $r(t)$ plays the role of a Lagrange multiplier associated with the constraint that the main unknown $p(t)$ is an adapted process.

11.2 Stochastic Dynamic Programming

11.2.1 Preliminaries; Optimality Principle

We make the assumptions (11.1.1), (11.1.2), and (11.1.5). We consider a family of stochastic control problems

$$dx = g(x(s), v(s))ds + \sigma(x(s), v(s))dw,\ t \leq s \leq T, \tag{11.2.1}$$
$$x(t) = x,$$

$$J_{x,t}(v(.)) = E[\int_t^T l(x(s), v(s))ds + h(x(T))]. \tag{11.2.2}$$

However, the situation is more complex than in the deterministic case. First we note that the initial conditions are deterministic, whereas in the case $t = 0$, we had allowed a random initial state; see (11.1.3). This difference obviously did not occur in the deterministic case. We next have to clarify the filtration with respect to which the control is adapted. We introduce $\mathcal{F}_t^s = \sigma(w(\tau) - w(t),\ t \leq \tau \leq s)$, and we require

$$v(.) \in L^2_{\mathcal{F}}(t, T; R^m),\ \ v(s) \in \mathcal{U}_{\text{ad}},\ \text{a.e a.s.}$$

The notation means that $v(s)$ is $\mathcal{F}_t^s$-measurable, $\forall s \in (t, T)$. Naturally, the control may depend on the initial conditions, which are deterministic.

We define the value function as

$$\Phi(x, t) = \inf_{v(.)} J_{x,t}(v(.)). \tag{11.2.3}$$

The optimality principle reads as follows:

$$\Phi(x, t) = \inf_{v(.)} E[\int_t^{t+\epsilon} l(x(s), v(s))ds + \Phi(x(t+\epsilon), t+\epsilon)],\ \forall t \leq T - \epsilon. \tag{11.2.4}$$

This is not as easy to check as in the deterministic case. In fact, this property is linked to the fact that an optimal control will satisfy a time consistency property, related to the Markov property. Take, to fix the ideas, $t = 0$ and $x = x_0$. The control satisfies the time consistency property if $\forall t$, the part of the control in the interval t, T depends on $\mathcal{F}^t$ only through the state $x(t)$. If we restrict the controls to satisfy this property, then the optimality principle can be checked as in the deterministic case.

But one has to prove that it is not a restriction. To prove it directly requires some delicate probabilistic arguments. In fact, it is much easier to proceed indirectly, as we shall see below, using fully the idea that dynamic programming is a sufficient condition theory.

11.2.2 HJB Equation

As we have seen in the deterministic case, a natural question is to derive from the optimality principle (11.2.4) an HJB equation. If one proceeds formally, it is not difficult to obtain the following equation:

$$\frac{\partial \Phi}{\partial t} + \inf_{v \in \mathcal{U}_{\text{ad}}} [l(x, v) + D\Phi.g(x, v) + \frac{1}{2}\text{tr}D^2\Phi\sigma\sigma^*(x, v)] = 0,$$
$$\Phi(x, T) = h(x). \tag{11.2.5}$$

From the point of view of partial differential equation (PDE), this is a difficult problem, which we shall consider only in particular cases. We refer to [29] and [33] for significant results for this fully nonlinear HJB equation (fully nonlinear refers to the fact that there is a control in the function σ).

However, a general theory is possible only when strong smoothness assumptions are made on the data, and one looks for a smooth solution $\Phi(x, t)$. This is a serious limitation that reduces the interest of a general theory.

EXERCISE 11.2. Apply the HJB equation (11.2.5) to the problem of Exercise 11.1. Compare the results with the stochastic maximum principle.

Equation (11.2.1) models the speed of evolution of the state of the system. The **drift term** $g(x, v)$ is similar to the deterministic case, and $\sigma(x, v)dw$ represents an uncertainty. The function $\sigma(x, v)$ itself models the variability of the uncertainty. It is called the **diffusion term**. In practice, the uncertainty is generally external to the system, and therefore its variability cannot be controlled. It therefore makes sense to assume that

$$\sigma(x, v) = \sigma(x). \tag{11.2.6}$$

We set

$$a(x) = \frac{1}{2}\sigma(x)\sigma^*(x),$$

and we shall make the assumption (coercivity assumption)

$$\alpha I \leq a(x) \leq MI, \ \alpha > 0. \tag{11.2.7}$$

The reason is purely technical. A rich mathematical theory is available in this case. However, it is meaningful in practice. It means that there is no degeneracy in the uncertainties, and the variability is bounded. However, this assumption is not verified in the case of the problem of Exercise 11.1.

Consider the second-order differential operator

$$A\varphi(x) = -\sum_{i,j=1}^{n} a_{ij}(x)\frac{\partial^2\varphi}{\partial x_i \partial x_j}(x). \tag{11.2.8}$$

Introduce also the Hamiltonian, see (10.2.9),

$$H(x, q) = \inf_{v \in \mathcal{U}_{\mathrm{ad}}} (l(x, v) + q.g(x, v)). \tag{11.2.9}$$

Then the HJB equation (11.2.5) can be written as follows:

$$\begin{aligned} -\frac{\partial \Phi}{\partial t} + A\Phi &= H(x, D\Phi), \\ \Phi(x, T) &= h(x). \end{aligned} \tag{11.2.10}$$

This is where we benefit considerably from the theory of parabolic PDEs. Such an equation is called quasilinear, because the term involving the second derivative is linear ($A\Phi$). The nonlinearity affects only the first-order derivative. In can be also viewed as a regularization of the deterministic HJB equation, which corresponds to the particular case $\sigma = 0$, a case that is excluded by the coercivity assumption.

In contrast to the case of the deterministic HJB equation, the approach of (11.2.10) is not to prove directly that the value function satisfies the PDE. The approach is to study it directly by PDE techniques, independently of its interpretation. We shall also make the following assumption:

$$h_{xx}(x) \text{ is measurable and bounded.} \tag{11.2.11}$$

This assumption is not indispensable, but it facilitates the treatment. In fact, it allows one to reduce the problem to the case $h = 0$. Indeed, considering $\Phi(x, t) - h(x)$, it is the solution of an HJB equation with the same $g(x, v), h(x) = 0$, and $l(x, v)$ changed to

$$l(x, v) + Dh(x).g(x, v) - Ah(x),$$

which has the same properties as $l(x, v)$.

We have the following theorem.

THEOREM 11.2. *We assume (11.1.1), (11.1.2), (11.1.5), (11.2.7), (10.2.10), and (11.2.11). Then there is a unique solution of (11.2.10) in the functional space*

$$\Phi, \frac{\partial \Phi}{\partial t}, \frac{\partial \Phi}{\partial x_i}, \frac{\partial^2 \Phi}{\partial x_i \partial x_j} \; L^p(0, T; L^p_{loc}(R^n)) \tag{11.2.12}$$
$$\forall p, \; 2 \le p < +\infty,$$

and

$$|\Phi(x, t)| \le C(1 + |x|^2), \tag{11.2.13}$$
$$|D\Phi(x, t)| \le C(1 + |x|).$$

The space $L^p_{\text{loc}}(R^n)$ is the space of functions z such that $z\varphi \in L^p(R^n), \forall \varphi \in C_0^\infty(R^n)$ (infinitely differentiable with compact support).

This unique solution is the value function (11.2.3). We emphasize the uniqueness. We also note that $\Phi(x, t)$ is continuous.

We can define a measurable feedback, as in the deterministic case. Let

$$|\bar{v}(x, q)| \le C(1 + |x| + |q|), \tag{11.2.14}$$

which minimizes the Lagrangian

$$L(x, v, q) = l(x, v) + q.g(x, v).$$

The feedback is defined by

$$\hat{v}(x,t) = \bar{v}(x, D\Phi(x,t)), \tag{11.2.15}$$

and it satisfies the bound

$$|\hat{v}(x,t)| \le C(1+|x|). \tag{11.2.16}$$

We note that with this feedback, we have

$$-\frac{\partial \Phi}{\partial t}(x,t) + A\Phi(x,t) = l(x, \hat{v}(x,t)) + D\Phi(x,t).g(x, \hat{v}(x,t)). \tag{11.2.17}$$

We do not give details of the proof of the existence and uniqueness of a solution of (11.2.10), (11.2.12), (11.2.13). It uses PDE techniques; see, for instance, [5]. We shall only discuss formally the interpretation as the value function. If $\Phi(x,t)$ were $C^{2,1}$ and $\hat{v}(x,t)$ Lipschitz continuous in x, then we could define the solution of

$$dy = g(y(s), \hat{v}(y(s),s))ds + \sigma(y(s))dw, \; t \le s \le T, \tag{11.2.18}$$
$$y(t) = x.$$

We can then set

$$u(s) = \hat{v}(y(s), s), \tag{11.2.19}$$

which is an admissible control.

We can then apply Itô's formula to compute the Itô differential of $\Phi(y(s), s)$. Thanks to (11.2.17), one checks easily that

$$\Phi(x,t) = J_{x,t}(u(.)). \tag{11.2.20}$$

On the other hand,

$$-\frac{\partial \Phi}{\partial t}(x,s) + A\Phi(x,s) \le l(x,v) + D\Phi(x,s).g(x,v), \;\; \forall x, s, v \in \mathcal{U}_{\text{ad}}.$$

If $v(.)$ is admissible and $x(.)$ is the corresponding state, then we can replace x by $x(s)$ in the preceding inequality and v by $v(s)$. Applying again Itô's formula, we obtain

$$\Phi(x,t) \le J_{x,t}(v(.)).$$

Therefore, $\Phi(x,t)$ is the value function, and $u(.)$ defined by (11.2.19) is the optimal control. So the interpretation is straightforward when sufficient smoothness is available. This reasoning constitutes the *verification principle*, as in Proposition 10.3.

The smoothness we have is substantial. However, it is insufficient to allow the use of Itô's formula as above. Nevertheless, we can proceed by smoothing approximations and use Itô's formula on smooth approximations. With the smoothness, passing to the limit is possible.

In this way, one can prove that $\Phi(x,t)$ is the value function. However, we cannot prove that there exists an optimal control.

Note that the estimates (11.2.13) can be derived from the interpretation of Φ as the value function. The proof is similar to the deterministic case; see Theorem 10.2.

The optimality principle can also be understood in the following way. Let us define

$$\Phi_{xt}(s) = \Phi(y(s), s). \tag{11.2.21}$$

We check easily, applying Itô's calculus, that

$$\Phi_{xt}(s) = E[h(y(T)) + \int_s^T l(y(\tau), u(\tau))d\tau | \mathcal{F}^s]. \tag{11.2.22}$$

On the other hand, for any adapted control $v(.)$, we define the trajectory $x_{y(s),s}^{v(.)}(\tau)$ as the solution of the equation

$$\begin{aligned} dx &= g(x(\tau), v(\tau))d\tau + \sigma(x(\tau), v(\tau))dw, \ s \le \tau \le T, \\ x(s) &= y(s). \end{aligned} \tag{11.2.23}$$

Then we have

$$\Phi_{xt}(s) \le E[h(x_{y(s),s}^{v(.)}(T)) + \int_s^T l(x_{y(s),s}^{v(.)}(\tau), v(\tau))d\tau | \mathcal{F}^s],$$

and therefore,

$$\Phi_{xt}(s) = \inf_{v(.)} E[h(x_{y(s),s}^{v(.)}(T)) + \int_s^T l(x_{y(s),s}^{v(.)}(\tau), v(\tau))d\tau | \mathcal{F}^s]. \tag{11.2.24}$$

Since the left- and right-hand sides of (11.2.24) are random, the infimum holds a.e. So it is an essential infimum. We call it inf to simplify notation.

REMARK 11.2. As in the deterministic case, the optimal control is obtained through a feedback on the state. Controls that are stochastic processes adapted to the fixed filtration $\mathcal{F}^t$ are called *open-loop controls*. Controls defined by a feedback are called *closed-loop controls*. As in the deterministic case, the optimal control in the class of open-loop controls and the optimal control in the class of closed-loop controls coincide. In fact, as discussed in the deterministic case, the two classes coincide; see Remark 10.6. However, the open-loop controls cannot be limited to deterministic functions, which would be too restrictive. On the other hand, the feedback functions are deterministic functions of the state. The final randomness will come from the state. Note that the feedback in the deterministic case and in the stochastic case are different, since the two HJB equations are different. ■

Next comes an idea that has no equivalent in the deterministic case that will allow to prove the existence of an optimal control. We are going to relax slightly the concept of optimal control using Girsanov's theorem. We shall call this new formulation the **weak formulation** of the stochastic control problem, whereas the formulation (11.2.2), (11.2.3) is called the **strong formulation** of the stochastic control problem. We begin by considering weak solutions of SDEs.

11.3 Weak Solution of Stochastic Differential Equations

11.3.1 The Concept

We introduced stochastic differential equations (SDEs) in Chapter 6, Section 6.4. We assumed in Theorem 6.1 that the drift term and the diffusion term were Lipschitz in x. The theory follows closely that of the deterministic case. The formulation is called the **strong formulation**, and the solution is called the **strong solution**. It applies with no coerciveness assumptions, and we may have $\sigma = 0$ (deterministic case). However, σ must be Lipschitz in x. A large mathematical theory is devoted to extending the concept of solution to situations in which g, σ are only continuous in x. For our objective, when we control the drift but not the diffusion term, it makes sense to consider a bisymmetric assumption on g, σ.

In fact, we want to study the SDE

$$\begin{aligned} dy &= g(y(t), t)dt + \sigma(y(t), t)dw(t), \\ y(0) &= y_0, \end{aligned} \tag{11.3.1}$$

with the assumptions

$$n = m, \ \sigma(x, t), \ \sigma^{-1}(x, t) \quad \text{continuous and bounded,} \tag{11.3.2}$$

$$\begin{aligned} |\sigma(x, t) - \sigma(x', t)| &\leq K|x - x'|, \\ g(x, t) &\text{ measurable,} \\ |g(x, t)| &\leq C(1 + |x|). \end{aligned} \tag{11.3.3}$$

With respect to the strong formulation, we have strengthened considerably the assumptions on σ, invertibility and boundedness, while keeping the Lipschitz regularity. However, we have relaxed considerably the regularity assumption of the drift term: not even continuity is required. Of course, the deterministic case cannot be a particular case, and such an extension will not be possible for deterministic differential equations.

However, we have to extend the concept of the solution of (11.3.1). The idea is to make use of Girsanov's theorem.

We consider first a probability space $\Omega, \mathcal{A}, \tilde{P}$ and a filtration $\mathcal{F}^t$. We assume that there exists an n-dimensional standard $\tilde{P}, \mathcal{F}^t$ Wiener process, denoted by $\tilde{w}(.)$. We also assume that y_0 is an $\mathcal{F}^0$-measurable, square integrable random variable. We can then solve in a strong sense the SDE without drift,

$$\begin{aligned} dy &= \sigma(y(t), t)d\tilde{w}(t), \\ y(0) &= y_0. \end{aligned} \tag{11.3.4}$$

The solution $y(.)$ is a continuous $\tilde{P}, \mathcal{F}^t$ martingale, and we have

$$E \sup_{0 \leq t \leq T} |y(t)|^2 \leq C_T(1 + E|y_0|^2). \tag{11.3.5}$$

We want to solve (11.3.1) on the interval $[0, T]$. Taking $T = +\infty$ is slightly more technical, but possible. However, there is no restriction on T, as soon as it is finite. We introduce the process

$$\alpha(t) = \sigma^{-1}(y(t), t)g(y(t), t). \tag{11.3.6}$$

The important fact is the property

$$E \exp[\int_0^T \alpha(t).d\tilde{w}(t) - \frac{1}{2}\int_0^T |\alpha(t)|^2 dt] = 1. \tag{11.3.7}$$

This is a technical result, for which we also need the property

$$E \exp k_0|y_0|^2 < +\infty. \tag{11.3.8}$$

EXERCISE 11.3. Replace $\alpha(.)$ by a piecewise approximation. Set $T = Nk$, and define

$$\begin{aligned} \alpha^j &= \frac{1}{j}\int_{(j-1)k}^{jk} \alpha(t)dt, \;\; j = 1, \cdots N, \\ \alpha^0 &= 0. \end{aligned}$$

We then set

$$\alpha_k(t) = \alpha^j, \;\; \forall t \in (jk, (j+1)k).$$

Show that (11.3.7) is true when $\alpha(.)$ is replaced with $\alpha_k(.)$. ■

If we accept (11.3.7), then we can define a new probability P by the Radon–Nikodym derivative

$$\frac{dP}{d\tilde{P}} = \exp[\int_0^T \alpha(t).d\tilde{w}(t) - \frac{1}{2}\int_0^T |\alpha(t)|^2 dt], \tag{11.3.9}$$

and we define the stochastic process

$$w(t) = \tilde{w}(t) - \int_0^t \alpha(s)ds. \tag{11.3.10}$$

According to Girsanov's theorem, see Chapter 6, Section 6.5, on the system $\Omega, \mathcal{A}, P$ the process $w(.)$ is a standard $P, \mathcal{F}^t$ Wiener process, and from the definition of $w(.)$, equation (11.3.4) reads like (11.3.1). However, we see that the pair $P, w(.)$ is ad hoc, and cannot be fixed a priori. It is in this sense that the concept of solution has been extended. Note that we do not have uniqueness, although we shall revisit the concept of uniqueness below. We say that $y(.)$ is a weak solution of the SDE (11.3.1).

REMARK 11.3. For $T = +\infty$, we define a family of probabilities on $\Omega, \mathcal{F}^t$, and using Kolmogorov's theorem, this family originates from a single P on the σ-algebra $\mathcal{A} = \cup_{t\geq 0}\mathcal{F}^t$ satisfying

$$\frac{dP}{d\tilde{P}}|_{\mathcal{F}^t} = \exp[\int_0^t \alpha(s).d\tilde{w}(s) - \frac{1}{2}\int_0^t |\alpha(s)|^2 ds].$$

11.3.2 Generalization of Itô's Formula

We shall use a generalization of Itô's formula adapted to weak solutions of the SDE (11.3.1), under the assumptions (11.3.2), (11.3.3). We first consider a function $f(x,t)$ that is in $L^p(0,T;L^p_{\text{loc}}(R^n)),\ p > \frac{n}{2}+1$. Let

$$B_R = \{x \in R^n |\ |x| \leq R\},$$

the ball with center 0 and radius R in R^n. Consider two stopping times with respect to the filtration $\mathcal{F}^t$, $\theta \leq \theta'$. Consider a weak solution $y(.)$ of the SDE (11.3.1). Let τ_R be the exit time of $y(.)$ from the ball B_R. So

$$\tau_R = \inf\{t|\ |y(t)| \geq R\}.$$

It is a stopping time with respect to the filtration $\mathcal{F}^t$. Consequently, $\theta \wedge \tau_R$ and $\theta' \wedge \tau_R$ are also stopping times. We consider the integral

$$\int_{\theta\wedge\tau_R}^{\theta'\wedge\tau_R} f(y(t),t)dt. \tag{11.3.11}$$

Since $f(x,t)$ is not a continuous function, one needs to give a precise meaning to this integral. In fact, since f is locally L^p integrable, it is defined as an element of L^p and not pointwise. This means that it is an equivalence class of functions defined pointwise, equal a.e. For each function $f(x,t)$ defined pointwise (even not continuous), we can give a meaning to the integral (11.3.11). The important result is that if we choose a representative, then the following estimate holds:

$$|E\int_{\theta\wedge\tau_R}^{\theta'\wedge\tau_R} f(y(t),t)dt| \leq C||f||_{L^p(B_R\times(0,T))}, \tag{11.3.12}$$

with a constant independent of R and of the representative f. This is a standard result; see, for instance, [5]. Therefore, the random variables constructed on the equivalence class are also equal a.s. They form an equivalence class of random variables that are equal a.s. There is compatibility of the two measures: the probability measure and the Lebesgue measure.

This property allows us to obtain a generalization of Itô's formula.

Consider a function $\Psi(x,t)$ that is continuous on $R^n\times[0,T]$ such that

$$\Psi, \frac{\partial\Psi}{\partial t}, \frac{\partial\Psi}{\partial x_i}, \frac{\partial^2\Psi}{\partial x_i\partial x_j}\ L^p(0,T;L^p_{\text{loc}}(R^n)), \tag{11.3.13}$$

$$p>\frac{n}{2}+1.$$

Then we have the formula

$$E\Psi(y(\theta'\wedge\tau_R),\theta'\wedge\tau_R)-E\Psi(y(\theta\wedge\tau_R),\theta\wedge\tau_R)$$

$$=E\int_{\theta\wedge\tau_R}^{\theta'\wedge\tau_R}(\frac{\partial\Psi}{\partial t}-A\Psi+g.D\Psi)(y(t),t)dt, \tag{11.3.14}$$

which is an integrated form of Itô's formula, more general than what can be obtained from the classical Itô's formula, since the function Ψ is not $C^{2,1}$.

11.3.3 Interpretation of Solutions of Linear PDEs

We consider the linear parabolic PDE

$$-\frac{\partial\Psi}{\partial t}+A\Psi-g.D\Psi=l(x,t), \tag{11.3.15}$$

$$\Psi(x,T)=h(x),$$

in which the function $g(x,t)$ satisfies (11.3.3), $\sigma(x,t)$ satisfies (11.3.2), and $l(x,t),\ h(x)$ are measurable functions such that

$$|l(x,t)|, |h(x)| \leq C(1+|x|^2), \tag{11.3.16}$$

$$|h_x(x)| \leq C(1+|x|),$$

$$|h_{xx}(x)| \leq C.$$

We see that we keep a smoothness assumption on h, whereas the smoothness on l has been removed. This asymmetry is due to the fact that we want to reduce (11.3.15) to the case $h = 0$ by simply considering $\Psi - h$ and replacing l with $l + Dh.g - Ah$.

Note that the matrix $a(x,t) = \frac{1}{2}\sigma\sigma^*(x,t)$ satisfies (11.2.7). We next consider the SDE (11.3.1) in the weak sense. So we have

$$dy = \sigma(y(s), s)d\tilde{w}(s), \tag{11.3.17}$$
$$y(t) = x,$$

and

$$w(s) = \tilde{w}(s) - \tilde{w}(t) - \int_t^s \sigma^{-1}g(y(\tau), \tau)d\tau, \tag{11.3.18}$$

which implies

$$dy = g(y(s), s)ds + \sigma(y(s), s)dw(s), \tag{11.3.19}$$
$$y(t) = x,$$

for a system $\Omega, \mathcal{A}_t, P_t, \mathcal{F}_t^s$, and $w(s)$ is a standard $P, \mathcal{F}_t^s$ Wiener process, with values in R^n, with $w(t) = 0$. We then begin with a system $\Omega, \mathcal{A}, \tilde{P}, \mathcal{F}^t$, and $\tilde{w}(.)$ is a $\tilde{P}, \mathcal{F}^t$ standard Wiener process with values in R^n. We set

$$\mathcal{F}_t^s = \sigma(\tilde{w}(\tau) - \tilde{w}(t),\ t \le \tau \le s),$$
$$\mathcal{A}_t = \cup_{s=t}^{+\infty}\mathcal{F}_t^s,$$

and P_t is derived from $\tilde{P}$ by the Radon–Nikodym derivative

$$\frac{dP_t}{d\tilde{P}}|_{\mathcal{F}_t^s} = \exp[\int_t^s \sigma^{-1}g(y)d\tilde{w} - \frac{1}{2}|\sigma^{-1}g(y)|^2 d\tau]. \tag{11.3.20}$$

We then have the following result.

PROPOSITION 11.2. *We assume (11.3.3), (11.3.2), and (11.3.16). Then there exists a unique solution of (11.3.15) such that*

$$\Psi, \frac{\partial\Psi}{\partial t}, \frac{\partial\Psi}{\partial x_i}, \frac{\partial^2\Psi}{\partial x_i \partial x_j}\ L^2(0,T; L^2_{loc}(R^n)), \tag{11.3.21}$$

$$|\Psi(x,t)| \le C(1 + |x|^2), \tag{11.3.22}$$

and

$$\Psi(x,t) = E_t[\int_t^T l(y(s), s)ds + h(y(T))]. \tag{11.3.23}$$

PROOF. We only sketch the proof. Note that if we return to the probability $\tilde{P}$, formula (11.3.23) can be written

$$\Psi(x,t) = \tilde{E}[\int_t^T l(y(s))\exp(\int_t^s \sigma^{-1}g(y)d\tilde{w} - \frac{1}{2}|\sigma^{-1}g(y)|^2 d\tau)ds +$$

$$+ h(y(T))\exp(\int_t^T \sigma^{-1}g(y)d\tilde{w} - \frac{1}{2}|\sigma^{-1}g(y)|^2 d\tau)]. \tag{11.3.24}$$

The technique is to find a priori estimates. Suppose we know that the function $\Psi(x,t)$ defined by (11.3.23) or (11.3.24) satisfies (11.3.15) with sufficient regularity. Can we obtain estimates on the norms of the function and derivatives? If this is possible, then one tries to find an approximation of the function Ψ that satisfies the estimate and that in the limit satisfies the equation and formulas (11.3.23) or (11.3.24). This is the basic concept of the a priori estimates approach. Now, from (11.3.19), one first checks that

$$E_t \sup_{t\le s\le T} |y(s)|^2 \le C(1+|x|^2),$$

and therefore, from (11.3.23) and the assumptions on l and h, see (11.3.16), we have

$$|\Psi(x,t)| \le C(1+|x|^2). \tag{11.3.25}$$

We now reduce the problem (11.3.15) to the case $h = 0$. We consider the equation

$$-\frac{\partial \Psi}{\partial t} + A\Psi - g.D\Psi = l, \tag{11.3.26}$$
$$\Psi(x,T) = 0.$$

Let $L^2_{\pi_\rho}(R^n)$ denote the space of functions φ such that $\varphi\pi_\rho^{\frac{1}{2}} \in L^2(R^n)$, with $\pi_\rho(x) = \dfrac{1}{(1+|x|^2)^\rho}$. If $\rho > \frac{n}{2}+2$, then l belongs to $L^2_{\pi_\rho}(R^n)$ for all t, with norm uniform in t. If a solution of (11.3.26) satisfies (11.3.25), then by multiplying the equation by $\pi_\rho\Psi$ and integrating in x, we obtain

$$-\tfrac{1}{2}\tfrac{d}{dt}\int \Psi^2\pi_\rho dx + \alpha\int |D\Psi|^2\pi_\rho dx \le \int l\Psi\pi_\rho dx + \int \Psi \sum_{j=1}^n \tfrac{\partial\Psi}{\partial x_j}[\sum_{i=1}^n(-\tfrac{\partial a_{ij}}{\partial x_i}$$
$$+2\rho a_{ij}\tfrac{x_i}{1+|x|^2}) + g_j]\pi_\rho dx.$$

We easily obtain

$$\int_0^T\int |D\Psi|^2\pi_\rho dxdt \le C \tag{11.3.27}$$

for $\rho > \frac{n}{2} + \frac{5}{2}$. If we multiply the equation by $\dfrac{\partial \Psi}{\partial t}\pi_\rho$ and integrate first in x, then in t, we obtain

$$\int_0^T \int |\frac{\partial \Psi}{\partial t}|^2 \pi_\rho dxdt \leq \sum_{j=1}^n \int_0^T \int \frac{\partial \Psi}{\partial t}\frac{\partial \Psi}{\partial x_j}[\sum_{i=1}^n (\frac{\partial a_{ij}}{\partial x_i} - 2\rho a_{ij}\frac{x_i}{1+|x|^2}) - g_j]\pi_\rho dxdt$$
$$-\frac{1}{2}\int_0^T \int \sum_{i,j=1}^n \frac{\partial a_{ij}}{\partial t}\frac{\partial \Psi}{\partial x_j}\frac{\partial \Psi}{\partial x_i}\pi_\rho dxdt - \int_0^T \int l \frac{\partial \Psi}{\partial t}\pi_\rho dxdt,$$

and it follows that

$$\int_0^T \int |\frac{\partial \Psi}{\partial t}|^2 \pi_\rho dxdt \leq C\int_0^T \int |D\Psi|^2 \pi_{\rho-1} dxdt + CT\int \pi_{\rho-2}dx,$$

and we obtain

$$\int_0^T \int |\frac{\partial \Psi}{\partial t}|^2 \pi_\rho dxdt \leq C, \tag{11.3.28}$$

for $\rho > \frac{n}{2} + \frac{7}{2}$. So we can write from (11.3.26) that

$$A\Psi = h$$

with

$$h = l + \frac{\partial \Psi}{\partial t} + g.D\Psi,$$

and h belongs to $L^2(0, T; L^2_{\pi_\rho}(R^n))$ with $\rho > \frac{n}{2} + \frac{7}{2}$. This implies

$$\frac{\partial^2 \Psi}{\partial x_i \partial x_j} \in L^2(0, T; L^2_{\pi_\rho}(R^n)),$$

and the estimates (11.3.21) follow.

To construct the solution, we consider for approximation the Dirichlet problem on a ball B_R, namely

$$-\frac{\partial \Psi_R}{\partial t} + A\Psi_R - g.D\Psi_R = l, \; x \in B_R, \tag{11.3.29}$$
$$\Psi_R|_{\Gamma_R} = 0,$$
$$\Psi(x, T) = 0,$$

where $\Gamma_R = \partial B_R$ is equal to the boundary of B_R. Note that l, g are bounded on B_R, although not uniformly with respect to R. We can find a solution Ψ_R satisfying

$$\Psi_R, \frac{\partial \Psi_R}{\partial x_i}, \frac{\partial^2 \Psi_R}{\partial x_i \partial x_j}, \frac{\partial \Psi_R}{\partial t} \in L^p(0, T; L^p(B_R))$$
$$\forall\, p\ 2 \leq p < +\infty.$$

In particular, $\Psi_R \in C^0(\bar{B}_R \times [0, T])$. Therefore, we can apply the integrated Itô's formula (11.3.14), which yields

$$\Psi_R(x, t) = E_t \int_t^{T \wedge \tau_R} l(y(t), t)dt, \ x \in B_R, \tag{11.3.30}$$

from which we get the estimate

$$|\Psi_R| \leq C(1 + |x|^2).$$

We use the fact that

$$\tau_R \uparrow +\infty, \text{ a.s. as } R \uparrow +\infty.$$

The estimates on the derivatives of Ψ_R are obtained from the equation (11.3.29), in a way similar to that done to obtain (11.3.27), (11.3.28). Note that from (11.3.30), Ψ_R can be extended by 0 outside B_R and thus can be defined on the whole space R^n. We can then let R tend to $+\infty$ and obtain the convergence to the single value

$$\Psi(x, t) = E_t[\int_t^T l(y(s), s)ds],$$

which concludes the proof. □

If g is bounded, the result of Proposition 11.2 can be improved.

PROPOSITION 11.3. *We make the assumptions of Proposition 11.2 and assume as well that g is bounded. Then we can replace (11.3.21) by*

$$\Psi, \frac{\partial \Psi}{\partial t}, \frac{\partial \Psi}{\partial x_i}, \frac{\partial^2 \Psi}{\partial x_i \partial x_j}\ L^p(0, T; L^p_{loc}(R^n)) \tag{11.3.31}$$
$$\forall p,\ 2 \leq p < +\infty.$$

PROOF. Define by $W^{1,p,\rho}(R^n)$, the space of functions φ such that $\varphi \pi_\rho^{\frac{1}{2}}$, $\frac{\partial \varphi}{\partial x_i} \pi_\rho^{\frac{1}{2}} \in L^p(R^n)$, with the norm

$$||\varphi||_{W^{1,p,\rho}(R^n)} = [\int |\varphi|^p (\pi_\rho)^{\frac{p}{2}} dx + \sum_{i=1}^n \int |\frac{\partial \varphi}{\partial x_i}|^p (\pi_\rho)^{\frac{p}{2}} dx]^{\frac{1}{p}},$$

and similarly, $W^{2,p,\rho}(R^n)$ is the space of functions φ such that $\varphi\pi_\rho^{\frac{1}{2}}$, $\frac{\partial\varphi}{\partial x_i}\pi_\rho^{\frac{1}{2}}$, $\frac{\partial^2\varphi}{\partial x_i\partial x_j}\pi_\rho^{\frac{1}{2}} \in L^p(R^n)$, with the norm

$$||\varphi||_{W^{2,p,\rho}(R^n)} = [\int|\varphi|^p(\pi_\rho)^{\frac{p}{2}}dx + \sum_{i=1}^n\int|\frac{\partial\varphi}{\partial x_i}|^p(\pi_\rho)^{\frac{p}{2}}dx$$
$$+\sum_{i,j=1}^n\int|\frac{\partial^2\varphi}{\partial x_i\partial x_j}|^p(\pi_\rho)^{\frac{p}{2}}dx]^{\frac{1}{p}}.$$

We note also that $W^{0,p,\rho}(R^n) = L^p_{\pi_\rho}(R^n)$. The function $l(x,t)$ belongs to $L^p_{\pi_\rho}(R^n)$, for $\rho > \frac{n}{2}+p$, for all t, with a uniform bound on the norm.

We use then the two following facts. First, if Ψ satisfies

$$-\frac{\partial\Psi}{\partial t} + A\Psi = f, \qquad (11.3.32)$$
$$\Psi(x,T) = 0,$$

with $f \in L^p(0,T;W^{0,p,\rho}(R^n))$, then $\Psi \in L^p(0,T;W^{2,p,\rho}(R^n))$, $\frac{\partial\Psi}{\partial t} \in L^p(0,T;W^{0,p,\rho}(R^n))$, with the estimate ($L^p$ theory of parabolic PDEs)

$$||\Psi||_{L^p(0,T;W^{2,p,\rho}(R^n))} + ||\frac{\partial\Psi}{\partial t}||_{L^p(0,T;W^{0,p,\rho}(R^n))} \le C||f||_{L^p(0,T;W^{0,p,\rho}(R^n))}. \qquad (11.3.33)$$

The second fact is

$$||\Psi||_{L^p(0,T;W^{1,p,\rho}(R^n))} \le C||\Psi||^{\frac{1}{2}}_{L^p(0,T;W^{2,p,\rho}(R^n))}||\Psi||^{\frac{1}{2}}_{L^p(0,T;W^{0,p,\rho}(R^n))}. \qquad (11.3.34)$$

Equation (11.3.26) now corresponds to

$$f = l + g.D\Psi.$$

Using the fact that g is bounded, we deduce from (11.3.33), that for $\rho > \dfrac{n}{2}+p$, we have

$$||\Psi||_{L^p(0,T;W^{2,p,\rho}(R^n))} + ||\frac{\partial\Psi}{\partial t}||_{L^p(0,T;W^{0,p,\rho}(R^n))} \le C + C||\Psi||_{L^p(0,T;W^{1,p,\rho}(R^n))}.$$

Taking account of (11.3.34) and of the fact that $||\Psi||_{L^p(0,T;W^{0,p,\rho}(R^n))}$ is bounded, we obtain

$$||\Psi||_{L^p(0,T;W^{2,p,\rho}(R^n))} + ||\frac{\partial\Psi}{\partial t}||_{L^p(0,T;W^{0,p,\rho}(R^n))} \le C,$$

and thus the result follows. □

11.4 Weak Formulation of Stochastic Control

11.4.1 Setting of the Problem

We assume

$$n = m, \ \sigma(x,t), \ \sigma^{-1}(x,t) \quad \text{continuous and bounded,} \qquad (11.4.1)$$
$$|\sigma(x,t) - \sigma(x',t)| \leq K|x - x'|.$$

We next assume

$$g(x,v) \text{ measurable and bounded,} \qquad (11.4.2)$$
$$l(x,v) \text{ measurable,}$$

$$|l(x,v)| \leq \bar{l}(1 + |x|^2 + |v|^2), \qquad (11.4.3)$$
$$l(x,v) \geq l_0|v|^2 - c_0,$$
$$|h(x)| \leq C(1 + |x|^2), \qquad (11.4.4)$$
$$|h_x(x)| \leq C(1 + |x|),$$
$$|h_{xx}(x)| \leq C,$$

$$\mathcal{U}_{\mathrm{ad}} = \text{nonempty closed convex subset of } R^m. \qquad (11.4.5)$$

We recall the definition of the Lagrangian:

$$L(x,v,q) = l(x,v) + q.g(x,v)$$

There exists a measurable map $\bar{v}(x,q)$ that achieves the minimum

$$L(x,\bar{v}(x,q),q) = \inf_{v \in \mathcal{U}_{\mathrm{ad}}} L(x,v,q), \ \text{a.e. } x, q \qquad (11.4.6)$$
$$= H(x,q).$$

The function $H(x,q)$ is the Hamiltonian. Thanks to the fact that g is bounded, we show easily that

$$|\bar{v}(x,q)| \leq C(1 + |x| + |q|^{\frac{1}{2}}); \qquad (11.4.7)$$

hence the properties

$$|H(x,q)| \leq C(1+|x|^2+|q|), \tag{11.4.8}$$
$$|H(x,q_1)-H(x,q_2)| \leq C|q_1-q_2|.$$

This important simplification explains our restriction to g being bounded.

11.4.2 The HJB Equation

We consider the HJB equation

$$-\frac{\partial \Psi}{\partial t}+A\Psi = H(x,D\Psi), \tag{11.4.9}$$
$$\Psi(x,T)=h(x),$$

with the regularity properties

$$|\Psi| \leq C(1+|x|^2), \tag{11.4.10}$$

$$\Psi, \frac{\partial \Psi}{\partial t}, \frac{\partial \Psi}{\partial x_i}, \frac{\partial^2 \Psi}{\partial x_i \partial x_j} \; L^p(0,T;L^p_{\text{loc}}(R^n)) \tag{11.4.11}$$
$$\forall p,\ 2 \leq p < +\infty.$$

THEOREM 11.3. *We make the assumptions (11.4.1), (11.4.2), (11.4.3), (11.4.4). Then there exists a unique solution of (11.4.9), (11.4.10), (11.4.11).*

PROOF. We use use the technique of a priori estimates. First, define

$$H_0(x,q)=\frac{H(x,q)+|q|^2H(x,0)}{1+|q|^2},$$
$$\Gamma(x,q)=q\,\frac{H(x,q)-H(x.0)}{1+|q|^2}.$$

Clearly,

$$H(x,q)=H_0(x,q)+q.\Gamma(x,q)$$

From the Lipschitz assumption (11.4.8) we have

$$|H(x,q)-H(x.0)| \leq C|q|;$$

hence $|\Gamma(x,q)| \leq C$ and

$$|H_0(x,q)| \leq C(1+|x|^2).$$

Therefore, the HJB equation reads

$$-\frac{\partial \Psi}{\partial t} + A\Psi = H_0(x, D\Psi) + D\Psi.\Gamma(x, D\Psi). \tag{11.4.12}$$

The interest of this writeup is that we obtain a solution such that $D\Psi$ is measurable. Then by noting that

$$l(x,t) = H_0(x, D\Psi(x,t)),$$
$$g(x,t) = \Gamma(x, D\Psi(x,t)),$$

we see that Ψ appears as the solution of a linear equation, of the form (11.3.15). We can then apply Proposition 11.3 to claim (11.4.10), (11.4.11). It remains to define a solution. We can consider the approximation

$$-\frac{\partial \Psi_\epsilon}{\partial t} + A\Psi_\epsilon = H_\epsilon(x, D\Psi_\epsilon), \tag{11.4.13}$$
$$\Psi_\epsilon(x,T) = h(x),$$

with

$$H_\epsilon(x,q) = \frac{H(x,q)}{1+\epsilon|H(x,q)|}.$$

Since $H_\epsilon(x,q)$ is a bounded function, it is easy to show that (11.4.13) has a solution with the appropriate regularity and that Ψ_ϵ is bounded. We can use a fixed-point approach. We do not detail the convergence. The uniqueness will be proved by a probabilistic interpretation, although an analytic proof is possible. ■

The policy iteration technique:

We consider the following iteration:

$$-\frac{\partial \Psi_0}{\partial t} + A\Psi_0 = l(x, v_0) + D\Psi_0.g(x, v_0), \tag{11.4.14}$$
$$\Psi_0(x,T) = h(x),$$

where v_0 is a fixed control in $\mathcal{U}_{ad}$,

$$-\frac{\partial \Psi_{k+1}}{\partial t} + A\Psi_{k+1} = l(x, v_k) + D\Psi_{k+1}.g(x, v_k), \tag{11.4.15}$$
$$\Psi_{k+1}(x,T) = h(x),$$

in which

$$v_k = \bar{v}(x, D\Psi_k).$$

We use the spaces $W^{1,p,\rho}(R^n)$, $W^{2,p,\rho}(R^n)$. Since $|v_k| \leq C(1 + |x| + |D\Psi_k|^{\frac{1}{2}})$, we check sequentially that

$$\Psi_k \in L^p(0, T; W^{2,p,\rho}(R^n)), \ \frac{\partial \Psi_k}{\partial t} \in L^p(0, T; W^{0,p,\rho}(R^n)),$$

with the estimate

$$||\Psi_{k+1}||_{L^p(0,T;W^{2,p,\rho}(R^n))}+||\frac{\partial \Psi_{k+1}}{\partial t}||_{L^p(0,T;W^{0,p,\rho}(R^n))} \leq C+C||\Psi_k||_{L^p(0,T;W^{1,p,\rho}(R^n))}. \tag{11.4.16}$$

Furthermore, we note that

$$l(x, v_{k-1}) + D\Psi_k.g(x, v_{k-1}) \geq l(x, v_k) + D\Psi_k.g(x, v_k),$$

and therefore, we can state that

$$-\frac{\partial(\Psi_{k+1} - \Psi_k)}{\partial t} + A(\Psi_{k+1} - \Psi_k) \leq D(\Psi_{k+1} - \Psi_k).g(x, v_k),$$
$$(\Psi_{k+1} - \Psi_k)(x, T) = 0.$$

Here we use a comparison argument to claim that

$$\Psi_{k+1} - \Psi_k \leq 0.$$

It follows that

$$\Psi_k \leq \Psi_0 \leq C(1 + |x|^2).$$

Similarly, from the second assumption (11.4.3), we have

$$-\frac{\partial \Psi_{k+1}}{\partial t} + A\Psi_{k+1} \geq -C + D\Psi_{k+1}.g(x, v_k),$$
$$\Psi_{k+1}(x, T) = h(x),$$

and again, by a comparison argument we get

$$\Psi_k \geq -C(1 + |x|^2).$$

Therefore, in particular,

$$||\Psi_k||_{L^p(0,T;W^{0,p,\rho}(R^n))} \leq C.$$

With this estimate, (11.4.16), and the inequality (11.3.34), we obtain easily

$$||\Psi_k||_{L^p(0,T;W^{2,p,\rho}(R^n))} + ||\frac{\partial \Psi_k}{\partial t}||_{L^p(0,T;W^{0,p,\rho}(R^n))} \leq C. \tag{11.4.17}$$

We can extract a convergent subsequence

$$\Psi_k \to \Psi$$

weakly in the functional space arising in the estimate (11.4.17). From the Lipschitz property of the Hamiltonian, see (11.4.8), we can assert that

$$H(x, D\Psi_{k+1}) \to H(x, D\Psi), \text{ in } L^p(0, T; W^{0,p,\rho}(R^n)).$$

Since

$$-\frac{\partial \Psi_{k+1}}{\partial t} + A\Psi_{k+1} \geq H(x, D\Psi_{k+1}),$$

we deduce that as $k \to +\infty$,

$$-\frac{\partial \Psi}{\partial t} + A\Psi \geq H(x, D\Psi).$$

On the other hand, for every fixed control in $\mathcal{U}_{\text{ad}}$ we have

$$\begin{aligned}
-\frac{\partial \Psi_k}{\partial t} + A\Psi_k - l(x, v) - D\Psi_k.g(x, v) \leq \\
-\frac{\partial \Psi_k}{\partial t} + A\Psi_k \quad - \quad H(x, D\Psi_k) \\
= -\frac{\partial(\Psi_k - \Psi_{k+1})}{\partial t} + A(\Psi_k - \Psi_{k+1}) \quad - \quad D(\Psi_k - \Psi_{k+1}).g(x, v_k),
\end{aligned}$$

and using the fact that $g(x, v_k)$ is bounded, we conclude that the right-hand side tends to 0 in $L^p(0, T; W^{0,p,\rho}(R^n))$ weakly. So in the limit, we have

$$-\frac{\partial \Psi}{\partial t} + A\Psi - l(x, v) - D\Psi.g(x, v) \leq 0.$$

Since this inequality is true for every $v \in \mathcal{U}_{\text{ad}}$, we get

$$-\frac{\partial \Psi}{\partial t} + A\Psi \leq H(x, D\Psi).$$

Since the reverse inequality has already been proven, the function Ψ is a solution of the HJB equation. □

11.4.3 Stochastic Control

We want now to interpret (11.4.9) as the HJB equation of an optimal stochastic control problem. We consider a system $\Omega, \mathcal{A}, P, \mathcal{F}^t$, and $w(.)$ is a $P, \mathcal{F}^t$ standard Wiener process with values in R^n. We solve

$$\begin{aligned} dX &= \sigma(X(s), s)dw(s), \\ X(t) &= x. \end{aligned} \tag{11.4.18}$$

The solution is denoted by $X_{xt}(s)$. We next define an admissible control as a process $v(s)$ adapted to the filtration $\mathcal{F}^s$ with values in $\mathcal{U}_{\text{ad}}$. We define

$$M_{xt}^v(s) = \exp[\int_t^s \sigma^{-1}g(X_{xt}(\tau), v(\tau))dw - \frac{1}{2}\int_t^s |\sigma^{-1}g(X_{xt}(\tau), v(\tau))|^2 d\tau] \tag{11.4.19}$$

and the probability

$$\frac{dP_{xt}^{v(.)}}{dP}|_{\mathcal{F}^s} = M_{xt}^v(s). \tag{11.4.20}$$

Let next, for $s \geq t$,

$$w_{xt}^{v(.)}(s) = w(s) - w(t) - \int_t^s \sigma^{-1}g(X_{xt}(\tau), v(\tau))d\tau. \tag{11.4.21}$$

For the system $\Omega, \mathcal{A}, P_{xt}^{v(.)}, \mathcal{F}^s$, the process $w_{xt}^{v(.)}(s)$ is a standard $P_{xt}^{v(.)}, \mathcal{F}^s$ Wiener process with values in R^n. The process $X_{xt}(s)$ appears as the solution of

$$\begin{aligned} dX_{xt} &= g(X_{xt}(s), v(s))ds + \sigma(X_{xt}(s), s)dw_{xt}^{v(.)}(s), \\ X_{xt}(t) &= x. \end{aligned} \tag{11.4.22}$$

We define

$$\begin{aligned} J_{x,t}(v(.)) &= E^{v(.)}[\int_t^T l(X_{xt}(s), v(s))ds + h(X_{xt}(T))] \\ &= E\, M_{xt}^v(T)[\int_t^T l(X_{xt}(s), v(s))ds + h(X_{xt}(T))]. \end{aligned} \tag{11.4.23}$$

We then have the following theorem.

THEOREM 11.4. *We make the assumptions (11.4.1), (11.4.2), (11.4.3), (11.4.4). Then the unique solution of (11.4.9), (11.4.10), (11.4.11) is the value function*

$$\Phi(x, t) = \inf_{v(.)} J_{x,t}(v(.)).$$

Moreover, there exists an optimal control defined by a feedback.

PROOF. Choose a solution Ψ of (11.4.9), (11.4.10), (11.4.11). For a ball B_R, consider the exit time τ_R of the process $y(.)$ out of the ball. Consider now an admissible control $v(.)$. We can use the integrated Itô's formula (11.3.14) to write

$$E^{v(.)}\Psi(X(T\wedge\tau_R),T\wedge\tau_R)-\Psi(x,t)=E^{v(.)}\int_t^{T\wedge\tau_R}(\frac{\partial\Psi}{\partial s}-A\Psi+g.D\Psi)(X(s),v(s))ds.$$

However, from the HJB equation, we have

$$(\frac{\partial\Psi}{\partial s}-A\Psi+g.D\Psi)(X(s),v(s))\geq -l(X(s),v(s)).$$

Therefore,

$$\Psi(x,t)\leq E^{v(.)}[\int_t^{T\wedge\tau_R}l(X(s),v(s))ds+\Psi(X(T\wedge\tau_R),T\wedge\tau_R)]. \tag{11.4.24}$$

Using the estimates

$$E^{v(.)}\sup_{t\leq s\leq T}|X(s)|^2\leq C(1+|x|^2)$$

$$|\Psi(x,t)|\leq C(1+|x|^2),$$

it is possible to let $R\uparrow+\infty$ in (11.4.24) and obtain

$$\Psi(x,t)\leq J_{x,t}(v(.)).$$

Next, we consider the function $\bar{v}(x,q)$ defined in (11.4.6) and define a feedback associated to the function $\Psi(x,t)$ by the formula

$$\hat{v}(x,t)=\bar{v}(x,D\Psi).$$

With this feedback we define an admissible control by setting

$$u(s)=\hat{v}(X(s),s).$$

Applying the integrated Itô's formula as above, and thanks to the definition of $u(s)$, we can write this time

$$\Psi(x,t)=E^{u(.)}[\int_t^{T\wedge\tau_R}l(X(s),u(s))ds+\Psi(X(T\wedge\tau_R),T\wedge\tau_R)],$$

and letting $R\uparrow+\infty$, we obtain

$$\Psi(x,t)=J_{x,t}(u(.)).$$

Therefore,

$$\Psi(x,t) = \Phi(x,t)$$
$$= \text{value function.}$$

and the control $u(.)$ is optimal. This concludes the proof. □

So far, the difference between the strong and weak formulations lies only in the extension of the concept of optimal feedback. The value function is the same, and the Bellman equation the same. We shall see an additional difference when we consider the process

$$\Psi_{xt}(s) = \Phi(X_{xt}(s), s), \tag{11.4.25}$$

in which $\Phi(x, s)$ is the value function, the solution of the Bellman equation (11.2.10). Recalling (11.2.21), we see that the processes $\Phi_{xt}(s)$ and $\Psi_{xt}(s)$ are different, except at time $s = t$. This is because the processes $y(s)$ and $X_{xt}(s)$ are different except at the initial value t. To facilitate the comparison, we summarize it again. We recall the Bellman equation

$$-\frac{\partial \Phi}{\partial t} + A\Phi = H(x, D\Phi), \tag{11.4.26}$$
$$\Phi(x,T) = h(x).$$

We then consider the optimal state, in the strong formulation sense, the solution of

$$dy = g(y(s), u(s))ds + \sigma(y(s))dw, \tag{11.4.27}$$
$$y(t) = x,$$

in which $u(s)$ is the optimal control. Setting $\zeta_{xt}(s) = \sigma^* D\Phi(y(s), s)$, we see that the optimality of $u(s)$ is expressed by

$$H(y(s), (\sigma^*)^{-1}(y(s))\zeta_{xt}(s)) = l(y(s), u(s)) + (\sigma^*)^{-1}(y(s))\zeta_{xt}(s).g(y(s), u(s)). \tag{11.4.28}$$

Since $\Phi_{xt}(s) = \Phi(y(s), s)$, we check easily that

$$\Phi_{xt}(s) = h(y(T)) + \int_s^T H(y(\tau), (\sigma^*)^{-1}(y(\tau))\zeta_{xt}(\tau))d\tau - \tag{11.4.29}$$
$$- \int_s^T \zeta_{xt}(\tau).(\sigma^{-1} g(y(\tau), \tau)d\tau + dw(\tau)).$$

We now turn to $\Psi_{xt}(s)$. Using the Bellman equation, we can write the formula

$$\Psi_{xt}(s) = h(X_{xt}(T)) + \int_s^T H(X_{xt}(\tau), D\Phi(X_{xt}(\tau), \tau))d\tau$$
$$- \int_s^T D\Phi(X_{xt}(\tau), \tau).\sigma(X_{xt}(\tau))dw(\tau).$$

If we set $Z_{xt}(s) = \sigma^* D\Phi(X_{xt}(s), s)$, this equation becomes

$$\Psi_{xt}(s) = h(X_{xt}(T)) + \int_s^T H(X_{xt}(\tau), (\sigma^*)^{-1}(X_{xt}(\tau))Z_{xt}(\tau))d\tau \quad (11.4.30)$$
$$- \int_s^T Z_{xt}(\tau).dw(\tau), t \leq s \leq T.$$

If we consider next the control

$$u_{xt}(s) = \bar{v}(X_{xt}(s), (\sigma^*)^{-1}(X_{xt}(s))Z_{xt}(s)) \quad (11.4.31)$$

the probability $P_{xt}^{u(.)}$, and the process $w_{xt}^{u(.)}$ defined by formulas (11.4.19), (11.4.20), (11.4.21), we get

$$\Psi_{xt}(s) = h(X_{xt}(T)) + \int_s^T H(X_{xt}(\tau), (\sigma^*)^{-1}(X_{xt}(\tau))Z_{xt}(\tau))d\tau - \quad (11.4.32)$$
$$- \int_s^T Z_{xt}(\tau).(\sigma^{-1}g(X_{xt}(\tau), \tau)d\tau + dw_{xt}^{u(.)}(\tau)).$$

Comparing (11.4.29) and (11.4.32), we see that the pair $\Psi_{xt}(s)$, $Z_{xt}(s)$ satisfy the same equation as $\Phi_{xt}(s)$, $\zeta_{xt}(s)$ with the process $w(s)$ changed into $w_{xt}^{u(.)}(s)$, and $y(s)$ changed into $X_{xt}(s)$. But when we change the probability P into $P_{xt}^{u(.)}$, $w_{xt}^{u(.)}(s)$ becomes a standard Wiener process, and $y(s)$ looks like $X_{xt}(s)$. It follows that for $P_{xt}^{u(.)}$, $\Psi_{xt}(s)$ has the same probability as $\Phi_{xt}(s)$ for the original probability P, although they are different processes, just as $y(s)$ is different from $X_{xt}(s)$.

Equation (11.4.30) is a nonlinear BSDE. The unknowns are the processes $\Psi_{xt}(s)$ and $Z_{xt}(s)$, which must be adapted to $\mathcal{F}_t^s$. If we can solve directly the BSDE (11.4.30), we shall recover the deterministic function $\Phi(x, t)$ by $\Phi(x, t) = \Psi_{xt}(t)$. The advantage of this formulation is that it does not require regularity properties of $\Phi(x, t)$. Then we have

$$\Psi_{xt}(s) = h(X_{xt}(T)) + \int_s^T l(X_{xt}(\tau), u_{xt}(\tau))d\tau - \int_s^T Z_{xt}(\tau).(dw(\tau) - \sigma^{-1}g(X_{xt}(\tau), u_{xt}(\tau))d\tau).$$

Hence

$$\Psi_{xt}(s) = E^{u(.)}[h(X_{xt}(T)) + \int_s^T l(X_{xt}(\tau), u_{xt}(\tau))d\tau|\mathcal{F}_t^s].$$

Similarly, for any control $v(s)$ adapted to $\mathcal{F}^s$, we will have

$$\Psi_{xt}(s) \leq h(X_{xt}(T)) + \int_s^T l(X_{xt}(\tau), v(\tau))d\tau - \int_s^T Z_{xt}(\tau).(dw(\tau)$$
$$-\sigma^{-1}g(X_{xt}(\tau), v(\tau))d\tau),$$

which implies

$$\Psi_{xt}(s) \le E^{v(.)}[h(X_{xt}(T)) + \int_s^T l(X_{xt}(\tau), v(\tau))d\tau | \mathcal{F}_t^s].$$

Therefore, we can assert that

$$\Psi_{xt}(s) = \inf_{v(.)} E^{v(.)}[h(X_{xt}(T)) + \int_s^T l(X_{xt}(\tau), v(\tau))d\tau | \mathcal{F}_t^s], \quad (11.4.33)$$

which is a result similar to (11.2.24).

11.5 Connection Between FBSDE and Partial Differential Equations

Inspired by equation (11.4.30), we can show a more general connection between PDEs and SBDEs. Consider the SDE

$$dX = b(X)ds + \sigma(X)dw,$$
$$X(t) = x,$$

denoted by $X_{xt}(s)$. Consider next the BSDE

$$\Psi_{xt}(s) = h(X_{xt}(T)) + \int_s^T F(X_{xt}(\tau), \Psi_{xt}(\tau), Z_{xt}(\tau))d\tau - \int_s^T Z_{xt}(\tau).dw(\tau) \quad (11.5.1)$$

with two unknowns $\Psi_{xt}(s)$ and $Z_{xt}(s)$. We can look for a solution of the form $\Psi_{xt}(s) = \Phi(X_{xt}(s), s)$, and it is clear that $\Phi(x, s)$ must be chosen as the solution of the PDE

$$-\frac{\partial \Phi}{\partial s} + A\Phi - D\Phi.b(x) = F(x, \Phi, \sigma^* D\Phi), \quad (11.5.2)$$
$$\Phi(x, T) = h(x).$$

If we can solve the BSDE (11.5.1), then we have $\Phi(x, t) = \Psi_{xt}(t)$.

The systematic study of BSDEs was introduced by Pardoux and Peng [39]; see also the book of Pardoux and Rascanu [40]. A general form of BSDEs is given by

$$\Psi(s) = h_T + \int_s^T F(\Psi(\tau), Z(\tau), \tau)d\tau - \int_s^T Z(\tau).dw(\tau), \quad (11.5.3)$$

in which h_T is $\mathcal{F}^T$-measurable and for every scalar Ψ and vector Z in R^n, the process $F(\Psi, Z, s)$ is adapted to $\mathcal{F}^s$. We give here just the basic result, following [17]. We shall give more complete results in the next chapter.

THEOREM 11.5. *Assume that* $E|h_T|^2 < +\infty$, $E\int_0^T |F(0,0,s)|^2 ds < +\infty$, *and*

$$|F(\Psi_1, Z_1, t) - F(\Psi_2, Z_2, t)| \leq C(|\Psi_1 - \Psi_2| + |Z_1 - Z_2|). \tag{11.5.4}$$

Then there exists a unique solution of (11.5.3) such that

$$||(\Psi, Z)||^2 = E[\sup_{0\leq t\leq T} |\Psi(t)|^2 + \int_0^T |Z(t)|^2 dt] \leq \tag{11.5.5}$$

$$\leq CE[|h_T|^2 + \int_0^T |F(0,0,t)|^2 dt.$$

PROOF. We only sketch the proof. We have

$$d|\Psi(s)|^2 = -2\Psi(s)F(\Psi(s), Z(s), s)ds + 2\Psi(s)Z(s)dw(s) + |Z(s)|^2 ds,$$

and thus

$$E[|\Psi(t)|^2 + \int_t^T |Z(s)|^2 ds] = E[|h_T|^2 + 2\int_t^T \Psi(s)F(\Psi(s), Z(s), s)ds,$$

and from the assumptions it is easy to check the estimate (11.5.5). For the uniqueness, we consider two solutions $\Psi^1(s), Z^1(s)$ and $\Psi^2(s), Z^2(s)$. Setting $\Delta\Psi(s) = \Psi^1(s) - \Psi^2(s)$, $\Delta Z(s) = Z^1(s) - Z^2(s)$, we obtain

$$\Delta\Psi(t) = \int_t^T [F(\Psi^1(s), Z^1(s), s) - F(\Psi^1(s), Z^1(s), s)]ds - \int_t^T \Delta Z(s)dw(s),$$

and thus

$$E[|\Delta\Psi(t)|^2 + \int_t^T |\Delta Z(s)|^2 ds] = 2E\int_t^T \Delta\Psi(s)(F(\Psi^1(s), Z^1(s), s) - F(\Psi^1(s), Z^1(s), s))ds$$

$$\leq 2CE\int_t^T [|\Delta\Psi(s)|^2 + |\Delta\Psi(s)||\Delta Z(s)|]ds.$$

Since this inequality holds for all t, it is standard to obtain

$$E|\Delta\Psi(t)|^2 = 0, \forall t, \ E\int_0^T |\Delta Z(s)|^2 ds = 0,$$

and thus $\Delta\Psi(t) = 0, \Delta Z(t) = 0$, which proves the uniqueness. The existence is proved by a fixed-point argument. Take $\chi(t)$ and $U(t)$ such that $\exp\frac{\gamma}{2}t\,\chi(t)$ and

$\exp \frac{\gamma}{2} t\, U(t)$ are in $L^2_{\mathcal{F}}(0,T)$ and $L^2_{\mathcal{F}}(0,T;R^n)$ respectively, for a convenient γ to be chosen. We define a map by solving

$$\Psi(s) = h_T + \int_s^T F(\chi(\tau), U(\tau), \tau)d\tau - \int_s^T Z(\tau).dw(\tau).$$

Consider a second pair $\tilde{\chi}(t)$, $\tilde{U}(t)$ and the corresponding image $\tilde{\Psi}(t), \tilde{U}(t)$. We have

$$\exp \gamma t\, E|\Psi(t) - \tilde{\Psi}(t)|^2 + \int_t^T \exp \gamma s\, E|Z(s) - \tilde{Z}(s)|^2 ds + \gamma \int_t^T \exp \gamma s\, E|\Psi(s) - \tilde{\Psi}(s)|^2 ds =$$

$$= 2\int_t^T \exp \gamma s\, E(\Psi(s) - \tilde{\Psi}(s))(F(\chi(s), U(s), s) - F(\tilde{\chi}(s), \tilde{U}(s), s))ds.$$

After easy majorizations, we can choose γ sufficiently large that

$$\int_0^T \exp \gamma s\, E|Z(s) - \tilde{Z}(s)|^2 ds + \int_0^T \exp \gamma s\, E|\Psi(s) - \tilde{\Psi}(s)|^2 ds \leq$$

$$\frac{1}{2}[\int_0^T \exp \gamma s\, E|\chi(s) - \tilde{\chi}(s)|^2 ds + \int_0^T \exp \gamma s\, E|U(s) - \tilde{U}(s)|^2 ds.$$

Therefore, the map that we have defined is a contraction. This completes the proof. □

11.6 Links Between Dynamic Programming and the Stochastic Maximum Principle

We proceed here formally, assuming sufficient regularity. So we are considering the strong formulation of stochastic control. Let $u(.)$ be an optimal control and let $y(.)$ be the corresponding optimal state.

We set (as in the deterministic case; see (10.3.5))

$$p(t) = D\Phi(y(t), t), \tag{11.6.1}$$

whence

$$dp_\alpha(t) = \sum_{\beta=1}^n \frac{\partial^2 \Phi}{\partial x_\alpha \partial x_\beta}(y(t), t)(g_\beta(y(t), u(t))dt + \sum_{j=1}^k \sigma_\beta^j(y(t))dw_j) \tag{11.6.2}$$

$$\frac{\partial^2 \Phi}{\partial t \partial x_\alpha}(y(t), t)dt + \frac{1}{2}\sum_{\beta,\gamma=1}^n \frac{\partial^3 \Phi}{\partial x_\alpha \partial x_\beta \partial x_\gamma}(y(t), t)\sum_{j=1}^k \sigma_\beta^j(y(t))\sigma_\gamma^j(y(t))dt.$$

Writing the HJB equation (11.2.10) as

$$\frac{\partial \Phi}{\partial t}+l(x,\hat{v})+\sum_{\beta=1}^{n}\frac{\partial \Phi}{\partial x_\beta}(x,t)g_\beta(x,\hat{v})+\frac{1}{2}\sum_{\beta,\gamma=1}^{n}\frac{\partial^2 \Phi}{\partial x_\beta \partial x_\gamma}(x,t)\sum_{j=1}^{k}\sigma_\beta^j(x)\sigma_\gamma^j(x)=0$$

and differentiating in x_α, we obtain

$$\frac{\partial^2 \Phi}{\partial t \partial x_\alpha}+l_{x_\alpha}(x,\hat{v})+\sum_{\beta=1}^{n}\frac{\partial^2 \Phi}{\partial x_\alpha \partial x_\beta}(x,t)g_\beta(x,\hat{v})+\sum_{\beta=1}^{n}\frac{\partial \Phi}{\partial x_\beta}(x,t)g_{\beta,x_\alpha}(x,\hat{v})+$$
$$+\frac{1}{2}\sum_{\beta,\gamma=1}^{n}\frac{\partial^3 \Phi}{\partial x_\alpha \partial x_\beta \partial x_\gamma}(x,t)\sum_{j=1}^{k}\sigma_\beta^j(x)\sigma_\gamma^j(x)$$
$$+\sum_{\beta,\gamma=1}^{n}\frac{\partial^2 \Phi}{\partial x_\beta \partial x_\gamma}(x,t)\sum_{j=1}^{k}\sigma_{\beta,x_\alpha}^j(x)\sigma_\gamma^j(x)=0,$$

where we have used the fact that $\hat{v}$ is optimal (we also assume $\mathcal{U}_{\text{ad}} = R^m$). If we take $x = y(t)$, then $\hat{v}$ must be replaced by $u(t)$. Combining the above with (11.6.1) and (11.4.2), we obtain

$$-dp_\alpha = (l_{x_\alpha}(y,u)+\sum_{\beta=1}^{n}p_\beta g_{\beta,x_\alpha}(y,u))dt$$
$$+\sum_{\beta,\gamma=1}^{n}\frac{\partial^2 \Phi}{\partial x_\beta \partial x_\gamma}\sum_{j=1}^{k}\sigma_{\beta,x_\alpha}^j(y)\sigma_\gamma^j(y)dt-\sum_{\beta=1}^{n}\frac{\partial^2 \Phi}{\partial x_\beta \partial x_\alpha}\sum_{j=1}^{k}\sigma_\beta^j(y)dw_j.$$

If we set

$$r_\alpha^j(t)=\sum_{\beta=1}^{n}\frac{\partial^2 \Phi}{\partial x_\beta \partial x_\alpha}(y(t),t)\sigma_\beta^j(y(t)), \tag{11.6.3}$$

the preceding equation becomes

$$-dp_\alpha = (l_{x_\alpha}(y,u)+\sum_{\beta=1}^{n}p_\beta g_{\beta,x_\alpha}(y,u))dt$$
$$+\sum_{j=1}^{k}\sigma_{\beta,x_\alpha}^j(y)r_\beta^j(t)dt-\sum_{j=1}^{k}r_\alpha^j(t)dw_j,$$

which is also (in vector form)

$$
\begin{aligned}
-dp &= (l_x(y,u) + g_x^*(y,u)p)dt \\
&\quad + \sum_{j=1}^{k} (\sigma_x^j(y))^* r^j dt - \sum_{j=1}^{k} r^j(t) dw_j,
\end{aligned}
$$

with the final condition

$$p(T) = h_x(y(T)),$$

and we recover the adjoint state; see (11.1.21).

11.7 Calculus of Variations Approach

11.7.1 Markov Properties of Diffusions

We consider SDEs in the strong sense. Assume

$$
\begin{aligned}
|g(x,t) - g(x',t)| + |\sigma(x,t) - \sigma(x',t)| &\leq K|x - x'|, \\
|g(x,t)|^2 &\leq K_0^2(1 + |x|^2),
\end{aligned}
\tag{11.7.1}
$$

and also

$$\sigma(x,t),\ \sigma^{-1}(x,t),\ \text{bounded.} \tag{11.7.2}$$

We solve the SDE

$$
\begin{aligned}
dy &= g(y(t),t)dt + \sigma(y(t),t)dw, \\
y(0) &= \xi,
\end{aligned}
\tag{11.7.3}
$$

where $w(t)$ is a standard Wiener process with values in R^n, defined on a probability space $\Omega, \mathcal{A}, P$ and a filtration $\mathcal{F}^t$. We assume also

$$
\begin{gathered}
w(t) \text{ is independent of } \mathcal{F}^0, \\
\xi \text{ is } \mathcal{F}^0 \text{ measurable,} \\
\text{the probability distribution of } \xi \text{ has density } m_0(x).
\end{gathered}
\tag{11.7.4}
$$

Recall the second-order differential operator (depending on t)

$$A\varphi(x) = -\sum_{i,j=1}^{n} a_{ij}(x) \frac{\partial^2 \varphi}{\partial x_i \partial x_j}, \tag{11.7.5}$$

where the matrix $a = a_{ij}$ is defined by

$$a(x) = \frac{1}{2}\sigma\sigma^*(x).$$

A very important result is that for all t, the probability distribution of the random variable $y(t)$ has also a density-$m(x,t)$ solution of the parabolic PDE, called the Fokker–Planck equation:

$$\frac{\partial m}{\partial t} + A^* m + \text{div }(gm) = 0, \tag{11.7.6}$$
$$m(x,0) = m_0(x).$$

The operator A^* is the dual of A, given by

$$A^*\varphi = -\sum_{i,j=1}^{n} \frac{\partial^2}{\partial x_i \partial x_j}(a_{ij}\varphi) \tag{11.7.7}$$

and

$$\text{div }(gm) = \sum_{i=1}^{n} \frac{\partial (mg_i)}{\partial x_i}.$$

We proceed formally. In particular, we do not state explicitly the regularity properties of the function m, and assume that all operations (derivation and integration) are legitimate. Let us show that

$$E\varphi(y(t)) = \int \varphi(x) m(x,t) dx. \tag{11.7.8}$$

We replace, for convenience of notation, t by T in (11.7.8). Consider then the PDE

$$-\frac{\partial \Phi}{\partial t} + A\Phi - g.D\Phi = 0, \tag{11.7.9}$$
$$\Phi(x,T) = \varphi.$$

From Itô's formula, we have

$$d\Phi(y(t),t) = D\Phi(y(t),t).\sigma(y(t),t)dw,$$

and therefore,

$$E\Phi(y(T),T) = E\Phi(y(0),0),$$

whence

$$E\varphi(y(T)) = \int \Phi(x,0)m_0(x)dx. \tag{11.7.10}$$

Now, considering (11.7.9) and (11.7.6), we can check that

$$\frac{d}{dt}\int \Phi(x,t)m(x,t)dx = 0.$$

Therefore,

$$\int \Phi(x,T)m(x,T)dx = \int \Phi(x,0)m_0(x)dx,$$

$$\int \varphi(x)m(x,T)dx = E,\varphi(y(T))$$

which proves the result (11.7.8) at T.

Consider now the diffusion with initial conditions x, t, namely $y_{xt}(s)$, the solution of

$$\begin{aligned} dy &= g(y(s),s)ds + \sigma(y(s),s)dw, \\ y(t) &= x, \end{aligned} \tag{11.7.11}$$

and the PDE (11.7.6) with initial condition a Dirac measure at x, t, that is,

$$\begin{aligned} \frac{\partial m}{\partial s} + A^*m + \text{div}\,(gm) = 0,\ s > t,\ \xi \in R^n, \\ m(\xi,t) = \delta(\xi - x). \end{aligned} \tag{11.7.12}$$

Note that in equation (11.7.12), x, t are parameters, and the arguments are ξ, s. We denote this function by $m(\xi, s;\ x, t)$ to emphasize the dependence in the parameters x, t. The fact that we can solve (11.7.12) is a deep result in PDE whose proof goes beyond the scope of this material. It is important to have in mind that the initial condition is a Dirac measure, not a function. Nevertheless, for every time $s > t$, the solution is a function. So there is a **smoothing effect**, which is an important result of diffusion parabolic equations.

EXERCISE 11.4. Consider the case $\sigma(x,t) = \sigma$ and $g(x,t) = F(t)x + f(t)$. Find explicitly the function $m(\xi, s;\ x, t)$.

Considering the solution of (11.7.9), we see as above that

$$\begin{aligned} \Phi(x,t) &= E\varphi(y_{x,t}(T)) \\ &= \int \varphi(\xi)m(\xi,T;\ x,t)d\xi. \end{aligned} \tag{11.7.13}$$

We now go back to (11.7.3). Consider $t_1 < t_2 < T$ and let $\Phi(x, t;\ t_2),\ t \leq t_2$, be the solution of

$$-\frac{\partial \Phi}{\partial t} + A\Phi - g.D\Phi = 0,\ t < t_2,$$
$$\Phi(x, t_2) = \varphi(x).$$

Similarly, consider $m(\xi, t;\ x, t_1),\ t > t_1$. We have

$$\int \varphi(\xi) m(\xi, t_2;\ x, t_1) = \Phi(x, t_1;\ t_2). \tag{11.7.14}$$

Consider now instead of deterministic times $t_1 < t_2$, random times with respect to the filtration $\mathcal{F}^t$, denoted by $\tau_1,\ \tau_2$, and we assume $\tau_1 < \tau_2 < T$. The bound is not necessary, but it avoids technicalities. Since we have

$$d\Phi(y(t), t, t_2) = D\Phi(y(t), t, t_2).\sigma(y(t), t)dw,$$

we can write, by integrating between t_1 and t_2,

$$\begin{aligned}\varphi(y(t_2)) &= \Phi(y(t_1), t_1, t_2) + \int_{t_1}^{t_2} D\Phi(y(t), t, t_2).\sigma(y(t), t)dw \\ &= \Phi(y(t_1), t_1, t_2) + \int_0^T \mathbb{1}_{t>t_1}\mathbb{1}_{t<t_2} D\Phi(y(t), t, t_2).\sigma(y(t), t)dw.\end{aligned}$$

We can replace t_1, t_2 by τ_1, τ_2 in this equality. Indeed, the process $\mathbb{1}_{t>\tau_1}\mathbb{1}_{t<\tau_2}$ $D\Phi(y(t), t, \tau_2)\sigma(y(t), t)$ is adapted. So we can write

$$\varphi(y(\tau_2)) = \Phi(y(\tau_1), \tau_1, \tau_2) + \int_0^T \mathbb{1}_{t>\tau_1}\mathbb{1}_{t<\tau_2} D\Phi(y(t), t, \tau_2).\sigma(y(t), t)dw.$$

We can take the conditional expectation with respect to $\mathcal{F}^{\tau_1}$ and obtain

$$\begin{aligned}E[\varphi(y(\tau_2))|\mathcal{F}^{\tau_1}] &= E[\Phi(y(\tau_1), \tau_1, \tau_2)|\mathcal{F}^{\tau_1}] \\ &= \int \varphi(\xi) E[m(\xi, \tau_2;\ y(\tau_1), \tau_1)|\mathcal{F}^{\tau_1}]d\xi.\end{aligned}$$

Suppose τ_2 is $\mathcal{F}^{\tau_1}$-measurable. Then this equality yields

$$E[\varphi(y(\tau_2))|\mathcal{F}^{\tau_1}] = \int \varphi(\xi) m(\xi, \tau_2;\ y(\tau_1), \tau_1)d\xi. \tag{11.7.15}$$

We obtain in particular the relation

$$E[\varphi(y(t+h))|\mathcal{F}^{t}] = \int \varphi(\xi) m(\xi, t+h;\ y(t), t)d\xi. \tag{11.7.16}$$

This relation expresses the Markov property of the diffusion process $y(t)$. At each time $t+h > t$, the conditional probability of $y(t+h)$ given $\mathcal{F}^t$ depends only on $y(t)$. The past history of the process matters only through the present value (if we think of t as the present time). The function $m(\xi, s; x, t)$ is called the **transition probability**. In particular, if we look at the process $y_{xt}(s)$, the solution of (11.7.11), we know the initial conditions x, t, and $m(\xi, s; \ x, t)$ is simply the probability density of $y_{xt}(s)$.

11.7.2 Feedback Control of Probability Densities

Consider now a control problem. We assume

$$\sigma(x,t),\ \sigma^{-1}(x,t) \tag{11.7.17}$$

continuous and bounded,

$$\begin{aligned} |\sigma(x,t) - \sigma(x',t)| &\leq K|x - x'|, \\ |g(x,v) - g(x',v')| &\leq C(|x - x'| + |v - v'|), \\ |g(x,v)| &\leq C(1 + |x| + |v|), \end{aligned} \tag{11.7.18}$$

$$\mathcal{U}_{\text{ad}} = \text{nonempty closed convex subset of } R^m, \tag{11.7.19}$$

$$\begin{aligned} |l(x,v)| &\leq \bar{l}(1 + |x|^2 + |v|^2), \\ |l_x(x,v)|\ ,\ |l_v(x,v)| &\leq \bar{l}(1 + |x| + |v|), \\ l(x,v) &\geq l_0|v|^2 - c_0, \end{aligned} \tag{11.7.20}$$

$$\begin{aligned} |h(x)| &\leq \bar{h}(1 + |x|^2), \\ |h_x(x)| &\leq \bar{h}(1 + |x|), \\ |h_{xx}(x)| &\leq \bar{h}. \end{aligned} \tag{11.7.21}$$

We have studied in Section 11.1 the problem

$$\begin{aligned} dx &= g(x(t), v(t))dt + \sigma(x(t))dw, \\ x(0) &= \xi, \end{aligned} \tag{11.7.22}$$

$$J(v(.)) = E[\int_0^T l(x(t), v(t))dt + h(x(T))], \tag{11.7.23}$$

constructed on a probability space $\Omega, \mathcal{A}, P$ equipped with a filtration $\mathcal{F}^t$ and a standard $P, \mathcal{F}^t$ Wiener process $w(.)$ with values in R^n. We assume that

ξ is $\mathcal{F}^0$ measurable independent of the Wiener process

has a probability density $m_0(x)$.

We have seen in Theorem 11.2 that if we consider the family of problems

$$dx = g(x(s), v(s))ds + \sigma(x(s))dw, \qquad (11.7.24)$$
$$x(t) = x,$$

$$J_{x,t}(v(.)) = E[\int_t^T l(x(s), v(s))ds + h(x(T))], \qquad (11.7.25)$$

then the value function

$$\Phi(x, t) = \inf_{v(.)} J_{x,t}(v(.)) \qquad (11.7.26)$$

is the unique solution of the HJB equation

$$-\frac{\partial \Phi}{\partial t} + A\Phi = H(x, D\Phi), \qquad (11.7.27)$$
$$\Phi(x, T) = h(x),$$

with

$$H(x, q) = \inf_{v \in \mathcal{U}_{ad}} L(x, v, q), \qquad (11.7.28)$$
$$L(x, v, q) = l(x, v) + q.g(x, v).$$

Under additional assumptions, there exists an optimal feedback $\hat{v}(x, t)$ obtained by the formula

$$\hat{v}(x, t) = \bar{v}(x, D\Phi),$$

where $\bar{v}(x, q)$ achieves the infimum in (11.7.28).

This motivates the fact that we restrict ourselves to controls obtained from feedbacks. Suppose we have a feedback $v(x, t)$ with linear growth. Then we consider the controlled state equation

$$dx = g(x(t), v(x(t), t))dt + \sigma(x(t))dw, \qquad (11.7.29)$$
$$x(0) = \xi.$$

Unless $v(.)$ is more regular, we cannot define (11.7.29) in the strong sense. Instead of using the SDE in the weak sense, we can consider the density, denoted by $m_{v(.)}(x, t)$, the solution of the Fokker–Planck equation

$$\frac{\partial m_{v(.)}}{\partial t} + A^* m_{v(.)} + \text{div}\,(g(x, v(x))m_{v(.)}) = 0, \qquad (11.7.30)$$
$$m(x, 0) = m_0(x).$$

We have, of course, to provide a precise meaning to this equation. We shall proceed only formally. With a feedback control, the objective $J(v(.))$ defined in (11.7.23) can be written as

$$J(v(.)) = E[\int_0^T l(x(t), v((x(t)))dt + h(x(T))],$$

and from the definition of the probability density, we have

$$J(v(.)) = \int_0^T \int_{R^n} l(x, v(x))m_{v(.)}(x,t)dxdt + \int_{R^n} h(x)m_{v(.)}(x,T). \tag{11.7.31}$$

It is interesting to observe that the problem is now deterministic. The state is $m_{v(.)}(x,t)$, the probability density of the random state. Its evolution is governed by a PDE, the Fokker–Planck equation (11.7.30). The objective functional is defined by (11.7.31). However, it is important to note that controls are restricted to functions of x. For the state equation (11.7.30), these functions $v(x,t)$ are open-loop controls. A closed-loop control would be a functional $v(x,m,t)$. But by a similar argument to that of Remark 10.6, the two classes coincide. So we may consider only open-loop controls. However, in view of the origin of the problem, we still call $v(x,t)$ a feedback. It is a feedback on the random state, not on the state m.

We shall not try to prove that there exists an optimal feedback $\hat{v}(x,t)$. We shall rather write a necessary condition for optimality. We need some notation. To a feedback control $v(x,t)$ we associate the function $\Phi_{v(.)}(x,t)$, the solution of the linear PDE

$$-\frac{\partial \Phi_{v(.)}}{\partial t} + A\Phi_{v(.)} - g(x, v(x)).D\Phi_{v(.)} = l(x, v(x)), \tag{11.7.32}$$

$$\Phi_{v(.)}(x,T) = h(x).$$

We can then compute the Gâteaux differential of the functional $J(v(.))$ at the point $\hat{v}(.)$. We have the following result.

PROPOSITION 11.4. *The Gâteaux differential of $J(v(.))$ at the point $\hat{v}(.)$ is given by*

$$\frac{d}{d\theta}J(\hat{v}(.){+}\theta v(.))|_{\theta=0} = \int_0^T \int_{R^n} \frac{\partial L}{\partial v}(x, \hat{v}(x,t), D\Phi_{\hat{v}(.)}(x,t))v(x,t)m_{\hat{v}(.)}(x,t)dxdt, \tag{11.7.33}$$

with $L(x,v,q) = l(x,v) + q.g(x,v)$.

PROOF. We provide only a formal proof. We shall also write $v(x)$ for $v(x,t)$ and $\hat{v}(x)$ for $\hat{v}(x,t)$. The first step is to check that

$$\frac{d}{d\theta}m_{\hat{v}(.)+\theta v(.)}(x,t)|_{\theta=0} = \tilde{m}(x,t),$$

where $\tilde{m}(x,t)$, depending on $\hat{v}(.)$, $v(.)$, is the solution of

$$\frac{\partial \tilde{m}}{\partial t} + A^*\tilde{m} + \text{div}(g(x,\hat{v}(x))\tilde{m}) + \tag{11.7.34}$$

$$+\text{div}(\frac{\partial}{\partial v}g(x,\hat{v}(x))v(x)m_{\hat{v}}(.)) = 0,$$

$$\tilde{m}(x,0) = 0.$$

But then

$$\frac{d}{d\theta}J(\hat{v}(.)+\theta v(.))|_{\theta=0} = \int_0^T \int_{R^n} l(x,\hat{v}(x))\tilde{m}(x,t)dxdt +$$

$$\int_0^T \int_{R^n} \frac{\partial l}{\partial v}(x,\hat{v}(x))v(x,t)m_{\hat{v}}(x,t)dxdt + \int_{R^n} h(x)\tilde{m}(x,T)dx.$$

Next,

$$\int_0^T \int_{R^n} l(x,\hat{v}(x))\tilde{m}(x,t)dxdt = \int_0^T \int_{R^n} (-\frac{\partial \Phi_{\hat{v}(.)}}{\partial t} + A\Phi_{\hat{v}(.)} - g(x,\hat{v}(x)).$$

$$D\Phi_{\hat{v}(.)})\tilde{m}(x,t)dxdt.$$

Integrating by parts and using the equation of $\tilde{m}$, see (11.7.34), we obtain

$$\int_0^T \int_{R^n} l(x,\hat{v}(x))\tilde{m}(x,t)dxdt = -\int_{R^n} h(x)\tilde{m}(x,T)dx +$$

$$-\int_0^T \int_{R^n} \text{div}(\frac{\partial}{\partial v}g(x,\hat{v}(x))v(x)m_{\hat{v}}(.))$$

$$\Phi_{\hat{v}(.)}(x,t)dxdt,$$

and thus

$$\frac{d}{d\theta}J(\hat{v}(.)+\theta v(.))|_{\theta=0} = \int_0^T \int_{R^n} \frac{\partial l}{\partial v}(x,\hat{v}(x))v(x,t)m_{\hat{v}}(x,t)dxdt$$

$$-\int_0^T \int_{R^n} \text{div}(\frac{\partial}{\partial v}g(x,\hat{v}(x))v(x)m_{\hat{v}}(.))\Phi_{\hat{v}(.)}(x,t)dxdt,$$

and we obtain (11.7.33). This concludes the proof. □

We can then state the following formal theorem.

THEOREM 11.6 (FORMAL). *In order that $\hat{v}(x,t)$ be optimal for the problem (11.7.30), (11.7.31) it is necessary that*

$$\hat{v}(x,t) = \bar{v}(x, D\Phi), \tag{11.7.35}$$

where $\Phi(x,t)$ is the value function.

PROOF. If $\hat{v}(x,t)$ is optimal, then for every admissible $v(x,t)$ we must have, using the Gâteaux differential (11.7.33) and the convexity of $\mathcal{U}_{\text{ad}}$,

$$\int_0^T \int_{R^n} \frac{\partial L}{\partial v}(x, \hat{v}(x,t), D\Phi_{\hat{v}(.)}(x,t))(v(x,t) - \hat{v}(x,t))\, m_{\hat{v}(.)}(x,t)dxdt \geq 0. \tag{11.7.36}$$

The density $m_{\hat{v}(.)}(x,t)$ is positive. We admit that it is strictly positive a.e. It follows that

$$\frac{\partial L}{\partial v}(x, \hat{v}(x,t), D\Phi_{\hat{v}(.)}(x,t))(v(x,t) - \hat{v}(x,t)) \geq 0, \text{ a.e.}$$

By localizing, we get

$$\frac{\partial L}{\partial v}(x, \hat{v}(x,t), D\Phi_{\hat{v}(.)}(x,t))(v - \hat{v}(x,t)) \geq 0, \text{ a.e.}$$
$$\forall v \in \mathcal{U}_{\text{ad}}.$$

This is a necessary condition for $\hat{v}(x,t)$ to minimize $L(x, v, D\Phi_{\hat{v}(.)}(x,t))$ in v. Therefore,

$$L(x, \hat{v}(x,t), D\Phi_{\hat{v}(.)}(x,t)) = H(x, D\Phi_{\hat{v}(.)}).$$

Returning to (11.7.32), we obtain

$$\Phi_{\hat{v}(.)}(x,t) = \Phi(x,t),$$

the value function, and

$$\hat{v}(x,t) = \bar{v}(x, D\Phi).$$

This concludes the formal proof. □

REMARK 11.4. We can now interpret the HJB equation as an adjoint equation for the control problem (11.7.30), (11.7.31). It is interesting to see that the optimal feedback is not obtained through a functional of m, so it is directly an open-loop control and not a closed-loop control on m. The reason lies in the condition (11.7.36), in which $m_{\hat{v}}(.)$ appears only as a multiplicative term, and since it is positive, it plays no role in the optimality condition. Therefore, the adjoint equation and the optimal state equation are completely decoupled.

11.8 Viscosity Theory

We rely considerably on the presentations of Crandall, Ishii, Lions [16], Fleming, Soner [22], and Nisio, [37]. We begin with the basic tools of the theory.

11.8.1 Second-Order Sub- and Superdifferentials

We denote by $\mathcal{S} = \mathcal{S}(n)$ the set of $n \times n$ symmetric matrices.

Let $\Phi \in C^0(R^n)$. We define the **second-order superdifferential** of Φ at x by

$$D^{+2}\Phi(x) = \{q \in R^n, M \in \mathcal{S} \mid \limsup_{y \to x} \frac{\Phi(y) - \Phi(x) - q.(y-x) - \frac{1}{2}(y-x)^* M(y-x)}{|y-x|^2} \leq 0\}, \tag{11.8.1}$$

and the **second-order subdifferential** of Φ at x by

$$D^{-2}\Phi(x) = \{q \in R^n, M \in \mathcal{S} \mid \liminf_{y \to x} \frac{\Phi(y) - \Phi(x) - q.(y-x) - \frac{1}{2}(y-x)^* M(y-x)}{|y-x|^2} \geq 0\}. \tag{11.8.2}$$

If $D^{+2}\Phi(x) \cap D^{-2}\Phi(x) \neq \emptyset$, then the function Φ is twice differentiable at x. Note that in this case,

$$D^{+2}\Phi(x) = \{D\Phi(x), M \geq D^2\Phi(x)\},\ D^{-2}\Phi(x) = \{D\Phi(x), M \leq D^2\Phi(x)\},$$

and the only common point is $D\Phi(x), D^2\Phi(x)$.

We will need the closure of these sets:

$$\overline{D^{+2}}\Phi(x) = \{q, M | \exists x_n, q_n, M_n \to x, q, M \text{ such that } q_n, M_n \in D^{+2}\Phi(x_n)\}, \tag{11.8.3}$$

$$\overline{D^{-2}}\Phi(x) = \{q, M | \exists x_n, q_n, M_n \to x, q, M \text{ such that } q_n, M_n \in D^{-2}\Phi(x_n)\}.$$

We can also give a very useful characterization of these sets as follows:

$$D^{+2}\Phi(x) = \{q, M | \exists \varphi_x(y) \in C^2 \text{such that } \Phi(y) \leq \varphi_x(y), \Phi(x) = \varphi_x(x),$$
$$D\varphi_x(x) = q, \quad D^2\varphi_x(x) = M\},$$
$$D^{-2}\Phi(x) = \{q, M | \exists \varphi_x(y) \in C^2 \text{such that } \Phi(y) \geq \varphi_x(y), \Phi(x) = \varphi_x(x),$$
$$D\varphi_x(x) = q, \quad D^2\varphi_x(x) = M\}.$$

We simply construct a function that can be associated to a pair $(q, M) \in D^{+2}\Phi(x)$. It is

$$\varphi_x(y) = \Phi(x) + q.(y-x) + \frac{1}{2}(y-x)^* M(y-x) + F(|y-x|^2), \qquad (11.8.4)$$

with

$$F(r) = \frac{2}{3r}\int_r^{2r} d\xi \int_\xi^{2\xi} h(\rho)d\rho,\ r > 0 \qquad (11.8.5)$$

and

$$h(\rho) = \sup_{|y-x|^2 \le \rho} \{\frac{(\Phi(y) - \Phi(x) - q.(y-x) - \frac{1}{2}(y-x)^* M(y-x)\}^+}{|y-x|^2}\}.$$

The function $h(\rho)$ is monotonically nondecreasing, and since $(q, M) \in (q, M) \in D^{+2}\Phi(x)$, we have $h(\rho) \to 0$ as $\rho \to 0$. We check easily that

$$F(r) \ge rh(r),\ |F'(r)| \le \frac{8}{3}h(4r),\ |F"(r)| \le \frac{28}{3}\frac{h(4r)}{r},$$
$$F(0) = 0,\ F'(0) = 0,$$

and

$$D\varphi_x(y) = q + (M + 2F'(|y-x|^2)I)(y-x),$$
$$D^2\varphi_x(y) = M + 4F"(|y-x|^2)(y-x)(y-x)^* + 2F'(|y-x|^2)I.$$

All required properties can be easily verified.

In the case of functions of t and x, $\Phi(x, t) \in C^0(R^n \times [0, T])$, the definitions extend as follows:

$$D^{+2,1}\Phi(x,t) = \{q \in R^n, \rho \in R, M \in \mathcal{S} \mid$$
$$\limsup_{y \to x, s \to t} \frac{\Phi(y,s) - \Phi(x,t) - q.(y-x) - \rho(s-t) - \frac{1}{2}(y-x)^* M(y-x)}{|y-x|^2 + |s-t|} \le 0\},$$
$$D^{-2,1}\Phi(x,t) = \{q \in R^n, \rho \in R, M \in \mathcal{S} \mid$$
$$\liminf_{y \to x, s \to t} \frac{\Phi(y,s) - \Phi(x,t) - q.(y-x) - \rho(s-t) - \frac{1}{2}(y-x)^* M(y-x)}{|y-x|^2 + |s-t|} \ge 0\}.$$

$$(11.8.6)$$

If $q, \rho, M \in D^{+2,1}\Phi(x,t)$, then we can construct a function $\varphi_{x,t}(y,s)$ such that $\Phi(y,s) \le \varphi_{x,t}(y,s)$, $\Phi(x,t) = \varphi_{x,t}(x,t)$ and $D\varphi_{x,t}(x,t) = q$, $\dfrac{\partial \varphi_{x,t}(x,t)}{\partial s} = \rho$, $D^2\varphi_{x,t}(x,t) = M$. It is given by

$$\varphi_{x,t}(y,s) = \Phi(x,t) + \rho(s-t) + q.(y-x) + \frac{1}{2}(y-x)^*M(y-x) + F((|y-x|^4 + (s-t)^2)^{\frac{1}{2}}), \tag{11.8.7}$$

where $F(r)$ is given by (11.8.5).

11.8.2 The Crandall–Ishii Lemma

We begin with a very useful result called the Crandall–Ishii lemma. One uses the following deep result, a consequence of Alexandrov's theorem and Jensen's lemma (see [16] for a proof).

THEOREM 11.7. *Let f be in $C^0(R^N)$, with*

$$f(0) = 0, \ f(\xi) + \frac{\lambda}{2}|\xi|^2 \textit{ is convex } \lambda \ge 0, \tag{11.8.8}$$

$$f(\xi) \le \frac{1}{2}\xi^* B\xi, \ B \in \mathcal{S}(N).$$

Then there exists a sequence $\xi_n \to 0$ such that f is twice differentiable in ξ_n with $Df(\xi_n) \to 0$, $D^2 f(\xi_n) \to M \in \mathcal{S}(N)$, with $-\lambda I \le M \le B$.

From the first assumption, one says that $f(\xi)$ is semiconvex. The result means that $(0, M)$ belongs to $\overline{D^{+2}f(0)} \cap \overline{D^{-2}f(0)}$. If f is C^2, this implies that $Df(0) = 0$, $-\lambda I \le D^2 f(0) \le B$.

We formulate the Crandall–Ishii lemma in the following particular form.

LEMMA 11.2. *Let $w(x) \in C^0(R^N)$. We assume that $w(0) = 0$ and*

$$w(x) \le \frac{1}{2}x^* Ax, \ A \in \mathcal{S}(N). \tag{11.8.9}$$

Then for every $\epsilon > 0$, there exists $M \in \mathcal{S}(N)$ such that

$$(0, M) \in \overline{D^{+2}w(0)}, \tag{11.8.10}$$

$$-(\frac{1}{\epsilon} + ||A||)I \le M \le A + \epsilon A^2. \tag{11.8.11}$$

PROOF. We see easily that if $w(x)$ is twice differentiable at 0, then $Dw(0) = 0$, $D^2w(0) \le A$. So we get the result with $M = D^2w(0)$ and $\epsilon = 0$. For the nondifferentiable case, we use the dual function. First, we state the easily obtained inequality

$$(Ax, x) \leq ((A + \epsilon A^2)\xi, \xi) + (\frac{1}{\epsilon} + ||A||)|x - \xi|^2, \forall x, \xi \in R^N. \tag{11.8.12}$$

We set $\lambda = \frac{1}{\epsilon} + ||A||$. Then we can write

$$w(x) - \frac{\lambda}{2}|x - \xi|^2 \leq \frac{1}{2}((A + \epsilon A^2)\xi, \xi).$$

We next introduce the dual function

$$\hat{w}(\xi) = \sup_x (w(x) - \frac{\lambda}{2}|x - \xi|^2),$$

so

$$\hat{w}(0) = 0, \ \hat{w}(\xi) \leq \frac{1}{2}((A + \epsilon A^2)\xi, \xi). \tag{11.8.13}$$

Moreover, $\hat{w}(\xi) + \frac{\lambda}{2}|\xi|^2$ is convex. According to Theorem 11.7, there exists a matrix $M \in \mathcal{S}(N)$ such that $(0, M) \in \overline{D^{+2}}\hat{w}(0) \cap \overline{D^{-2}}\hat{w}(0)$, and $-\lambda I \leq M \leq A + \epsilon A^2$. If we show that it belongs to $\overline{D^{+2}}w(0)$, then the results (11.8.10), (11.8.11) follow, where we recall the definition of λ. To establish this property, we study the dual function in greater detail. In fact, there are interesting correspondence formulas between functions and their duals. We will make them explicit in a smooth situation, to simplify the presentation. Consider $\hat{w}(\xi)$. The supremum is attained at $\hat{x}(\xi)$, and we have

$$\begin{aligned} Dw(\hat{x}(\xi)) &= \lambda(\hat{x}(\xi) - \xi), \\ D\hat{w}(\xi) &= \lambda(\hat{x}(\xi) - \xi). \end{aligned} \tag{11.8.14}$$

The first relation is simply the first-order condition, and the second is a consequence of the so-called envelope theorem. Recall that $\hat{x}(0) = 0$. Hence $D\hat{w}(0) = Dw(0) = 0$. Now

$$w(x) - \frac{\lambda}{2}|\hat{x}(\xi) - \xi|^2 \leq \hat{w}(\xi + x - \hat{x}(\xi))$$

$$= \hat{w}(\xi) + D\hat{w}(\xi).(x - \hat{x}(\xi)) + \frac{1}{2}(D^2\hat{w}(\xi)(x - \hat{x}(\xi)), x - \hat{x}(\xi)) + o(|x - \hat{x}(\xi)|^2),$$

but $\hat{w}(\xi) = w(\hat{x}(\xi)) - \frac{\lambda}{2}|\hat{x}(\xi) - \xi|^2$. Therefore, we can write

$$w(x) \leq w(\hat{x}(\xi)) + D\hat{w}(\xi).(x - \hat{x}(\xi)) + \frac{1}{2}(D^2\hat{w}(\xi)(x - \hat{x}(\xi)), x - \hat{x}(\xi)) + o(|x - \hat{x}(\xi)|^2).$$

From (11.8.14), it follows that

$$w(x) \leq w(\hat{x}(\xi))+Dw(\hat{x}(\xi)).(x-\hat{x}(\xi))+\frac{1}{2}(D^2\hat{w}(\xi)(x-\hat{x}(\xi)),x-\hat{x}(\xi))+o(|x-\hat{x}(\xi)|^2).$$

But

$$w(x) = w(\hat{x}(\xi))+Dw(\hat{x}(\xi)).(x-\hat{x}(\xi))+\frac{1}{2}(D^2w(\hat{x}(\xi))(x-\hat{x}(\xi)),x-\hat{x}(\xi))+o(|x-\hat{x}(\xi)|^2).$$

Combining the two preceding relations, we obtain

$$D\hat{w}(\xi) = Dw(\hat{x}(\xi)),\ D^2\hat{w}(\xi) \geq D^2w(\hat{x}(\xi)). \tag{11.8.15}$$

In particular,

$$D\hat{w}(0) = Dw(0),\ D^2\hat{w}(0) \geq D^2w(0). \tag{11.8.16}$$

So $(0, D^2\hat{w}(0))$ belongs to $D^{+2}w(0)$. Therefore, if the dual function $\hat{w}(\xi)$ is C^2at 0, the properties (11.8.10), (11.8.11) are valid. To remove the smoothness assumption on $\hat{w}(\xi)$, we make a useful observation, which we first explain in the smooth case.

Set $q = D\hat{w}(\xi) = Dw(\hat{x}(\xi))$. From (11.8.14), we have $\hat{x}(\xi) = \xi+\frac{q}{\lambda}$. Therefore, (11.8.15) implies $M \geq D^2\hat{w}(\xi) \Rightarrow M \geq D^2w(\xi+\frac{q}{\lambda})$. This means in the smooth case that

$$(q, M) \in D^{+2}\hat{w}(\xi) \Rightarrow (q, M) \in D^{+2}w(\xi+\frac{q}{\lambda}). \tag{11.8.17}$$

It is interesting to see that this relation is also valid in the nonsmooth case. The reasoning does not need smoothness. It is sufficient to know that the function $\hat{x}(\xi)$ is well defined. In that case, we shall have necessarily $q = \lambda(\hat{x}(\xi) - \xi)$. See also (11.8.14). Indeed, assume $(q, M) \in D^{+2}\hat{w}(\xi)$. We write

$$\begin{aligned} w(x) - \frac{\lambda}{2}|x-\eta|^2 &\leq \hat{w}(\eta) \\ &\leq \hat{w}(\xi)+q.(\eta-\xi)+\frac{1}{2}(M(\eta-\xi),\eta-\xi))+o(|\eta-\xi|^2). \end{aligned} \tag{11.8.18}$$

Take first $\eta = x+\xi-\hat{x}(\xi)$, and note that $\hat{w}(\xi)+\frac{\lambda}{2}|\xi-\hat{x}(\xi)|^2 = w(\hat{x}(\xi))$. Therefore,

$$\begin{aligned} w(x) \leq w(\hat{x}(\xi))+q.(x-\hat{x}(\xi))+\frac{1}{2}(M(x-\hat{x}(\xi)),x-\hat{x}(\xi))+ \\ +o(|x-\hat{x}(\xi)|^2). \end{aligned}$$

This implies that $(q, M) \in D^{+2}w(\hat{x}(\xi))$. It remains to prove the relation $\hat{x}(\xi) = \xi + \frac{q}{\lambda}$. Returning to (11.8.18), we take $x = \hat{x}(\xi),\ \eta = \xi + \theta(q + \lambda(\xi - \hat{x}(\xi))$. We obtain, after rearranging,

$$0 \leq \theta |q + \lambda(\xi - \hat{x}(\xi))|^2 + O(\theta^2 |q + \lambda(\xi - \hat{x}(\xi))|^2).$$

Since θ is arbitrary, we can take $\theta < 0$ and divide by $-\theta$, and then let $\theta \to 0$. It follows immediately that $q + \lambda(\xi - \hat{x}(\xi)) = 0$.

We then consider an element $(0, M) \in \overline{D^{+2}\hat{w}(0)}$, which exists according to the beginning of the proof. There exists a sequence ξ_n, q_n, M_n such that $\xi_n \to 0,\ q_n \to 0,\ M_n \to M$ and $(q_n, M_n) \in D^{+2}\hat{w}(\xi_n)$. Therefore, also $(q_n, M_n) \in D^{+2}w(\xi_n + \frac{q_n}{\lambda})$, which proves the property

$$(0, M) \in \overline{D^{+2}\hat{w}(0)} \Rightarrow (0, M) \in \overline{D^{+2}w(0)}. \tag{11.8.19}$$

This completes the proof of (11.8.10), (11.8.11). □

One can extend the result (11.8.10), (11.8.11) immediately to the case of a general quadratic form: Let $\varphi(x) = b.x + \frac{1}{2}x^* Ax$ and $w(x) \in C^0(R^N)$. Let $\hat{x}$ be a maximizer of the function $w(x) - \varphi(x)$. Then for every $\epsilon > 0$, there exists $M \in \mathcal{S}(N)$ such that

$$(D\varphi(\hat{x}), M) \in \overline{D^{+2}w(\hat{x})} \tag{11.8.20}$$
$$-(\frac{1}{\epsilon} + ||A||)I \leq M \leq A + \epsilon A^2.$$

We next follow M. Nisio [37] to extend the result to the time-dependent case. Let $\varpi_i(x_i, t), i = 1, 2,\ x_1, x_2 \in R^n, t \in R$, by continuous functions such that

$$\varpi_i(x_i, t) < -1, \forall x_i, t,\ |x_i|, |t| \geq a_0 > 0, \tag{11.8.21}$$
$$\varpi_i(0, 0) = 0,$$
$$\varpi_1(x_1, t) + \varpi_2(x_2, t) < \frac{\lambda}{2}|x_1 - x_2|^2, \forall x_1, x_2, t \neq 0,$$

and such that they satisfy the property

$$\forall q, \rho, M \in \overline{D^{+2,1}\varpi_i(x_i, t)} \tag{11.8.22}$$
$$|x_i| + |t| + |q| + ||M|| \leq h \Rightarrow \rho \geq -c(h).$$

PROPOSITION 11.5. *We assume (11.8.21), (11.8.22). Then* $\forall \epsilon > 0$, *there exist* $Z_i \in \mathcal{S}(n)$ *such that*

$$-(\frac{1}{\epsilon}+2\lambda)I_{2n} \leq \begin{pmatrix} Z_1 & 0 \\ 0 & Z_2 \end{pmatrix} \leq \lambda(1+2\lambda\epsilon)\hat{A} \tag{11.8.23}$$

$$\hat{A} = \begin{pmatrix} I_n & -I_n \\ -I_n & I_n \end{pmatrix}$$

and $\rho \in R$ *such that* $(0,\rho,Z_1) \in \overline{D^{+2,1}}\varpi_1(0,0)$, $(0,-\rho,Z_2) \in \overline{D^{+2,1}}\varpi_2(0,0)$.

PROOF. We set $z = (x_1,t_1;x_2,t_2)$, $\varpi(z) = \varpi_1(x_1,t_1)+\varpi_2(x_2,t_2)$, $\varphi_\delta(z) = \frac{\lambda}{2}|x_1-x_2|^2+\frac{1}{2\delta}|t_1-t_2|^2$, and $\Phi_\delta(z) = \varpi(z)-\varphi_\delta(z)$. The function $\Phi_\delta(z)$ attains its maximum at a point z_δ. We note that $\Phi_\delta(z_\delta) \geq 0$, and thus from the first assumption (11.8.21), the sequence z_δ remains bounded as $\delta \to 0$. We can use the Crandall–Ishii lemma, the result (11.8.20), to claim that for every ϵ, there exists $M_{\delta\epsilon} \in \mathcal{S}(2n+2)$ such that $(D_z\varphi_\delta(z_\delta), M_{\delta\epsilon}) \in \overline{D^{+2}}\varpi(z_\delta)$. The matrix $M_{\delta\epsilon}$ satisfies $-(\frac{1}{\epsilon}+||A_\delta||)I_{2n+2} \leq M_{\delta\epsilon} \leq A_\delta+\epsilon||A_\delta||^2$. The matrix A_δ is given by

$$A_\delta = \begin{pmatrix} \lambda I_n & 0 & -\lambda I_n & 0 \\ 0 & \frac{1}{\delta} & 0 & \frac{1}{\delta} \\ -\lambda I_n & 0 & \lambda I_n & 0 \\ 0 & -\frac{1}{\delta} & 0 & \lambda I_n \end{pmatrix}.$$

We can split the result as follows: there exist $Z_{1,\delta\epsilon}, Z_{2,\delta\epsilon} \in \mathcal{S}(n)$ such that

$$\begin{pmatrix} \lambda(x_{1\delta}-x_{2\delta}) \\ \frac{1}{\delta}(t_{1\delta}-t_{2\delta}) \end{pmatrix},\ Z_{1,\delta\epsilon} \in \overline{D^{+2,1}}\varpi_1(x_{1\delta},t_{1\delta}), \tag{11.8.24}$$

$$-\begin{pmatrix} \lambda(x_{1\delta}-x_{2\delta}) \\ \frac{1}{\delta}(t_{1\delta}-t_{2\delta}) \end{pmatrix},\ Z_{2,\delta\epsilon} \in \overline{D^{+2,1}}\varpi_2(x_{2\delta},t_{2\delta}),$$

and

$$-(\frac{1}{\epsilon}+\lambda||\hat{A}||)I_{2n} \leq \begin{pmatrix} Z_{1,\delta\epsilon} & 0 \\ 0 & Z_{2,\delta\epsilon} \end{pmatrix} \leq (1+\lambda\epsilon)\hat{A}. \tag{11.8.25}$$

We know that $x_{1\delta}, t_{1\delta}, x_{2\delta}, t_{2\delta}$ remains bounded as $\delta \to 0$. From (11.8.25) we see that $||Z_{1,\delta\epsilon}||, ||Z_{2,\delta\epsilon}||$ are also bounded. Using the assumption (11.8.22), we obtain

$$\frac{1}{\delta}(t_{1\delta}-t_{2\delta}) \geq -c_0,\ -\frac{1}{\delta}(t_{1\delta}-t_{2\delta}) \geq -c_0.$$

Therefore, $\frac{1}{\delta}|t_{1\delta}-t_{2\delta}| \leq c_0$, whence $t_{1\delta}-t_{2\delta} \to 0$ as $\delta \to 0$. For a subsequence, we claim that

$$t_{i\delta} \to \theta,\ x_{i\delta} \to \xi_i,\ Z_{i,\delta\epsilon} \to Z_{i\epsilon},\ \frac{1}{\delta}(t_{1\delta}-t_{2\delta}) \to \rho.$$

Using

$$\varpi_1(x_{1\delta}, t_{1\delta}) + \varpi_2(x_{2\delta}, t_{2\delta}) \geq \frac{\lambda}{2}|x_{1\delta} - x_{2\delta}|^2 + \frac{1}{2\delta}|t_{1\delta} - t_{2\delta}|^2,$$

we get

$$\varpi_1(\xi_1, \theta) + \varpi_2(\xi_2, \theta) \geq \frac{\lambda}{2}|\xi_1 - \xi_2|^2,$$

and from the last assumption (11.8.21), we necessarily have $\xi_1 = \xi_2 = 0, \theta = 0$. This concludes the proof. □

REMARK 11.5. Concerning the assumption (11.8.22), we see from the proof that an opposite inequality will lead to the same result. The assumption is made in this form because it is the way in which we are going to apply it below.

One can apply Proposition 11.5 to get a result appropriate to our applications. Let $u(x,t),\ v(x,t) \in C^0(R^{n+1})$. We set

$$\varphi(x, y, t) = \frac{\lambda}{2}|x - v|^2 + \beta(T - t), \tag{11.8.26}$$

and we assume that there exists a maximizer of $u(x,t) - v(y,t) - \varphi(x,y,t)$ in the three variables x, y, t, denoted by $\hat{x}, \hat{y}, \hat{t}$ with $\hat{t} \in (0, T)$. We define

$$\varpi_1(x_1, t) = u(\hat{x} + x_1, \hat{t} + t) - u(\hat{x}, \hat{t}) + \frac{\beta}{2}t$$
$$-\lambda(\hat{x} - \hat{y})x_1 - \frac{r}{2}(|x_1|^4 + \frac{|t|^4}{2})$$
$$\varpi_2(x_2, t) = -v(\hat{y} + x_2, \hat{t} + t) + v(\hat{y}, \hat{t}) + \frac{\beta}{2}t$$
$$+\lambda(\hat{x} - \hat{y})x_2 - \frac{r}{2}(|x_2|^4 + \frac{|t|^4}{2}).$$

It is clear that the functions $\varpi_1(x_1, t), \varpi_2(x_2, t)$ satisfy (11.8.21). Moreover, if $q_1, \rho_1, M_1 \in \overline{D^{+2,1}}\varpi_1(x_1, t)$, then

$$q_1 + \lambda(\hat{x} - \hat{y}) + 2r|x_1|^2 x_1,\ \rho_1 - \frac{\beta}{2} + rt^3,\ M_1 + 2r[|x_1|^2 I + 2x_1 x_1^*] \in \overline{D^{+2,1}}u(\hat{x} + x_1, \hat{t} + t),$$

and if $q_2, \rho_2, M_2 \in \overline{D^{+2,1}}\varpi_2(x_2, t)$, then

$$-q_2 + \lambda(\hat{x} - \hat{y}) - 2r|x_2|^2 x_2, -\rho_2 + \frac{\beta}{2} - rt^3, -M_2 - 2r[|x_2|^2 I + 2x_2 x_2^*] \in \overline{D^{-2,1}}v(\hat{y} + x_2, \hat{t} + t),$$

and the assumption (11.8.22) is satisfied if we make appropriate assumptions on $u(x,t)$, $v(y,t)$, namely

$$\forall q, \rho, M \in \overline{D^{+2,1}}u(x,t) \tag{11.8.27}$$
$$|x| + |t| + |q| + ||M|| \leq h \Rightarrow \rho \geq -c(h),$$
$$\forall q, \rho, M \in \overline{D^{-2,1}}v(y,t) \tag{11.8.28}$$
$$|y| + |t| + |q| + ||M|| \leq h \Rightarrow \rho \leq c(h).$$

We consider a maximizer $\hat{x}, \hat{y}, \hat{t}$ of $u(x,t) - v(y,t) - \varphi(x,y,t)$, with $\varphi(x,y,t)$ given by (11.8.26). We can apply Proposition 11.5 to state that there exist $\rho_i \in R$, $X_i \in \mathcal{S}(n)$ such that

$$(\lambda(\hat{x}-\hat{y}), \rho_1, X_1) \in \overline{D^{+2,1}}u(\hat{x},\hat{t}), \tag{11.8.29}$$
$$(\lambda(\hat{x}-\hat{y}), \rho_2, X_2) \in \overline{D^{-2,1}}v(\hat{y},\hat{t}),$$
$$\rho_1 - \rho_2 = -\beta,$$
$$-(\frac{1}{\epsilon} + 2\lambda)I_{2n} \leq \begin{pmatrix} X_1 & 0 \\ 0 & -X_2 \end{pmatrix} \leq \lambda(1+2\epsilon\lambda)\hat{A}.$$

Since ϵ is arbitrary, we can take $\epsilon = \dfrac{1}{\lambda}$ to assert that

$$-3\lambda I_{2n} \leq \begin{pmatrix} X_1 & 0 \\ 0 & -X_2 \end{pmatrix} \leq 3\lambda\hat{A}. \tag{11.8.30}$$

We are ready to define viscosity solutions.

11.8.3 Viscosity Solutions

We return to the value function $\Phi(x,t)$ of the stochastic control problem (11.2.1), (11.2.2) (we assume that σ does not depend on the control v). It satisfies the optimality principle

$$\Phi(x,t) = \inf_{v(.)} E[\int_t^{t+\epsilon} l(x(s), v(s))ds + \Phi(x(t+\epsilon), t+\epsilon)],\ \forall t \leq T-\epsilon, \tag{11.8.31}$$
$$\Phi(x,T) = h(x).$$

A subsolution $\Psi(x,t)$ of the Bellman equation (11.8.31) satisfies the inequalities

$$\Psi(x,t) \leq \inf_{v(.)} E[\int_t^{t+\epsilon} l(x(s),v(s))ds + \Psi(x(t+\epsilon),t+\epsilon)],\ \forall t \leq T-\epsilon, \tag{11.8.32}$$

$$\Psi(x,T) \leq h(x).$$

As in Theorem10.3, Chapter 10, we can take $v(s) = v$ constant in the inequality (11.8.32).

Similarly, a supersolution $\chi(x,t)$ of the Bellman equation (11.8.31) satisfies the inequalities

$$\chi(x,t) \geq \inf_{v(.)} E[\int_t^{t+\epsilon} l(x(s),v(s))ds + \chi(x(t+\epsilon),t+\epsilon)],\ \forall t \leq T-\epsilon, \tag{11.8.33}$$

$$\chi(x,T) \geq h(x).$$

The value function is the largest of the subsolutions and the smallest of the supersolutions. We next consider the HJB equation (11.2.10):

$$-\frac{\partial \Phi}{\partial t} + A\Phi = H(x,t,D\Phi) \tag{11.8.34}$$

$$\Phi(x,T) = h(x),$$

with

$$A\Phi(x) = -\text{tr}\,(a(x)D^2\Phi(x)),$$

$$H(x,t,D\Phi) = \inf_v(l(x,v) + D\Phi.g(x,v)).$$

A subsolution that is twice differentiable will satisfy the inequality

$$\frac{\partial \Psi}{\partial t} - A\Psi + H(x,t,D\Psi), \geq 0 \tag{11.8.35}$$

and a supersolution will satisfy

$$\frac{\partial \Psi}{\partial t} - A\Psi + H(x,t,D\Psi) \leq 0. \tag{11.8.36}$$

This leads to the following natural definition:

A **viscosity subsolution** of the HJB equation is a function $\Psi(x,t)$ that satisfies

$$\rho + \text{tr}\,(a(x)M) + H(x,t,q) \geq 0,\ \forall (q,\rho,M) \in \overline{D^{+2,1}}\Psi(x,t), \tag{11.8.37}$$

$$\Psi(x,T) \leq h(x).$$

A **viscosity supersolution** of the HJB equation is a function $\chi(x,t)$ that satisfies

$$\rho + \text{tr}\,(a(x)M) + H(x,t,q) \leq 0,\ \forall(q,\rho,M) \in \overline{D^{-2,1}}\chi(x,t), \quad (11.8.38)$$
$$\chi(x,T) \geq h(x).$$

We check immediately that a subsolution of the Bellman equation is a viscosity subsolution of the HJB equation, and a supersolution of the Bellman equation is a viscosity supersolution of the HJB equation. We note indeed that we can replace the closure of the super- and subdifferentials by the super- and subdifferentials themselves. We can then use the characterization of super- and subdifferentials by the property that for each element of a superdifferential, at a given point we can construct a smooth function whose first and second derivatives coincide with this element and that dominates the function and is equal to the function at the point. Smoothness allows us to use Itô's calculus and to conclude easily as in the deterministic case; see Proposition 10.5. So the value function is a viscosity solution of the HJB equation.

We now formulate the main result, which is a comparison between viscosity subsolutions and viscosity supersolutions, similar to the deterministic case; see Proposition 10.6.

Maximum Principle:

THEOREM 11.8. *We assume that $a(x)$ is uniformly Lipschitz and that the Hamiltonian satisfies*

$$|H(x,t,q) - H(x',t,q')| \leq c(|x-x'| + |q-q'|) \tag{11.8.39}$$

and $h(x) \in C^0(R^n)$. Let $\Psi(x,t)$, $\chi(x,t)$ be bounded continuous functions that are respectively a viscosity subsolution and a viscosity supersolution of the HJB equation (11.8.34). Then $\Psi(x,t) \leq \chi(x,t)$.

As a consequence, the value function is the unique viscosity solution of the HJB equation.

PROOF. We first define K such that

$$|\Psi(x,t)| + |\chi(x,t)| \leq K. \tag{11.8.40}$$

Suppose the property is not true. Then there exist (z,θ) such that

$$\Psi(z,\theta) - \chi(z,\theta) = 4\Delta > 0,\ \theta \in [0,T]. \tag{11.8.41}$$

We define

$$u(x,t) = \Psi(x,t) - \frac{\gamma}{t} - \epsilon|x|^2,\ v(x,t) = \chi(x,t) + \epsilon|x|^2,$$

$$\varphi(x,y,t) = \frac{\mu}{2}|x-y|^2 + \beta(T-t),$$
$$\Phi(x,y,t) = u(x,t) - v(y,t) - \varphi(x,y,t).$$

The number μ will tend to $+\infty$. The numbers γ and β will be chosen, subject to restrictions, but fixed. The number ϵ is fixed, subject to restrictions. After μ has gone to $+\infty$, it will converge to 0. The first point is to see that there exists a maximizer $\hat{x}, \hat{y}, \hat{t}$ of $\Phi(x, y, t)$ such that $\hat{t} \in (0, T)$. We consider the domain in which we look for the maximizer. The triple (x, y, t) must satisfy $\Phi(x, y, t) \geq \Phi(z, z, \theta)$. If we choose β, γ, ϵ such that

$$\beta T \leq \Delta,\ \gamma \leq \theta\Delta,\ 2\epsilon|z|^2 \leq \Delta, \tag{11.8.42}$$

then $\Phi(z, z, \theta) \geq \theta$. Therefore, the set of points in which to look for a maximum is contained in the set $\{x, y, t|\ \Phi(x, y, t) \geq \Delta\}$. So it is contained in the set of points such that

$$\frac{\gamma}{t} + \epsilon(|x|^2 + |y|^2) + \frac{\mu}{2}|x - y|^2 + \beta(T - t) \leq \Psi(x, t) - \chi(y, t) - \Delta \leq 2K. \tag{11.8.43}$$

Therefore, we get

$$t \geq t_0 = \frac{\gamma}{2K},\ |x|^2 + |y|^2 \leq \frac{2K}{\epsilon}. \tag{11.8.44}$$

Also,

$$\begin{aligned} 0 &\leq \Psi(x, t) - \Psi(x, T) - (\chi(x, t) - \chi(x, T)) \\ &+ h(x) - h(y) - \Delta, |x - y| \leq 2\sqrt{\frac{K}{\mu}}. \end{aligned} \tag{11.8.45}$$

The functions Ψ and χ are uniformly continuous on $B(\sqrt{\frac{2K}{\epsilon}}) \times [0, T]$, and h is uniformly continuous on $B(\sqrt{\frac{2K}{\epsilon}})$, where $B(r)$ is the ball with center 0 and radius r. Therefore, it follows from (11.8.45) that

$$0 \leq l_\epsilon(T - t) + m_\epsilon(2\sqrt{\frac{K}{\mu}}) - \Delta,$$

in which l_ϵ is a common modulus of continuity for Ψ and χ on $B(\sqrt{\frac{2K}{\epsilon}}) \times [0, T]$, and m_ϵ is a modulus of continuity for h on $B(\sqrt{\frac{2K}{\epsilon}})$. We can choose $\mu_0(\epsilon)$ such that for $\mu > \mu_0(\epsilon)$, we have $m_\epsilon(2\sqrt{\frac{K}{\mu}}) < \frac{\Delta}{2}$. We define $t_{1\epsilon}$ such that for $T - t < T - t_{1\epsilon}$, we have $l_\epsilon(T - t) < \frac{\Delta}{2}$. The inequality shows that a time t such that $T - t < T - t_{1\epsilon}$ cannot be part of a maximizer. Therefore, we have established that

$$T - \hat{t} \geq T - t_{1\epsilon} > 0. \tag{11.8.46}$$

Therefore, we can guarantee that $\hat{t} \in (0, T)$. We are almost ready to apply the result (11.8.29), (11.8.30), provided we check that the assumptions (11.8.27) and (11.8.28) are satisfied. But

$$(q, \rho, M) \in \overline{D^{+2,1}}u(x,t) \Rightarrow (q + 2\epsilon x, \rho - \frac{\gamma}{t^2}, M + 2\epsilon I) \in \overline{D^{+2,1}}\Psi(x,t),$$

$$(q, \rho, M) \in \overline{D^{-2,1}}v(x,t) \Rightarrow (q - 2\epsilon x, \rho, M - 2\epsilon I) \in \overline{D^{-2,1}}\chi(x,t).$$

Since $\Psi(x,t)$ and $\chi(x,t)$ are a viscosity subsolution and a viscosity supersolution of the HJB equation, we can write

$$(q, \rho, M) \in \overline{D^{+2,1}}u(x,t) \Rightarrow \rho - \frac{\gamma}{t^2} + \text{tr}\,(a(x)M) + 2\epsilon\text{tr}\,(a(x)) + \\ + H(x, t, q + 2\epsilon x) \geq 0,$$

and from the assumption (11.8.39), it is easy to check that the property (11.8.27) is satisfied. Similar reasoning holds for (11.8.28). We can then apply the result (11.8.29), (11.8.30). There exist real numbers ρ_1, ρ_2 and matrices $X_1, X_2 \in \mathcal{S}(n)$ such that

$$(\mu(\hat{x} - \hat{y}), \rho_1, X_1) \in \overline{D^{+2,1}}u(\hat{x}, \hat{t}), \\ (\mu(\hat{x} - \hat{y}), \rho_2, X_2) \in \overline{D^{-2,1}}v(\hat{y}, \hat{t}), \\ \rho_1 - \rho_2 = -\beta, \\ -3\mu I_{2n} \leq \begin{pmatrix} X_1 & 0 \\ 0 & -X_2 \end{pmatrix} \leq 3\mu\hat{A}. \tag{11.8.47}$$

This means also, returning to the superdifferential of Ψ and subdifferential of χ, that

$$\rho_1 - \frac{\gamma}{(\hat{t})^2} + \text{tr}(a(\hat{x})X_1) + 2\epsilon\text{tr}\,(a(\hat{x})) + H(\hat{x}, \hat{t}, \mu(\hat{x} - \hat{y}) + 2\epsilon\hat{x}) \geq 0,$$

$$\rho_2 + \text{tr}(a(\hat{y})X_2) - 2\epsilon\text{tr}\,(a(\hat{y})) + H(\hat{y}, \hat{t}, \mu(\hat{x} - \hat{y}) - 2\epsilon\hat{y}) \geq 0,$$

and on subtracting, we get

$$-\beta + \text{tr}(a(\hat{x})X_1) - \text{tr}(a(\hat{y})X_2) + 2\epsilon\text{tr}\,(a(\hat{x}) + a(\hat{y})) + \\ + H(\hat{x}, \hat{t}, \mu(\hat{x} - \hat{y}) + 2\epsilon\hat{x}) - H(\hat{y}, \hat{t}, \mu(\hat{x} - \hat{y}) - 2\epsilon\hat{y}) \geq 0. \tag{11.8.48}$$

Now letting e_k denote the coordinate vectors of R^n, we check easily that

$$\text{tr}(a(\hat{x})X_1) - \text{tr}(a(\hat{y})X_2) = \sum_{k=1}^{n} (e_k^*\sigma(\hat{x}), e_k^*\sigma(\hat{y})) \begin{pmatrix} X_1 & 0 \\ 0 & -X_2 \end{pmatrix} \begin{pmatrix} \sigma^*(\hat{x})e_k \\ \sigma^*(\hat{y})e_k \end{pmatrix},$$

and from the second property (11.8.47), we obtain

$$\mathrm{tr}(a(\hat{x})X_1) - \mathrm{tr}(a(\hat{y})X_2) \leq 3\mu \sum_{k=1}^{n} |(\sigma^*(\hat{x}) - \sigma^*(\hat{y}))e_k|^2.$$

Also, from the assumption (11.8.39), we have

$$|H(\hat{x}, \hat{t}, \mu(\hat{x} - \hat{y}) + 2\epsilon\hat{x}) - H(\hat{y}, \hat{t}, \mu(\hat{x} - \hat{y}) - 2\epsilon\hat{y})| \leq c(|\hat{x} - \hat{y}| + 2\epsilon|\hat{x}| + |\hat{y}|).$$

Collecting results, we write

$$\begin{aligned} -\beta + 3\mu \sum_{k=1}^{n} |(\sigma^*(\hat{x}_{\mu\epsilon}) - \sigma^*(\hat{y}_{\mu\epsilon}))e_k|^2 + c|\hat{x}_{\mu\epsilon} - \hat{y}_{\mu\epsilon}| + \\ +2\epsilon[(|\hat{x}_{\mu\epsilon}| + |\hat{y}_{\mu\epsilon}|) + \mathrm{tr}\,(a(\hat{x}_{\mu\epsilon}) + a(\hat{y}_{\mu\epsilon}))] \geq 0. \end{aligned} \tag{11.8.49}$$

We have reinstated the indices μ, ϵ, because we are going to let first $\mu \to +\infty$, then $\epsilon \to 0$. The other parameters β, γ are fixed. Moreover, by writing $\Phi(\hat{x}, \hat{y}, \hat{t}) \geq \Phi(\hat{y}, \hat{y}, \hat{t})$, we obtain the inequality

$$\frac{\mu}{2}|\hat{x}_{\mu\epsilon} - \hat{y}_{\mu\epsilon}|^2 \leq \Psi(\hat{x}_{\mu\epsilon}, \hat{t}_{\mu\epsilon}) - \Psi(\hat{y}_{\mu\epsilon}, \hat{t}_{\mu\epsilon}) + \epsilon(|\hat{y}_{\mu\epsilon}|^2 - |\hat{x}_{\mu\epsilon}|^2). \tag{11.8.50}$$

On the other hand, taking account of (11.8.43) and recalling that ϵ is fixed, we can extract a subsequence in μ such that

$$\hat{x}_{\mu\epsilon} \to \hat{x}_\epsilon,\ \hat{y}_{\mu\epsilon} \to \hat{y}_\epsilon,\ \hat{t}_{\mu\epsilon} \to \hat{t}_\epsilon, \text{ as } \mu \to +\infty$$

and $\hat{x}_{\mu\epsilon} - \hat{y}_{\mu\epsilon} \to 0$, so $\hat{x}_\epsilon = \hat{y}_\epsilon$. But then, the right-hand side of (11.8.50) tends to 0. Therefore, we also have $\mu|\hat{x}_{\mu\epsilon} - \hat{y}_{\mu\epsilon}|^2 \to 0$. Returning to (11.8.49) and using the fact that σ is uniformly Lipschitz, we let $\mu \to +\infty$, to obtain

$$-\beta + 4\epsilon(|\hat{x}_\epsilon| + \mathrm{tr}\, a(\hat{x}_\epsilon)) \geq 0.$$

But we have also $\epsilon|\hat{x}_\epsilon|^2 \leq K$. Letting $\epsilon \to 0$, we can assert that $\epsilon|\hat{x}_\epsilon| \to 0$. Using the fact that $a(x)$ is uniformly Lipschitz, we also have $\epsilon \mathrm{tr}\, a(\hat{x}_\epsilon) \to 0$. We obtain eventually $-\beta \geq 0$, which is a contradiction. Hence the initial conjecture is wrong, which proves the theorem. □

REMARK 11.6. We emphasize the method over the result. With the regularity assumptions that we have made, the value function will be differentiable a.e., and the HJB equation will have a solution in the usual sense; see Theorem 10.2, Chapter 10.

11.8.4 Existence; Perron's Method

We know that the value function is a viscosity solution of the HJB equation. We know also that the value function is the greatest element of the set of subsolutions of the Bellman equation. Perron's method indicates that the same property is valid for the set of viscosity subsolutions. In this way, one provides an existence result of viscosity solutions of the HJB equation that does not rely on the probabilistic interpretation.

PROPOSITION 11.6. *Assume that there exists one viscosity subsolution $\underline{\Psi}(x,t)$ and one viscosity supersolution $\overline{\chi}(x,t)$ of the HJB equation (11.8.34) such that $\underline{\Psi}(x,T)=\overline{\chi}(x,T)=h(x)$. Let*

$$\Phi(x,t)=\sup\{\Psi(x,t) \text{ continuous viscosity subsolution } |\underline{\Psi}(x,t)\leq\Psi(x,t)\leq\overline{\chi}(x,t)\}. \tag{11.8.51}$$

We assume that the function $\Phi(x,t)$ is continuous. Then it is a viscosity solution of the HJB equation.

Before giving the proof, we begin with important properties of viscosity subsolutions of the HJB equation

LEMMA 11.3. *We assume $a(x)$ continuous and $H(x,t,q)$ continuous. If $\Psi_n(x,t)$ is a monotonically increasing sequence of viscosity subsolutions converging pointwise to $\Psi(x,t)$, then $\Psi(x,t)$ is also a viscosity subsolution. Also, if $\Psi_1(x,t)$ and $\Psi_2(x,t)$ are two viscosity subsolutions, then $\Psi(x,t)=\max(\Psi_1(x,t),\Psi_2(x,t))$ is also a viscosity subsolution.*

PROOF. Let $(q,\rho,M)\in D^{+2,1}\Psi(x,t)$. Then on can write

$$\Psi(y,s)-\Psi(x,t)-\rho(s-t)-q.(y-x)-\frac{1}{2}(y-x)^*M(y-x)\leq$$

$$\delta(|s-t|+|y-x|^2)\forall(y,s) \text{ such that } |s-t|+|y-x|^2\leq r(\delta).$$

The function $r(\delta)$ is monotonically nondecreasing and goes to 0 as $\delta\to0$. Let $\hat{x}_n,\hat{t}_n$ be a maximum point for the function

$$\Psi_n(y,s)-\rho(s-t)-q.(y-x)-\frac{1}{2}(y-x)^*M(y-x)-2\delta(\sqrt{|s-t|^2+1}+|y-x|^2)$$

on $B_{r(\delta)}(x,t)=|s-t|+|y-x|^2\leq r(\delta)$. So for all $(y,s)\in B_{r(\delta)}(x,t)$, we can write

$$\Psi_n(y,s)\leq\Psi_n(\hat{x}_n,\hat{t}_n)+\rho(s-\hat{t}_n)+q.(y-\hat{x}_n)+$$

$$+\tfrac{1}{2}(y-x)^*M(y-x)-\frac{1}{2}(\hat{x}_n-x)^*M(\hat{x}_n-x)+$$

$$+2\delta((\sqrt{|s-t|^2+1}+|y-x|^2)-(\sqrt{|\hat{t}_n-t|^2+1}+|\hat{x}_n-x|^2)).$$

In particular, for $y = x$, $s = t$, we obtain

$$\Psi_n(x,t) \le \Psi_n(\hat{x}_n, \hat{t}_n) + \rho(t - \hat{t}_n) + q.(x - \hat{x}_n) - \tag{11.8.52}$$
$$-\frac{1}{2}(\hat{x}_n - x)^* M(\hat{x}_n - x) - 2\delta(\sqrt{|\hat{t}_n - t|^2 + 1} - 1 + |\hat{x}_n - x|^2).$$

We note that $\Psi_n(\hat{x}_n, \hat{t}_n) \le \Psi(\hat{x}_n, \hat{t}_n)$. Moreover, by taking a subsequence, we can say that $\hat{x}_n \to \hat{x}$, $\hat{t}_n \to \hat{t}$. Using the continuity of Ψ and collecting results, we can assert that

$$\Psi(x,t) \le \Psi(\hat{x}, \hat{t}) + \rho(t - \hat{t}) + q.(x - \hat{x}) -$$
$$-\frac{1}{2}(\hat{x} - x)^* M(\hat{x} - x) - 2\delta(\sqrt{|\hat{t} - t|^2 + 1} - 1 + |\hat{x} - x|^2).$$

On the other hand, since $(\hat{x}, \hat{t}) \in B_{r(\delta)}(x,t)$, we have

$$\Psi(\hat{x}, \hat{t}) - \Psi(x,t) - \rho(\hat{t} - t) - q.(\hat{x} - x) - \frac{1}{2}(\hat{x} - x)^* M(\hat{x} - x) \le \delta(|\hat{t} - t| + |\hat{x} - x|^2)$$
$$\le \delta(\sqrt{|\hat{t} - t|^2 + 1} - 1 + |\hat{x} - x|^2).$$

Adding these two inequalities, we get immediately $\hat{t} = t$, $\hat{x} = x$. So $\hat{x}_n \to \hat{x}$ and $\hat{t}_n \to \hat{t}$, and from (11.8.52), we obtain

$$\Psi(x,t) \le \liminf \Psi_n(\hat{x}_n, \hat{t}_n).$$

Since also $\Psi_n(\hat{x}_n, \hat{t}_n) \le \Psi(\hat{x}_n, \hat{t}_n)$, we also have $\limsup \Psi_n(\hat{x}_n, \hat{t}_n) \le \Psi(x,t)$. Therefore, $\Psi_n(\hat{x}_n, \hat{t}_n) \to \Psi(x,t)$. Now, the function

$$\varphi(y,s) = \rho(s-t) + q.(y-x) + \frac{1}{2}(y-x)^* M(y-x) + 2\delta(\sqrt{|s-t|^2+1} + |y-x|^2)$$

is smooth, and

$$D\varphi(y,s) = q + (M + 2\delta I)(y-x), \frac{\partial \varphi}{\partial s}(y,s) = \rho + 2\delta\frac{s-t}{\sqrt{|s-t|^2+1}}, D^2\varphi(y,s) = M + 4\delta I.$$

Since $\Psi_n(y,s) - \varphi(y,s)$ attains its maximum over $B_{r(\delta)}(x,t)$ at a point $(\hat{x}_n, \hat{t}_n)$ that is necessarily in the open ball $B_{r(\delta)}(x,t)$, we get that

$$(q + (M + 4\delta I)(\hat{x}_n - x),\ \rho + 2\delta\frac{\hat{t}_n - t}{\sqrt{|\hat{t}_n - t|^2 + 1}},\ M + 4\delta I) \in D^{+2,1}\Psi_n(\hat{x}_n, \hat{t}_n)$$

for n large enough. Since $\Psi_n(x,t)$ is a viscosity subsolution, we can assert that

$$\rho + 2\delta \frac{\hat{t}_n - t}{\sqrt{|\hat{t}_n - t|^2 + 1}} + \mathrm{tr}(a(\hat{x}_n)(M + 4\delta I)) + \\ + H(\hat{x}_n, \hat{t}_n, q + (M + 4\delta I)(\hat{x}_n - x) \geq 0.$$

Letting $n \to +\infty$, we obtain immediately, using the continuity of the Hamiltonian and of the function a,

$$\rho + 2\delta + \mathrm{tr}(a(\hat{x})(M + 4\delta I)) + H(x,t,q) \geq 0,$$

and since δ is arbitrarily small, we get $\rho + \mathrm{tr}(a(\hat{x})(M) + H(x,t,q) \geq 0$, which proves that Ψ is also a viscosity subsolution.

Now let $\Psi(x,t) = \max(\Psi_1(x,t), \Psi_2(x,t))$, where $\Psi_1(x,t)$, $\Psi_2(x,t)$ are two viscosity subsolutions. Let $(q,\rho,M) \in D^{+2,1}\Psi(x,t)$. Then

$$\limsup_{y\to x, s\to t} \frac{\Psi(y,s) - \Psi(x,t) - q.(y-x) - \rho(s-t) - \frac{1}{2}(y-x)^* M(y-x)}{|y-x|^2 + |s-t|} \leq 0.$$

Suppose $\Psi(x,t) = \Psi_1(x,t)$. Then since we have $\Psi(y,s) \geq \Psi_1(y,s)$, we have also necessarily

$$\limsup_{y\to x, s\to t} \frac{\Psi_1(y,s) - \Psi_1(x,t) - q.(y-x) - \rho(s-t) - \frac{1}{2}(y-x)^* M(y-x)}{|y-x|^2 + |s-t|} \leq 0,$$

which implies $(q,\rho,M) \in D^{+2,1}\Psi_1(x,t)$. Similarly, if $\Psi(x,t) = \Psi_2(x,t)$, then $(q,\rho,M) \in D^{+2,1}\Psi_2(x,t)$. In both cases, since $\Psi_1(x,t)$ and $\Psi_2(x,t)$ are both viscosity subsolutions, we have $\rho + \mathrm{tr}\,(a(x)M) + H(x,t,q) \geq 0$, and thus $\Psi(x,t)$ is also a viscosity subsolution. ■

It is a classical result that in the definition of $\Phi(x,t)$, see (11.8.51), we can replace the full set of continuous viscosity subsolutions by a countable subset. Since the maximum of two continuous viscosity subsolutions is a continuous viscosity subsolution, we can obtain that $\Phi(x,t)$ is the limit of an increasing sequence of continuous viscosity subsolutions. □

Proof of Proposition 11.6

PROOF. The first point is that $\Phi(x,t)$ is a viscosity subsolution. This comes from the fact that one can find a sequence of viscosity subsolutions that converges monotonically to $\Phi(x,t)$, and applying the preceding lemma, the function $\Phi(x,t)$ is then the greatest viscosity subsolution. We are going to prove that if it is not also a viscosity supersolution, then it cannot be the greatest viscosity subsolution. From this contradiction, the function $\Phi(x,t)$ is seen to be a viscosity solution. Suppose $\Phi(x,t)$ is not a viscosity supersolution. Since from the assumptions we have $\Phi(x,T)=h(x)$, the assertion comes from the fact that there exist $x_0, t_0 < T$, and a triple $(q_0,\rho_0,M_0) \in D^{-2,1}\Phi(x_0,t_0)$ such that

$$\rho_0 + \text{tr}\,(a(x_0)M_0) + H(x_0,t_0,q_0) > 0. \tag{11.8.53}$$

We then build the function

$$\Phi_{\gamma\delta\epsilon}(x,t) = \Phi(x_0,t_0) + \delta + \rho_0(t-t_0) + q_0.(x-x_0) +$$
$$+\frac{1}{2}(x-x_0)^* M_0(x-x_0) - \gamma(|x-x_0|^2 + \sqrt{(t-t_0)^2+\epsilon}\,).$$

We have

$$D\Phi_{\gamma\delta\epsilon}(x,t) = q_0+(M_0-2\gamma I)(x-x_0), \frac{\partial \Phi_{\gamma\delta\epsilon}}{\partial t}(x,t) = \rho_0 - \gamma\frac{t-t_0}{\sqrt{(t-t_0)^2+\epsilon}}, D^2\Phi_{\gamma\delta\epsilon}(x,t) = M_0-2\gamma I,$$

and from the assumption (11.8.53), we get

$$\rho_0-\gamma\frac{t-t_0}{\sqrt{(t-t_0)^2+\epsilon}} + \text{tr}(a(x)(M_0-2\gamma I)) + H(x,t,q_0+(M_0-2\gamma I)(x-x_0)) \geq 0 \tag{11.8.54}$$

for $\gamma \leq \gamma_0$ and $|x-x_0|^2+|t-t_0| \leq r_0$. Since $(q_0,\rho_0,M_0) \in D^{-2,1}\Phi(x_0,t_0)$, we can assert that

$$\liminf_{y\to x_0, s\to t_0} \frac{\Phi(x,t)-\Phi(x_0,t_0)-q_0.(x-x_0)-\rho_0(t-t_0)-\frac{1}{2}(t-x_0)^*M_0(t-x_0)}{|x-x_0|^2+|t-t_0|} \geq 0.$$

So we can define a monotonically nondecreasing function $\lambda(\gamma)$ such that $\lambda(\gamma) \to 0$ as $\gamma \to 0$ and

$$\Phi(x,t)-\Phi(x_0,t_0)-q_0.(x-x_0)-\rho_0(t-t_0)-\frac{1}{2}(x-x_0)^*M_0(x-x_0 \geq -\frac{\gamma}{2}(|t-t_0|+|x-x_0|^2)$$

as soon as $|x-x_0|^2+|t-t_0| \leq \lambda(\gamma)$. Introducing $\Phi_{\gamma\delta\epsilon}(x,t)$, we see easily that

$$\Phi(x,t) - \Phi_{\gamma\delta\epsilon}(x,t) + \delta \geq \frac{\gamma}{2}(|t-t_0|+|x-x_0|^2) \tag{11.8.55}$$

as soon as $|x-x_0|^2+|t-t_0| \leq \lambda(\gamma)$. But also if $\frac{\lambda(\gamma)}{2} \leq |x-x_0|^2+|t-t_0| \leq \lambda(\gamma)$, we have $\Phi(x,t)-\Phi_{\gamma\delta\epsilon}(x,t)+\delta \geq \frac{\gamma\lambda(\gamma)}{4}$. If we take $\delta = \frac{\gamma\lambda(\gamma)}{8}$, we obtain $\Phi(x,t)-\Phi_{\gamma\delta\epsilon}(x,t) \geq \frac{\gamma\lambda(\gamma)}{8}$.

We now define γ and r such that $\frac{\lambda(\gamma)}{2} \leq r \leq \lambda(\gamma) \leq r_0$ and $\gamma \leq \gamma_0$. We also take $\epsilon = (\frac{\lambda(\gamma)}{16})^2$. We then consider the function

$$\chi_{\gamma r}(x,t) = \begin{cases} \max(\Phi(x,t), \Phi_{\gamma\delta\epsilon}(x,t)) \text{ if } |x-x_0|^2 + |t-t_0| \leq r. \\ \\ \Phi(x,t) \text{ if } |x-x_0|^2 + |t-t_0| \leq r \end{cases}$$

This function is continuous. Indeed, if $|x-x_0|^2 + |t-t_0| = r$, we have, by construction, $\frac{\lambda(\gamma)}{2} \leq |x-x_0|^2 + |t-t_0| \leq \lambda(\gamma)$. Hence $\Phi(x,t) - \Phi_{\gamma\delta}(x,t) \geq \frac{\gamma\lambda(\gamma)}{8} > 0$. This ensures the continuity. We have clearly $\chi_{\gamma r}(x,t) \geq \Phi(x,t)$ and $\Phi_{\gamma\delta\epsilon}(x_0,t_0) = \Phi(x_0,t_0) + \delta - \gamma\sqrt{\epsilon} = \Phi(x_0,t_0) + \gamma\frac{\lambda(\gamma)}{16}$. Therefore, $\chi_{\gamma r}(x_0,t_0) = \Phi_{\gamma\delta\epsilon}(x_0,t_0) > \Phi(x_0,t_0)$. Therefore, $\chi_{\gamma r}(x,t)$ dominates $\Phi(x,t)$ and does not coincide with $\Phi(x,t)$.

If we show that $\chi_{\gamma r}(x,t)$ is a viscosity subsolution, we shall get a contradiction with the fact that $\Phi(x,t)$ is the greatest viscosity subsolution. The fact that $\chi_{\gamma r}(x,t)$ is a subsolution is naturally related to the property that the maximum of two viscosity subsolutions is also a viscosity subsolution. Let $(q,\rho,M) \in D^{+2,1}\chi_{\gamma r}(x,t)$. We have

$$\limsup_{y\to x, s\to t} \frac{\chi_{\gamma r}(y,s) - \chi_{\gamma r}(x,t) - q.(y-x) - \rho(s-t) - \frac{1}{2}(y-x)^* M(y-x)}{|y-x|^2 + |s-t|} \leq 0.$$

If $\chi_{\gamma r}(x,t) = \Phi(x,t)$, then using $\chi_{\gamma r}(y,s) \geq \Phi(y,s)$, we obtain immediately

$$\limsup_{y\to x, s\to t} \frac{\Phi(y,s) - \Phi(x,t) - q.(y-x) - \rho(s-t) - \frac{1}{2}(y-x)^* M(y-x)}{|y-x|^2 + |s-t|} \leq 0,$$

which proves that $(q,\rho,M) \in D^{+2,1}\Phi(x,t)$. Suppose now that $\chi_{\gamma r}(x,t) = \Phi_{\gamma\delta\epsilon}(x,t) > \Phi(x,t)$. Necessarily $|x-x_0|^2 + |t-t_0| \leq r \leq r_0$. Since $\Phi_{\gamma\delta\epsilon}(x,t) - \Phi(x,t) > 0$, for y,s sufficiently close to x,t we have $\Phi_{\gamma\delta\epsilon}(y,s) - \Phi(y,s) \geq 0$. Hence for y,s sufficiently close to x,t, we have $\chi_{\gamma r}(y,s) = \Phi_{\gamma\delta\epsilon}(y,s)$. As in the previous case, we can assert that $(q,\rho,M) \in D^{+2,1}\Phi_{\gamma\delta\epsilon}(x,t)$, so

$$q = q_0 + (M_0 - 2\gamma I)(x-x_0),\ \rho = \rho_0 - \gamma\frac{t-t_0}{\sqrt{(t-t_0)^2+\epsilon}},\ M \geq M_0 - 2\gamma I.$$

Since $|x-x_0|^2 + |t-t_0| \leq r \leq r_0$, it follows from (11.8.54) that $\rho + \text{tr}\,(a(x)M) + H(x,t,q) \geq 0$. In all cases, this inequality is satisfied, which implies that $\chi_{\gamma r}(x,t)$ is a viscosity subsolution. This concludes the proof. □

11.8.5 Stochastic Perron's Method

Bayraktar and Sîrbu [2] have introduced a different idea to show the existence of a viscosity solution, which they call the stochastic Perron's method. We present the approach with some differences, in a way that builds on the Perron's method discussed in the previous section. We consider the set of subsolutions of the Bellman equation

$$\Psi(x,t) \leq \inf_v E[\int_t^{t+\epsilon} l(x(s),v)ds + \Psi(x(t+\epsilon),t+\epsilon)],\ \forall t \leq T-\epsilon,$$

$$\Psi(x,T) \leq h(x). \tag{11.8.56}$$

Note that we take v constant in the infimum on the right-hand side. The supersolutions are defined by

$$\Psi(x,t) \geq \inf_{v(.)} E[\int_t^{t+\epsilon} l(x(s),v(s))ds + \Psi(x(t+\epsilon),t+\epsilon)],\ \forall t \leq T-\epsilon,$$

$$\Psi(x,T) \geq h(x). \tag{11.8.57}$$

We want to prove the following analogue of Proposition 11.6.

PROPOSITION 11.7. . *Assume that there exists one subsolution $\underline{\Psi}(x,t)$ (11.8.56) and one supersolution $\overline{\chi}(x,t)$ (11.8.57) of the Bellman equation such that $\underline{\Psi}(x,T) = \overline{\chi}(x,T) = h(x)$. Let*

$$\Phi(x,t) = \sup\{\Psi(x,t) \text{ continuous subsolution } | \underline{\Psi}(x,t) \leq \Psi(x,t) \leq \overline{\chi}(x,t)\}. \tag{11.8.58}$$

We assume that the function $\Phi(x,t)$ is continuous. Then it is a viscosity solution of the HJB equation.

PROOF. Note that we are replacing the set of viscosity subsolutions of the HJB equation by the set of subsolutions of the Bellman equation. The claim is that the result is the same. The set of subsolutions is contained in the set of viscosity subsolutions. We begin by stating properties of the set of subsolutions. If a sequence Ψ_n of subsolutions is increasing and converges monotonically to Ψ, then Ψ is also a subsolution. Also, the maximum of two subsolutions is a subsolution. Note that for $t \leq s_1 \leq s_2 \leq T$ and for fixed v, if $\Psi(x,t)$ is a subsolution, then

$$\Psi(x(s_1),s_1) \leq E[\int_{s_1}^{s_2} l(x(\tau),v)d\tau + \Psi(x(s_2),s_2)|\mathcal{F}_t^{s_1}].$$

This follows from the Markov property of the diffusion (with fixed control)

$$\begin{aligned} dx &= g(x(s), v)ds + \sigma(x(s))dw(s), \\ x(t) &= x. \end{aligned}$$

Therefore,

$$\Psi(x^v_{xt}(s), s) + \int_t^s l(x^v_{xt}(\tau), v)d\tau \text{ is a } P, \mathcal{F}^s_t \text{ submartingale,} \tag{11.8.59}$$

where we have written $x^v_{xt}(s)$ for the solution of the diffusion, emphasizing the initial conditions and the control, and $\mathcal{F}^s_t = \sigma(w(\tau) - w(t),\ t \leq \tau \leq s)$.

Consider the function $\Phi(x, t)$, assumed to be continuous. It is also a subsolution, hence a viscosity subsolution, and $\Phi(x, T) = h(x)$. It is the greatest subsolution, but not necessarily the greatest viscosity subsolution. Suppose it is not a viscosity supersolution. Then we are in the situation described in Proposition 11.6. There exist $x_0, t_0 < T$ and a triple $(q_0, \rho_0, M_0) \in D^{-2,1}\Phi(x_0, t_0)$ such that

$$\rho_0 + \text{tr}\,(a(x_0)M_0) + H(x_0, t_0, q_0) > 0. \tag{11.8.60}$$

We thus construct the same function

$$\begin{aligned} \Phi_{\gamma\delta\epsilon}(x, t) &= \Phi(x_0, t_0) + \delta + \rho_0(t - t_0) + q_0.(x - x_0) + \\ &+ \frac{1}{2}(x - x_0)^* M_0(x - x_0) - \gamma(|x - x_0|^2 + \sqrt{(t - t_0)^2 + \epsilon}\,). \end{aligned}$$

We recall the facts related to $\Phi_{\gamma\delta\epsilon}(x, t)$. We have

$$\begin{aligned} &\frac{\partial \Phi_{\gamma\delta\epsilon}}{\partial t} + \text{tr}\,(a(x)D^2\Phi_{\gamma\delta\epsilon}) + H(x, t, D\Phi_{\gamma\delta\epsilon}) \geq 0 \\ &\qquad \text{if } \gamma \leq \gamma_0,\ |x - x_0|^2 + |t - t_0| \leq r_0. \end{aligned} \tag{11.8.61}$$

We define the function $\lambda(\gamma)$ as in Proposition 11.6. If we choose γ, δ, ϵ, and r as

$$\delta = \frac{\gamma\lambda(\gamma)}{8},\ \epsilon = \epsilon = (\frac{\lambda(\gamma)}{16})^2,\ \frac{\lambda(\gamma)}{2} \leq r \leq \lambda(\gamma) \leq r_0,$$

then we obtain that for $\frac{\lambda(\gamma)}{2} \leq |x - x_0|^2 + |t - t_0| \leq \lambda(\gamma)$, we have $\Phi(x, t) - \Phi_{\gamma\delta}(x, t) \geq \frac{\gamma\lambda(\gamma)}{8} > 0$. In particular, this is true when $|x - x_0|^2 + |t - t_0| = r$. Moreover, $\Phi_{\gamma\delta\epsilon}(x_0, t_0) = \Phi(x_0, t_0) + \gamma\frac{\lambda(\gamma)}{16} > \Phi(x_0, t_0)$. We next consider the function

$$\chi_{\gamma r}(x,t) = \begin{cases} \max(\Phi(x,t), \Phi_{\gamma\delta\epsilon}(x,t)) \text{ if } |x-x_0|^2 + |t-t_0| \leq r, \\ \\ \Phi(x,t) \text{ if } |x-x_0|^2 + |t-t_0| \leq r. \end{cases}$$

Then $\chi_{\gamma r}(x,t) \geq \Phi(x,t)$ and $\chi_{\gamma r}(x,t) \not\equiv \Phi(x,t)$. We know that $\chi_{\gamma r}(x,t)$ is a continuous viscosity subsolution, but we need to show that it is a subsolution. At a point where $\chi_{\gamma r}(x,t) = \Phi(x,t)$, we have, for all v,

$$\chi_{\gamma r}(x,t) = \Phi(x,t) \leq E[\int_t^{t+\epsilon} l(x(s),v)ds + \Phi(x(t+\epsilon), t+\epsilon)]$$

$$\leq E[\int_t^{t+\epsilon} l(x(s),v)ds + \chi_{\gamma r}(x(t+\epsilon), t+\epsilon)].$$

Consider now a point such that $\chi_{\gamma r}(x,t) = \Phi_{\gamma\delta\epsilon}(x,t) > \Phi(x,t)$. Necessarily, $|x-x_0|^2 + |t-t_0| \leq r$. Since when $|x-x_0|^2 + |t-t_0| = r$, we have $\Phi(x,t) > \Phi_{\gamma\delta\epsilon}(x,t)$, we can define a stopping time τ_{xt}^v by

$$\tau_{xt}^v = \inf\{s > t | \Phi_{\gamma\delta\epsilon}(x_{xt}^v(s), s) = \Phi(x_{xt}^v(s), x_{xt}^v(s))\}.$$

We can state that

$$|x_{xt}^v(s) - x_0|^2 + |s - t_0| < r, \ \forall s \text{ such that } t \leq s \leq \tau_{xt}^v.$$

Since $r < r_0$, we have from (11.8.61),

$$\frac{\partial \Phi_{\gamma\delta\epsilon}}{\partial t}(x_{xt}^v(s), s) + \text{tr }(a(x_{xt}^v(s))D^2\Phi_{\gamma\delta\epsilon}(x_{xt}^v(s), s))$$
$$+H(x_{xt}^v(s), s, D\Phi_{\gamma\delta\epsilon}(x_{xt}^v(s), s)) \geq 0$$

for $t \leq s \leq \tau_{xt}^v$. Therefore, also

$$\frac{\partial \Phi_{\gamma\delta\epsilon}}{\partial t}(x_{xt}^v(s), s) + \text{tr }(a(x_{xt}^v(s))D^2\Phi_{\gamma\delta\epsilon}(x_{xt}^v(s), s)) +$$
$$+l(x_{xt}^v(s), v) + D\Phi_{\gamma\delta\epsilon}(x_{xt}^v(s), s).g(x_{xt}^v(s), v) \geq 0.$$

From Itô's formula we obtain

$$\Phi_{\gamma\delta\epsilon}(x,t) \leq E[\int_t^{(t+\epsilon)\wedge\tau} l(x_{xt}^v(s), v)ds + \Phi_{\gamma\delta\epsilon}(x_{xt}^v((t+\epsilon)\wedge\tau), (t+\epsilon)\wedge\tau)]$$

$$\leq E[\int_t^{(t+\epsilon)\wedge\tau} l(x_{xt}^v(s), v)ds + \Phi_{\gamma\delta\epsilon}(x_{xt}^v(t+\epsilon), t+\epsilon)\mathbb{1}_{\{t+\epsilon<\tau\}} +$$
$$+ \Phi(x_{xt}^v(\tau), \tau)\mathbb{1}_{\{\tau\leq t+\epsilon\}}]. \qquad (11.8.62)$$

Now using the property (11.8.59) for the function $\Phi(x,t)$ and the optional sampling theorem, we can assert

$$\Phi(x_{xt}^v(\tau),\tau) \leq E[\int_\tau^{\tau\vee(t+\epsilon)} l(x_{xt}^v(s),v)ds + \Phi(x_{xt}^v(\tau\vee(t+\epsilon)),\tau\vee(t+\epsilon))|\mathcal{F}_t^\tau],$$

and since $\mathbb{1}_{\{\tau\leq t+\epsilon\}}$ is $\mathcal{F}_t^\tau$ measurable, it follows that

$$E\Phi(x_{xt}^v(\tau),\tau)\mathbb{1}_{\{\tau\leq t+\epsilon\}} = E[\mathbb{1}_{\{\tau\leq t+\epsilon\}}(\int_\tau^{t+\epsilon} l(x_{xt}^v(s),v)ds+\Phi(x_{xt}^v(t+\epsilon),t+\epsilon))].$$

Inserting this relation in (11.8.62) and noting that $\chi_{\gamma r}(x,t) = \Phi_{\gamma\delta\epsilon}(x,t)$ and $\chi_{\gamma r}(x_{xt}^v(t+\epsilon),(t+\epsilon))\mathbb{1}_{\{t+\epsilon<\tau\}} = \Phi_{\gamma\delta\epsilon}(x_{xt}^v(t+\epsilon),(t+\epsilon))\mathbb{1}_{\{t+\epsilon<\tau\}}$, we get

$$\chi_{\gamma r}(x,t) \leq E[\int_t^{(t+\epsilon)\wedge\tau} l(x_{xt}^v(s),v)ds + \chi_{\gamma r}(x_{xt}^v(t+\epsilon),(t+\epsilon))\mathbb{1}_{\{t+\epsilon<\tau\}} +$$
$$+ \mathbb{1}_{\{\tau\leq t+\epsilon\}}(\int_\tau^{t+\epsilon} l(x_{xt}^v(s),v)ds + \chi_{\gamma r}(x_{xt}^v(t+\epsilon),t+\epsilon))],$$

where we have used the fact that $\Phi(x_{xt}^v(t+\epsilon),t+\epsilon) \leq \chi_{\gamma r}(x_{xt}^v(t+\epsilon),t+\epsilon)$, We conclude again

$$\chi_{\gamma r}(x,t) \leq E[\int_t^{t+\epsilon} l(x(s),v)ds + \chi_{\gamma r}(x(t+\epsilon),t+\epsilon)].$$

Therefore, $\chi_{\gamma r}(x,t)$ is a subsolution of the Bellman equation. We get a contradiction with the fact that $\Phi(x,t)$ is maximal. Necessarily, $\Phi(x,t)$ is a viscosity supersolution. Hence it is a viscosity solution of the HJB equation. □

REMARK 11.7. The fact that the function $\Phi(x,t)$ defined by (11.8.58) is the value function can be seen in a way similar to the deterministic case; see Theorem 10.3 of Chapter 10. We deal with stochastic processes, which requires some care, but not basic changes. This provides a proof that the value function is a viscosity solution that does not rely on the optimality principle. One objective of Bayraktar and Sîrbu [2] is to prove the optimality principle in this way. That requires proving that the greatest subsolution (or the value function) is a supersolution, and not just a viscosity supersolution. One difficulty is that we can no longer use constant controls, and thus we normally lose the Markov property, unless we are working with feedback controls, but that is very restrictive. It is necessary to go to weak solutions, as done in [2].

Chapter 12
Additional Results for BSDE

12.1 Solutions of Parabolic PDEs

12.1.1 General Comments

This chapter uses many ideas of P. Briand, Y. Hu, F. Da Lio, O. Ley, M. Kobylanski, N. El Karoui, S. Hamadene, and A. Matoussi. The presentation is slightly innovative. We try to present the basic concepts, and we do so for PDEs first. Parabolic quasilinear equations have been, of course, at the foundation of functional analysis and its development. However, in physics and in mechanics, one deals mostly with boundary value problems. The boundary is often taken to be bounded, and the solution is bounded. Stochastic control leads mostly to problems in R^n. Moreover, the functions are unbounded, and the Hamiltonian may have quadratic growth. There may be conflicts that prevent solutions from existing.

We try in this chapter to clarify the various relevant concepts and methods.

12.1.2 General Result

We are considering the following problem:

$$
\begin{aligned}
-\frac{\partial u}{\partial t} - \frac{1}{2}\Delta u &= H(x, u, Du), \qquad (12.1.1)\\
u(x, T) &= h(x).
\end{aligned}
$$

We have taken $-\frac{1}{2}\Delta u$ instead of a general coercive operator A, to simplify calculations. The argument x is in R^n. The Hamiltonian $H(x, y, z)$ is a measurable function on $R^n \times R \times R^n$ such that

A. Bensoussan, *Estimation and Control of Dynamical Systems*, Interdisciplinary Applied Mathematics 48, https://doi.org/10.1007/978-3-319-75456-7_12

$$y, z \to H(x, y, z) \text{ is continuous,} \tag{12.1.2}$$

$$|H(x, y, z)| \leq \lambda(x) + k|y| + \frac{\gamma}{2}|z|^2. \tag{12.1.3}$$

The functions $\lambda(x)$ and $|h(x)|$ are unbounded, but they cannot have a growth rate that is more than quadratic, with limited growth rates depending on the size of T. If T is sufficiently small, then general quadratic growth is possible. We shall make this more precise in the next section. The key issue is that there is a conflict between the growth of the functions $\lambda(x)$, $|h(x)|$ and the quadratic growth of the Hamiltonian in z. When the Hamiltonian has growth rate less than quadratic, things become much easier.

We shall prove the existence of a solution of (12.1.1). More conditions are needed to obtain uniqueness.

12.1.3 Growth Discussion

To understand the issue of growth, we use the inequality (12.1.3) in equation (12.1.1) to obtain

$$-\frac{\partial u}{\partial t} - \frac{1}{2}\Delta u \leq \lambda(x) + k|u| + \frac{\gamma}{2}|Du|^2.$$

This leads to introducing the PDE

$$\begin{aligned} -\frac{\partial \bar{u}}{\partial t} - \frac{1}{2}\Delta \bar{u} - k\bar{u} - \frac{\gamma}{2}|D\bar{u}|^2 &= \lambda(x), \\ \bar{u}(x, T) &= |h(x)|, \end{aligned} \tag{12.1.4}$$

and we expect

$$|u(x, t)| \leq \bar{u}(x, t). \tag{12.1.5}$$

The issue is that the PDE (12.1.4) may fail to have a solution. The growth conditions guarantee the solvability of this equation. In fact, by simple calculations the function $\bar{v}(x, t) = \exp(\gamma \bar{u}(x, t) \exp kt)$ satisfies the inequality

$$\begin{aligned} -\frac{\partial \bar{v}}{\partial t} - \frac{1}{2}\Delta \bar{v} &\leq \gamma \lambda(x) \exp kt\, \bar{v}, \\ \bar{v}(x, T) &= \exp(\gamma |h(x)| \exp kT). \end{aligned} \tag{12.1.6}$$

This introduces naturally the linear equation

$$\begin{aligned} -\frac{\partial v^*}{\partial t} - \frac{1}{2}\Delta v^* &= \gamma \lambda(x) \exp kt\, v^*, \\ v^*(x, T) &= \exp(\gamma |h(x)| \exp kT), \end{aligned} \tag{12.1.7}$$

and $\bar{u}(x,t) \leq \dfrac{1}{\gamma}(\exp -kt)\log v^*(x,t)$. The problem reduces to the solvability of (12.1.7), which is not guaranteed, in spite of the linearity. One good way to capture the solvability issue is to use the probabilistic interpretation of $v^*(x,t)$. Indeed, we have

$$v^*(x,t) = E\exp\gamma[|h|(w_{xt}(T))\exp kT + \int_t^T \lambda(w_{xt}(s))\exp ks\,ds], \tag{12.1.8}$$

in which $w_{xt}(s) = x + w(s) - w(t)$, where $w(s)$ is a standard Wiener process in R^n. It is clear that the expected value on the right-hand side of (12.1.8) can be $+\infty$ unless the growth rates of λ and $|h|$ are restricted.

12.1.4 Sufficient Condition

We introduce a function $\zeta(x)$ such that

$$(\mu - k)\zeta - \frac{1}{2}\Delta\zeta - \frac{\gamma}{2}|D\zeta|^2 \geq \lambda(x)\exp\mu T, \tag{12.1.9}$$
$$\zeta(x) \geq |h(x)|\exp\mu T,$$

for a convenient $\mu > k$. We have the following result.

LEMMA 12.1.

$$\bar{u}(x,t) \leq \zeta(x). \tag{12.1.10}$$

PROOF. Let $\bar{\zeta}(x,t) = \exp\gamma\zeta(x)\exp(-(\mu-k)t)$. We get

$$\begin{aligned}
-\frac{\partial\bar{\zeta}}{\partial t} - \frac{1}{2}\Delta\bar{\zeta} &= \gamma\bar{\zeta}\exp(-(\mu-k)t)[(\mu-k)\zeta - \frac{1}{2}\Delta\zeta - \\
&\quad - \frac{\gamma}{2}\exp -(\mu-k)t\,|D\zeta|^2 \\
&\geq \gamma\bar{\zeta}\lambda(x)\exp kt\,\exp\mu(T-t) \\
&\geq \gamma\bar{\zeta}\lambda(x)\exp kt
\end{aligned}$$

and

$$\bar{\zeta}(x,T) \geq \exp\gamma|h(x)|\exp kT.$$

Comparing with the inequalities (12.1.6), we see immediately that $\bar{\zeta}(x,t) \geq \bar{v}(x,t)$. Hence $\exp[\gamma\zeta(x)\exp kt\,\exp(-\mu t)] \geq \exp\gamma\bar{u}(x,t)\exp kt$, which implies immediately the result (12.1.10). □

Therefore, a sufficient condition for the solvability of (12.1.4) is to find a function $\zeta(x)$ satisfying (12.1.9). We can then state the following.

LEMMA 12.2. *We assume that* $\lambda(x)$, $|h(x)|$ *satisfy the growth condition*

$$\lambda(x) \leq |x|^2 \frac{\exp -(2+kT)}{2\gamma T^2} + C, \tag{12.1.11}$$

$$|h(x)| \leq |x|^2 \frac{\exp -(2+kT)}{2\gamma T^2} + C.$$

Then the function

$$\zeta(x) = \frac{|x|^2}{2T\gamma} + B \tag{12.1.12}$$

and $\mu - k = \dfrac{2}{T}$ *satisfy (12.1.9), for a constant B sufficiently large (compared to C).*

PROOF. The result is obtained by direct calculation. We must take

$$B \geq C\exp(2+kT),\ B \geq \frac{n}{4\gamma} + C\frac{T}{2}\exp(2+kT).$$

□

12.1.5 Main Result

We have the following theorem.

THEOREM 12.1. *We assume (12.1.2), (12.1.3), and (12.1.11). Then there exists a solution of (12.1.1).*

The proof will be given in the next section.

We can give a uniqueness result, with additional assumptions. We assume in addition

$$|H(x,y,z) - H(x,y',z)| \leq k|y-y'|, \tag{12.1.13}$$

$$z \to H(x,y,z) \text{ is convex.} \tag{12.1.14}$$

We then have the following result.

THEOREM 12.2. *With the assumptions of Theorem 12.1, (12.1.13), (12.1.14), and the growth condition (12.2.35), the solution of (12.1.1) is unique.* ■

12.2 Methodology

In this section, we prove Theorems 12.1 and 12.2. We proceed by first making more assumptions, which will permit us to construct an approximation. We begin with the standard case, that is, the Lipschitz case.

PROPOSITION 12.1. *We assume that*

$$|H(x, y, z) - H(x, y', z')| \le C(|y - y'| + |z - z'|), \tag{12.2.1}$$

$$|H(x, y, z)| \le \lambda(x) + k(|y| + |z|), \tag{12.2.2}$$

with $\lambda(x)$, $|h(x)|$ polynomial growth. There exists a unique solution of (12.1.1) in a Sobolev space with weights.

PROOF. Let $\Phi(x)$ be a positive function of the type $\Phi(x) = \dfrac{1}{(1+|x|^2)^{\frac{m}{2}}}$, with m a positive integer. We assume $|D\Phi(x)| \le c_0\Phi(x)$. This is true in the example. The assumption on $\lambda(x)$ and $|h|(x)$ is made precise as follows:

$$\int_{R^n} \lambda^2(x)\Phi^2(x)dx < +\infty, \int_{R^n} h^2(x)\Phi^2(x)dx < +\infty. \tag{12.2.3}$$

We note that $\lambda, h \in L^2_\Phi(R^n)$.

We consider the Sobolev space

$$H^1_\Phi(R^n) = \{v| \int_{R^n} v^2(x)\Phi^2(x)dx < +\infty, \int_{R^n} |Dv|^2(x)\Phi^2(x)dx < +\infty\}$$

and the space $\mathcal{H} = L^2(0, T; H^1_\Phi(R^n))$ with the norm

$$||v||^2 = \int_0^T \int_{R^n} (|v|^2 + |Dv|^2)\Phi^2 \exp \mu t \, dxdt,$$

in which μ is a positive constant, defined below. We will obtain the solution as the fixed point of a contraction map. For $v \in \mathcal{H}$, we define $\Theta(v) = u$ as the solution of

$$-\frac{\partial u}{\partial t} - \frac{1}{2}\Delta u = H(x, v, Dv),$$
$$u(x, T) = h(x).$$

Note that $\dfrac{\partial u}{\partial t} \in L^2(0, T; (H^1_\Phi(R^n))')$, where $(H^1_\Phi(R^n))'$ is the dual space of $H^1_\Phi(R^n)$. Therefore, $u(., T)$ is defined as an element of $L^2_\Phi(R^n)$. Thanks to the assumptions (12.2.3) and (12.2.2), we see easily that the map Θ is well defined and maps $\mathcal{H}$ into itself. To prove that for some choice of μ it is a contraction, consider two functions v_1, v_2 and the images $\Theta(v_1) = u_1$ and $\Theta(v_2) = u_2$. We also set $\tilde{v} = v_1 - v_2$, $\tilde{u} = u_1 - u_2$. We have

$$-\frac{\partial \tilde{u}}{\partial t} - \frac{1}{2}\Delta \tilde{u} = H(x, v_1, Dv_1) - H(x, v_2, Dv_2),$$
$$\tilde{u}(x, T) = 0.$$

Testing this equation with $\tilde{u}\Phi^2$ and integrating, we obtain

$$\frac{1}{2}\int_{R^n} \tilde{u}^2(x,0)\Phi^2(x)dx + \frac{\mu}{2}\int_0^T\int_{R^n} \tilde{u}^2\Phi^2 \exp \mu t\, dxdt +$$
$$+\frac{1}{2}\int_0^T\int_{R^n} |D\tilde{u}|^2\Phi^2 \exp \mu t\, dxdt + \int_0^T\int_{R^n} D\tilde{u}.D\Phi\tilde{u}\Phi \exp \mu t dxdt \leq$$
$$\leq C\int_0^T\int_{R^n} (|\tilde{v}| + |D\tilde{v}|)|\tilde{u}|\Phi^2 \exp \mu t\, dxdt.$$

Using the condition on $|D\Phi|$, this implies

$$\frac{\mu}{2}\int_0^T\int_{R^n} \tilde{u}^2\Phi^2 \exp \mu t\, dxdt + \frac{1}{2}\int_0^T\int_{R^n} |D\tilde{u}|^2\Phi^2 \exp \mu t\, dxdt \leq$$
$$c_0\int_0^T\int_{R^n} |D\tilde{u}||\tilde{u}|\Phi^2 \exp \mu t\, dxdt + C\int_0^T\int_{R^n} (|\tilde{v}| + |D\tilde{v}|)|\tilde{u}|\Phi^2 \exp \mu t\, dxdt.$$

By easy majorizations, we obtain

$$\frac{1}{4}\int_0^T\int_{R^n} |D\tilde{u}|^2\Phi^2 \exp \mu t\, dxdt + (\frac{\mu}{2} - 4C^2 - c_0^2)\int_0^T\int_{R^n} \tilde{u}^2\Phi^2 \exp \mu t\, dxdt \leq \frac{1}{8}||\tilde{v}||^2.$$

By choosing μ such that $\frac{\mu}{2} - 4C^2 - c_0^2 \geq \frac{1}{4}$, it follows that $\frac{1}{4}||\tilde{u}||^2 \leq \frac{1}{8}||\tilde{v}||^2$, which proves that Θ is a contraction. This completes the proof. □

We next state an important comparison result. Consider another Hamiltonian $H'(x, y, z)$ and another initial condition h'. We assume that

$$H'(x, y, z) \geq H(x, y, z),\ h'(x) \geq h(x). \tag{12.2.4}$$

We do not make the same assumptions on $H'(x, y, z)$ as for $H(x, y, z)$. However, we assume that there exists a solution $u'(x, t)$ of the equation

$$-\frac{\partial u'}{\partial t} - \frac{1}{2}\Delta u' = H(x, u', Du'),$$
$$u'(x, T) = h'(x),$$

in the space $\mathcal{H}$. We may now state the following proposition.

PROPOSITION 12.2. *We have the comparison property* $u'(x,t) \geq u(x,t)$.

PROOF. Setting $\tilde{u} = u' - u$, we have

$$-\frac{\partial \tilde{u}}{\partial t} - \frac{1}{2}\Delta \tilde{u} = H'(x, u', Du') - H(x, u, Du)$$
$$\geq H(x, u', Du') - H(x, u, Du)$$

and $\tilde{u}(x, T) \geq 0$, and $\tilde{u}$ belongs to $\mathcal{H}$. We test the inequality with $(\tilde{u})^- \Phi^2 \exp \mu t$. We note that

$$(H(x, u', Du') - H(x, u, Du))(\tilde{u})^- \geq -C(|(\tilde{u})^-|^2 + (\tilde{u})^- |D(\tilde{u})^-|).$$

After integration, we get easily

$$(\frac{\mu}{2} - C)\int_0^T \int_{R^n} |(\tilde{u})^-|^2 \Phi^2 \exp \mu t \, dxdt + \frac{1}{2}\int_0^T \int_{R^n} |D(\tilde{u})^-|^2 \Phi^2 \exp \mu t \, dxdt \leq$$
$$(C + c_0)\int_0^T \int_{R^n} |D(\tilde{u})^-||(\tilde{u})^-| \Phi^2 \exp \mu t \, dxdt,$$

and from the choice of μ, we necessarily have $(\tilde{u})^- = 0$, whence the result. □

12.2.1 Continuous Hamiltonian

Our objective now is to relax the assumption that the Hamiltonian is Lipschitz. We still assume (12.2.2), but we replace (12.2.1) by (12.1.2), with the same growth conditions on $\lambda(x), |h(x)|$. We then have the following.

PROPOSITION 12.3. *We assume (12.1.2), (12.2.2), and that* $\lambda(x), |h(x)|$ *have polynomial growth. Then the set of solutions of (12.1.1) in* $\mathcal{H}$ *is not empty and has a minimum and a maximum solution.*

PROOF. We first check that the solutions, if they exist, lie in an interval. Indeed, we define the unique solution $\bar{u}(x,t)$ of

$$-\frac{\partial \bar{u}}{\partial t} - \frac{1}{2}\Delta \bar{u} = \lambda(x) + k(|\bar{u}| + |D\bar{u}|), \tag{12.2.5}$$
$$\bar{u}(x, T) = |h(x)|.$$

The existence and uniqueness of $\bar{u}$ in the Hilbert space $\mathcal{H}$ is a simple application of Proposition 12.1. The next step is to prove that if u is a solution of (12.1.1) in the Hilbert space $\mathcal{H}$, then we have

$$|u(x,t)| \leq \bar{u}(x,t). \tag{12.2.6}$$

Let us consider $\tilde{u}(x,t) = u(x,t)+\bar{u}(x,t)$ and let us prove that it is positive. Indeed, $\tilde{u}$ satisfies

$$\begin{aligned} -\frac{\partial \tilde{u}}{\partial t} - \frac{1}{2}\Delta \tilde{u} &= \lambda(x) + k(|\bar{u}| + |D\bar{u}|) + H(x,u,Du) \\ &\geq k(|\bar{u}| + |D\bar{u}|) - k(|u| + |Du|) \\ &\geq -k|\tilde{u}| - k|D\tilde{u}|, \\ \tilde{u}(x,T) &\geq 0. \end{aligned}$$

We test with $(\tilde{u})^- \Phi^2 \exp \mu t$. After integration and majorizations that we have already seen we can write

$$0 \geq (\frac{\mu}{2} - k) \int_0^T \int_{R^n} |(\tilde{u})^-|^2 \, \Phi^2 \exp \mu t \, dxdt + \frac{1}{2} \int_0^T \int_{R^n} |D(\tilde{u})^-|^2 \Phi^2 \exp \mu t \, dxdt -$$
$$- (c_0 + k) \int_0^T \int_{R^n} |D(\tilde{u})^-||(\tilde{u})^-|\Phi^2 \exp \mu t \, dxdt,$$

and for a convenient choice of μ, we obtain $(\tilde{u})^- = 0$. Therefore, $u(x,t) \geq -\bar{u}(x,t)$. Similarly, we prove $u(x,t) \leq \bar{u}(x,t)$. Hence the result (12.2.6) is proved. We now construct a minimum solution. We define the sequence of Hamiltonians

$$H_N(x,y,z) = \inf_{\xi,\eta}(H(x,\xi,\eta) + N(|\xi - x| + |\eta - y|)),$$

and N will tend to $+\infty$. We have $H_N(x,y,z) \leq H(x,y,z)$. Also,

$$\begin{aligned} H(x,\xi,\eta) + N(|\xi - x| + |\eta - y|) &\geq -\lambda(x) - k|\xi| - k|\eta| \\ &\quad + N(|\xi - x| + |\eta - y|) \\ &\geq -\lambda(x) - k|x| - k|y| + (N-k)(|\xi - x| + |\eta - y|). \end{aligned}$$

So for $N > k$, this expression goes to $+\infty$ as $|\xi|, |\eta| \to +\infty$. Therefore, since $H(x,\xi,\eta)$ is continuous in ξ, η, the infimum is attained at a point ξ_N, η_N. Also,

$$H(x,y,z) \geq -\lambda(x) - k|x| - k|y| + (N-k)(|\xi_N - x| + |\eta_N - y|).$$

This implies $\xi_N \to x$, $\eta_N \to y$ as $N \to +\infty$. Now

$$H_N(x,y,z) \geq H(x,\xi_N,\eta_N)$$

and $\liminf H_N(x,y,z) \geq H(x,y,z)$. Therefore,

$$H_N(x,y,z) \uparrow H(x,y,z). \tag{12.2.7}$$

Since also from above, $H_N(x, y, z) \geq -\lambda(x) - k|x| - k|y|$, we have the estimate

$$|H_N(x, y, z)| \leq \lambda(x) + k(|y| + |z|). \tag{12.2.8}$$

On the other hand,

$$|H_N(x, y, z) - H_N(x, y', z')| \leq N(|y - y'| + |z - z'|),$$

so H_N is Lipschitz. Therefore, from Proposition 12.1, there exists a unique solution $u_N(x, t)$ of the equation

$$-\frac{\partial u_N}{\partial t} - \frac{1}{2}\Delta u_N = H_N(x, u_N, Du_N),$$
$$u_N(x, T) = h(x),$$

in the space $\mathcal{H}$. Although we should take a norm in $\mathcal{H}$ depending on N, we can see easily that we can take a norm not depending on N. Indeed, testing with $u_N \Phi^2 \exp \mu t$, we obtain the inequality

$$(\frac{\mu}{2} - k)\int_0^T \int_{R^n} (u_N)^2 \Phi^2 \exp \mu t \, dxdt + \frac{1}{2}\int_0^T \int_{R^n} |Du_N|^2 \Phi^2 \exp \mu t \, dxdt \leq$$
$$(c_0 + k)\int_0^T \int_{R^n} |u_N||Du_N|\Phi^2 \exp \mu t \, dxdt +$$
$$+\int_0^T \int_{R^n} \lambda(x)|u_N|\Phi^2 \exp \mu t \, dxdt + \exp \mu T \int_{R^n} h^2(x)\Phi^2(x)dx,$$

and we see that we can take the norm in $\mathcal{H}$ independent of N. We have also

$$|u_N(x, t)| \leq \bar{u}(x, t), \ ||u_N||_{\mathcal{H}} \leq C,$$

and since $H_N(x, y, z)$ is monotonically increasing in N, we can assert from the comparison property, Proposition 12.2, that $u_N(x, t) \leq u_{N+1}(x, t)$. Therefore, the sequence $u_N(x, t)$ converges pointwise to $u(x, t) \leq \bar{u}(x, t)$. Also from the estimate in $\mathcal{H}$ we can assert for a subsequence

$$u_N \rightharpoonup u, \text{ weakly in } \mathcal{H}.$$

We want to prove that u_N converges strongly in $\mathcal{H}$. Let $M > N$. We are going to let M tend to $+\infty$ for fixed N. Set $\tilde{u}_M = u_M - u_N$. We have

$$-\frac{\partial \tilde{u}_M}{\partial t} - \frac{1}{2}\Delta \tilde{u}_M = H_M(x, u_M, Du_M) - H_N(x, u_N, Du_N),$$
$$\tilde{u}_M(x, T) = 0.$$

We shall use the inequality

$$|H_M(x,u_M,Du_M)-H_N(x,u_N,Du_N)|\leq k|\tilde{u}_M|+k|D\tilde{u}_M|+ \\ +2\lambda(x)+k(|u_N|+|Du_N|).$$

We test the equation with $\tilde{u}_M\Phi^2\exp\mu t$, and after integration and majorizations already seen, we obtain

$$(\frac{\mu}{2}-k)\int_0^T\int_{R^n}(\tilde{u}_M)^2\Phi^2\exp\mu t\,dxdt+\frac{1}{2}\int_0^T\int_{R^n}|D\tilde{u}_M|^2\Phi^2\exp\mu t\,dxdt\leq \\ (c_0+k)\int_0^T\int_{R^n}|\tilde{u}_M||D\tilde{u}_M|\Phi^2\exp\mu t\,dxdt+ \\ +2\int_0^T\int_{R^n}(\lambda(x)+k(|u_N|+|Du_N|)|\tilde{u}_M|\Phi^2\exp\mu t\,dxdt,$$

and for μ fixed, sufficiently large, it follows that

$$\frac{1}{4}\int_0^T\int_{R^n}|D\tilde{u}_M|^2\Phi^2\exp\mu t\,dxdt \\ \leq 2\int_0^T\int_{R^n}(\lambda(x)+k(|u_N|+|Du_N|)|\tilde{u}_M|\Phi^2\exp\mu t\,dxdt.$$

We now let $M\to+\infty$. From the weak convergence in $\mathcal{H}$, the weak lower semicontinuity of the norm, and the strong pointwise convergence, we can assert that

$$\frac{1}{4}\int_0^T\int_{R^n}|D(u-u_N)|^2\Phi^2\exp\mu t\,dxdt \\ \leq 2\int_0^T\int_{R^n}(\lambda(x)+k(|u_N|+|Du_N|)|u-u_N|\Phi^2\exp\mu t\,dxdt. \quad (12.2.9)$$

As $N\to+\infty$, the right-hand side goes to 0. This proves that $u_N\to u$ in $\mathcal{H}$ strongly.

We can then extract a subsequence, also denoted by u_N, such that $Du_N(x,t)\to Du(x,t)$ a.e. From this subsequence we can also extract a new one such that

$$\int_0^T\int_{R^n}\sup_N|Du_N|^2\Phi^2dxdt<+\infty. \quad (12.2.10)$$

This is a classical result. It comes from the fact that Du_N is also a Cauchy sequence in $L^2(0,T;L^2_\Phi(R^n))$. Therefore, we can find a subsequence Du_{N_j} such that

$$(\int_0^T\int_{R^n}|Du_{N_{j+1}}-Du_{N_j}|^2\Phi^2dxdt)^{\frac{1}{2}}\leq\frac{1}{2^j},$$

and therefore, the function

$$g = |Du_{N_0}| + \sum_{j=0}^{+\infty} |Du_{N_{j+1}} - Du_{N_j}|$$

is in $L^2(0, T; L^2_\Phi(R^n))$, and $|Du_{N_j}| \leq g, \forall j$. Therefore, $\sup_j |Du_{N_j}| \leq g$, and (12.2.10) follows for this subsequence. The next claim is for this subsequence, still denoted by u_N, that

$$H_N(x, u_N, Du_N) \to H(x, u, Du) \text{ a.e.} \tag{12.2.11}$$

This is done in a way very similar to (12.2.7), together with the pointwise convergence of u_N, Du_N. Also,

$$\sup_N |H_N(x, u_N, Du_N)| \leq \lambda(x) + k\bar{u}(x,t) + k \sup_N |Du_N(x,t)|,$$

and the right-hand side belongs to $L^2(0, T; L^2_\Phi(R^n))$. Therefore, we can apply Lebesgue's theorem to conclude from (12.2.11) that

$$\int_0^T \int_{R^n} |H_N(x, u_N, Du_N)|^2 \Phi^2 dxdt \to \int_0^T \int_{R^n} |H(x, u, Du)|^2 \Phi^2 dxdt.$$

This implies that $H_N(x, u_N, Du_N) \to H(x, u, Du)$ in $L^2(0, T; L^2_\Phi(R^n))$. This proves that u is a solution of (12.1.1). This solution is the minimum one. Indeed, if we have another one in $\mathcal{H}$, called u^*, then

$$-\frac{\partial u^*}{\partial t} - \frac{1}{2}\Delta u* = H(x, u*, Du*),$$
$$u*(x, T) = h(x).$$

But $H(x, u*, Du*) \geq H_N(x, u*, Du*)$, and from the comparison result, Proposition 12.2, we conclude that $u* \geq u_N$. Therefore, also $u* \geq u$. This proves the assertion.

The maximum solution is constructed by considering the approximation

$$H^N(x, y, z) = \sup_{\xi,\eta}(H(x, \xi, \eta) - N(|\xi - x| + |\eta - y|)), \tag{12.2.12}$$

and the proof is complete. □

We can also give a comparison result. Consider two Hamiltonians $H(x, y, z)$, $H'(x, y, z)$ satisfying

$$H(x, y, z),\ H'(x, y, z) \text{ are continuous in } y, z, \tag{12.2.13}$$
$$|H(x, y, z)|, |H'(x, y, z)| \leq \lambda(x) + k(|y| + |z|),$$

and

$$H(x,y,z) \geq H'(x,y,z),\ \ h(x) \geq h'(x). \tag{12.2.14}$$

We may state the following result.

PROPOSITION 12.4. *Let u, u' be the minimum (or maximum) solutions of CX*

$$-\frac{\partial u}{\partial t} - \frac{1}{2}\Delta u = H(x,u,Du), \tag{12.2.15}$$
$$u(x,T) = h(x),$$

and

$$-\frac{\partial u'}{\partial t} - \frac{1}{2}\Delta u' = H(x,u',Du'), \tag{12.2.16}$$
$$u'(x,T) = h(x).$$

Then we have $u \geq u'$.

PROOF. Considering the approximations $H_N(x,y,z)$ and $H'_N(x,y,z)$, we have $H_N(x,y,z) \geq H'_N(x,y,z)$. Therefore, the corresponding solutions $u_N(x,t)$ and $u'_N(x,t)$ satisfy $u_N \geq u'_N$. Hence for the limit, $u \geq u'$. This implies the result. □

12.2.2 Proof of Theorem 12.1

We construct an approximation of H as follows:

$$H_{\epsilon\eta}(x,y,z) = \frac{H^+(x,y,z)}{1+\epsilon|z|^2} - \frac{H^-(x,y,z)}{1+\eta|z|^2}. \tag{12.2.17}$$

Clearly,

$$|H_{\epsilon\eta}(x,y,z)| \leq \lambda(x) + k|y| + \frac{\gamma}{2}|z|^2, \tag{12.2.18}$$

$$H_{\epsilon\eta}(x,y,z) \text{ is decreasing in } \epsilon \text{ and increasing in } \eta,$$

$$-\lambda(x) - k|y| - \frac{\gamma}{2\eta} \leq H_{\epsilon\eta}(x,y,z) \leq \lambda(x) + k|y| + \frac{\gamma}{2\epsilon}.$$

To simplify notation we set $\rho = (\epsilon, \eta)$. The Hamiltonian $H_\rho(x,y,z)$ satisfies the assumptions of Proposition 12.3. So considering the problem

$$-\frac{\partial u_\rho}{\partial t} - \frac{1}{2}\Delta u_\rho = H_\rho(x,u_\rho,Du_\rho), \tag{12.2.19}$$
$$u_\rho(x,T) = h(x),$$

there exists a minimum solution in the space $\mathcal{H}$. Consider now equation (12.1.4). From the growth conditions, we can assert that

$$|u_\rho(x,t)| \le \bar{u}(x,t) \le \zeta(x). \tag{12.2.20}$$

Indeed,

$$H_\rho(x,y,z) \le \lambda(x) + k|y| + \frac{\gamma}{2}\frac{|z|^2}{1+\epsilon|z|^2},$$

and by the comparison property, we get $u_\rho(x,t) \le \bar{u}_\epsilon(x,t)$, the solution of

$$-\frac{\partial \bar{u}_\epsilon}{\partial t} - \frac{1}{2}\Delta\bar{u}_\epsilon = \lambda(x) + k|\bar{u}_\epsilon| + \frac{\gamma}{2}\frac{|D\bar{u}_\epsilon|^2}{1+\epsilon|D\bar{u}_\epsilon|^2},$$
$$\bar{u}_\epsilon(x,T) = |h(x)|.$$

The solution is unique and positive, so we do not need the absolute value. We then see that $\bar{u}_\epsilon(x,t) \le \bar{u}(x,t)$. Similarly, we check that $u_\rho(x,t) \ge -\bar{u}(x,t)$.

Because of the quadratic growth of the Hamiltonian, we cannot use the same Hilbert space $\mathcal{H}$ as before. We shall need to use a weight function that goes rapidly to 0 at infinity. Let $\delta > 0$, to be chosen later. We take the weight

$$\Phi(x) = \exp -\frac{\delta}{2}\zeta^2(x). \tag{12.2.21}$$

The property $|D\Phi| \le c_0\Phi$ does not hold, but we have $|D\zeta| \le c_1\zeta$, since ζ is quadratic. We can assert that

$$\int_{R^n} \lambda^2(x)\Phi^2(x)\exp\frac{\delta}{2}\zeta^2(x)dx < +\infty, \int_{R^n} \zeta^2(x)\Phi^2(x)\exp\frac{\delta}{2}\zeta^2(x)dx < +\infty$$
$$\int_{R^n} |D\Phi|^2 \exp\frac{\delta}{2}\zeta^2(x)dx < +\infty. \tag{12.2.22}$$

To save notation, we still write $L^2_\Phi(R^n) = \{v| \int_{R^n} v^2(x)\Phi^2(x)dx\}$ and the Sobolev space $H^1_\Phi(R^n) = \{v|v \in L^2_\Phi(R^n), Dv \in (L^2_\Phi(R^n))^n\}$. We now test (12.2.19) with $u_\rho\Phi^2\exp\frac{\delta}{2}(u_\rho)^2$. We obtain

$$\frac{1}{2}\int_0^T\int_{R^n} |Du_\rho|^2(1+\delta u_\rho^2)\Phi^2\exp\frac{\delta}{2}(u_\rho)^2dxdt$$
$$+\int_0^T\int_{R^n} Du_\rho.D\Phi u_\rho\Phi\exp\frac{\delta}{2}(u_\rho)^2dxdt \le$$
$$\int_0^T\int_{R^n}(\lambda(x)+k|u_\rho|+\frac{\gamma}{2}|Du_\rho|^2)|u_\rho|\Phi^2\exp\frac{\delta}{2}(u_\rho)^2dxdt$$
$$+\frac{1}{\delta}\int_{R^n}\Phi^2\exp\frac{\delta}{2}h^2dx.$$

For all $\beta > 0$, we can then write

$$\frac{1}{2}\int_0^T\int_{R^n}|Du_\rho|^2(1+(\delta-\beta)u_\rho^2-\gamma|u_\rho|)\Phi^2\exp\frac{\delta}{2}(u_\rho)^2dxdt \leq$$

$$\frac{1}{2\beta}\int_0^T\int_{R^n}|D\Phi|^2\exp\frac{\delta}{2}(u_\rho)^2dxdt$$

$$+\int_0^T\int_{R^n}(\lambda(x)+k|u_\rho|)|u_\rho|\Phi^2\exp\frac{\delta}{2}(u_\rho)^2dxdt+\frac{1}{\delta}\int_{R^n}\Phi^2\exp\frac{\delta}{2}h^2dx.$$

Since $|u_\rho| \leq \zeta$, we majorize the right-hand side by replacing u_ρ by ζ. We next choose β and δ such that $\delta > 2\beta + \frac{\gamma^2}{4}$. Defining $a_0 = \min(\beta, 1-\frac{\gamma^2}{4(\delta-2\beta)})$, we finally obtain the inequality

$$\frac{a_0}{2}\int_0^T\int_{R^n}|Du_\rho|^2(1+u_\rho^2)\Phi^2\exp\frac{\delta}{2}(u_\rho)^2dxdt \leq \frac{T}{2}\int_{R^n}\lambda^2\Phi^2\exp\frac{\delta}{2}\zeta^2(x)dx+ \tag{12.2.23}$$

$$+T(k+\frac{1}{2})\int_{R^n}\Phi^2\zeta^2\exp\frac{\delta}{2}\zeta^2dx+\frac{T}{2\beta}\int_{R^n}|D\Phi|^2\exp\frac{\delta}{2}\zeta^2dx+\frac{1}{\delta}\int_{R^n}\Phi^2\exp\frac{\delta}{2}h^2dx.$$

Thanks to (12.2.22), the right-hand side is bounded, and therefore u_ρ remains in a bounded set of $L^2(0,T;H^1_\Phi(R^n))$.

Next, we proceed in two steps. In $\rho = (\epsilon,\eta)$, we fix η, and we let $\epsilon \to 0$. To simplify notation, we simply write $\rho = \epsilon$. We have the following properties:

$$H_\epsilon(x,y,z) \uparrow H^+(x,y,z) - \frac{H^-(x,y,z)}{1+\eta|z|^2}. \tag{12.2.24}$$

From the comparison property, we can assert that

$$u_\epsilon \uparrow u \leq \zeta, \tag{12.2.25}$$

and from the bound (12.2.23), we get also, for a subsequence,

$$Du_\epsilon \rightharpoonup Du, \text{ in } L^2(0,T;L^2_\Phi(R^n)) \text{ weakly}. \tag{12.2.26}$$

We are going to show strong convergence. However, we shall need to use a weight Ψ more stringent than Φ. Let $\epsilon' < \epsilon$. We test equation (12.2.19) with $(u_\epsilon - u_{\epsilon'})\Psi^2\exp\frac{\delta}{2}(u_\epsilon-u_{\epsilon'})^2$. Recall that ρ has been replaced with ϵ (η is fixed). After integration and rearrangements, we can state the inequality

$$\frac{1}{2}\int_0^T\int_{R^n}|D(u_\epsilon-u_{\epsilon'})|^2[1-\beta+\delta(u_\epsilon-u_{\epsilon'})^2-\gamma|u_\epsilon-u_{\epsilon'}|]\Psi^2\exp\frac{\delta}{2}(u_\epsilon-u_{\epsilon'})^2dxdt \leq$$

$$\frac{1}{2\beta}\int_0^T\int_{R^n}|D\Psi|^2(u_\epsilon-u_{\epsilon'})^2\exp\frac{\delta}{2}(u_\epsilon-u_{\epsilon'})^2dxdt+$$

$$+\int_0^T\int_{R^n}[2\lambda+2k\zeta+3\gamma|Du_\epsilon-Du|^2+3\gamma|Du|^2]|u_\epsilon-u_{\epsilon'}|\Psi^2\exp\frac{\delta}{2}(u_\epsilon-u_{\epsilon'})^2dxdt.$$

We choose $\beta<\frac{1}{2},\delta>\frac{\gamma^2}{2(\frac{1}{2}-\beta)}$. Then the preceding inequality implies

$$\frac{1}{4}\int_0^T\int_{R^n}|D(u_\epsilon-u_{\epsilon'})|^2[1+\delta(u_\epsilon-u_{\epsilon'})^2]\Psi^2\exp\frac{\delta}{2}(u_\epsilon-u_{\epsilon'})^2dxdt\leq \tag{12.2.27}$$

$$\frac{1}{2\beta}\int_0^T\int_{R^n}|D\Psi|^2(u_\epsilon-u_{\epsilon'})^2\exp\frac{\delta}{2}(u_\epsilon-u_{\epsilon'})^2dxdt+$$

$$+\int_0^T\int_{R^n}[2\lambda+2k\zeta+3\gamma|Du_\epsilon-Du|^2+3\gamma|Du|^2]|u_\epsilon-u_{\epsilon'}|\Psi^2\exp\frac{\delta}{2}(u_\epsilon-u_{\epsilon'})^2dxdt.$$

We want to let ϵ' tend to 0 while ϵ remains fixed. Recalling that $|u_\epsilon|,|u_{\epsilon'}|\leq\zeta$, and setting

$$F_{\epsilon'}=\Bigg[\frac{|D\Psi|^2}{2\beta}(u_\epsilon-u_{\epsilon'})^2+[2\lambda+2k\zeta+$$

$$+3\gamma|Du_\epsilon-Du|^2+3\gamma|Du|^2]|u_\epsilon-u_{\epsilon'}|\Psi^2\Bigg]\exp\frac{\delta}{2}(u_\epsilon-u_{\epsilon'})^2,$$

we see that

$$F_{\epsilon'}\to\Bigg[\frac{|D\Psi|^2}{2\beta}(u_\epsilon-u)^2+[2\lambda+2k\zeta+$$

$$+3\gamma|Du_\epsilon-Du|^2+3\gamma|Du|^2]|u_\epsilon-u|\Psi^2\Bigg]\exp\frac{\delta}{2}(u_\epsilon-u)^2$$

pointwise. Also,

$$F_{\epsilon'}\leq 2\frac{|D\Psi|^2}{\beta}\zeta^2\exp 2\delta\zeta^2+2[2\lambda+2k\zeta+$$

$$+3\gamma|Du_\epsilon-Du|^2+3\gamma|Du|^2]\Psi^2\zeta\exp 2\delta\zeta^2.$$

So if we choose $\Psi(x)=\exp-2\delta\zeta^2(x)$, then we get

$$F_{\epsilon'}\leq\frac{32\delta^2}{\beta}\zeta^2|D\zeta|^2\exp-2\delta\zeta^2(x)+2[2\lambda+2k\zeta+$$

$$+3\gamma|Du_\epsilon-Du|^2+3\gamma|Du|^2]\zeta\exp-2\delta\zeta^2(x),$$

which is a function independent of ϵ' that is integrable, since $\int_0^T \int_{R^n} |Du_\epsilon|^2 \exp -\delta\zeta^2 dxdt < +\infty$, $\int_0^T \int_{R^n} |Du|^2 \exp -\delta\zeta^2 dxdt < +\infty$. So we can pass to the limit on the right-hand side of (12.2.27). For the left-hand side, we first consider

$$X_{\epsilon'\alpha} = D(u_\epsilon - u_{\epsilon'}) \frac{[1+\delta(u_\epsilon - u_{\epsilon'})^2]^{\frac{1}{2}} \exp \frac{\delta}{4}(u_\epsilon - u_{\epsilon'})^2}{1+\alpha(1+2\sqrt{\delta}\zeta)\exp\delta\zeta^2}.$$

Since $|u_\epsilon|, |u_{\epsilon'}| \leq \zeta$, the quantity $\dfrac{[1+\delta(u_\epsilon - u_{\epsilon'})^2]^{\frac{1}{2}} \exp \frac{\delta}{4}(u_\epsilon - u_{\epsilon'})^2}{1+\alpha(1+2\sqrt{\delta}\zeta)\exp\delta\zeta^2}$ is bounded by $\dfrac{1}{\alpha}$ and converges pointwise as $\epsilon' \to 0$ to $\dfrac{[1+\delta(u_\epsilon - u)^2]^{\frac{1}{2}} \exp \frac{\delta}{4}(u_\epsilon - u)^2}{1+\alpha(1+2\sqrt{\delta}\zeta)\exp\delta\zeta^2}$. Since $D(u_\epsilon - u_{\epsilon'})$ converges weakly to $D(u_\epsilon - u)$ in $L^2(0,T; L^2_\Phi(R^n))$, the same is true for $X_{\epsilon'\alpha}$. Therefore,

$$\int_0^T \int_{R^n} |D(u_\epsilon - u)|^2 \frac{[1+\delta(u_\epsilon - u)^2]\exp\frac{\delta}{2}(u_\epsilon - u)^2}{[1+\alpha(1+2\sqrt{\delta}\zeta)\exp\delta\zeta^2]^2}\Psi^2 dxdt \leq$$

$$\liminf_{\epsilon'\to 0} \int_0^T \int_{R^n} |D(u_\epsilon - u_{\epsilon'})|^2 \frac{[1+\delta(u_\epsilon - u_{\epsilon'})^2]\exp\frac{\delta}{2}(u_\epsilon - u_{\epsilon'})^2}{[1+\alpha(1+2\sqrt{\delta}\zeta)\exp\delta\zeta^2]^2}\Psi^2 dxdt$$

$$\leq \liminf_{\epsilon'\to 0} \int_0^T \int_{R^n} |D(u_\epsilon - u_{\epsilon'})|^2 [1+\delta(u_\epsilon - u_{\epsilon'})^2]\exp\frac{\delta}{2}(u_\epsilon - u_{\epsilon'})^2 \Psi^2 dxdt,$$

and from the above considerations regarding $F_{\epsilon'}$, we can state that

$$\leq \frac{2}{\beta}\int_0^T \int_{R^n} |D\Psi|^2 (u_\epsilon - u)^2 \exp\frac{\delta}{2}(u_\epsilon - u)^2 dxdt +$$

$$+4\int_0^T \int_{R^n} [2\lambda + 2k\zeta + 3\gamma|Du_\epsilon - Du|^2 + 3\gamma|Du|^2]|u_\epsilon - u|\Psi^2 \exp\frac{\delta}{2}(u_\epsilon - u)^2 dxdt.$$

Letting $\alpha \to 0$, we conclude that

$$\int_0^T \int_{R^n} |D(u_\epsilon - u)|^2 [1+\delta(u_\epsilon - u)^2]\exp\frac{\delta}{2}(u_\epsilon - u)^2 \Psi^2 dxdt \leq \tag{12.2.28}$$

$$\frac{2}{\beta}\int_0^T \int_{R^n} |D\Psi|^2 (u_\epsilon - u)^2 \exp\frac{\delta}{2}(u_\epsilon - u)^2 dxdt +$$

$$+4\int_0^T \int_{R^n} [2\lambda + 2k\zeta + 3\gamma|Du_\epsilon - Du|^2 + 3\gamma|Du|^2]|u_\epsilon - u|\Psi^2 \exp\frac{\delta}{2}(u_\epsilon - u)^2 dxdt.$$

So we also have

$$\int_0^T \int_{R^n} |D(u_\epsilon - u)|^2[1 + \delta(u_\epsilon - u)^2 - 12\gamma|u_\epsilon - u|] \exp \frac{\delta}{2}(u_\epsilon - u)^2\Psi^2 dxdt \leq$$

$$\frac{2}{\beta}\int_0^T \int_{R^n} |D\Psi|^2(u_\epsilon - u)^2 \exp \frac{\delta}{2}(u_\epsilon - u)^2 dxdt$$

$$+4\int_0^T \int_{R^n} [2\lambda + 2k\zeta + 3\gamma|Du|^2]|u_\epsilon - u|\Psi^2 \exp \frac{\delta}{2}(u_\epsilon - u)^2 dxdt.$$

We recall that β and δ were chosen so that

$$\beta < \frac{1}{2}, \delta > \frac{\gamma^2}{2(\frac{1}{2} - \beta)}.$$

By choosing $\beta > \frac{1}{2} - \frac{1}{288}$, we can ensure that $\delta > 144\gamma^2$, so that $\frac{1}{2} + \frac{\delta}{2}(u_\epsilon - u)^2 - 12\gamma|u_\epsilon - u| \geq 0$. Therefore,

$$\frac{1}{2}\int_0^T \int_{R^n} |D(u_\epsilon - u)|^2[1 + \delta(u_\epsilon - u)^2] \exp \frac{\delta}{2}(u_\epsilon - u)^2\Psi^2 dxdt \leq \tag{12.2.29}$$

$$\frac{2}{\beta}\int_0^T \int_{R^n} |D\Psi|^2(u_\epsilon - u)^2 \exp \frac{\delta}{2}(u_\epsilon - u)^2 dxdt$$

$$+8\int_0^T \int_{R^n} (\lambda + k\zeta)|u_\epsilon - u|\Psi^2 \exp \frac{\delta}{2}(u_\epsilon - u)^2 dxdt.$$

The right-hand side tends to 0 as $\epsilon \to 0$. Therefore, $u_\epsilon \to u$ in $L^2(0, T; H^1_\Psi(R^n))$ strongly. We can choose a subsequence such that

$$u_\epsilon \uparrow u, \; Du_\epsilon \to Du, \text{ a.e.}, \int_0^T \int_{R^n} \sup_\epsilon |Du_\epsilon|^2\Psi^2 dxdt < +\infty.$$

One easily checks that

$$H_\epsilon(x, u_\epsilon, Du_\epsilon) \to H^+(x, u, Du) - \frac{H^-(x, u, Du)}{1 + \eta|Du|^2}, \text{ a.e.}$$

and in $L^1(0, T; L^1_\Psi(R^n))$. Therefore, u is a solution of

$$-\frac{\partial u}{\partial t} - \frac{1}{2}\Delta u = H^+(x, u, Du) - \frac{H^-(x, u, Du)}{1 + \eta|Du|^2}, \tag{12.2.30}$$

$$u(x, T) = h(x).$$

Of course, this solution depends on η, called u_η (not to be mistaken for u_ϵ, which was shorthand for u_ρ), and we have the property that u_η is monotonically increasing in η. This is because $u_{\epsilon,\eta_1}(x,t) \leq u_{\epsilon,\eta_2}(x,t)$ if $\eta_1 < \eta_2$, with fixed ϵ. We can then let η tend to 0 monotonically and operate in a way similar to that for ϵ. This leads to a solution of (12.1.1). This concludes the proof of Theorem 12.1. ■

12.2.3 Formal Proof of Theorem 12.2

We turn now to the uniqueness property. We make the assumptions (12.1.13), (12.1.14). We use a trick introduced by Da Lio and Ley [18], which takes advantage of the convexity assumption. Suppose we have two solutions u_1, u_2 in $L^2(0,T; H^1_\Psi(R^n))$. In fact, we are going to assume sufficient smoothness of these functions. This is the formal aspect. For $0 < \theta < 1$, we consider $\tilde{u} = u_1 - \theta u_2$. Eventually θ will tend to 1. We have

$$-\frac{\partial \tilde{u}}{\partial t} - \frac{1}{2}\Delta\tilde{u} = H(x,u_1,Du_1) - \theta H(x,u_2,Du_2),$$
$$\tilde{u}(x,T) = (1-\theta)h(x).$$

From the convexity assumption we have

$$\begin{aligned} H(x,u_2,Du_1) &= H(x,u_2,\theta Du_2 + (1-\theta)\frac{Du_1 - \theta Du_2}{1-\theta}) \\ &\leq \theta H(x,u_2,Du_2) + (1-\theta)H(x,u_2,\frac{D\tilde{u}}{1-\theta}); \end{aligned}$$

hence

$$\begin{aligned} H(x,u_1,Du_1) - \theta H(x,u_2,Du_2) \leq H(x,u_1,Du_1) - H(x,u_2,Du_1) + \\ +(1-\theta)H(x,u_2,\frac{D\tilde{u}}{1-\theta}). \end{aligned}$$

Also,

$$\begin{aligned} H(x,u_1,Du_1) - H(x,u_2,Du_1) &\leq k|u_1 - u_2| \\ &\leq k|\tilde{u}| + k(1-\theta)|u_2|. \end{aligned}$$

Collecting estimates, we obtain

$$-\frac{\partial \tilde{u}}{\partial t} - \frac{1}{2}\Delta\tilde{u} - k|\tilde{u}| - \frac{\gamma}{2(1-\theta)}|D\tilde{u}|^2 \leq (1-\theta)(\lambda(x) + 2k|u_2|), \quad (12.2.31)$$
$$\tilde{u}(x,T) \leq (1-\theta)|h(x)|.$$

But then $\tilde{v} = \dfrac{\tilde{u}}{1-\theta}$ satisfies

$$-\frac{\partial \tilde{v}}{\partial t} - \frac{1}{2}\Delta\tilde{v} - k|\tilde{v}| - \frac{\gamma}{2}|D\tilde{v}|^2 \leq \lambda(x) + 2k|u_2|, \tag{12.2.32}$$

$$\tilde{v}(x,T) \leq |h(x)|, \tag{12.2.33}$$

and $(\tilde{v})^+(x,t) \leq z(x,t)$, the solution of

$$-\frac{\partial z}{\partial t} - \frac{1}{2}\Delta z - kz - \frac{\gamma}{2}|Dz|^2 \exp kt = \lambda(x) + 2k\zeta(x), \tag{12.2.34}$$
$$z(x,T) = |h(x)|.$$

Here we make the growth condition

$$\text{The function } z(x,t) \text{ is well defined.} \tag{12.2.35}$$

The inequality can be seen as follows. Define $\chi(x,t) = \exp(\gamma(\tilde{v})^+(x,t)\exp kt)$ and $\zeta(x,t) = \exp(\gamma z(x,t)\exp kt)$. It suffices to show that $\chi(x,t) \leq \zeta(x,t)$. Note first that ζ is the solution of the linear equation

$$-\frac{\partial \zeta}{\partial t} - \frac{1}{2}\Delta\zeta - \zeta\gamma\exp kt(\lambda(x) + 2k\zeta(x)) = 0, \tag{12.2.36}$$
$$\zeta(x,T) = \exp(\gamma|h(x)|\exp kT).$$

When $(\tilde{v})^+(x,t) = 0$, we have $\chi(x,t) = 1 \leq \zeta(x,t)$, since $z(x,t) \geq 0$. Also, $\chi(x,T) \leq \exp(\gamma|h(x)|\exp kT) = \zeta(x,T)$. Here comes the formal argument: On the domain $(\tilde{v})^+(x,t) > 0$, $\tilde{v}(x,t)$ satisfies the inequality

$$-\frac{\partial \tilde{v}}{\partial t} - \frac{1}{2}\Delta\tilde{v} - k\tilde{v} - \frac{\gamma}{2}|D\tilde{v}|^2 \leq \lambda(x) + 2k\zeta,$$

and $\chi(x,t) = \exp(\gamma\tilde{v}(x,t)\exp kt)$ satisfies

$$-\frac{\partial \chi}{\partial t} - \frac{1}{2}\Delta\chi - \chi\gamma\exp kt(\lambda(x) + 2k\zeta) \leq 0.$$

The formal aspect lies in the definition of the domain $(\tilde{v})^+(x,t) > 0$, which requires some smoothness of the solutions u_1, u_2. To conclude, we note that we have proved that $(\tilde{u})^+(x,t) \leq (1-\theta)z(x,t)$, or $(u_1 - \theta u_2)^+(x,t) \leq (1-\theta)z(x,t)$. But $z(x,t)$ is a fixed function, not depending on θ. So we can let $\theta \to 1$, which leads to $u_1 - u_2 \leq 0$. Reversing the roles of u_1, u_2, we obtain the opposite inequality, whence $u_1 = u_2$, and the uniqueness follows.

12.3 Solutions of BSDEs

12.3.1 Statement of the Problem

We shall give here results of BSDEs, which are similar to those expressed in previous sections regarding parabolic PDEs in R^n. We have a probability space $\Omega, \mathcal{A}, P$, a standard Wiener process $w(t)$ in R^n, and the filtration $\mathcal{F}^t$ generated by the Wiener process. We next define a family $H(t, y, z)$, where $y \in R, z \in R^n$, of scalar stochastic processes adapted to $\mathcal{F}^t$. We assume that

$$H(t, y, z) \text{ is measurable and continuous in } y, z, \tag{12.3.1}$$

$$|H(t, y, z)| \leq \lambda(t) + k|y| + \frac{\gamma}{2}|z|^2, \tag{12.3.2}$$

in which $\lambda(t)$ is positive adapted, and $E\int_0^T \lambda^2(t)dt$. Additional integrability conditions will be given below. We also define a random variable ξ in $L^2(\Omega, \mathcal{A}, P)$ that is $\mathcal{F}^T$-measurable. We introduce the BSDE

$$Y(t) = \xi + \int_t^T H(s, Y(s), Z(s))ds - \int_t^T Z(s).dw(s), \tag{12.3.3}$$

in which $Y(t)$, $Z(t)$ are adapted processes. We look for solutions such that

$$\sup_{0 \leq t \leq T} |Y(t)| < +\infty, \int_0^T |Z(t)|^2 dt < +\infty, \text{ a.s.} \tag{12.3.4}$$

12.3.2 Integrability Condition and Existence

We shall need the integrability condition that plays the role of the growth condition of parabolic PDEs:

$$E[\exp \gamma(|\xi| \exp kT + \int_0^T \lambda(s) \exp ks\, ds)] < +\infty. \tag{12.3.5}$$

Then we shall prove the following result (equivalent of Theorem 12.1).

THEOREM 12.3. *We assume (12.3.1), (12.3.2), (12.3.5). Then there exists a solution of (12.3.3) such that (12.3.4) holds.* ■

Consider the process

$$\varphi(t) = E[\exp \gamma(|\xi| \exp kT + \int_t^T \lambda(s) \exp ks\, ds)|\mathcal{F}^t]$$

and

$$\bar{Y}(t) = \frac{\exp -kt}{\gamma} \log \varphi(t). \tag{12.3.6}$$

We may introduce the monotonically increasing sequence of stopping times $\tau_N = \inf\{t \in [0, T] | \bar{Y}(t) \geq N\}$. We have

$$\{\tau_N < T\} \subset \{ \sup_{0 \leq t \leq T} \bar{Y}(t) \geq N\} \subset \{ \sup_{0 \leq t \leq T} \varphi(t) \geq \exp \gamma N\}$$

$$\subset \{ \sup_{0 \leq t \leq T} E[\exp \gamma(|\xi| \exp kT + \int_0^T \lambda(s) \exp ks\, ds)|\mathcal{F}^t] \geq \exp \gamma N\}.$$

Hence by classical martingale properties, it follows that

$$P(\{\tau_N < T\}) \leq \frac{\varphi(0)}{\exp \gamma N}.$$

Therefore, we also have $P(\cap_N\{\tau_N < T\}) = 0,\ P(\cup_N\{\tau_N = T\}) = 1$. It follows that a.s. $\exists N$ such that $\tau_N = T$. We look for solutions such that

$$|Y(t)| \leq \bar{Y}(t),\ \mathbb{1}_{t<\tau_N} Z(t) \in L^2_{\mathcal{F}}(0, T; R^n) < +\infty, \forall N. \tag{12.3.7}$$

This implies (12.3.4). We can prove uniqueness under additional assumptions. Namely, we assume

$$|H(t, y, z) - H(t, y', z)| \leq k|y - y'|, \tag{12.3.8}$$

$$z \to H(t, y, z) \text{ is convex,} \tag{12.3.9}$$

$$E[\exp \gamma(|\xi| \exp kT + \int_0^T (\lambda(s) + 2k\bar{Y}(s)) \exp ks\, ds)] < +\infty. \tag{12.3.10}$$

We may then state the following theorem.

THEOREM 12.4. *We make the assumptions of Theorem 12.3 and (12.3.8), (12.3.9), (12.3.10). Then the solution of (12.3.3), (12.3.4) is unique.* ■

12.3.3 Methodology

We first assume the Lipschitz condition

$$|H(t, y, z) - H(t, y', z)| \leq C(|y - y'| + |z - z'|) \tag{12.3.11}$$

and the integrability condition

$$|H(t,0,0)| \leq \lambda(t) \in L^2_{\mathcal{F}}(0,T). \tag{12.3.12}$$

Then we state the following.

PROPOSITION 12.5. *We assume (12.3.11), (12.3.12). For $\xi \in L^2(\Omega,\mathcal{F}^T,P)$, there exists a unique solution of (12.3.3) such that $Y \in L^2_{\mathcal{F}}(0,T)$, $Z \in L^2_{\mathcal{F}}(0,T;R^n)$.* ■

This is the result already stated in Theorem 11.5. We have also the following comparison result.

PROPOSITION 12.6. *Consider another Hamiltonian $H'(t,y,z)$ such that $H'(t,y,z) \geq H(t,y,z)$. Suppose that for $\xi' \in L^2(\Omega,\mathcal{A},P)$, $\xi' \geq \xi$, there exists a solution $Y'(t) \in L^2_{\mathcal{F}}(0,T)$, $Z'(t) \in L^2_{\mathcal{F}}(0,T;R^n)$ of*

$$Y'(t) = \xi' + \int_t^T H(s,Y'(s),Z'(s))ds - \int_t^T Z'(s).dw(s).$$

Then we have $Y'(t) \geq Y(t)$ $\forall t$ a.s.

PROOF. We define $\tilde{Y}(t) = Y'(t) - Y(t)$, $\tilde{Z}(t) = Z'(t) - Z(t)$. We compute

$$\frac{1}{2}d((\tilde{Y})^-)^2 = 1\!\!1_{\tilde{Y}(t)<0}(\tilde{Y}d\tilde{Y} + \frac{1}{2}|\tilde{Z}|^2dt)$$

$$\geq 1\!\!1_{\tilde{Y}(t)<0}[-\tilde{Y}(H(t,Y',Z') - H(t,Y,Z)) + \frac{1}{2}|\tilde{Z}|^2]dt$$
$$- (\tilde{Y})^-\tilde{Z}dw.$$

Therefore, using $(\tilde{Y})^-(T) = 0$, we obtain

$$0 \geq E\int_0^T \exp\mu t\, 1\!\!1_{\tilde{Y}(t)<0}[\mu(\tilde{Y})^2 - \tilde{Y}(H(t,Y',Z') - H(t,Y,Z)) + \frac{1}{2}|\tilde{Z}|^2]dt$$
$$\geq E\int_0^T \exp\mu t\, 1\!\!1_{\tilde{Y}(t)<0}[\mu(\tilde{Y})^2 - C|\tilde{Y}|(|\tilde{Y}| + |\tilde{Z}|) + \frac{1}{2}|\tilde{Z}|^2]dt.$$

Since μ can be chosen sufficiently large, we obtain $1\!\!1_{\tilde{Y}(t)<0} = 0$ a.s. This proves the result. □

We remove now the assumption (12.3.11). We assume (12.3.1) and

$$|H(t,y,z)| \leq \lambda(t) + k(|y| + |z|), \tag{12.3.13}$$
$$\lambda(t) \in L^2_{\mathcal{F}}(0,T).$$

We then state the following.

PROPOSITION 12.7. *We assume (12.3.1) and (12.3.13). For* $\xi \in L^2(\Omega, \mathcal{F}^T, P)$, *the set of solutions of (12.3.3) such that* $Y \in L^2_{\mathcal{F}}(0, T)$, $Z \in L^2_{\mathcal{F}}(0, T; R^n)$ *is not empty and has a minimum and a maximum element. Moreover, the comparison principle of Proposition 12.6 holds between minimum solutions.*

PROOF. Let us consider the BSDE □

$$\bar{Y}(t) = |\xi| + \int_t^T (\lambda(s) + k(|\bar{Y}(s) + |\bar{Z}(s)|))ds - \int_t^T \bar{Z}(s).dw(s). \qquad (12.3.14)$$

It enters in the framework of Proposition 12.5. Therefore, there exists a unique solution of (12.3.14). By comparison arguments similar to those used in the proof of Proposition 12.6, it is easy to check that every solution of (12.3.3) such that $Y \in L^2_{\mathcal{F}}(0, T)$, $Z \in L^2_{\mathcal{F}}(0, T; R^n)$ will satisfy

$$|Y(t)| \leq \bar{Y}(t). \qquad (12.3.15)$$

We use the same method as for parabolic PDEs; see Proposition 12.3. We introduce the approximation

$$H_N(t, y, z) = \inf_{\xi,\eta}(H(t, \xi, \eta) + N|\xi - y| + N|\eta - z|).$$

Then for $N > k$, we have

$$|H_N(t, y, z)| \leq \lambda(t) + k(|y| + |z|), \qquad (12.3.16)$$
$$|H_N(t, y, z) - H_N(t, y', z')| \leq N(|y' - y| + |z' - z|).$$

Moreover,

$$H_N(t, y, z) \uparrow H(t, y, z), \ \forall y, z, \text{ a.s.} \qquad (12.3.17)$$

We can then solve the BSDE

$$Y_N(t) = \xi + \int_t^T H_N(s, Y_N(s), Z_N(s))ds - \int_t^T Z_N(s).dw(s), \qquad (12.3.18)$$

whose solution $Y_N(t), Z_N(t)$ is uniquely defined. We see easily that $Y_N(t)$ is increasing in N and $Y_N(t) \leq \bar{Y}(t)$. Therefore, $Y_N(t) \uparrow Y(t) \leq \bar{Y}(t)$. Next, considering the process $\frac{1}{2}|Y_N(t)|^2 \exp \mu t$ and using Itô's formula, we obtain easily

$$\frac{1}{2}E|\xi|^2 \exp \mu T \geq E\int_0^T \exp \mu t[(\frac{\mu}{2} - k)|Y_N(t)|^2 - |Y_N(t)|(\lambda(t) + k|Z_N(t)) +$$
$$+ \frac{1}{2}|Z_N(t)|^2]dt,$$

from which it follows easily that Z_N remains bounded in $L^2_{\mathcal{F}}(0, T; R^n)$. We can thus extract a subsequence such that $Z_N \rightharpoonup Z$ weakly in $L^2_{\mathcal{F}}(0, T; R^n)$. We prove now that the convergence is strong. Consider $M > N$ and set $\tilde{Y}_M(t) = Y_M(t) - Y_N(t)$, $\tilde{Z}_M(t) = Z_M(t) - Z_N(t)$. The fact that we use only the index M is due to the fact that N will be fixed, and we shall let $M \to +\infty$. We make calculations that have similarities to those made in the proof of Proposition 12.3. We consider the process $\frac{(\tilde{Y}_M(t))^2}{2}$. After use of Itô's formula and assumption (12.3.13), we can write the inequality (note that $\tilde{Y}_M(t) > 0$)

$$0 \geq E \int_0^T \exp \mu t \left(\mu \frac{(\tilde{Y}_M(t))^2}{2} + \frac{(\tilde{Z}_M(t))^2}{2} - \right.$$
$$\left. - \tilde{Y}_M(t)[2\lambda(t) + k(|Y_M(t)| + |Y_N(t)| + |Z_M(t)| + |Z_N(t)|)] \right) dt.$$

For $\frac{\mu}{2} > k + k^2$ and by standard majorizations, we can assert that

$$E \int_0^T \exp \mu t \, |\tilde{Z}_M(t)|^2 dt \leq 8E \int_0^T \exp \mu t (\lambda(t) + k(|Y_N(t)| + |Z_N(t)|))\tilde{Y}_M(t) dt.$$

Passing to the lim inf as $M \to +\infty$, we obtain

$$E \int_0^T \exp \mu t |Z(t) - Z_N(t)|^2 dt \leq 8E \int_0^T \exp \mu t (\lambda(t) + k(|Y_N(t)| + |Z_N(t)|))(Y(t) - Y_N(t)) dt.$$

In this inequality, we can now let $N \to +\infty$. The right-hand side tends to 0. This proves the strong convergence of Z_N to Z in $L^2_{\mathcal{F}}(0, T; R^n)$. So we can extract a subsequence such that

$$Y_N(t) \uparrow Y(t), \text{ a.e., a.s., } Z_N(t) \to Z(t), \text{ a.e., a.s.,}$$
$$\sup_N |Z_N(t)| \in L^2_{\mathcal{F}}(0, T; R^n).$$

We check easily that

$$H_N(t, Y_N(t), Z_N(t)) \to H(t, Y(t), Z(t)), \text{ a.e., a.s.,}$$
$$\sup_N |H_N(t, Y_N(t), Z_N(t))| \leq \lambda(t) + k\bar{Y}(t) + k \sup_N |Z_N(t)|;$$

hence

$$H_N(t, Y_N(t), Z_N(t)) \to H(t, Y(t), Z(t) \in L^2_{\mathcal{F}}(0, T; R^n).$$

It follows immediately that the pair $Y(t), Z(t)$ is a solution of (12.3.3). We claim that if $Y^*(t), Z^*(t)$ is another solution belonging to $L^2_{\mathcal{F}}(0, T) \times L^2_{\mathcal{F}}(0, T; R^n)$, then $Y^*(t) \geq Y(t)$. This is proved by checking that $Y^*(t) \geq Y_N(t)$. One considers the process $\frac{1}{2}((Y^*(t) - Y_N(t))^-)^2 \exp \mu t$ and proceeds as in Proposition 12.6. One recalls that $H(t, Y^*(t), Z^*(t)) \geq H_N(t, Y^*(t), Z^*(t))$ and that H_N(t,y,z) is Lipschitz. Details are left to the reader. Since N is arbitrary, we have $Y^*(t) \geq Y(t)$. We finally prove the comparison property. If we have two Hamiltonians $H(t, y, z)$ and $H'(t, y, z)$ such that $H(t, y, z) \geq H'(t, y, z)$ and $\xi \geq \xi'$, both satisfying (12.3.1), (12.3.13), then the solutions $Y(t), Z(t)$ and $Y'(t), Z'(t)$ are limits of $Y_N(t), Z_N(t)$ and $Y_N^{'}(t), Z_N^{'}(t)$. But $H_N(t, y, z) \geq H_N'(t, y, z)$. Therefore, from the comparison result for Lipschitz Hamiltonians, we can assert that $Y_N(t) \geq Y_N'(t)$. This implies $Y(t) \geq Y'(t)$, which proves the comparison principle. The maximum solution is obtained by considering the approximation of the Hamiltonian defined by (12.2.12). This completes the proof. ■

12.3.4 Quadratic Growth Case

We turn now to the proof of Theorem 12.3. Setting as in the proof of Theorem 12.1 $\rho = (\epsilon, \eta)$ and

$$H_\rho(t, y, z) = \frac{H^+(t, y, z)}{1 + \epsilon|z|^2} - \frac{H^-(t, y, z)}{1 + \eta|z|^2}, \tag{12.3.19}$$

we see that there exists a minimum solution of

$$Y_\rho(t) = \xi + \int_t^T H_\rho(s, Y_\rho(s), Z_\rho(s))ds - \int_t^T Z_\rho(s)dw(s), \tag{12.3.20}$$

$$Y_\rho \in L^2_{\mathcal{F}}(0, T), Z_\rho \in L^2_{\mathcal{F}}(0, T; R^n).$$

Consider next the solutions of

$$\bar{Y}_\epsilon(t) = |\xi| + \int_t^T [\lambda(s) + k\bar{Y}_\epsilon(s) + \frac{\gamma}{2}\frac{|\bar{Z}_\epsilon(s)|^2}{1 + \epsilon|\bar{Z}_\epsilon(s)|^2}]ds - \int_t^T \bar{Z}_\epsilon dw, \tag{12.3.21}$$

$$\bar{Y}_\eta(t) = |\xi| + \int_t^T [\lambda(s) + k\bar{Y}_\eta(s) + \frac{\gamma}{2}\frac{|\bar{Z}_\eta(s)|^2}{1 + \eta|\bar{Z}_\eta(s)|^2}]ds - \int_t^T \bar{Z}_\eta dw. \tag{12.3.22}$$

The solutions $\bar{Y}_\epsilon, \bar{Z}_\epsilon$ and $\bar{Y}_\eta, \bar{Z}_\eta$ are uniquely defined, since the Hamiltonians are Lipschitz and have linear growth. We set $Y_\eta^*(t) = -\bar{Y}_\eta(t),\ Z_\eta^*(t) = -\bar{Z}_\eta(t)$. Then equation (12.3.22) becomes

$$Y_\eta^*(t) = -|\xi| + \int_t^T [-\lambda(s) + kY_\eta^*(s) - \frac{\gamma}{2}\frac{|Z_\eta^*(s)|^2}{1 + \eta|Z_\eta^*(s)|^2}]ds - \int_t^T Z_\eta^* dw. \tag{12.3.23}$$

Noting that $\bar{Y}_\epsilon(t) > 0,\ Y^*_\eta(t) < 0$, we see that the Hamiltonians related to (12.3.21), (12.3.23) can be written as

$$\bar{H}_\epsilon(t, y, z) = \lambda(t) + k|y| + \frac{\gamma}{2}\frac{|z|^2}{1+\epsilon|z|^2},$$

$$H^*_\eta(t, y, z) = -\lambda(t) - k|y| - \frac{\gamma}{2}\frac{|z|^2}{1+\eta|z|^2}.$$

We notice immediately that

$$H^*_\eta(t, y, z) \leq H_\rho(t, y, z) \leq \bar{H}_\epsilon(t, y, z).$$

From the comparison property of Proposition 12.7, we can assert that

$$Y^*_\eta(t) \leq Y_\rho(t) \leq \bar{Y}_\epsilon(t).$$

Hence $|Y_\rho(t)| \leq \max(\bar{Y}_\epsilon(t), \bar{Y}_\eta(t))$. If we define $\varphi_\epsilon(t) = \exp(\gamma \bar{Y}_\epsilon(t) \exp kt)$, then from Itô's formula, we obtain

$$d\varphi_\epsilon(t) \geq \gamma\varphi_\epsilon(t) \exp kt[-\lambda(t)dt + \bar{Z}_\epsilon dw].$$

From the assumption (12.3.5) it is easy to check that $\varphi_\epsilon(t) \leq \varphi(t)$. Therefore, $\bar{Y}_\epsilon(t) \leq \bar{Y}(t)$. Since also $\bar{Y}_\eta(t) \leq \bar{Y}(t)$, we obtain the estimate

$$|Y_\rho(t)| \leq \bar{Y}(t). \tag{12.3.24}$$

From the comparison properties, we note that $Y_\rho(t)$ is decreasing in ϵ and increasing in η. We now obtain additional estimates. Let δ be defined later. We compute the Itô differential $\frac{1}{\delta} d \exp \frac{\delta}{2} Y^2_\rho(t)$. After easy majorizations, we obtain

$$\frac{1}{\delta} d \exp \frac{\delta}{2} Y^2_\rho(t) \geq \exp \frac{\delta}{2} Y^2_\rho(t)[(-|Y_\rho(t)|(\lambda(t) + k|Y_\rho(t)| + \frac{\gamma}{2}|Z_\rho(t)|^2) +$$
$$+ \frac{1}{2}(1 + \delta Y^2_\rho(t))|Z_\rho(t)|^2)dt + Y_\rho(t) Z_\rho(t) dw(t)].$$

Using the estimate (12.3.24), we further estimate, choosing $\delta > \frac{\gamma^2}{4}$,

$$\frac{1}{\delta} d \exp \frac{\delta}{2} Y^2_\rho(t) \geq -\exp \frac{\delta}{2}(\bar{Y}(t))^2(\lambda(t)\bar{Y}(t) + k(\bar{Y}(t))^2)dt +$$
$$+ \frac{1}{2}(1 - \frac{\gamma^2}{4\delta})|Z_\rho(t)|^2 dt + \exp \frac{\delta}{2} Y^2_\rho(t)\, Y_\rho(t) Z_\rho(t) dw(t).$$

Recalling the definition $\tau_N = \inf\{t \in [0, T] | \bar{Y}(t) \geq N\}$, we deduce

$$\frac{1}{2}(1 - \frac{\gamma^2}{4\delta})E \int_0^T 1\!\!1_{t<\tau_N} |Z_\rho(t)|^2 dt \leq \exp \frac{\delta}{2} N^2 [(\frac{1}{\delta} + kN^2 T) + NE \int_0^T \lambda(t) dt].$$

So

$$E \int_0^T 1\!\!1_{t<\tau_N} |Z_\rho(t)|^2 dt \leq C_N. \tag{12.3.25}$$

In the pair $\rho = (\epsilon, \eta)$ we fix ϵ and let $\eta \to 0$. Since ϵ is fixed, we identify ρ with η. We note that $Y_\rho(t) = Y_\eta(t) \downarrow Y(t)$ and

$$H_\rho(t, y, z) \downarrow \frac{H^+(t, y, z)}{1 + \epsilon |z|^2} - H^-(t, y, z).$$

We can extract a subsequence such that

$$1\!\!1_{t<\tau_N} Z_\eta(t) \rightharpoonup 1\!\!1_{t<\tau_N} Z(t), \text{ in } L^2_{\mathcal{F}}(0, T; R^n), \text{ weakly}. \tag{12.3.26}$$

The process $Z(t)$ does not depend on N, but it is not in $L^2_{\mathcal{F}}(0, T; R^n)$. Indeed, if we write $1\!\!1_{t<\tau_N} Z_\eta(t) \rightharpoonup Z^N(t)$, we check immediately that $Z^N(t) = 1\!\!1_{t<\tau_N} Z^{N'}(t)$, $\forall N < N'$, a.s., a.e. However, a.s. there exists N_T such that a.s. $\tau_{N_T} = T$. It follows that $Z^{N'}(t) = Z^{N_T}(t) \, \forall N' > N_T$. Writing $Z(t) = Z^{N_T}(t)$, we also have $Z^N(t) = 1\!\!1_{t<\tau_N} Z(t)$.

We then prove strong convergence. We fix η and consider a subsequence $\eta' < \eta$ that tends to 0. So $Y_{\eta'} \downarrow Y < Y_\eta$. We consider the process $\frac{1}{\delta} \exp \frac{\delta}{2}(Y_{\eta'} - Y_\eta)^2(t)$ for a coefficient δ more stringent than what was used above. We note the inequality

$$-H_{\eta'}(t, Y_{\eta'}, Z_{\eta'}) + H_\eta(t, Y_\eta, Z_\eta) \leq 2\lambda(t) + k|Y_{\eta'}| + k|Y_\eta| + 3\gamma |Z|^2 + \\ + 3\gamma |Z_\eta - Z|^2 + \gamma |Z_{\eta'} - Z_\eta|^2.$$

After some now standard majorizations, we claim that for $\delta > 4\gamma^2$, we have

$$\frac{1}{\delta} d \exp \frac{\delta}{2}(Y_{\eta'} - Y_\eta)^2(t) \geq \exp \frac{\delta}{2}(Y_{\eta'} - Y_\eta)^2(t) \Big\{ [(Y_{\eta'} - Y_\eta)(t)(2\lambda(t) + 2\bar{Y}(t)$$
$$+3\gamma |Z(t)|^2 + 3\gamma |Z_\eta(t) - Z(t)|^2) + + \frac{1}{4}(1 + \delta (Y_{\eta'} - Y_\eta)^2(t)) |Z_{\eta'}(t) - Z_\eta(t)|^2] dt$$
$$+(Y_{\eta'} - Y_\eta)(t)(Z_{\eta'}(t) - Z_\eta(t)) dw(t) \Big\}.$$

Integrating between 0 and τ_N and taking the expectation yields

$$\frac{1}{\delta}E[\exp\frac{\delta}{2}(Y_{\eta'}-Y_\eta)^2(\tau_N)1\!\!1_{\tau_N<T}]+$$

$$+E\int_0^{\tau_N}\exp\frac{\delta}{2}(Y_{\eta'}-Y_\eta)^2(t)(Y_\eta-Y_{\eta'})(t)(2\lambda(t)+2\bar{Y}(t)+3\gamma|Z(t)|^2+3\gamma|Z_\eta(t)-Z(t)|^2)dt$$

$$\geq\frac{1}{4}E\int_0^{\tau_N}\exp\frac{\delta}{2}(Y_{\eta'}-Y_\eta)^2(t)(1+\delta(Y_{\eta'}-Y_\eta)^2(t))|Z_{\eta'}(t)-Z_\eta(t)|^2dt. \tag{12.3.27}$$

Now

$$1\!\!1_{t<\tau_N}(Z_{\eta'}-Z_\eta)\rightharpoonup 1\!\!1_{t<\tau_N}(Z-Z_\eta)\text{ in }L^2_{\mathcal{F}}(0,T;R^n)\text{ weakly}$$

$$1\!\!1_{t<\tau_N}[(1+\delta(Y_{\eta'}-Y_\eta)^2(t))]^{\frac{1}{2}}\exp\frac{\delta}{4}(Y_{\eta'}-Y_\eta)^2(t)\to$$

$$1\!\!1_{t<\tau_N}[(1+\delta(Y-Y_\eta)^2(t))]^{\frac{1}{2}}\exp\frac{\delta}{4}(Y-Y_\eta)^2(t)$$

pointwise and is bounded.

Therefore, we can assert that

$$1\!\!1_{t<\tau_N}[(1+\delta(Y_{\eta'}-Y_\eta)^2(t))]^{\frac{1}{2}}\exp\frac{\delta}{4}(Y_{\eta'}-Y_\eta)^2(t)\,(Z_{\eta'}-Z_\eta)(t)\rightharpoonup$$

$$1\!\!1_{t<\tau_N}[(1+\delta(Y-Y_\eta)^2(t))]^{\frac{1}{2}}\exp\frac{\delta}{4}(Y-Y_\eta)^2(t)\,(Z-Z_\eta)(t)\text{ in }L^2_{\mathcal{F}}(0,T;R^n)\text{.weakly}$$

From (12.3.27) it then follows, from weak lower semicontinuity, that

$$\frac{1}{\delta}E\exp\frac{\delta}{2}(Y-Y_\eta)^2(\tau_N)+E\int_0^{\tau_N}\exp\frac{\delta}{2}(Y-Y_\eta)^2(t)(Y_\eta-Y)(t)(2\lambda(t)+2\bar{Y}(t)$$
$$+3\gamma|Z(t)|^2+3\gamma|Z_\eta(t)-Z(t)|^2)dt$$
$$\geq\frac{1}{4}E\int_0^{\tau_N}\exp\frac{\delta}{2}(Y-Y_\eta)^2(t)(1+\delta(Y-Y_\eta)^2(t))|Z(t)-Z_\eta(t)|^2dt.$$

Therefore, we have also

$$\frac{1}{\delta}E\exp\frac{\delta}{2}(Y-Y_\eta)^2(\tau_N)+E\int_0^{\tau_N}\exp\frac{\delta}{2}(Y-Y_\eta)^2(t)(Y_\eta-Y)(t)(2\lambda(t)$$
$$+2\bar{Y}(t)+3\gamma|Z(t)|^2)dt$$
$$\geq E\int_0^{\tau_N}\exp\frac{\delta}{2}(Y-Y_\eta)^2(t)[\frac{1}{4}(1+\delta(Y-Y_\eta)^2(t))$$
$$-3\gamma(Y_\eta-Y)(t)]|Z_\eta(t)-Z(t)|^2dt$$

$$\geq (\frac{1}{4} - \frac{9\gamma^2}{\delta}) E \int_0^{\tau_N} |Z_\eta(t) - Z(t)|^2 dt.$$

We can take $\delta > 36\gamma^2$. We then let $\eta \to 0$. We obtain that $\mathbb{1}_{t<\tau_N} Z_\eta(t) \to \mathbb{1}_{t<\tau_N} Z(t)$ in $L^2_{\mathcal{F}}(0, T; R^n)$ strongly. Also, for a subsequence, $\sup_\eta \mathbb{1}_{t<\tau_N} |Z_\eta(t)| \in L^2_{\mathcal{F}}(0, T; R^n)$. Then

$$\sup_\eta |H_\eta(t, Y_\eta(t), Z_\eta(t))| \mathbb{1}_{t<\tau_N} \leq (\lambda(t) + kN + \frac{\gamma}{2} \sup_\eta |Z_\eta(t)|^2) \mathbb{1}_{t<\tau_N} \in L^1_{\mathcal{F}}(0, T; R^n). \tag{12.3.28}$$

We can consider a new subsequence such that $Y_\eta(t), \mathbb{1}_{t<\tau_N} Z_\eta(t) \to Y(t), \mathbb{1}_{t<\tau_N} Z(t)$ pointwise, a.e, a.s. (recall that this limit depends on ϵ, which is fixed). Thanks to the assumption (12.3.2), it is easy to check that

$$H_\eta(t, Y_\eta(t), Z_\eta(t)) \mathbb{1}_{t<\tau_N} \to H_{\epsilon 0}(t, Y(t), Z(t)) \mathbb{1}_{t<\tau_N}, \text{ pointwise,a.e., a.s.}$$

Thanks to the estimate (12.3.28), we can assert that

$$H_\eta(t, Y_\eta(t), Z_\eta(t)) \mathbb{1}_{t<\tau_N} \rightharpoonup H_{\epsilon 0}(t, Y(t), Z(t)) \mathbb{1}_{t<\tau_N} \text{weakly in } L^1_{\mathcal{F}}(0, T; R^n).$$

Let ξ_t be a random variable, $\mathcal{F}^t$ measurable and bounded. We can then claim that

$$E\xi_t \mathbb{1}_{t<\tau_N} \int_t^{\tau_N} H_\eta(s, Y_\eta(s), Z_\eta(s)) ds \to E\xi_t \mathbb{1}_{t<\tau_N} \int_t^{\tau_N} H_{\epsilon 0}(s, Y(s), Z(s)) ds. \tag{12.3.29}$$

Now, from the BSDE (12.3.20), we can write

$$\mathbb{1}_{t<\tau_N} Y_\eta(t) = \mathbb{1}_{t<\tau_N} Y_\eta(\tau_N) + \mathbb{1}_{t<\tau_N} \int_t^{\tau_N} H_\eta(s, Y_\eta(s), Z_\eta(s)) ds - $$
$$- \mathbb{1}_{t<\tau_N} \int_t^{\tau_N} Z_\eta(s).dw.$$

Since $E\xi_t \mathbb{1}_{t<\tau_N} Y_\eta(t) \downarrow E\xi_t \mathbb{1}_{t<\tau_N} Y(t)$, $E\xi_t \mathbb{1}_{t<\tau_N} Y_\eta(\tau_N) \downarrow E\xi_t \mathbb{1}_{t<\tau_N} Y(\tau_N)$, $E\xi_t \mathbb{1}_{t<\tau_N} \int_t^{\tau_N} Z_\eta(s).dw \to E\xi_t \mathbb{1}_{t<\tau_N} \int_t^{\tau_N} Z(s).dw$, we obtain from (12.3.29) that

$$\mathbb{1}_{t<\tau_N} Y(t) = \mathbb{1}_{t<\tau_N} Y(\tau_N) + \mathbb{1}_{t<\tau_N} \int_t^{\tau_N} H_{\epsilon 0}(s, Y(s), Z(s)) ds - $$
$$- \mathbb{1}_{t<\tau_N} \int_t^{\tau_N} Z(s).dw.$$

This equality holds a.s. for all integers N. Since a.s. $\tau_N = T$ for some N, we can write, recalling the dependence in ϵ and writing $Y(t) = Y_{\epsilon 0}(t)$, $Z(t) = Z_{\epsilon 0}(t)$,

$$Y_{\epsilon 0}(t) = \xi + \int_t^T H_{\epsilon 0}(s, Y(s), Z(s))ds - \int_t^T Z_{\varepsilon 0}(s).dw(s), \tag{12.3.30}$$

in which we have used the fact that $Y_\eta(T) = \xi$ and thus $Y(T) = \xi$. A similar argument is made letting $\epsilon \to 0$. We obtain a solution of (12.3.3) such that (12.3.7) holds. This completes the proof of Theorem 12.3∎

REMARK 12.1. The assumption (12.3.5) is sufficient but not necessary. Suppose, for instance, that we assume (12.3.1), and instead of (12.3.2),

$$\xi \geq 0, \; -\frac{\gamma}{2}|z|^2 \leq H(t, y, z) \leq \lambda(t), \; \lambda(t) \geq 0, \tag{12.3.31}$$

$$\xi \in L^2(\Omega, \mathcal{F}^T, P), \; \lambda(t) \in L^2_{\mathcal{F}}(0, T).$$

Then there exists a solution of (12.3.3) such that $Y(t) \in L^\infty_{\mathcal{F}}(0, T; L^2(\Omega, \mathcal{A}, P)$ and $Z(t) \in L^2_{\mathcal{F}}(0, T; R^n)$. Indeed, we first check that $Y(t) \geq 0$. This is proved by computing

$$d \exp -rY(t) = \exp -rY(t)[(\frac{1}{2}r^2|Z(t)|^2 + rH(t, Y(t), Z(t)))dt - rZ(t).dw(t)].$$

Since $\frac{1}{2}r^2|Z(t)|^2 + rH(t, Y(t), Z(t)) \geq (\frac{1}{2}r^2 - r\frac{\gamma}{2})|Z(t)|^2 \geq 0$ for $r \geq \gamma$, we obtain

$$d \exp -rY(t) \geq -r \exp -rY(t)Z(t).dw(t).$$

Hence $\exp -rY(t) \leq E[\exp -r\xi|\mathcal{F}^t] \leq 1$. This implies $Y(t) \geq 0$. We also have

$$Y(t) \leq E[\xi + \int_t^T \lambda(s)ds|\mathcal{F}^t],$$

which is much less stringent than (12.3.5). The rest of the proof of Theorem 12.3 is similar to that above for the more general case. ∎

12.3.5 Proof of Uniqueness

We now proceed with the proof of Theorem 12.4. Consider two solutions $Y_1(t), Z_1(t)$ and $Y_2(t), Z_2(t)$ such that

$$|Y_i(t)| \leq \bar{Y}(t), \; \mathbb{1}_{t<\tau_N} Z_i(t) \in L^2_{\mathcal{F}}(0, T; R^n), \forall N.$$

We define $Y_\theta(t) = Y_1(t) - \theta Y_2(t)$, $Z_\theta(t) = Z_1(t) - \theta Z_2(t)$. We next define

$$W_\theta(t) = \exp(\frac{\gamma}{1-\theta} Y_\theta^+(t) \exp kt).$$

If $Y_\theta(t) < 0$, we have $W_\theta(t) = 1$, If $Y_\theta(t) > 0$,

$$dW_\theta(t) = W_\theta(t)\frac{\gamma}{1-\theta}\exp kt\left\{[kY_\theta(t) + \frac{1}{2}\frac{\gamma}{1-\theta}\exp kt - \right.$$

$$(H(t, Y_1(t), Z_1(t)) - \theta H(t, Y_2(t), Z_2(t)))]dt + Z_\theta dw(t)\}.$$

As we have seen in the proof of Theorem 12.2, we can write, using convexity (12.3.9) and (12.3.8),

$$H(t, Y_1(t), Z_1(t)) - \theta H(t, Y_2(t), Z_2(t)) \leq k|Y_\theta(t)| +$$

$$+(1-\theta)[k|Y_2(t)| + H(t, Y_2(t), \frac{Z_\theta(t)}{1-\theta})$$

$$\leq k|Y_\theta(t)| + (1-\theta)(\lambda(t) + 2k\bar{Y}(t)) + \frac{\gamma}{2(1-\theta)}|Z_\theta(t)|^2.$$

Finally,

$$dW_\theta(t) \geq -W_\theta(t)\gamma \exp kt(\lambda(t) + 2k\bar{Y}(t))dt + W_\theta(t)\frac{\gamma}{1-\theta}\exp kt \; Z_\theta dw(t) \tag{12.3.32}$$

and

$$d[W_\theta(t) \exp \gamma \int_0^t (\lambda(s) + 2k\bar{Y}(s)) \exp ks \, ds] \geq$$

$$W_\theta(t) \exp \gamma \int_0^t (\lambda(s) + 2k\bar{Y}(s)) \exp ks \, ds \, \frac{\gamma}{1-\theta} \exp kt \; Z_\theta dw(t). \tag{12.3.33}$$

Let $\tau^\theta = \inf\{t | Y_\theta(t) \leq 0\}$. We can integrate the previous inequality between $t \wedge \tau_N \wedge \tau^\theta$ and $\tau_N \wedge \tau^\theta$, and we obtain

$$W_\theta(t \wedge \tau_N \wedge \tau^\theta)$$

$$= E[W_\theta(\tau_N \wedge \tau^\theta) \exp \gamma \int_{t\wedge\tau_N\wedge\tau^\theta}^{\tau_N\wedge\tau^\theta} (\lambda(s) + 2k\bar{Y}(s)) \exp ks \, ds | \mathcal{F}^{t\wedge\tau_N\wedge\tau^\theta}],$$

$$\mathbb{1}_{t<\tau_N\wedge\tau^\theta} W_\theta(t)$$

$$= \mathbb{1}_{t<\tau_N\wedge\tau^\theta} E[W_\theta(\tau_N \wedge \tau^\theta) \exp \gamma \int_t^{\tau_N\wedge\tau^\theta} (\lambda(s) + 2k\bar{Y}(s)) \exp ks \, ds | \mathcal{F}^t]$$

$$\leq \mathbb{1}_{t<\tau_N\wedge\tau^\theta} E[W_\theta(\tau_N) \exp \gamma \int_t^{\tau_N} (\lambda(s) + 2k\bar{Y}(s)) \exp ks \, ds | \mathcal{F}^t].$$

We note that $W_\theta(T) = \exp(\gamma\xi^+ \exp kT)$. Since $\tau_N \uparrow T$ a.s., we obtain, as $N \uparrow +\infty$,

$$\mathbb{1}_{t<\tau^\theta} W_\theta(t) \leq \mathbb{1}_{t<\tau^\theta} E[\exp\gamma(\xi^+ \exp kT + \int_t^T (\lambda(s) + 2k\bar{Y}(s)) \exp ks\, ds)|\mathcal{F}^t],$$

which is a.s. finite, from the assumption (12.3.10), $\forall t$. Therefore, also

$$\mathbb{1}_{t<\tau^\theta} W_\theta(t) \leq E[\exp\gamma(\xi^+ \exp kT + \int_t^T (\lambda(s) + 2k\bar{Y}(s)) \exp ks\, ds)|\mathcal{F}^t]. \tag{12.3.34}$$

Let $\tau^1 = \inf\{t|Y_1(t) - Y_2(t) \leq 0\}$. If $\tau^1 > 0$, then for $t < \tau^1$, we have $\mathbb{1}_{t<\tau^\theta} W_\theta(t) \to +\infty$ as $\theta \to 1$, which is impossible, from (12.3.34). Therefore, $\tau^1 = 0$, which means that $Y_1(0)-Y_2(0) \leq 0$. Initial time 0 can now be replaced with any time $t_0 < T$. Therefore, $Y_1(t) - Y_2(t) \leq 0, \forall t$, a.s. Since the roles of $Y_1(t)$ and $Y_2(t)$ are symmetric, we infer that $Y_1(t) = Y_2(t)$. This implies also $Z_1(t) = Z_2(t)$. The proof is complete. ■

Chapter 13
Stochastic Control Problems in Finance

13.1 General Description

13.1.1 Financial Markets

We consider a financial market with n risky assets. The price of asset i at time t is denoted by $Y_i(t)$. It is a random process, defined on a probability space $\Omega, \mathcal{A}, P$. We consider a filtration $\mathcal{F}^t$ and a standard $P, \mathcal{F}^t$ Wiener process with values in R^n. The components of the Wiener process are denoted by $w_i(t)$.

The Wiener process represents the only source of uncertainties in the financial market. In fact, each component is an uncertainty, and the number of uncertainties is equal to the number of risky assets. These uncertainties are external and influence the evolution of the price of risky assets. The fact that the number of risky assets and the number of uncertainties are the same is a part of the assumption of complete markets, which will be sufficient in our context; see below for further discussion.

We assume that the evolution of $Y_i(t)$ is governed by the following equation:

$$dY_i = Y_i(\alpha_i(t)dt + \sum_{j=1}^{n} \sigma_{ij}(t)dw_j), \tag{13.1.1}$$

$$Y_i(0) = Y_i^0.$$

We assume, in order to simplify, that $\alpha_i(t)$ and $\sigma_{ij}(t)$ are deterministic functions. Then the process $Y_i(t)$ is distributed according to a log normal distribution. The matrix $\sigma(t)$ is invertible. We assume

$$\alpha(t), \sigma(t), \sigma^{-1}(t) \text{ bounded.} \tag{13.1.2}$$

A. Bensoussan, *Estimation and Control of Dynamical Systems*, Interdisciplinary Applied Mathematics 48, https://doi.org/10.1007/978-3-319-75456-7_13

The invertibility of the matrix $\sigma(t)$ is the assumption of **complete markets** in the finance literature. This corresponds to the idea that the prices of risky assets provide the economic agent with sufficient information to mitigate the risks due to the uncertainties. We shall not develop these considerations, since we are more interested in the stochastic control problems arising in this context. Nevertheless, a very important concept is that of indicator of the market. We need first to state that besides the risky assets, there is one safe asset (cash). Its value evolves in a deterministic manner. One dollar at time 0 becomes $\exp rt$ at time t. The number r is the return per unit of time.

We introduce the vector (the vector Sharpe ratio in finance)

$$\theta(t) = \sigma^{-1}(t)(\alpha(t) - r\mathbb{1}) \tag{13.1.3}$$

and the stochastic process $Z(t)$ defined by the relation

$$\begin{aligned} dZ &= -Z\theta(t).dw(t), \\ Z(0) &= 1. \end{aligned} \tag{13.1.4}$$

Clearly, $Z(t)$ is a $P, \mathcal{F}^t$ martingale, satisfying

$$Z(t) \geq 0, \;\; EZ(t) = 1.$$

This process is called a **martingale measure**, and it constitutes the *market indicator*. A major attribute is the following fact:

$$Z(t)Y_i(t) \exp -rt \text{ is a } P, \mathcal{F}^t \text{ martingale.} \tag{13.1.5}$$

This property has major importance in finance and also for the stochastic control to be discussed below.

EXERCISE 13.1. Prove property (13.1.5).

13.1.2 Optimal Consumption and Investment Problem

One of the basic problems in mathematical finance is that of allocating one's wealth between consumption and investment in the financial market. This is the consumer–investor problem. The agent has utility functions to take his decisions.

Let us denote by $X(t)$ the wealth at time t. In terms of system theory, it represents the state of the system. Formally, there are also the $Y_i(t)$, but in view of our simple model, we will not need to introduce them explicitly, as we shall see below. The notation $X(t)$ is traditional in finance, so we shall keep it, although it is not customary in control theory.

Following the invariant embedding principle, we shall consider initial conditions x, t and denote by $X_{x,t}(s)$ the evolution of the wealth, $s \geq t$, with $X_{x,t}(t) = x$. We now describe the model.

In our model, except for the initial endowment x, there is no extra source of income (such as wages). The income comes from gains on investments in the market. At time s, we can write

$$X(s) = \pi_0(s) \exp r(s-t) + \sum_{i=1}^{n} \pi_i(s) Y_i(s), \; s > t,$$

where $\pi_0(s)$ represents the amount of cash, and $\pi_i(s)$ the amount invested in the risky asset i, at time s. The processes $\pi_0(s)$ and $\pi(s)$ (vector of components $\pi_i(s)$) are control variables. The vector $\pi(s)$ is called the portfolio of investments. The wealth is self-financing. This is expressed mathematically by saying that $X(s)$ has an Itô differential given by

$$dX = r\pi_0(s) \exp r(s-t)ds + \sum_{i=1}^{n} \pi_i(s) dY_i(s) - C(s)ds.$$

We have introduced a new control variable, the consumption rate $C(s)$. The interpretation of the preceding relation is clear when one replaces ds by a small interval of time Δs. Just before $s + \Delta s$, the amounts of cash and assets are still unchanged with respect to time s; they are $\pi_0(s)$ and $\pi(s)$. However, the value of these assets has changed, because the value of cash has become $\exp r(s-t)(1+r\Delta s)$ and the value of asset i has become $Y_i(s) + \Delta Y_i(s)$. During this interval of time, an amount of wealth has been devoted to consumption.

After easy manipulations we can eliminate $\pi_0(s)$ to obtain

$$\begin{aligned} dX &= rXds + \sum_{i=1}^{n} \pi_i Y_i ((\alpha_i - r)ds + \sum_{j=1}^{n} \sigma_{ij} dw_j - Cds \\ &= rXds + \sum_{i=1}^{n} \pi_i Y_i \sum_{j=1}^{n} \sigma_{ij}(\theta_j ds + dw_j) - Cds. \end{aligned}$$

At this stage, it is customary to perform a change of control variables. We define

$$\varpi_i(s) = \frac{\pi_i(s) Y_i(s)}{X(s)}. \tag{13.1.6}$$

This, of course, is legitimate only when $X(s)$ does not vanish. With this new control, the wealth evolution becomes

$$\begin{aligned} dX &= rXds + X\sigma^* \varpi.(\theta ds + dw) - Cds, \qquad (13.1.7) \\ X(t) &= x. \end{aligned}$$

In this equation the control variables are the processes $\varpi(s)$ and $C(s)$. They are processes adapted to the filtration $\mathcal{F}_t^s = \sigma(w(\tau) - w(t), t \leq \tau \leq s)$. We can then write explicitly the solution $X(s)$ of equation (13.1.7).

EXERCISE 13.2. Write explicitly the solution of (13.1.7). Begin with the case $C(s) = 0$.

We consider processes $\varpi(.)$ and $C(.)$ such that

$$E\int_0^T X(s)^2|\varpi(s)|^2 ds,\ E\int_0^T C(s)^2 ds < +\infty. \tag{13.1.8}$$

This constitutes the class of admissible controls. There are also the implicit constraints $C(s) > 0$, $X(s) > 0$. We shall discard them, and obtain that thanks to the assumptions and the setup of the model, they will be automatically satisfied by the optimal controls.

We see from the expression (13.1.8) that the processes $Y_i(s)$ do not appear explicitly in the evolution of the wealth. This is, of course, due to our simplified model and the change of control variables. We write $X_{xt}(s; \varpi(.), C(.))$ for the wealth at time s, mentioning explicitly the dependence with respect to the initial conditions and to the control variables.

We now define the payoff function. The agent has utility functions $U_1(c)$ for his consumption and $U_2(x)$ for his final wealth (think of T as the retirement age). We assume that

$$\begin{gathered} U_1(c), U_2(x) : R^+ \rightarrow R^+ \\ \text{concave } C^2, \\ U_i'(0) = +\infty \qquad U_i'(+\infty) = 0, i = 1, 2. \end{gathered} \tag{13.1.9}$$

Examples are

$$U(x) = \frac{x^{1-\gamma}}{1-\gamma},\ 0 < \gamma < 1,$$

and $U(x) = \log x$. We can now define the payoff functional by

$$J_{x,t}(\varpi(.), C(.)) = E[\int_t^T U_1(C(s))\exp(-r(s-t))ds + U_2(X(T))\exp(-r(T-t))], \tag{13.1.10}$$

and the value function is defined by

$$\Phi(x,t) = \sup_{\varpi(.),C(.)} J_{x,t}(\varpi(.), C(.)). \tag{13.1.11}$$

Clearly,

$$\Phi(x,T) = U_2(x). \tag{13.1.12}$$

REMARK 13.1. We discount the consumptions and the final wealth with the same discount as the cash return. This is purely a matter of convenience.

13.2 Dynamic Equation Approach

13.2.1 HJB Equation

We use the standard approach of dynamic programming. However, we note that the problem involves controls on the diffusion term for which we have not developed any theory. So we cannot use results of previous chapters, and we shall proceed formally.

The HJB equation is given formally as follows:

$$\frac{\partial \Phi}{\partial t}+rx\frac{\partial \Phi}{\partial x}+\sup_c(U_1(c)-c\frac{\partial \Phi}{\partial x})+\sup_{\varpi}(x\sigma^*\varpi.\theta\frac{\partial \Phi}{\partial x}\ +\frac{1}{2}x^2|\sigma^*\varpi|^2\frac{\partial^2 \Phi}{\partial x^2})-r\Phi=0,$$

$$\Phi(x,T)=U_2(x). \tag{13.2.1}$$

We can perform formally the optimization in the operators of the left-hand side. We introduce the dual problem

$$\sup_c(U_1(c)-c\lambda),$$

where λ is a positive parameter. The solution of this optimization is given by the equation

$$U_1'(\hat{c})=\lambda. \tag{13.2.2}$$

According to the assumption, $U_1'(c)$ is a continuous decreasing function from $+\infty$ to 0. Therefore, for $\lambda > 0$, there exists a unique solution of the equation (13.2.2). Since the same is true for $U_2(x)$, we shall call $\beta_i(x)$ the unique solution of

$$U_i'(\beta_i(x))=x, \tag{13.2.3}$$

where $\beta_i(x)$ is the inverse function of $U_i'(x)$.

Therefore, we get the first feedback

$$\hat{c}(x,t)=\beta_1(\frac{\partial \Phi}{\partial x}(x,t)). \tag{13.2.4}$$

Similarly, we obtain easily

$$\hat{\varpi}(x,t) = -\frac{\frac{\partial \Phi}{\partial x}(x,t)}{x\frac{\partial^2 \Phi}{\partial x^2}(x,t)}(\sigma^*)^{-1}\theta(t), \tag{13.2.5}$$

provided, of course, that this expression has meaning. With these feedbacks, the HJB equation (13.2.1) reads

$$\frac{\partial \Phi}{\partial t} + rx\frac{\partial \Phi}{\partial x} + U_1(\beta_1(\frac{\partial \Phi}{\partial x})) - \beta_1(\frac{\partial \Phi}{\partial x})\frac{\partial \Phi}{\partial x} - \frac{1}{2}|\theta|^2\frac{(\frac{\partial \Phi}{\partial x})^2}{\frac{\partial^2 \Phi}{\partial x^2}} - r\Phi = 0,$$

$$\Phi(x,T) = U_2(x). \tag{13.2.6}$$

This expression is formal. In particular, the functional space in which the solution is looked for is not made precise.

13.2.2 Solution of the HJB Equation

Although the HJB equation (13.2.6) looks quite complex, it turns out that it has a simple explicit solution. To figure it out, we consider the derivative

$$\lambda(x,t) = \frac{\partial \Phi}{\partial x}(x,t). \tag{13.2.7}$$

We note that the HJB equation can be written as

$$\frac{\partial \Phi}{\partial t} + rx\lambda + U_1(\beta_1(\lambda)) - \lambda\beta_1(\lambda) - \frac{1}{2}|\theta|^2\frac{\lambda^2}{\frac{\partial \lambda}{\partial x}} - r\Phi = 0.$$

We differentiate this equation in x, to obtain

$$\frac{\partial \lambda}{\partial t} + \frac{\partial \lambda}{\partial x}(rx - \beta_1(\lambda)) + \frac{1}{2}|\theta|^2\lambda^2\frac{\frac{\partial^2 \lambda}{\partial x^2}}{(\frac{\partial \lambda}{\partial x})^2} - |\theta|^2\lambda = 0, \tag{13.2.8}$$

$$\lambda(x,T) = U_2'(x).$$

This equation looks simpler than the HJB equation, although it is still rather complex. It turns out that it can be solved simply by considering the inverse

$$x = G(\lambda,t). \tag{13.2.9}$$

The intuition will come from the probabilistic interpretation; see below. We can write the derivatives as

$$\frac{\partial \lambda}{\partial t} = -\frac{\frac{\partial G}{\partial t}}{\frac{\partial G}{\partial \lambda}},$$
$$\frac{\partial \lambda}{\partial x} = \frac{1}{\frac{\partial G}{\partial \lambda}},$$
$$\frac{\partial^2 \lambda}{\partial x^2} = -\frac{\frac{\partial^2 G}{\partial \lambda^2}}{(\frac{\partial G}{\partial \lambda})^3},$$

and using them in (13.2.8), we obtain a linear equation

$$-\frac{\partial G}{\partial t} - |\theta|^2 \lambda(\frac{\partial G}{\partial \lambda} + \frac{1}{2}\lambda \frac{\partial^2 G}{\partial \lambda^2}) + rG = \beta_1(\lambda), \tag{13.2.10}$$
$$G(\lambda, T) = \beta_2(\lambda).$$

To show that this equation has a solution, necessarily unique, we need estimates. Since $\beta_1(\lambda)$ and $\beta_2(\lambda)$ are unbounded, because the value at 0 is $+\infty$, we make the following additional assumption:

$$0 < \delta \leq \lambda^m \beta_i(\lambda) \leq C, \ m > 0. \tag{13.2.11}$$

This assumption is satisfied in the examples above.

We consider then

$$G_m(\lambda, t) = \lambda^m G(\lambda, t), \tag{13.2.12}$$

which is the solution of

$$-\frac{\partial G_m}{\partial t} - |\theta|^2 \lambda((1-m)\frac{\partial G_m}{\partial \lambda} + \frac{1}{2}\lambda \frac{\partial^2 G_m}{\partial \lambda^2}) + (r + \frac{|\theta|^2}{2} m(1-m))G_m$$
$$= \lambda^m \beta_1(\lambda), G_m(\lambda, T) = \lambda^m \beta_2(\lambda). \tag{13.2.13}$$

From the assumption (13.2.11) and the classical theory of linear parabolic PDEs, there exists a unique bounded solution of (13.2.13). There exists consequently a unique solution of (13.2.10) with the growth condition

$$\frac{\delta_0}{\lambda^m} \leq G(\lambda, t) \leq \frac{C_0}{\lambda^m}. \tag{13.2.14}$$

Since $\beta_1(\lambda)$ and $\beta_2(\lambda)$ are decreasing, the function $G(\lambda, t)$ is also decreasing in λ. Now $G(0, t)$ is the solution of

$$-\frac{\partial G(0,t)}{\partial t} + rG(0,t) = \beta_1(0),$$
$$G(0,T) = \beta_2(0).$$

Since $\beta_i(0) = +\infty$, we get $G(0,t) = +\infty$. From (13.2.14) we also get $G(\infty,t) = 0$. Since $G(\lambda,t)$ is continuous in λ, we can take the inverse $\lambda(x,t)$, the solution of (13.2.9), which has also a unique solution, which is decreasing and satisfies the growth condition (an immediate consequence of (13.2.4))

$$\frac{\delta_1}{x^{\frac{1}{m}}} \leq \lambda(x,t) \leq \frac{C}{x^{\frac{1}{m}}}. \tag{13.2.15}$$

So we have defined in a unique way the derivative $\frac{\partial \Phi}{\partial x}$, with the growth condition

$$\frac{\delta_1}{x^{\frac{1}{m}}} \leq \frac{\partial \Phi}{\partial x} \leq \frac{C}{x^{\frac{1}{m}}}. \tag{13.2.16}$$

It remains to find Φ. In fact, we shall check that

$$\Phi(x,t) = \Psi(\lambda(x,t),t), \tag{13.2.17}$$

where $\Psi(\lambda,t)$ is the solution of a linear PDE, which is

$$-\frac{\partial \Psi}{\partial t} - \frac{1}{2}|\theta|^2\lambda^2\frac{\partial^2 \Psi}{\partial \lambda^2} + r\Psi = U_1(\beta_1(\lambda)), \tag{13.2.18}$$
$$\Psi(\lambda,T) = U_2(\beta_2(\lambda)).$$

To solve equation (13.2.18), we assume

$$\lambda^p|U_i(\beta_i(\lambda))| \leq C_0 + C_1\lambda^q, \; p,q \geq 0. \tag{13.2.19}$$

EXERCISE 13.3. Show that this assumption is satisfied for the examples.

To solve (13.2.18), we consider

$$\Psi_p(\lambda,t) = \lambda^p\Psi(\lambda,t),$$

which is a solution of

$$-\frac{\partial \Psi_p}{\partial t} - \frac{1}{2}|\theta|^2\lambda^2\frac{\partial^2 \Psi_p}{\partial \lambda^2} + |\theta|^2\lambda p\frac{\partial \Psi_p}{\partial \lambda} + (r - \frac{1}{2}|\theta|^2 p(p+1))\Psi_p$$
$$= \lambda^p U_1(\beta_1(\lambda)), \; \Psi_p(\lambda,T) = \lambda^p U_2(\beta_2(\lambda)). \tag{13.2.20}$$

We then look for a solution of (13.2.20) such that

$$|\Psi_p(\lambda,t)| \leq c_0 + c_1\,\lambda^q, \tag{13.2.21}$$

and there is a unique solution of (13.2.20) with the growth condition (13.2.21). Therefore, there is a unique solution of (11.2.18) with the growth condition

$$\lambda^p|\Psi(\lambda,t)| \leq c_0 + c_1\,\lambda^q. \tag{13.2.22}$$

To prove (13.2.17), we consider the solution $\Psi(\lambda,t)$ of (13.2.18), (13.2.22), the function $G(\lambda,t)$ defined by (13.2.10), its inverse $\lambda(x,t)$, and we set

$$\tilde{\Phi}(x,t) = \Psi(\lambda(x,t),t).$$

We are going to prove that $\tilde{\Phi}(x,t)$ is the solution of the HJB equation and also that

$$\frac{\partial\tilde{\Phi}}{\partial x} = \lambda(x,t) = \frac{\partial\Phi}{\partial x}, \tag{13.2.23}$$
$$\tilde{\Phi}(x,T) = \Phi(x,T).$$

Then from the HJB equation, we obtain immediately

$$\frac{\partial(\tilde{\Phi}-\Phi)}{\partial t} - r(\tilde{\Phi}-\Phi) = 0,$$
$$(\tilde{\Phi}-\Phi)(x,T) = 0,$$

which implies $\tilde{\Phi}-\Phi = 0$.

To save notation, we write directly Φ instead of $\tilde{\Phi}$.

We first prove the following equality:

$$\frac{\partial\Psi}{\partial\lambda} = \lambda\frac{\partial G}{\partial\lambda}. \tag{13.2.24}$$

EXERCISE 13.4. Prove (13.2.24) by showing that the functions $\dfrac{\partial\Psi}{\partial\lambda}$ and $\lambda\dfrac{\partial G}{\partial\lambda}$ satisfy the same equation and the same initial condition at time T.

It follows from (13.2.24) that

$$\frac{\partial\Phi}{\partial x} = \frac{\partial\Psi}{\partial\lambda}\frac{\partial\lambda}{\partial x} =$$

$$= \frac{\dfrac{\partial \Psi}{\partial \lambda}}{\dfrac{\partial G}{\partial \lambda}} = \lambda,$$

so the first property (13.2.23) holds.

Now compute

$$\frac{\partial \Phi}{\partial t} = \frac{\partial \Psi}{\partial \lambda}\frac{\partial \lambda}{\partial t} + \frac{\partial \Psi}{\partial t}.$$

Using the equations of Ψ and λ, see (13.2.18) and (13.2.8), we obtain

$$\frac{\partial \Phi}{\partial t} = \frac{\partial \Phi}{\partial x}(\beta_1(\lambda) + |\theta|^2 \frac{\lambda}{\dfrac{\partial \lambda}{\partial x}} - rx)$$

$$- \frac{1}{2}|\theta|^2 \lambda^2 \frac{\dfrac{\partial^2 \Phi}{\partial x^2}}{(\dfrac{\partial \lambda}{\partial x})^2} + r\Phi - U_1(\beta_1(\lambda)),$$

$$\Phi(x, T) = U_2(x).$$

Using in this relation the fact that $\dfrac{\partial \Phi}{\partial x} = \lambda$, we obtain easily that Φ satisfies the HJB equation.

We have proved the following result.

THEOREM 13.1. *We assume (13.1.2), (13.1.9), (13.2.11),(13.2.19). The HJB equation (13.2.6) has a unique solution, which is C^2 in x, with the growth condition (13.2.16) and* $\dfrac{\partial^2 \Phi}{\partial x^2} < 0$.

It is given explicitly by the formula (13.2.17).

13.3 Solution of the Consumer–Investor Problem

We consider the feedbacks (13.2.4), (13.2.5). The corresponding state, denoted by $\hat{X}(s)$, the solution of (13.1.7) with initial conditions x at time 0, appears as the solution of

$$d\hat{X} = (r\hat{X} - |\theta|^2 \frac{\dfrac{\partial \Phi}{\partial x}}{\dfrac{\partial^2 \Phi}{\partial x^2}}(\hat{X}) - \beta_1(\frac{\partial \Phi}{\partial x}(\hat{X})))ds - \frac{\dfrac{\partial \Phi}{\partial x}}{\dfrac{\partial^2 \Phi}{\partial x^2}}(\hat{X})\theta.dw, \quad (13.3.1)$$

$$\hat{X}(0) = x.$$

A direct study of the SDE (13.3.1) appears to be out of reach. Fortunately, it has an explicit solution. Indeed, we introduce the martingale

$$\begin{aligned} d\zeta &= -\zeta\theta.dw, \\ \zeta(0) &= \lambda(x,0). \end{aligned} \tag{13.3.2}$$

Then we claim that

$$\hat{X}(s) = G(\zeta(s), s). \tag{13.3.3}$$

We define $\hat{X}(s)$ by (13.3.3). At time 0, we indeed have

$$x = G(\lambda(x,0),0).$$

Let us check that the Itô differential satisfies (13.3.1). In fact, we have

$$\begin{aligned} \zeta(s) &= \lambda(\hat{X}(s), s) \\ &= \frac{\partial\Phi}{\partial x}(\hat{X}(s), s). \end{aligned}$$

Also,

$$\begin{aligned} \frac{\partial G}{\partial\lambda}(\zeta(s), s) &= \frac{\partial G}{\partial\lambda}(\lambda(\hat{X}(s), s), s) \\ &= \frac{1}{\frac{\partial\lambda}{\partial x}(\hat{X}(s), s)} = \\ &= \frac{1}{\frac{\partial^2\Phi}{\partial x^2}(\hat{X}(s), s)}. \end{aligned}$$

Now from (13.3.3), using Itô's formula, we get

$$d\hat{X} = (\frac{\partial G}{\partial s}(\zeta(s), s) + \frac{1}{2}|\theta|^2\zeta^2(s)\frac{\partial^2 G}{\partial\lambda^2}(\zeta(s), s))ds - \frac{\partial G}{\partial\lambda}(\zeta(s), s)\zeta(s)\theta.dw.$$

Using the PDE (13.2.10) and the formulas above, we obtain (13.3.1). The corresponding controls obtained through the feedbacks become

$$\begin{aligned} \hat{C}(s) &= \beta_1(\zeta(s)), \\ \hat{\varpi}(s)\hat{X}(s) &= -\frac{\partial G}{\partial\lambda}(\zeta(s), s)\zeta(s)(\sigma^*)^{-1}\theta(s) \\ &= -\frac{\partial\Psi}{\partial\lambda}(\zeta(s), s)(\sigma^*)^{-1}\theta(s). \end{aligned} \tag{13.3.4}$$

We have to guarantee that these controls are admissible. The second condition (13.1.8) reads $E\int_0^T(\beta_1(\zeta(s)))^2ds < +\infty$. From the assumption (13.2.11), it is sufficient to check

$$E\int_0^T \frac{ds}{(\zeta(s))^{2m}} < +\infty. \tag{13.3.5}$$

However, $\eta(s) = \frac{1}{\zeta(s)}$ satisfies the stochastic differential equation

$$\begin{aligned} d\eta &= \eta|\theta|^2 ds + \eta\theta dw(s), \\ \eta(0) &= \frac{1}{\lambda(x,0)}, \end{aligned}$$

and (13.3.5) follows easily. To check the first condition (13.1.8), we need to show that

$$E\int_0^T (\frac{\partial\Psi}{\partial\lambda}(\zeta(s)))^2 ds < +\infty, \tag{13.3.6}$$

for which we need an estimate on $\chi(\lambda) = \frac{\partial\Psi(\lambda)}{\partial\lambda}$. We check easily that $\chi(\lambda)$ is the solution of the PDE

$$\begin{aligned} -\frac{\partial\chi}{\partial t} - \frac{1}{2}|\theta|^2\lambda^2\frac{\partial^2\chi}{\partial\lambda^2} - |\theta|^2\lambda\frac{\partial\chi}{\partial\lambda} + r\chi &= \lambda\beta_1'(\lambda), \qquad (13.3.7)\\ \chi(\lambda, T) &= \lambda\beta_2'(\lambda). \end{aligned}$$

So we are in a situation similar to that of $G(\lambda, t)$, (13.2.10). So we assume that

$$\lambda^{m+1}|\beta_i'(\lambda)| \leq C, \tag{13.3.8}$$

which implies the growth property $\lambda^m|\chi(\lambda)| \leq C$. The property (13.3.6) follows easily.

We next consider

$$\Phi(\hat{X}(s), s)\exp -r(s-t) = \Psi(\zeta(s), s)\exp -r(s-t).$$

Therefore,

$$\begin{aligned} d(\Phi(\hat{X}(s), s)\exp -r(s-t)) &= d(\Psi(\zeta(s), s)\exp -r(s-t)) \\ = -U_1(\beta_1(\zeta(s)))\exp -r(s-t)ds &- \frac{\partial\Psi}{\partial\lambda}(\zeta(s), s)\zeta(s)\theta(s).dw. \end{aligned}$$

Hence

$$\Phi(x,t) = E[\int_0^T U_1(\hat{C}(s)) \exp -r(s-t)ds + U_2(\hat{X}(T)) \exp -r(T-t)]$$
$$= J_{x,t}(\hat{\varpi}(.), \hat{C}(.)).$$

We can also prove that $\Phi(x,t) \geq J_{x,t}(\varpi(.), C(.))$, which implies that $\hat{\varpi}(.), \hat{C}(.)$ are optimal. So we can state our next theorem.

THEOREM 13.2. *We make the assumptions of Theorem 13.1 and (13.3.8). The processes $\hat{\varpi}(.), \hat{C}(.)$ obtained from the feedbacks $\hat{\varpi}(x,t)$, $\hat{c}(x,t)$ defined by equations (13.2.4), (13.2.5) provide a solution of the consumption–investment problem.*

13.4 Entrepreneur Decision-Making

13.4.1 The Model

We consider here a model introduced in [6]. It concerns an entrepreneur who makes decisions on projects and on the level of her efforts. The projects are scalable, with increasing expected return, but also with increasing risks. The evolution of the outcome depends also on the personal efforts of the entrepreneur. These efforts have a positive deterministic impact on the wealth. However, they are integrated in the objective functional negatively. The objective functional on a horizon T represents the final wealth reduced by the reimbursement of a loan. We will consider afterward the problem of the bank that grants this loan. The entrepreneur begins her activity with initial conditions x, t. She decides two control processes, denoted by $u(s)$ and $v(s)$, $s > t$. We suppose that we have an underlying probability space $(\Omega, \mathcal{A}, P)$ on which is defined a scalar standard Wiener process $w(t)$. We denote by $\mathcal{F}^t$ the filtration generated by $w(t)$. The processes $u(s), v(s)$ are adapted to $\mathcal{F}^s$. The process $u(s)$ represents the level of the entrepreneur's effort. It takes positive values. The process $v(s)$ represents the scale of the projects. With these two processes, one can define the wealth process $X(s)$ as follows:

$$dX = \delta u(s)ds + v(s)X(s)(\alpha ds + dw(s)), \tag{13.4.1}$$
$$X(0) = x,$$

and $\delta, \alpha > 0$. We will impose the following restrictions on the controls:

$$u(s) > 0,\ E\int_0^T u(s)^2 ds < +\infty,\ v(s)X(s) > 0,\ X(T) > 0. \tag{13.4.2}$$

We can see that the effort $u(s)$ has a direct deterministic positive impact on the wealth evolution. An increase of the scale of projects has a positive impact on the expected return, but it may increase the risk. We note that in contrast to the consumer–investor problem, we do not require square integrability of the process $v(s)X(s)$. That will facilitate the definition of optimal controls.

The objective function is defined as follows:

$$J(u(.), v(.)) = E[F(X(T)) - \frac{1}{2}\int_0^T u(s)^2 ds], \tag{13.4.3}$$

with

$$F(x) = U(w + (x - K)^+) \tag{13.4.4}$$

and the assumptions

$$\begin{aligned} U : R^+ &\to R, \text{ monotone increasing and concave,} && (13.4.5)\\ U(0) = -\infty \quad & U(+\infty) = +\infty,\ U'(0) = +\infty,\ U'(+\infty) = 0, \\ U(x) - xU'(x) \quad & \to +\infty, \text{ as } x \to +\infty. \end{aligned}$$

The number $w \geq 0$. This number represents a minimum survival income. Indeed, if the wealth is x at time T, the remaining income is $(x - K)^+$, since there is a reimbursement of value K. Because of the term w, the function $F(x)$ is not globally concave, unless $w = 0$. The functional $J(u(.), v(.))$ is well defined, but may take the value $+\infty$.

Following the invariant embedding principle, we introduce a family of problems, indexed by initial conditions $X(t) = x$.

We denote by $X_{xt}(s)$ the solution, referring to the initial conditions. The payoff becomes

$$J_{xt}(u(.), v(.)) = E[F(X_{xt}(T)) - \frac{1}{2}\int_t^T u(s)^2 ds]. \tag{13.4.6}$$

As usual, we are looking for the value function

$$\Phi(x, t) = \sup_{u(.),v(.)} J_{xt}(u(.), v(.)), \tag{13.4.7}$$

in which the controls must be admissible, i.e., satisfying (13.4.2).

13.4.2 Dynamic Programming

We can write easily the Bellman equation for $t < T$, assuming sufficient smoothness of the value function. We get

$$\frac{\partial\Phi}{\partial t} + \sup_{u>0}(-\frac{1}{2}u^2 + \delta u\frac{\partial\Phi}{\partial x}) + \sup_{\{v|vx>0\}}(\alpha vx\frac{\partial\Phi}{\partial x} + \frac{1}{2}x^2v^2\frac{\partial^2\Phi}{\partial x^2}) = 0.$$

If the constraints are not binding, we get also

$$\frac{\partial\Phi}{\partial t} + \frac{\delta^2}{2}(\frac{\partial\Phi}{\partial x})^2 - \frac{1}{2}\alpha^2\frac{(\frac{\partial\Phi}{\partial x})^2}{\frac{\partial^2\Phi}{\partial x^2}} = 0, \tag{13.4.8}$$

with the feedbacks

$$u(x,t) = \delta\frac{\partial\Phi}{\partial x}, \; xv(x,t) = -\alpha\frac{\frac{\partial\Phi}{\partial x}}{\frac{\partial^2\Phi}{\partial x^2}}, \tag{13.4.9}$$

and we see that the constraints are satisfied if

$$\frac{\partial\Phi}{\partial x} > 0, \; \frac{\partial^2\Phi}{\partial x^2} < 0. \tag{13.4.10}$$

We of course cannot request these properties to be true up to time T. Indeed, from the definition of the value function, we should set $\Phi(x,T) = F(x)$, and we know that this is not a globally concave function. The discussion of what happens as $t \to T$ is one of the the major difficulties of this problem.

Nevertheless, if the smoothness and (13.4.10) are true for $t < T$, we may introduce $\lambda(x,t) = \frac{\partial\Phi}{\partial x}(x,t)$, which is positive and decreasing. Note that the Bellman equation (13.4.8) can be written in the form

$$\frac{\partial\Phi}{\partial t} + \frac{\delta^2}{2}\lambda^2 - \frac{1}{2}\alpha^2\frac{\lambda^2}{\frac{\partial\lambda}{\partial x}} = 0, \tag{13.4.11}$$

and if we postulate that λ is C^2 in x, then we can differentiate in x to get a PDE for $\lambda(x,t)$, namely

$$\frac{\partial\lambda}{\partial t} + \delta^2\lambda\frac{\partial\lambda}{\partial x} - \alpha^2\lambda + \frac{1}{2}\alpha^2\lambda^2\frac{\frac{\partial^2\lambda}{\partial x^2}}{(\frac{\partial\lambda}{\partial x})^2} = 0. \tag{13.4.12}$$

Since $\lambda(x,t)$ is monotonically decreasing in x, we can look at its inverse $G(\lambda,t)$, so that $\lambda(x,t)$ is the solution of

$$x = G(\lambda,t). \tag{13.4.13}$$

Using

$$1 = \frac{\partial G}{\partial \lambda}\frac{\partial \lambda}{\partial x},$$

$$0 = \frac{\partial^2 G}{\partial \lambda^2}(\frac{\partial \lambda}{\partial x})^2 + \frac{\partial G}{\partial \lambda}\frac{\partial^2 \lambda}{\partial x^2},$$

we obtain easily a PDE for G, namely

$$\frac{\partial G}{\partial t} + \alpha^2(\lambda \frac{\partial G}{\partial \lambda} + \frac{1}{2}\lambda^2 \frac{\partial^2 G}{\partial \lambda^2}) = \delta^2 \lambda. \tag{13.4.14}$$

We can then express $\frac{\partial \Phi}{\partial t}$ and $\frac{\partial \lambda}{\partial t}$ in terms of λ, using (13.4.11) and (13.4.12). We can write

$$\frac{\partial \Phi}{\partial t} = -\frac{\delta^2}{2}\lambda^2 + \frac{1}{2}\alpha^2 \lambda^2 \frac{\partial G}{\partial \lambda}, \tag{13.4.15}$$

$$\frac{\partial \lambda}{\partial t} = -\frac{\delta^2 \lambda}{\frac{\partial G}{\partial \lambda}} + \alpha^2 \lambda + \frac{1}{2}\alpha^2 \lambda^2 \frac{\frac{\partial^2 G}{\partial \lambda^2}}{\frac{\partial G}{\partial \lambda}}. \tag{13.4.16}$$

Of course, on the right-hand side, λ must be replaced by $\lambda(x, t)$. We can next express Φ itself as a function of λ, namely $\Phi(x, t) = \Psi(\lambda(x, t), t)$. Necessarily we have

$$\frac{\partial \Psi}{\partial \lambda} = \lambda \frac{\partial G}{\partial \lambda}, \tag{13.4.17}$$

so

$$\lambda^2 \frac{\partial^2 G}{\partial \lambda^2} = \lambda \frac{\partial^2 \Psi}{\partial \lambda^2} - \frac{\partial \Psi}{\partial \lambda}$$

and

$$\frac{\partial \Phi}{\partial t} = -\frac{\delta^2}{2}\lambda^2 + \frac{1}{2}\alpha^2 \lambda \frac{\partial \Psi}{\partial \lambda},$$

$$\frac{\partial \lambda}{\partial t} = -\frac{\delta^2 \lambda^2}{\frac{\partial \Psi}{\partial \lambda}} + \frac{1}{2}\alpha^2 \lambda + \frac{1}{2}\alpha^2 \lambda^2 \frac{\frac{\partial^2 \Psi}{\partial \lambda^2}}{\frac{\partial \Psi}{\partial \lambda}}.$$

Then combining these expressions in the formula

$$\frac{\partial \Phi}{\partial t} = \frac{\partial \Psi}{\partial t} + \frac{\partial \Psi}{\partial \lambda}\frac{\partial \lambda}{\partial t},$$

we get the PDE

$$\frac{\partial \Psi}{\partial t} + \frac{1}{2}\alpha^2 \lambda^2 \frac{\partial^2 \Psi}{\partial \lambda^2} - \frac{\delta^2}{2}\lambda^2 = 0. \tag{13.4.18}$$

13.4.3 Choice of Initial Conditions

It is of particular interest to decide first on the initial condition for $G(\lambda, t)$, i.e., fixing the value $G(\lambda, T)$. Indeed, since $G(\lambda, t)$ is the solution of a linear PDE, we derive explicitly its value for $t < T$, once $G(\lambda, T)$ is known. It will be a smooth function, even though $G(\lambda, T)$ may not be. If we can solve the equation $G(\lambda, t) = x$ for every x in a unique way, we will obtain the function $\lambda(x, t)$ for $t < T$. We then obtain $\lambda(x, T)$ by letting $t \uparrow T$, and not by solving $G(\lambda, T) = x$, since $G(\lambda, T)$ may not be invertible. This fixes $\frac{\partial \Phi(x,t)}{\partial x} = \lambda(x, t)$, but not completely $\Phi(x, t)$. We recall that we want $\Phi(x, t) = \Psi(\lambda(x, t), t)$, hence also $\Psi(\lambda, t) = \Phi(G(\lambda, t), t)$. Since $\Psi(\lambda, t)$ is also the solution of a linear PDE, fixing $\Psi(\lambda, T)$ will permit us to give an explicit formula for $\Psi(\lambda, t)$. Since we have fixed $G(\lambda, T)$, it is natural to fix $\Psi(\lambda, T) = F(G(\lambda, T))$. With these steps, we obtain a function $\Phi(x, t)$ for all $t < T$. By construction, it satisfies the Bellman equation. By considering its limit as $t \uparrow T$, we can define the value $\Phi(x, T)$. This is the road map we intend to follow. It remains to perform the various steps and to check that we have indeed obtained the value function. We proceed with choosing $G(\lambda, T)$. We note that we can interpret $G(\lambda, t), t < T$, as the value of x that maximizes $\Phi(x, t) - \lambda x$. Since it is positive, it also solves the problem $\sup_{x \geq 0}(\Phi(x, t) - \lambda x)$. Also, $\Psi(\lambda, t) = \Phi(G(\lambda, t), t)$. So it is natural to fix the initial value $G(\lambda, T)$ at the point $s(\lambda)$ that maximizes $F(x) - \lambda x$ over $x \geq 0$. Let us show that this point is uniquely defined. Indeed, set $L(x, \lambda) = F(x) - \lambda x$. We have

$$\frac{\partial L}{\partial x} = F'(x) - \lambda = \left| \begin{array}{c} -\lambda, \text{ if } x < K, \\ U'(w + x - K) - \lambda, \text{ if } x \geq K. \end{array} \right.$$

Note that for $x \geq K$, we have $U'(w + x - K) \leq U'(w)$. Therefore, if $U'(w) \leq \lambda$, we get $\frac{\partial L(x,\lambda)}{\partial x} \leq 0$. Hence $s(\lambda) = 0$ if $U'(w) \leq \lambda$. Let us then consider the situation in which $\lambda < U'(w)$. We then have

$$\frac{\partial L(x, t)}{\partial x} = \left| \begin{array}{c} < 0, \text{ if } x < K, \\ > 0, \text{ if } K \leq x < K - w + U'^{-1}(\lambda), \\ \leq 0, \text{ if } x \geq K - w + U'^{-1}(\lambda). \end{array} \right.$$

If we set $\theta(\lambda) = K - w + U'^{-1}(\lambda)$, then we can state that

$$s(\lambda) = \left| \begin{array}{c} \theta(\lambda) \text{ if } F(\theta(\lambda)) - \lambda\theta(\lambda) \geq F(0), \\ 0 \text{ if } F(\theta(\lambda)) - \lambda\theta(\lambda) < F(0). \end{array} \right.$$

Note that $F'(\theta(\lambda)) = \lambda$, since $\theta(\lambda) > K$ (recall that $\lambda < U'(w)$). Therefore, the condition is equivalent to $F(\theta(\lambda)) - \theta(\lambda)F'(\theta(\lambda)) \geq F(0)$. However, the function $F(x) - xF'(x)$ is monotonically increasing on $[K, +\infty)$. Also, $F(K) - KF'(K) = U(w) - KU'(w) < F(0) = U(w)$, and

$$F(x) - xF'(x) = U(w + x - K) - xU'(w + x - K) \to +\infty, \text{ as } x \to +\infty.$$

Therefore, there exists a single value $\zeta > K$ such that

$$F(\zeta) - \zeta F'(\zeta) = F(0), \tag{13.4.19}$$

and we can state that

$$s(\lambda) = \begin{vmatrix} \theta(\lambda) \text{ if } \theta(\lambda) \geq \zeta, \\ 0 \text{ if } \theta(\lambda) < \zeta. \end{vmatrix}$$

Finally, $s(\lambda)$ is uniquely defined by the formula

$$s(\lambda) = (K - w + U'^{-1}(\lambda))\mathbb{1}_{\lambda \leq U'(\zeta + w - K)}. \tag{13.4.20}$$

This is not a continuous function of λ. The value $U'(\zeta + w - K)$ is a discontinuity. The left limit is ζ, and the right limit is 0. The function $s(\lambda)$ is invertible only for $\lambda \leq U'(\zeta + w - K)$, and its inverse is then $\lambda(x) = U'(w + x - K) = F'(x)$. Note that we need $x \geq \zeta$.

13.4.4 Solution of the Bellman Equation

Since we have fixed the value of $G(\lambda, T) = s(\lambda)$, we can give an explicit formula for $G(\lambda, t)$. We first consider $\tilde{G}(\lambda, t) = \lambda G(\lambda, t)$. It is the solution of the PDE

$$\frac{\partial \tilde{G}}{\partial t} + \frac{1}{2}\lambda^2\alpha^2\frac{\partial^2 \tilde{G}}{\partial \lambda^2} = \delta^2\lambda^2, \tag{13.4.21}$$
$$\tilde{G}(\lambda, T) = \lambda s(\lambda).$$

Set $\rho = U'(\zeta + w - K)$. After some tedious calculations, one can obtain the explicit formula

$$G(\lambda, t) = \frac{1}{\sqrt{2\pi}} \int_{-\infty}^{\frac{1}{\alpha\sqrt{T-t}}\log\frac{\rho}{\lambda} - \frac{1}{2}\alpha\sqrt{T-t}} (K - w + \tag{13.4.22}$$

$$U'^{-1}(\lambda \exp(\alpha\xi\sqrt{T-t} + \frac{1}{2}\alpha^2(T-t)))\Big) \exp -\frac{1}{2}\xi^2 d\xi - \delta^2\lambda\tilde{T}_t,$$

with

$$\tilde{T}_t = \frac{\exp \alpha^2(T-t) - 1}{\alpha^2}. \tag{13.4.23}$$

This formula is valid for $t < T$.It is valid as well for $t = T$, provided that for $\rho = \lambda$, we interpret the ratio $\frac{1}{\sqrt{T-t}} \log \frac{\rho}{\lambda}$ as $+\infty$ and not 0. The function $G(\lambda, t)$ is smooth for $t < T$. We have $G(0, t) = +\infty$ and $G(+\infty, t) = -\infty$. Moreover,

$$\frac{\partial G(\lambda, t)}{\partial \lambda} = -\frac{\zeta \exp -\frac{1}{2}(\frac{1}{\alpha\sqrt{T-t}} \log \frac{\rho}{\lambda} - \frac{1}{2}\alpha\sqrt{T-t})^2}{\lambda\alpha\sqrt{2\pi(T-t)}} - \delta^2 \tilde{T}_t +$$

$$+ \int_{-\infty}^{\frac{1}{\alpha\sqrt{T-t}} \log \frac{\rho}{\lambda} - \frac{1}{2}\alpha\sqrt{T-t}} \frac{(U'^{-1})'}{\sqrt{2\pi}} (\lambda \exp(\alpha\xi\sqrt{T-t} + \frac{1}{2}\alpha^2(T-t))) \times$$

$$\exp[-\frac{1}{2}\xi^2 + \alpha\xi\sqrt{T-t} + \frac{1}{2}\alpha^2(T-t)] d\xi.$$

It is clear that $\frac{\partial G(\lambda,t)}{\partial \lambda} < 0$. Therefore, $G(\lambda, t)$ is decreasing in λ. It follows that there exists a unique $\lambda(x, t)$ such that

$$G(\lambda, t) = x. \tag{13.4.24}$$

This property is lost for $t = T$. So an important question concerns the limit of $\lambda(x, t)$ as $t \to T$. The function $\lambda(x, t)$ is decreasing with x. Let us check the property

$$\lambda(x, t) \to \lambda(x, T) = \left| \begin{array}{c} U'(w + x - K) = F'(x) \text{ if } x > \zeta, \\ \rho \text{ if } x \leq \zeta. \end{array} \right. \tag{13.4.25}$$

Indeed, we first prove that $\limsup_{t \to T} \lambda(x, t) \leq \rho$. We prove it by contradiction. If $\limsup_{t \to T} \lambda(x, t) > \rho$, then there exists a subsequence that converges to a number, possibly $+\infty$, at a rate strictly greater than ρ. In that case, it easy to check that the corresponding subsequence $G(\lambda(x, t), t)$ approaches zero as $t \to T$, which is impossible, since it is equal to x. Therefore, $\lambda(x, t)$ remains bounded. If we extract a subsequence converging to λ^*, then

$$G(\lambda(x, t), t) \to \left| \begin{array}{c} K - w + U'^{-1}(\lambda^*), \text{ if } \lambda^* < \rho, \\ 0, \text{ if } \lambda^* > \rho. \end{array} \right.$$

The second case is impossible, because $G(\lambda(x, t), t) = x$. So $\lambda^* \leq \rho$. If $\lambda^* < \rho$, then $K - w + U'^{-1}(\lambda^*) = x$, which implies $\lambda^* = U'(w + x - K)$. This requires $x > \zeta$. If $x \leq \zeta$, this cannot be true, and hence $\lambda^* = \rho$. Since the number λ^* is uniquely defined, the whole sequence converges, and thus we have proved (13.4.25). We next recall that $\Psi(\lambda, t) = \Phi(G(\lambda, t), t)$. Hence from the choice of $G(\lambda, T)$, we must have

$$\Psi(\lambda, T) = F(s(\lambda)) = \left| \begin{array}{c} U(U'^{-1}(\lambda)) \text{ if } \lambda \leq \rho, \\ U(w) \text{ if } \lambda > \rho. \end{array} \right. \tag{13.4.26}$$

From (13.4.18) and (13.4.26) we can give the explicit value of $\Psi(\lambda, t)$. By comparison with $\tilde{G}(\lambda, t)$, we shall write

$$\Psi(\lambda, t)=\frac{1}{\sqrt{2\pi}}\int_{-\infty}^{\frac{1}{\alpha\sqrt{T-t}}\log\frac{\rho}{\lambda}+\frac{1}{2}\alpha\sqrt{T-t}} U(U'^{-1}(\lambda\exp(\alpha\xi\sqrt{T-t}-\frac{1}{2}\alpha^2(T-t))))$$

$$\exp-\frac{1}{2}\xi^2 d\xi+\frac{U(w)}{\sqrt{2\pi}}\int_{\frac{1}{\alpha\sqrt{T-t}}\log\frac{\rho}{\lambda}+\frac{1}{2}\alpha\sqrt{T-t}}^{+\infty}\exp-\frac{1}{2}\xi^2 d\xi-\frac{1}{2}\delta^2\lambda^2\tilde{T}_t. \quad (13.4.27)$$

For $x > \zeta$, we have $\lambda(x, t) \to \lambda(x, T) < \rho$; hence

$$\Phi(x, t) = \Psi(\lambda(x, t), t) \to U(U'^{-1}(\lambda(x, T))) = U(w + x - K) = F(x).$$

Therefore,

$$\Phi(x, T) = F(x), \ x > \zeta. \quad (13.4.28)$$

We cannot use (13.4.27) when $x \leq \zeta$, since $\lambda(x, t) \to \rho$. This implies that we cannot decide about the limit of $\frac{1}{\sqrt{T-t}}\log\frac{\rho}{\lambda}$. We proceed differently. We consider the concave hull of $F(x)$, namely the function

$$cF(x) = \left| \begin{array}{l} F(x) \text{ if } x \geq \zeta, \\ F(0) + x\rho \text{ if } x < \zeta. \end{array} \right. \quad (13.4.29)$$

Recall that $\rho = F'(\zeta)$. The function $cF(x)$ is concave, monotonically increasing, and C^1. From (13.4.25) we have

$$\frac{\partial\Phi(x, t)}{\partial x} = \lambda(x, t) \to (cF(x))'. \quad (13.4.30)$$

Therefore, $\Phi(x, t)-cF(x) \to C$ as $t \to T$. Since we know that $\Phi(x, t)-cF(x) = \Phi(x, t) - F(x) \to 0$ as $t \to T$ for $x > \zeta$, the constant is necessarily 0. Therefore, we have proved the following theorem.

THEOREM 13.3. *We assume (13.4.5). Equation (13.4.8) with initial condition* $\Phi(x, T) = cF(x)$ *has a solution that is* $C^0(R^+\times[0, T])$, $\frac{\partial\Phi(x,t)}{\partial x}$ *is* $C^0(R^+\times[0, T])$, *while* $\frac{\partial\Phi(x,t)}{\partial t}$ *and* $\frac{\partial^2\Phi(x,t)}{\partial x^2}$ *are* $C^0(R^+ \times [0, T))$. *Moreover, the properties (13.4.10) are satisfied. The function* $\Phi(x, t)$ *can be given explicitly by its gradient* $\frac{\partial\Phi(x,t)}{\partial x} = \lambda(x, t)$, *the solution of (13.4.24), (13.4.22), and its value for* $x > \zeta$, *defined by* $\Phi(x, t) = \Psi(\lambda(x, t), t)$, *where* $\Psi(\lambda, t)$ *is given by (13.4.27).*

13.4.5 Solution of the Stochastic Control Problem

We note that we do not recover the initial condition $\Phi(x, T) = F(x)$, which could have been be expected. The initial condition is $\Phi(x, T) = cF(x)$. It is therefore legitimate to check whether the function Φ is indeed the value function. We first introduce feedbacks

$$\hat{u}(x,t) = \delta\lambda(x,t),\ \hat{v}(x,t) = -\alpha\frac{\lambda(x,t)}{x\frac{\partial\lambda(x,t)}{\partial x}}. \tag{13.4.31}$$

We see in particular that $\hat{u}(x,t) > 0$, $x\hat{v}(x,t) > 0$. Moreover, the Bellman equation (2.1.6) reads

$$\frac{\partial\Phi}{\partial t} - \frac{1}{2}\hat{u}^2 + \delta\hat{u}\frac{\partial\Phi}{\partial x} + \alpha\hat{v}x\frac{\partial\Phi}{\partial x} + \frac{1}{2}x^2\hat{v}^2\frac{\partial^2\Phi}{\partial x^2} = 0. \tag{13.4.32}$$

We proceed with a verification argument. Consider an admissible control $u(s)$, $v(s)$ for the problem (13.4.1), (13.4.2), (13.4.6). We take the initial condition at time 0 with initial value x. From Itô's formula we get

$$\begin{aligned} d\Phi(X(s),s) = [&\frac{\partial\Phi(X(s),s)}{\partial s} + \frac{\partial\Phi(X(s),s)}{\partial x}(\delta u(s) + \alpha v(s)X(s)) \\ &+ \frac{1}{2}\frac{\partial\Phi^2(X(s),s)}{\partial x^2}v(s)^2X(s)^2]ds + + \frac{\partial\Phi(X(s),s)}{\partial x}v(s)X(s)dw(s), \end{aligned}$$

and thus from the Bellman equation,

$$d\Phi(X(s),s) \leq \frac{\partial\Phi(X(s),s)}{\partial x}v(s)X(s)dw(s) + \frac{1}{2}u(s)^2ds.$$

Take $M > x$ and let $\tau_M = \inf_s\{|X(s)| \geq M\}$. We integrate between 0 and $T \wedge \tau_M$ and take the mathematical expectation. We get

$$E[\Phi(X(T\wedge\tau_M), T\wedge\tau_M) - \frac{1}{2}\int_0^{T\wedge\tau_M} u(s)^2ds] \leq \Phi(x,0).$$

In this inequality we let $M \to +\infty$, then $\tau_M \to +\infty$. Since Φ is continuous up to T, we obtain, by Fatou's lemma,

$$E[cF(X(T) - \frac{1}{2}\int_0^T u(s)^2ds] \leq \Phi(x,0).$$

However, since $cF(x) \geq F(x)$, we also have $\Phi(x,0) \geq J_{x0}(u(.), v(.))$, and since this inequality holds for every admissible $u(.)$, $v(.)$, we can write

$$\Phi(x,0) \geq \sup_{u(.),v(.)} J_{x0}(u(.), v(.)). \tag{13.4.33}$$

We then construct the trajectory linked with the feedbacks $\hat{u}(x,t)$, $\hat{v}(x,t)$. We define the process $\hat{Y}(s)$ by

$$\begin{aligned} d\hat{Y}(s) &= -\alpha \hat{Y}(s) dw(s), \\ \hat{Y}(0) &= \lambda(x,0), \end{aligned}$$

and we set $\hat{X}(s) = G(\hat{Y}(s), s)$. Applying Itô's calculus and the PDE (13.4.14), we check easily the following formula,

$$\begin{aligned} d\hat{X}(s) &= \delta \hat{u}(\hat{X}(s), s) ds + \hat{v}(\hat{X}(s), s)\hat{X}(s)(\alpha ds + dw(s)), \\ \hat{X}(0) &= x, \end{aligned}$$

and we define the stochastic controls $\hat{u}(s) = \hat{u}(\hat{X}(s), s) = \hat{Y}(s)$, $\hat{v}(s) = \hat{v}(\hat{X}(s), s)$. Therefore, $\hat{v}(\hat{X}(s), s)\hat{X}(s) = -\alpha \hat{Y}(s) \frac{\partial G(\hat{Y}(s), s)}{\partial \lambda}$. We have $\hat{u}(s) > 0$, $\hat{v}(s)\hat{X}(s) > 0$. In fact, $\hat{X}(s) > 0, \forall s$, whence also $\hat{X}(T) > 0$. It follows that the controls $\hat{u}(s)$, $\hat{v}(s)$ are admissible. Defining for $M > x$,

$$\hat{\tau}_M = \inf_s \{\hat{X}(s) \geq M\},$$

we obtain this time

$$E[\Phi(\hat{X}(T \wedge \hat{\tau}_M), T \wedge \hat{\tau}_M) - \frac{1}{2} \int_0^{T \wedge \hat{\tau}_M} \hat{u}(s)^2 ds] = \Phi(x, 0). \tag{13.4.34}$$

To proceed, we need some information about the growth of $G(\lambda, t)$ in λ. We assume, besides (13.4.5), that

$$\lambda^m U'^{-1}(\lambda) \leq C_0, \ m > 0. \tag{13.4.35}$$

As for (13.2.14), one can check easily that

$$-C_1 \lambda \leq G(\lambda, t) \leq C_0 + \frac{C}{\lambda^m}. \tag{13.4.36}$$

We next assume

$$\lambda^p |U(U'^{-1}(\lambda))| \leq c_0 + c_1 \lambda^q, \ p, q \geq 0, \tag{13.4.37}$$

from which we obtain

$$|\Psi(\lambda, t)| \leq \alpha_0 + \beta \lambda^{-p} + \alpha_1 \lambda^2 + \gamma \lambda^{q-p}. \tag{13.4.38}$$

Therefore,

$$(\hat{X}(s))^2 \leq C_1 + C_2 (Y(s))^2 + C_3 ((Y(s))^{-2m},$$

and it follows easily that

$$E \sup_{0 \le s \le T} (\hat{X}(s))^2 \le C_T(x), \forall T.$$

From this estimate we derive that $\hat{\tau}_M \uparrow +\infty$ a.s. Now

$$E\Phi(\hat{X}(T \wedge \hat{\tau}_M), T \wedge \hat{\tau}_M) = E[cF(\hat{X}(T))1\!\!1_{T \le \hat{\tau}_M}] + \\ + E[\Psi(\hat{Y}(\hat{\tau}_M), \hat{\tau}_M)1\!\!1_{T > \hat{\tau}_M}].$$

But

$$E|\Psi(\hat{Y}(\hat{\tau}_M), \hat{\tau}_M)|1\!\!1_{T > \hat{\tau}_M} \le E \sup_{0 \le t \le T} |\Psi(\hat{Y}(t), t)|1\!\!1_{T > \hat{\tau}_M}.$$

From the estimate (13.4.38), we can state that $E \sup_{0 \le t \le T} |\Psi(\hat{Y}(t), t)| < +\infty$, and therefore, $E \sup_{0 \le t \le T} |\Psi(\hat{Y}(t), t)|1\!\!1_{T > \hat{\tau}_M} \rightarrow 0$, as $M \rightarrow +\infty$. Also, $E|cF(\hat{X}(T), T)| < +\infty$. It follows that $E\Phi(\hat{X}(T \wedge \hat{\tau}_M), T \wedge \hat{\tau}_M) \rightarrow E[cF(\hat{X}(T))]$. But $\hat{X}(T) = s(\hat{Y}(T))$. We use the fact that

$$\sup_{x \ge 0}(F(x) - \lambda x) = \sup_{x \ge 0}(cF(x) - \lambda x), \tag{13.4.39}$$

and hence $F(s(\lambda)) = cF(s(\lambda))$. Therefore, $E\Phi(\hat{X}(T \wedge \hat{\tau}_M), T \wedge \hat{\tau}_M) \rightarrow E[F(\hat{X}(T))]$. It follows that

$$E[F(\hat{X}(T)] - \frac{1}{2}\int_0^T \hat{u}(s)^2 ds] = J_{x0}(\hat{u}(.), \hat{v}(.)) = \Phi(x, 0).$$

Comparing with (13.4.33), it follows that the pair $\hat{u}(.)$, $\hat{v}(.)$ is optimal. We thus have obtained the following theorem.

THEOREM 13.4. *We assume (13.4.5) and (13.4.35), (13.4.37). The stochastic control problem (13.4.1), (13.4.2), (13.4.3) has a solution given by $\hat{u}(.)$, $\hat{v}(.)$, defined by the feedbacks (13.4.31).*

REMARK 13.2. We see that taking $\Phi(x, T) = F(x)$ or $cF(x)$ does not change the solution of the optimal control problem. This can be seen right away in our method when we define the function $G(\lambda, t)$, which will be the same in both cases. Furthermore, when $w = 0$, we have $F(x) = cF(x)$, and $\zeta = K,\ \rho = +\infty$.

13.5 The Martingale Method

The martingale method has been proposed by Karatzas et al. [27] and Cox and Huang [15] to solve optimal consumption and portfolio selection problems. It is an attempt to solve directly the stochastic control problem without using dynamic programming. It is then possible to check that the solution obtained in this manner satisfies dynamic programming. So we want to solve problem (13.4.1), (13.4.2), (13.4.3). Since we want to recover dynamic programming eventually, it is convenient to consider the family of problems indexed by initial conditions, x, t, so $X_{xt}(s)$ is defined by

$$\begin{aligned} dX &= \delta u(s)ds + v(s)X(s)(\alpha ds + dw(s)), \\ X(t) &= x, \end{aligned} \tag{13.5.1}$$

and the payoff to maximize is given by

$$J_{xt}(u(.), v(.)) = E[F(X_{xt}(T)) - \frac{1}{2}\int_t^T u(s)^2 ds]. \tag{13.5.2}$$

The idea of the method is to take advantage of the fact that the functional $J_{xt}(u(.), v(.))$ does not depend directly on the second control $v(.)$. Moreover, a martingale process can be derived from the wealth process $X(s)$, as is customary in mathematical finance. Let us first introduce the process $Z_t(s)$, which is an $\mathcal{F}_t^s$ martingale, where $\mathcal{F}_t^s = \sigma(w(\tau) - w(t), \tau \leq s)$, defined by

$$\begin{aligned} dZ(s) &= -\alpha Z(s)dw(s), \\ Z(t) &= 1. \end{aligned} \tag{13.5.3}$$

It is then easy to check that the process $X(s)Z(s) - \delta \int_t^s Z(\tau)u(\tau)d\tau$ is an $\mathcal{F}_t^s$ martingale. Consequently, we have in particular

$$E[X(T)Z(T) - \delta \int_t^T Z(s)u(s)ds] = x. \tag{13.5.4}$$

This is called the budget equation. It can be interpreted as a relation between the controls $u(.), v(.)$, although the control $v(.)$ does not appear explicitly. It appears indirectly, through the final wealth $X(T)$. At this stage, it is important to stress the common structure between the payoff (13.5.2) and the budget equation. Only the final wealth and the effort $u(.)$ appear explicitly. The intermediary wealth values and the scale of projects $v(.)$ do not appear explicitly. This observation is at the origin of the method, called the martingale approach. The idea is to introduce a new problem. The link between the new problem and our problem will be addressed later. The new problem is to consider a random variable $X(T)$ simply $\mathcal{F}_t^T$ measurable (it is not the final wealth) and positive that is a decision variable, together with an adapted process $u(s)$, which is also a control. These two controls can be decided independently, except for the constraint

$$EX(T)Z_t(T) - \delta E\int_t^T Z_t(s)u(s)ds = x. \tag{13.5.5}$$

They must optimize the payoff

$$K_t(X(T), u(.)) = E[F(X(T)) - \frac{1}{2}\int_t^T u(s)^2 ds]. \tag{13.5.6}$$

The intuition behind this problem is easy to understand. Of course the optimal $X(T)$, assuming it exists, should correspond to the final wealth obtained after solving our original stochastic control problem (13.5.1), (13.5.2). It is an idealized final wealth, since it may not be realized via the stochastic differential equation (13.5.1). In that case, the problems are not connected. If, by luck, it can be realized via (13.5.1), then the two problems coincide. This issue is addressed afterward. The interesting point with (13.5.6) is that it is much simpler than the original problem. So we focus first on this problem. We will also denote the value function by

$$\Phi(x, t) = \sup_{\{X(T)\geq 0, u(.)|\text{B.E.}\}} K_t(X(T), u(.)), \tag{13.5.7}$$

where the symbol B.E. means that the budget equation is a constraint. We have used $\Phi(x, t)$ for the value function, which is the same notation as (13.4.7), to save notation. Indeed, the value functions are different, since the problems are a priori different. But we do not consider them simultaneously, so it is not an issue. Besides, we will show eventually that the two problems coincide.

Note also that the only constraint is the budget equation. We do not impose a positivity condition on $u(.)$. In fact, it will be automatically satisfied. So we omit it at this stage. Finally, the value function depends on x, because of the budget equation. Since the budget equation is a scalar equality constraint, we can address it by the Lagrange multiplier technique. Define

$$\begin{aligned} K_t(X(T), u(.); q) &= K_t(X(T), u(.)) \\ &-qEX(T)Z_t(T) + q\delta E\int_t^T Z_t(s)u(s)ds. \end{aligned} \tag{13.5.8}$$

We will limit q to positive numbers. We first find, for positive q, a solution to the problem

$$\sup_{X(T)\geq 0, u(.)} K_t(X(T), u(.); q). \tag{13.5.9}$$

If there exists a value of $q, q(x,t)$ such that the optimal solution for the problem $K_t(X(T), u(.); q(x,t))$ satisfies the budget equation, then this solution is also optimal for the problem (13.5.7). So we begin by solving (13.5.9) for positive q. In the problem (13.5.9), the decision $X(T)$ and the control $u(.)$ are no longer coupled. The optimal $u(s)$, denoted by $\hat{u}(s)$, is easily obtained, since it maximizes a quadratic concave function. We get

$$\hat{u}(s) = q\delta Z_t(s), \tag{13.5.10}$$

and we note immediately that this process is adapted and takes positive values. For $X(T)$ we have to maximize

$$E[F(X(T)) - qZ_t(T)X(T)].$$

Since we want $X(T)$ $\mathcal{F}_t^T$ measurable and positive, the solution of this problem is clearly $s(qZ_t(T))$, as seen in Section 13.4.3.

It follows immediately that the optimal $X(T)$, denoted by $\hat{X}(T)$, is given by

$$\hat{X}(T) = s(qZ_t(T)). \tag{13.5.11}$$

For all $q > 0$, we have solved the uncoupled problem. If we find q such that the optimal solution satisfies the budget equation, then it is also the solution of the problem (13.5.7). Using (13.5.10), (13.5.11) in the budget equation (13.5.5), we get

$$E[Z_t(T)s(qZ_t(T)) - \delta^2 q \int_t^T (Z_t(s))^2 ds] = x.$$

It is easy to compute

$$\tilde{T}_t = E\int_t^T (Z_t(s))^2 ds = \frac{\exp \alpha^2(T-t) - 1}{\alpha^2}.$$

Define the function

$$G(q,t) = E[Z_t(T)s(qZ_t(T))] - \delta^2 q\tilde{T}_t. \tag{13.5.12}$$

We leave to the reader to check that this function is exactly the function $G(\lambda, t)$ given by formula (13.4.12). So from now on, we call $\lambda(x,t)$ the Lagrange multiplier $q(x,t)$. We recall that in the Bellman equation (13.4.8), we have $\lambda(x,t) = \frac{\partial \Phi(x,t)}{\partial x}$.

It follows that the solution of problem (13.5.7) is given by the formulas

$$\hat{u}_{xt}(s) = \lambda(x,t)\delta Z_t(s), \quad \hat{X}_{xt}(T) = s(\lambda(x,t)Z_t(T)), \; t < T. \tag{13.5.13}$$

It remains to show that (13.5.7) is equivalent to our original problem (13.5.1), (13.5.2). For that we introduce $\hat{Y}(s) = \lambda(x,t)Z_t(s)$. From formulas (13.5.13), we can write

$$\hat{u}_{xt}(s) = \hat{u}(s) = \delta \hat{Y}(s), \tag{13.5.14}$$
$$\hat{X}_{xt}(T) = \hat{X}(T) = s(\hat{Y}(T)).$$

Consider the process

$$\hat{X}(s) = G(\hat{Y}(s), s). \tag{13.5.15}$$

For $s = T$, we recover $\hat{X}(T)$ as defined by (13.5.14). Also, $\hat{Y}(t) = \lambda(x, t)$. Therefore, $\hat{X}(t) = x$. We have

$$d\hat{Y}(s) = -\alpha \hat{Y}(s) dw(s),$$

and applying Itô's formula, we obtain easily that

$$d\hat{X}(s) = \delta^2 \hat{Y}(s) ds - \alpha \frac{\partial G}{\partial \lambda}(\hat{Y}(s), s)\hat{Y}(s)(\alpha ds + dw(s)).$$

We next define

$$\hat{v}_{xt}(s) = \hat{v}(s) = \frac{-\alpha \frac{\partial G}{\partial \lambda}(\hat{Y}(s), s)\hat{Y}(s)}{G(\hat{Y}(s), s)}, \tag{13.5.16}$$

and recalling (13.5.14), we can rewrite the Itô differential of $\hat{X}(s)$ as follows:

$$d\hat{X}(s) = \delta \hat{u}(s) ds + \hat{v}(s)\hat{X}(s)(\alpha ds + dw(s)), \tag{13.5.17}$$
$$\hat{X}(t) = x.$$

We note that $\hat{v}(s)\hat{X}(s) > 0$ and $E \int_t^T \hat{u}(s)^2 ds + \infty$. So $\hat{X}(T)$ can be viewed as the final value of a wealth process, as defined in the definition of problem (13.5.1), (13.5.2).

So we have solved completely the original problem. It remains to introduce the feedbacks

$$\hat{v}(x, t) = \frac{-\alpha \frac{\partial G}{\partial \lambda}(\lambda(x, t), t)\lambda(x, t)}{x} \tag{13.5.18}$$
$$= \frac{-\alpha \frac{\partial G}{\partial \lambda}(G^{-1}(x, t), t)G^{-1}(x, t)}{x},$$

and

$$\hat{u}(x, t) = \delta \lambda(x, t) \tag{13.5.19}$$

$$= \delta G^{-1}(x,t),$$

and to check that

$$\hat{v}(\hat{X}(s),s) = \hat{v}_{xt}(s),\ \ \hat{u}_{xt}(s) = \hat{u}(\hat{X}(s),s),$$

to recover what was obtained from dynamic programming.

REMARK 13.3. We have introduced the intermediate wealth $\hat{X}(s)$ by formula (13.5.15). A probabilistic approach would be to use the fact that $\hat{X}(s)Z(s) - \delta\int_t^s Z(\tau)\hat{u}(\tau)d\tau$ must be an $\mathcal{F}_t^s$ martingale. Hence

$$\hat{X}(s)Z(s) = E[\hat{X}(T)Z(T) - \delta\int_s^T Z(\tau)\hat{u}(\tau)d\tau|\mathcal{F}_t^s].$$

This formula can be used as a definition of $\hat{X}(s)$ for all s. One then uses the representation theorem of martingales of Kunita and Watanabe [31] to write

$$\hat{X}(s)Z(s) - \delta\int_t^s Z(\tau)\hat{u}(\tau)d\tau = \int_t^s \hat{X}(\tau)Z(\tau)(\hat{v}(\tau)-\alpha)dw(\tau),$$

which defines a unique adapted process $\hat{v}(s)$. This suffices to show the equivalence of problems (13.5.7) and (13.5.1), (13.5.2). However, this is not a constructive approach, and we do not get the feedback properties and the equivalence with dynamic programming.

13.5.1 The Case $\alpha = 0, \delta = 0$

From the preceding analysis, we obtain (formally) that for $\alpha = 0, \delta = 0$, the value function is independent of time and equal to $cF(x)$. We can check this property directly. Since the state equation reduces to

$$dX = v(s)X(s)dw(s), \tag{13.5.20}$$
$$X(t) = x,$$

so it is independent of $u(.)$, it is clear that we can assume $u(s) = 0$. So the problem reduces to maximizing

$$J_{xt}(v(.)) = E[F(X_{xt}(T;v(.)))]. \tag{13.5.21}$$

Let us define

$$\Phi(x,t) = \sup_{v(.)} J_{xt}(v(.)).$$

We want to prove the following result.

PROPOSITION 13.1. *We have* $\Phi(x,t) = cF(x), \forall t < T$.

PROOF. We first note that when F is concave (case $w = 0$), the result is immediate. Indeed, by Jensen's inequality, we have

$$\Phi(x,t) \leq \sup_{v(.)} E[cF(X_{xt}(T;v(.)))] \leq \sup_{v(.)} cF[EX_{xt}(T;v(.))] = cF(x).$$

But also, taking a control equal to 0 yields $\Phi(x,t) \geq F(x)$. In the concave case, we have $F(x) = cF(x)$, whence the result. So the control 0 is optimal. In general, we need to prove

$$\Phi(x,t) \geq cF(x). \tag{13.5.22}$$

Of course, if we know that $\Phi(x,t)$ is concave, the result is an immediate consequence of the definition of the concave hull. Otherwise, we first recall that $F(x)$ is bounded below. Let us consider $F_M(x) = F(x) \wedge M$. Define

$$\Phi_M(x,t) = \sup_{v(.)} E[F_M(X_{xt}(T;v(.)))].$$

It is clear that $\Phi_M(x,t) \uparrow \Phi(x,t)$ as $M \uparrow +\infty$. If we prove that $\Phi_M(x,t) \geq cF_M(x)$, then we will get (13.5.22) by letting $M \uparrow +\infty$. So we may assume that F is bounded. We then introduce a small α and define $w_\alpha(s)$ by $dw_\alpha(s) = dw(s) - \alpha ds$, $w_\alpha(t) = 0$, and consider the martingale

$$M_t^\alpha(s) = \exp[\alpha(w(s) - w(t)) - \frac{1}{2}\alpha^2(s-t)].$$

We make a change of probability by setting

$$\frac{dP^\alpha}{dP}|_{\mathcal{F}_t^s} = M_t^\alpha(s),$$

where $\mathcal{F}_t^s = \sigma(w(\tau) - w(t), \tau \leq s)$. For P^α, $w_\alpha(s)$ is a standard Wiener process, and the state $X_{xt}(s)$ appears as the solution of

$$\begin{aligned} dX &= v(s)X(s)(\alpha ds + dw_\alpha(s)), \\ X(t) &= x. \end{aligned} \tag{13.5.23}$$

We then define

$$\begin{aligned} \Phi^\alpha(x,t) &= \sup_{v(.)} E^\alpha[F(X_{xt}(T;v(.)))] \\ &= \sup_{v(.)} E[M_t^\alpha(s)F(X_{xt}(T;v(.)))]. \end{aligned}$$

For $\alpha > 0$, the theory of Section 13.4.5 is valid. Therefore, we can write

$$\Phi^\alpha(x,t) = \sup_{v(.)} E[M_t^\alpha(s) cF(X_{xt}(T; v(.)))].$$

On the other hand, we can also see, using the fact that F is bounded, that

$$\Phi^\alpha(x,t) \to \Phi(x,t),$$
$$\Phi^\alpha(x,t) \to \sup_{v(.)} E[cF(X_{xt}(T; v(.)))].$$

Therefore,

$$\Phi(x,t) = \sup_{v(.)} E[cF(X_{xt}(T; v(.)))].$$

We are now in the concave case, which implies $\Phi(x,t) = cF(x)$, and the proof is complete. □

REMARK 13.4. The result of Proposition 13.4 completes a result mentioned in [14].

13.6 Optimal Loan

The number K represents the debt to be paid by the entrepreneur. In practice, the contract is such that the bank grants a loan L and asks for K as reimbursement. The bank problem is to define K as a function of L. In case of no risk on reimbursement, we have $L = K \exp -r(T-t)$. Assuming that the bank is risk-neutral, the relation with risk is

$$\begin{aligned} L &= \exp -r(T-t) E \min(K, \hat{X}_{xt}(T)) \\ &= \exp -r(T-t) E \min(K, s(\lambda(x,t) Z_t(T))) \\ &= K \exp -r(T-t) E \mathbb{1}_{\lambda(x,t) Z_t(T) \leq \rho}. \end{aligned} \tag{13.6.1}$$

The initial wealth is $x = x_0 + L$. Recalling formula (13.5.12), we can also write, to separate the term involving K,

$$G(\lambda,t) = G_K(\lambda,t) = KG_1(\lambda,t) + G_2(\lambda,t),$$

with

$$G_1(\lambda,t) = E[Z_t(T)\mathbb{1}_{\lambda \leq \rho}],$$
$$G_2(\lambda,t) = E[Z_t(T)(U'^{-1}(\lambda Z_t(T)) - w)\mathbb{1}_{\lambda \leq \rho}] - \delta^2 \lambda \tilde{T}_t.$$

In the equation $x = G(\lambda, t)$ that defines $\lambda(x, t)$, we make use of (13.6.1) to write $G(\lambda, t) = G_L(\lambda, t)$, with

$$G_L(\lambda, t) = \frac{L \exp r(T-t) E[Z_t(T) \mathbb{1}_{\lambda \leq \rho}]}{E \mathbb{1}_{\lambda Z_t(T) \leq \rho}} + G_2(\lambda, t), \tag{13.6.2}$$

which expresses the function $G(\lambda, t)$ in terms of L instead of K. The function $G_2(\lambda, t)$ is clearly decreasing in λ on R^+. Set

$$\Gamma(\lambda, t) = \frac{E[Z_t(T) \mathbb{1}_{\lambda \leq \rho}]}{E \mathbb{1}_{\lambda Z_t(T) \leq \rho}}.$$

Then we write explicitly

$$\Gamma(\lambda, t) = \frac{N(-d_1(\lambda, t))}{N(-d_2(\lambda, t))}, \tag{13.6.3}$$

with

$$\begin{aligned} d_1(\lambda, t) &= \frac{1}{\alpha\sqrt{T-t}} \log \frac{\lambda}{\rho} + \frac{\alpha}{2}\sqrt{T-t}, \\ d_2(\lambda, t) &= d_1(\lambda, t) - \alpha\sqrt{T-t}, \end{aligned} \tag{13.6.4}$$

and $N(x)$ is the cumulative distribution function of the standard Gaussian random variable.

Let us check that the function $\Gamma(\lambda, t)$ decreases in λ from the value 1 at $\lambda = 0$ to the value 0 at $\lambda = +\infty$. Since

$$-d_2 = -d_1 + \alpha\sqrt{T-t},$$

we have

$$N(-d_2) = N(-d_1) + \frac{1}{\sqrt{2\pi}} \int_{-d_1}^{-d_2} \exp -\frac{1}{2}x^2 dx$$

and

$$\Gamma(\lambda, t) = \frac{1}{1 + \frac{\frac{1}{\sqrt{2\pi}} \int_{-d_1}^{-d_2} \exp -\frac{1}{2}x^2 dx}{N(-d_1)}}.$$

Therefore, to prove the assertion on $\Gamma(\lambda, t)$, we have to show that the ratio

$$X = \frac{\frac{1}{\sqrt{2\pi}} \int_{-d_1}^{-d_2} \exp -\frac{1}{2}x^2 dx}{N(-d_1)}$$

is increasing in λ from 0 to $+\infty$. But

$$X = \frac{\int_{-d_1}^{-d_2} \exp -\frac{1}{2}x^2 dx}{\int_{-\infty}^{-d_1} \exp -\frac{1}{2}x^2 dx},$$

and thus

$$X = \frac{\int_0^{\alpha\sqrt{T-t}} \exp -\frac{1}{2}(-d_1 + y)^2 dy}{\int_0^{+\infty} \exp -\frac{1}{2}(d_1 + y)^2 dy},$$

$$X = \frac{\int_0^{\alpha\sqrt{T-t}} \exp -\frac{1}{2}y^2 \exp y d_1 \, dy}{\int_0^{+\infty} \exp -\frac{1}{2}y^2 \exp -y d_1 dy}.$$

This expression is clearly increasing in d_1, and hence increasing in λ, from formula (13.6.4). For $\lambda = 0$, we have $d_1 = -\infty$, whence $X = 0$, and for $\lambda = +\infty$, we have $d_1 = +\infty$ and $X = +\infty$. This completes the proof of the assertion. Now $G_2(\lambda, t)$ decreases from $+\infty$ at 0 to $-\infty$ at $+\infty$. So the same is true for $G_L(\lambda, t)$.

If we consider the equation

$$x_0 + L = G(\lambda, t) = G_L(\lambda, t)$$

for L fixed > 0, then it has a unique solution $\lambda(L, t)$. Using finally

$$K = \frac{L \exp r(T - t)}{N(-d_2(\lambda, t))},$$

we obtain a map from L to K. This defines the level of the debt K in terms of the loan L. Note that if $w = 0$, then $\rho = \infty$, whence $d_2(\lambda, t) = -\infty$, for all $\lambda > 0$. There is no risk for the bank, since the final wealth is always greater than K, so we get $K = L \exp r(T - t)$.

13.7 Duality

13.7.1 Consumer–Investor Problem

Consider the HJB equation (13.2.6). We consider the Legendre–Fenchel transform of the value function $\Phi(x, t)$, defined for $\lambda \geq 0$ by

$$\tilde{\Phi}(\lambda, t) = \sup_{x \geq 0}(\Phi(x, t) - x\lambda). \tag{13.7.1}$$

We shall call $\tilde{\Phi}(\lambda, t)$ the dual function of Φ. In the context of the utility functions $U_1(c)$ and $U_2(x)$, we have introduced the functions β_i; see (13.2.3). So we can write

$$\tilde{U}_i(\lambda) = U_i(\beta_i(\lambda)) - \lambda\beta_i(\lambda). \tag{13.7.2}$$

We shall see that the simplifications obtained in solving the HJB equation can be understood in the context of duality. In fact, $\tilde{\Phi}(\lambda, t)$ will be the solution of a linear PDE. This is not very surprising, since the value of x, which achieves its supremum in (13.7.1), is simply $G(\lambda, t)$. Indeed, it is the solution of

$$\frac{\partial \Phi(x,t)}{\partial x} = \lambda,$$

which is the inverse of the function $\lambda(x, t)$. It follows from (13.7.1) that

$$\frac{\partial \tilde{\Phi}(\lambda, t)}{\partial \lambda} = -G(\lambda, t). \tag{13.7.3}$$

Looking at equation (13.2.10) and noting that $\beta_i(\lambda) = -\frac{\partial \tilde{U}_i(\lambda)}{\partial \lambda}$, we can write the relation

$$\frac{\partial}{\partial \lambda}[\frac{\partial \tilde{\Phi}}{\partial t} + \frac{|\theta(t)|^2}{2}\lambda^2\frac{\partial^2 \tilde{\Phi}}{\partial \lambda^2} - r\tilde{\Phi} + \tilde{U}_1] = 0,$$
$$\frac{\partial}{\partial \lambda}[\tilde{\Phi}(\lambda, T) - \tilde{U}_2(\lambda)] = 0.$$

Since $\Phi(x, T) = U_2(x)$, we have $\tilde{\Phi}(\lambda, T) - \tilde{U}_2(\lambda) = 0$. Therefore, the above relation implies

$$\frac{\partial \tilde{\Phi}}{\partial t} + \frac{|\theta(t)|^2}{2}\lambda^2\frac{\partial^2 \tilde{\Phi}}{\partial \lambda^2} - r\tilde{\Phi} + \tilde{U}_1 = c(t), \tag{13.7.4}$$
$$\tilde{\Phi}(\lambda, T) - \tilde{U}_2(\lambda) = 0.$$

It remains to show that $c(t) = 0$. We first note that $\tilde{U}_1(\lambda) = U_1(\beta_1(\lambda)) - \lambda\beta_1(\lambda)$. We next write

$$\frac{\partial \tilde{\Phi}(\lambda, t)}{\partial t} = \frac{\partial \Phi(G(\lambda, t), t)}{\partial t}, \; \lambda = \frac{\partial \Phi(G(\lambda, t), t)}{\partial x}.$$

Differentiating the second relation in λ, we have

$$1 = \frac{\partial^2 \Phi(G(\lambda, t), t)}{\partial x^2}\frac{\partial G(\lambda, t)}{\partial \lambda},$$

$$1 = -\frac{\partial^2 \Phi(G(\lambda, t), t)}{\partial x^2} \frac{\partial^2 \tilde{\Phi}(\lambda, t)}{\partial \lambda^2},$$

so

$$\frac{\partial^2 \tilde{\Phi}(\lambda, t)}{\partial \lambda^2} = -\frac{1}{\frac{\partial^2 \Phi(G(\lambda,t),t)}{\partial x^2}}.$$

We also note that

$$\begin{aligned}\tilde{\Phi}(\lambda, t) &= \Phi(G(\lambda, t), t) - \lambda G(\lambda, t) \\ &= \Phi(G(\lambda, t), t) - G(\lambda, t) \frac{\partial \Phi(G(\lambda, t), t)}{\partial x},\end{aligned}$$

$$\tilde{U}_1(\lambda) = U_1(\beta_1(\frac{\partial \Phi(G(\lambda, t), t)}{\partial x})) - \frac{\partial \Phi(G(\lambda, t), t)}{\partial x} \beta_1(\frac{\partial \Phi(G(\lambda, t), t)}{\partial x}).$$

Collecting results, we obtain from (13.7.4) that

$$\frac{\partial \Phi(G(\lambda, t), t)}{\partial t} - \frac{|\theta(t)|^2}{2} \frac{(\frac{\partial \Phi(G(\lambda,t),t)}{\partial x})^2}{\frac{\partial^2 \Phi(G(\lambda,t),t)}{\partial x^2}} - r\Phi(G(\lambda, t), t) +$$

$$+ rG(\lambda, t) \frac{\partial \Phi(G(\lambda, t), t)}{\partial x} + U_1(\beta_1(\frac{\partial \Phi(G(\lambda, t), t)}{\partial x})) - \frac{\partial \Phi(G(\lambda, t), t)}{\partial x} \beta_1(\frac{\partial \Phi(G(\lambda, t), t)}{\partial x}) = c(t).$$

But applying equation (13.2.6) at $x = G(\lambda, t)$, we immediately obtain $c(t) = 0$. Therefore, the dual function $\tilde{\Phi}(\lambda, t)$ is the solution of the linear PDE

$$\begin{aligned}\frac{\partial \tilde{\Phi}}{\partial t} + \frac{|\theta(t)|^2}{2} \lambda^2 \frac{\partial^2 \tilde{\Phi}}{\partial \lambda^2} - r\tilde{\Phi} + \tilde{U}_1 &= 0, \\ \tilde{\Phi}(\lambda, T) - \tilde{U}_2(\lambda) &= 0.\end{aligned} \tag{13.7.5}$$

One can write an explicit formula for $\tilde{\Phi}(\lambda, t)$.

EXERCISE 13.5. Find the explicit formula for $\tilde{\Phi}(\lambda, t)$. Introduce the stochastic process

$$\begin{aligned}dY(s) &= -Y(s)\theta(s).dw(s), \\ Y(t) &= \lambda,\end{aligned}$$

denoted by $Y_{\lambda t}(s)$, and show the probabilistic interpretation

$$\tilde{\Phi}(\lambda, t) = E[\int_t^T \tilde{U}_1(Y_{\lambda t}(s)) \exp -r(s - t)\, ds + \tilde{U}_2(Y_{\lambda t}(T)) \exp -r(T - t)]$$

and derive the explicit solution.

We finally express $\Phi(x,t)$ in terms of the dual function. We have

$$\tilde{\Phi}(\lambda,t) = \Phi(G(\lambda,t),t) - \lambda G(\lambda,t),$$

and if we apply this formula with $\lambda = \lambda(x,t)$, we obtain

$$\Phi(x,t) = \tilde{\Phi}(\lambda(x,t),t) + x\lambda(x,t).$$

But from (13.7.3), we have

$$\frac{\partial \tilde{\Phi}(\lambda(x,t),t)}{\partial \lambda} = -x,$$

and thus we can write

$$\Phi(x,t) = \min_{\lambda \geq 0}(\tilde{\Phi}(\lambda,t) + \lambda x). \tag{13.7.6}$$

13.7.2 Entrepreneur Problem

Define again the dual function $\tilde{\Phi}(\lambda,t) = \sup_{x \geq 0}(\Phi(x,t) - x\lambda)$. The supremum is attained at $x = G(\lambda,t)$, and we have $\frac{\partial \tilde{\Phi}(\lambda,t)}{\partial \lambda} = -G(\lambda,t)$. Using equation (13.4.14), we can write, as in the preceding section,

$$\frac{\partial}{\partial \lambda}[\frac{\partial \tilde{\Phi}}{\partial t} + \frac{\alpha^2}{2}\lambda^2 \frac{\partial^2 \tilde{\Phi}}{\partial \lambda^2} + \frac{\delta^2 \lambda^2}{2}] = 0.$$

Hence

$$\frac{\partial \tilde{\Phi}}{\partial t} + \frac{\alpha^2}{2}\lambda^2 \frac{\partial^2 \tilde{\Phi}}{\partial \lambda^2} + \frac{\delta^2 \lambda^2}{2} = c(t).$$

By calculations identical to those of the preceding section, we have also

$$\frac{\partial \Phi(G(\lambda,t),t)}{\partial t} + \frac{\delta^2}{2}(\frac{\partial \Phi(G(\lambda,t),t)}{\partial x})^2 - \frac{1}{2}\alpha^2 \frac{(\frac{\partial \Phi(G(\lambda,t),t)}{\partial x})^2}{\frac{\partial^2 \Phi(G(\lambda,t),t)}{\partial x^2}} = c(t),$$

and from the HJB equation (13.4.8), necessarily $c(t) = 0$. Therefore, the dual function is the solution of the linear PDE

$$\begin{aligned} \frac{\partial \tilde{\Phi}}{\partial t} + \frac{\alpha^2}{2}\lambda^2 \frac{\partial^2 \tilde{\Phi}}{\partial \lambda^2} + \frac{\delta^2 \lambda^2}{2} &= 0, \\ \tilde{\Phi}(\lambda,T) = \widetilde{cF}(\lambda) &= \tilde{F}(\lambda), \end{aligned} \tag{13.7.7}$$

and we can write the probabilistic representation formula

$$\tilde{\Phi}(\lambda, t) = \frac{\delta^2}{2}\lambda^2\tilde{T}_t + E\tilde{F}(\lambda Z_t(T)). \tag{13.7.8}$$

We can next write

$$\Phi(x, t) = \min_{\lambda \geq 0}(\tilde{\Phi}(\lambda, t) + x\lambda). \tag{13.7.9}$$

13.8 Optimal Retirement

13.8.1 Setting of the Problem

We consider the problem of the consumer–investor with an additional decision: when to retire. This problem has been considered by Fahri and Panageas [21]. We consider a more general framework for the utilities, and use variational inequalities (VI). Before retirement, a person benefits from a fixed income, which will disappear when he retires. We consider a stationary problem, which is simpler. So there is no upper bound on the retirement time. The initial time is then 0, and the wealth evolution follows the dynamics

$$\begin{aligned} dX &= rXds + X\sigma^*\varpi.(\theta ds + dw) + (I - C)ds, \\ X(0) &= x, \end{aligned} \tag{13.8.1}$$

in which I denotes the fixed income by unit of time. As in the consumer–investor case, the processes $C(s)$ and $\varpi(s)$ are controls. In addition, there is also a stopping time with respect to the filtration $\mathcal{F}^t$, denoted by τ. The payoff is defined by the formula

$$J_x(C(.), \varpi(.), \tau) = E[\int_0^\tau \exp -rt\, U_1(C(t))dt + \exp -r\tau\, U_2(X_x(\tau))], \tag{13.8.2}$$

and the value function is defined by

$$\Phi(x) = \sup_{C(.),\varpi(.),\tau} J_x(C(.), \varpi(.), \tau). \tag{13.8.3}$$

13.8.2 Variational Inequality

We state directly the problem obtained from dynamic programming arguments. The function $\Phi(x)$ is the solution of a variational inequality (instead of an equation), namely

$$\frac{1}{2}|\theta|^2\frac{(\Phi')^2}{\Phi''} - U_1(\beta_1(\Phi')) + \Phi'\beta_1(\Phi') - \Phi'(rx+I) + r\Phi \geq 0, \quad (13.8.4)$$
$$\Phi(x) - U_2(x) \geq 0,$$
$$\text{Product } = 0.$$

The argument x is positive, which corresponds to the fact that it is a possible value of wealth. The inequalities must hold a.e. x. We look for a function that is C^1 and monotonically increasing and concave. When we write "product," it means the product of the two positive terms. So for all x, one of the two positive terms must be 0. We recall that β_1 represents the inverse of U_1'. Similarly, β_2 used later represents the inverse of U_2'. We try to solve the variational inequality by claiming that there exists $\hat{x} > 0$ such that

$$\frac{1}{2}|\theta|^2\frac{(\Phi')^2}{\Phi''} - U_1(\beta_1(\Phi')) + \Phi'\beta_1(\Phi') - \Phi'(rx+I) + r\Phi = 0, \ 0 < x < \hat{x},$$
$$\Phi(x) - U_2(x) = 0, \ x > \hat{x}. \quad (13.8.5)$$

We write the smooth matching conditions

$$\Phi(\hat{x}) - U_2(\hat{x}) = 0, \quad (13.8.6)$$
$$\Phi'(\hat{x}) - U_2'(\hat{x}) = 0,$$

which allows us to obtain $\hat{x}$. We consider $\lambda(x) = \Phi'(x)$, which we expect to be a continuous positive monotonically decreasing function satisfying the equation

$$\frac{1}{2}|\theta|^2\lambda^2\frac{\lambda''}{(\lambda')^2} + \lambda'(rx+I-\beta_1(\lambda)) - |\theta|^2\lambda \quad = \quad 0, \ 0 < x < \hat{x}, \quad (13.8.7)$$
$$\lambda(x) = U_2'(x) \ x \geq \hat{x}.$$

We set $\hat{\lambda} = \lambda(\hat{x}) = U_2'(\hat{x})$. Therefore, $\hat{x} = \beta_2(\hat{\lambda})$. At this stage we consider that $\hat{\lambda}$ is fixed and positive (hence $\hat{x}$ is also fixed). We try to define $\lambda(x)$ by inverting a function $G(\lambda)$. More precisely, we look for $\lambda(x)$ by writing

$$G(\lambda(x)) = x + \frac{I}{r}. \quad (13.8.8)$$

Since $\lambda(x)$ is decreasing, for $x < \hat{x}$ we have $\lambda(x) > \hat{\lambda}$. So $G(\lambda)$ is defined on $[\hat{\lambda}, +\infty)$ and satisfies the boundary condition

$$G(\hat{\lambda}) = \beta_2(\hat{\lambda}) + \frac{I}{r}. \quad (13.8.9)$$

Note that $G(\lambda)$ must be positive and decreasing. From (13.8.7), one obtains easily that $G(\lambda)$ satisfies the linear second-order differential equation

$$-|\theta|^2(\frac{1}{2}\lambda^2 G''(\lambda)+\lambda G'(\lambda))+rG(\lambda)=\beta_1(\lambda),\ \lambda>\hat{\lambda}. \tag{13.8.10}$$

13.8.3 Solution of the Linear Problem

We can find a function $G_0(\lambda)$ that satisfies (13.8.10) for all $\lambda > 0$. The following function has this property:

$$G_0(\lambda)=\int_0^{+\infty}\frac{\exp -rt}{|\theta|\sqrt{2\pi t}}\int_0^{+\infty}\beta_1(\lambda\eta)\exp[-\frac{1}{2}(\frac{1}{2}|\theta|\sqrt{t}+\frac{1}{|\theta|\sqrt{t}}\log\eta)^2]d\eta dt. \tag{13.8.11}$$

Although a direct calculation is possible, the best way to check that $G_0(\lambda)$ satisfies (13.8.10) for all positive λ is to use probabilistic techniques. First consider a function $\Gamma_0(\lambda)$, the solution of

$$-\frac{|\theta|^2}{2}\lambda^2\Gamma_0''(\lambda)+r\Gamma_0(\lambda)=\lambda\beta_1(\lambda). \tag{13.8.12}$$

Then $G_0(\lambda)=\dfrac{\Gamma_0(\lambda)}{\lambda}$ will satisfy (13.8.10). Consider the stochastic differential equation

$$d\zeta=\zeta|\theta|dB,$$
$$\zeta(0)=\lambda,$$

where $B(t)$ is a standard scalar Wiener process. By Itô's formula,

$$d(\Gamma_0(\zeta(t))\exp -rt)=-\zeta(t)\beta_1(\zeta(t))\exp -rt\,dt+$$
$$+\Gamma_0'(\zeta(t))\zeta(t)|\theta|\exp -rt\,dB,$$

and taking the mathematical expectation, we obtain easily

$$\Gamma_0(\lambda)=\int_0^{+\infty}\exp -rt\ E\zeta(t)\beta_1(\zeta(t))dt.$$

This result needs some assumptions on the function $\beta_1(\lambda)$ so that the expectation of the stochastic integral vanishes, and also $E\Gamma_0(\zeta(t))\exp -rt \to 0$ as $t\to+\infty$. Formula (13.8.11) follows easily.

We note that $G_0(\lambda)$ is positive and monotonically decreasing. Since $\beta_1(\lambda)\to 0$ as $\lambda\to+\infty$, we see that $G_0(\lambda)\to 0$ as $\lambda\to+\infty$. Now since (13.8.10) is linear, the solution is of the form

$$G(\lambda) = G_0(\lambda) + A\lambda^{-\gamma},$$

with

$$\gamma(\gamma - 1) = \frac{2r}{|\theta|^2} \tag{13.8.13}$$

and $\gamma > 1$. We obtain the constant A by matching the boundary condition. Hence

$$A = (\beta_2(\hat{\lambda}) + \frac{I}{r} - G_0(\hat{\lambda}))\hat{\lambda}^{\gamma}.$$

We need to have $A > 0$. So we make the assumption

$$\beta_2(\lambda) - G_0(\lambda) \geq 0. \tag{13.8.14}$$

We obtain the formula

$$G(\lambda) = G_0(\lambda) + (\beta_2(\hat{\lambda}) + \frac{I}{r} - G_0(\hat{\lambda}))(\frac{\lambda}{\hat{\lambda}})^{-\gamma}. \tag{13.8.15}$$

With this assumption we have obtained a positive monotonically decreasing solution of (13.8.9), (13.8.10) such that $G(\lambda) \to 0$, as $\lambda \to +\infty$. For $x < \hat{x}$, we have $x + \frac{I}{r} < G(\hat{\lambda})$, so there exists a single $\lambda(x) > \hat{\lambda}$ such that $G(\lambda) = x + \frac{I}{r}$. The function $\lambda(x)$ is positive, monotonically decreasing, and satisfies (13.8.10) and the boundary condition $\lambda(\hat{x}) = \hat{\lambda}$.

13.8.4 Equation for $\hat{\lambda}$

To obtain an equation for $\hat{\lambda}$, we match the values of Φat $\hat{x}$. On the one hand, $\Phi(\hat{x}) = U_2(\hat{x}) = U_2(\beta_2(\hat{\lambda}))$. On the other hand, by the HJB equation (13.8.5), we may write

$$r\Phi(x) = -\frac{1}{2}|\theta|^2\lambda^2 G'(\lambda) + U_1(\beta_1(\lambda)) - \lambda\beta_1(\lambda) + r\lambda G(\lambda), \tag{13.8.16}$$

in which λ must be replaced with $\lambda(x)$. Applying $\lambda = \hat{\lambda}$, we obtain

$$\begin{aligned} -\frac{1}{2}|\theta|^2(\hat{\lambda})^2 G'(\hat{\lambda}) + U_1(\beta_1(\hat{\lambda})) - \hat{\lambda}\beta_1(\hat{\lambda}) \\ -r(U_2(\beta_2(\hat{\lambda})) - \hat{\lambda}\beta_2(\hat{\lambda})) + \hat{\lambda}I = 0. \end{aligned} \tag{13.8.17}$$

From the formula (13.8.16) we get the equation

$$-\frac{1}{2}|\theta|^2[(\hat{\lambda})^2 G_0'(\hat{\lambda}) + \gamma\hat{\lambda}(G_0(\hat{\lambda}) - \beta_2(\hat{\lambda}))] + \hat{\lambda}I(1+\gamma\frac{|\theta|^2}{2r}) + \quad (13.8.18)$$
$$+U_1(\beta_1(\hat{\lambda})) - \hat{\lambda}\beta_1(\hat{\lambda}) - r(U_2(\beta_2(\hat{\lambda})) - \hat{\lambda}\beta_2(\hat{\lambda})) = 0.$$

Next, we define the function

$$F(\lambda) = -\frac{1}{2}|\theta|^2[(\lambda)^2 G_0'(\lambda) + \gamma\lambda(G_0(\lambda) - \beta_2(\lambda))] + \lambda I(1+\gamma\frac{|\theta|^2}{2r}) +$$
$$+U_1(\beta_1(\lambda)) - \lambda\beta_1(\lambda) - r(U_2(\beta_2(\lambda)) - \lambda\beta_2(\lambda)).$$

We compute $F'(\lambda)$. After some easy computations and rearrangements, using the equation of G_0 and the definition of γ, we obtain the formula

$$F'(\lambda) = \frac{r}{\gamma - 1}[\gamma\frac{I}{r} + \gamma(\beta_2(\lambda) - G_0(\lambda)) + \lambda(\beta_2'(\lambda) - G_0'(\lambda))].$$

To obtain $F'(\lambda) > 0$, it is sufficient to reinforce (13.8.14) by

$$\lambda^\gamma(\beta_2(\lambda) - G_0(\lambda)) \uparrow \geq 0. \quad (13.8.19)$$

We need also to assume

$$F(\lambda) \to -\infty, \text{ as } \lambda \to 0, \quad (13.8.20)$$
$$F(\lambda) \to +\infty, \text{ as } \lambda \to +\infty.$$

Since the term in square brackets in the definition of $F(\lambda)$ is negative, the second assumption is satisfied whenever $+U_1(\beta_1(\lambda)) - \lambda\beta_1(\lambda) - r(U_2(\beta_2(\lambda)) - \lambda\beta_2(\lambda))$ is bounded below as $\lambda \to +\infty$. So with assumptions (13.8.19) and (13.8.20), there exists a unique solution $\hat{\lambda} > 0$ of (13.8.18).

We now define the function $\Phi(x)$. We shall use formula (13.8.16) together with (13.8.15). Combining, we obtain $\Phi(x) = \Psi(\lambda(x))$ with

$$r\Psi(\lambda) = -\frac{1}{2}|\theta|^2\lambda^2 G_0'(\lambda) + \lambda r G_0(\lambda) + U_1(\beta_1(\lambda)) - \lambda\beta_1(\lambda) + \quad (13.8.21)$$
$$+ r(1+\gamma\frac{|\theta|^2}{2r})\frac{(\beta_2(\hat{\lambda}) + \frac{I}{r} - G_0(\hat{\lambda}))}{\hat{\lambda}^{-\gamma}}\lambda^{-\gamma+1}.$$

We want to check that Ψ is the solution of

$$-\frac{1}{2}|\theta|^2\lambda^2\Psi'' + r\Psi = U_1(\beta_1(\lambda)),\ \lambda > \hat{\lambda}, \quad (13.8.22)$$

$$\Psi(\hat{\lambda}) = U_2(\beta_2(\hat{\lambda})).$$

In contrast to $G(\lambda)$, $\Psi(\lambda)$ does not vanish at infinity. Its behavior is dictated by formula (13.8.21). To check (13.8.22), we first notice that $\psi(\lambda) = \lambda^{-\gamma+1}$ satisfies

$$-\frac{1}{2}|\theta|^2\lambda^2\psi'' + r\psi = 0$$

because of the choice of γ; see (13.8.13). So if we define $\Psi_1(\lambda)$ by the formula

$$r\Psi_1(\lambda) = -\frac{1}{2}|\theta|^2\lambda^2 G_0'(\lambda) + \lambda r G_0(\lambda) + U_1(\beta_1(\lambda)) - \lambda\beta_1(\lambda),$$

we must check that

$$-\frac{1}{2}|\theta|^2\lambda^2\Psi_1'' + r\Psi_1 = U_1(\beta_1(\lambda)).$$

This is easily checked, from the definition of $G_0(\lambda)$. We also can check that

$$\Psi'(\lambda) = \lambda G'(\lambda),\ \lambda \geq \hat{\lambda}. \tag{13.8.23}$$

The boundary condition at $\hat{\lambda}$ is also obtained directly from formula (13.8.22), using the definition of $\hat{\lambda}$; see (13.8.18). We now have completely defined $\Phi(x)$ by

$$\Phi(x) = \Psi(\lambda(x)),\ 0 < x < \hat{x},$$
$$\Phi(x) = U_2(x),\ x \geq \hat{x}. \tag{13.8.24}$$

We have $\Phi'(x) = \lambda(x)$, and $\Phi(x)$ is C^1. We claim that $\Phi(x)$ solves the variational inequality (13.8.4). From the construction of $\Phi(x)$, what remains to be done is to check

$$\Phi(x) \geq U_2(x),\ x < \hat{x}, \tag{13.8.25}$$
$$\frac{1}{2}|\theta|^2\frac{(\Phi')^2}{\Phi''} - U_1(\beta_1(\Phi')) + \Phi'\beta_1(\Phi') - \Phi'(rx + I) + r\Phi \geq 0,\ x > \hat{x}.$$

We begin with the first inequality. Set $\chi(x) = \Phi(x) - U_2(x)$. Since $\chi(\hat{x}) = 0$, it is sufficient to prove that $\chi'(x) \leq 0$ for $x < \hat{x}$. This means $\lambda(x) \leq U_2'(x)$ for $x < \hat{x}$ or $x \leq \beta_2(\lambda(x))$, which means also that $G(\lambda(x)) - \frac{I}{r} \leq \beta_2(\lambda(x))$, for $x < \hat{x}$. Therefore, it is sufficient to prove that

$$\Gamma(\lambda) = G(\lambda) - \frac{I}{r} - \beta_2(\lambda) < 0,\ \lambda > \hat{\lambda}. \tag{13.8.26}$$

We have $\Gamma(\hat{\lambda}) = 0$ and $\Gamma(+\infty) = -\frac{I}{r}$. Moreover, $\Gamma(\lambda)$ satisfies the equation

$$-|\theta|^2(\frac{1}{2}\lambda^2\Gamma''(\lambda) + \lambda\Gamma'(\lambda)) + r\Gamma(\lambda) = -I + \beta_1(\lambda) - r\beta_2(\lambda) + |\theta|^2(\frac{1}{2}\lambda^2\beta_2'' + \lambda\beta_2').$$

So if we assume

$$-|\theta|^2(\frac{1}{2}\lambda^2\beta_2'' + \lambda\beta_2') + r\beta_2(\lambda) \geq \beta_1(\lambda), \ \forall\lambda > 0, \tag{13.8.27}$$

it follows immediately that $\Gamma(\lambda) \leq 0$ for $\lambda \geq \hat{\lambda}$. We turn to the second inequality (13.8.25). Setting

$$L(x) = \frac{1}{2}|\theta|^2\frac{(\Phi')^2}{\Phi''} - U_1(\beta_1(\Phi')) + \Phi'\beta_1(\Phi') - \Phi'(rx + I) + r\Phi,$$

we have

$$L'(x) = |\theta|^2(\lambda - \frac{1}{2}\frac{\lambda^2\lambda''}{(\lambda')^2}) + \lambda'(\beta_1(\lambda) - rx - I).$$

But for $x > \hat{x}$, we have $\lambda(x) = U_2'(x)$. So $x = \beta_2(\lambda(x))$. Moreover,

$$L'(x) = |\theta|^2(U_2'(x) - \frac{1}{2}\frac{(U_2'(x))^2U_2'''(x)}{(U_2''(x))^2}) + U_2''(x)(\beta_1(U_2'(x)) - rx - I).$$

Let us check that $L'(\beta_2(\lambda)) \geq 0$. It will follow that $L'(x) = L'(\beta_2(\lambda(x))) \geq 0$ for $x > \hat{x}$. From $U_2'(\beta_2(\lambda)) = \lambda$, we deduce

$$U_2''(\beta_2(\lambda))\beta_2'(\lambda) = 1, \ U_2'''(\beta_2(\lambda))(\beta_2'(\lambda))^2 + U_2''(\beta_2(\lambda))\beta_2''(\lambda) = 0.$$

Using these formulas in the calculation of $L'(\beta_2(\lambda))$, we obtain the formula

$$L'(\beta_2(\lambda)) = \frac{1}{\beta_2'(\lambda)}[-I + \beta_1(\lambda) - r\beta_2(\lambda) + |\theta|^2(\frac{1}{2}\lambda^2\beta_2'' + \lambda\beta_2')].$$

From the assumption (13.8.27) we see immediately that $L'(\beta_2(\lambda)) \geq 0$. So we have proved that $L(x)$ is monotonically increasing for $x > \hat{x}$. To check that $L(x) > 0$, it suffices to check that $L(\hat{x}) \geq 0$. However, for $x < \hat{x}$, we have

$$L(x) = \frac{1}{2}|\theta|^2(\lambda(x))^2G'(\lambda(x)) - U_1(\beta_1(\lambda(x))) + \lambda(x)\beta_1(\lambda(x)) - \lambda(x)(rx + I) + r\Phi(x).$$

Taking $x = \hat{x}$, we obtain

$$L(\hat{x}) = \frac{1}{2}|\theta|^2\hat{\lambda}^2 G'(\hat{\lambda}) - U_1(\beta_1(\hat{\lambda})) + \hat{\lambda}\beta_1(\hat{\lambda}) - \hat{\lambda}(r\beta_2(\hat{\lambda}) + I) + rU_2(\beta_2(\hat{\lambda})),$$

and from the definition of $\hat{\lambda}$, see (13.8.17), it follows that $L(\hat{x}) = 0$. Finally, we note that $\Phi(x)$ is monotonically increasing, since $\Phi'(x) = \lambda(x) > 0$ and is concave, since it is concave on $[0, \hat{x}]$ and on $[\hat{x}, +\infty)$ and is C^1. We thus have proved the following result.

PROPOSITION 13.2. *We assume (13.8.19), (13.8.20), (13.8.27), and (13.1.9) for the utility functions. Then the function $\Phi(x)$ defined by (13.8.24) is the solution of the variational inequality (13.8.4). It is C^1, monotonically increasing, and concave.* ■

EXERCISE 13.6. Consider the case

$$U_1(x) = \frac{x^\delta}{\delta},\ U_2(x) = K\frac{x^\delta}{\delta},\ 0 \le \delta < 1;$$

the case $\delta = 0$ is interpreted as $\log x$. Check that the assumptions of Proposition 13.2 are satisfied when $\gamma > \dfrac{1}{1-\delta}$ and r is sufficiently large.

13.8.5 Solution of the Optimal Retirement Problem

We can now give the full solution of the problem. On $(0, \hat{x})$ we consider the same feedbacks as in the consumer–investor problem

$$\hat{c}(x) = \beta_1(\Phi'(x)), \tag{13.8.28}$$

$$\hat{\varpi}(x) = -\frac{\Phi'(x)}{x\Phi''(x)}(\sigma^*)^{-1}\theta, \tag{13.8.29}$$

and use them in the wealth equation (13.8.1) to obtain

$$d\hat{X} = (r\hat{X} - |\theta|^2\frac{\Phi'(\hat{X})}{\Phi''(\hat{X})} - \beta_1(\Phi'(\hat{X})) + I)ds - \frac{\Phi'(\hat{X})}{\Phi''(\hat{X})}\theta.dw, \tag{13.8.30}$$

$$\hat{X}0) = x.$$

If we introduce the process

$$d\zeta = -\zeta(s)\theta.dw, \tag{13.8.31}$$

$$\zeta(0) = \lambda(x),$$

then we get the important property

$$\hat{X}(s) = G(\zeta(s)) - \frac{I}{r}. \tag{13.8.32}$$

We define the stopping time

$$\hat{\tau} = \hat{\tau}_x = \inf\{s|\hat{X}(s) \geq \hat{x}\}. \tag{13.8.33}$$

So between 0 and $\hat{\tau}_x$, the process $\hat{X}(s)$ remains in the interval $[0, \hat{x}]$. We define the processes

$$\begin{aligned}\hat{C}(s) &= \beta_1(\zeta(s)), \qquad (13.8.34)\\ \hat{\varpi}(s)\hat{X}(s) &= -\zeta(s)G'(\zeta(s))(\sigma^*)^{-1}\theta\\ &= -\Psi'(\zeta(s))(\sigma^*)^{-1}\theta,\end{aligned}$$

where $\Psi(\lambda)$ has been defined in (13.8.21). We have $\Phi(\hat{X}(s)) = \Psi(\zeta(s))$. We note that $\lambda(x) \geq \hat{\lambda}$, and $\hat{\tau}_x$ is the first time that $\zeta(s)$ reaches $\hat{\lambda}$. Using (13.8.22), we get easily

$$\Phi(x) = \Psi(\lambda(x)) = J_x(\hat{C}(.), \hat{\varpi}(.), \hat{\tau}).$$

We can also prove that $\Phi(x) \geq J_x(C(.), \varpi(.), \tau)$ for every triple $C(.), \varpi(.), \tau$. This proves the optimality of the triple $\hat{C}(s)$, $\hat{\varpi}(s)$, $\hat{\tau}$.

Chapter 14
Stochastic Control for Non-Markov Processes

14.1 Statement of the Problem

We consider a probability space $\Omega, \mathcal{A}, P$ equipped with a filtration $\mathcal{F}^t$ and a standard $P, \mathcal{F}^t$ Wiener process with values in R^n. We consider

$$g(x, v) : R^n \times R^m \to \text{adapted process with values in } R^n, \tag{14.1.1}$$

$$\sigma(x) : R^n \to \text{adapted process with values in } \mathcal{L}(R^n; R^n), \text{ invertible.}$$

The two approaches of strong and weak formulation of SDEs will lead to interesting differences in the context of non-Markov processes. We begin with the weak formulation of the state equation as

$$\begin{aligned} dX &= \sigma(X(s))dw(s), \\ X(t) &= x. \end{aligned} \tag{14.1.2}$$

We have used capital letters to stress the fact that $X_{xt}(s)$ is a fixed adapted process, with parameters x, t representing the initial conditions. We next consider a control $v(s)$ that is adapted to the filtration $\mathcal{F}^t$ and define the process

$$\begin{aligned} M^v_{xt}(s) = \exp\Bigg(&\int_t^s \sigma^{-1}(X_{xt}(\tau))g(X_{xt}(\tau), v(\tau))dw(\tau) \\ &-\frac{1}{2}\int_t^s |\sigma^{-1}(X_{xt}(\tau))g(X_{xt}(\tau), v(\tau))|^2 d\tau\Bigg). \end{aligned} \tag{14.1.3}$$

This process satisfies the stochastic differential equation

$$dM^v_{xt}(s) = M^v_{xt}(s)\sigma^{-1}(X_{xt}(s))g(X_{xt}(s), v(s))dw(s), \; M^v_{xt}(t) = 1.$$

A. Bensoussan, *Estimation and Control of Dynamical Systems*, Interdisciplinary Applied Mathematics 48, https://doi.org/10.1007/978-3-319-75456-7_14

The process $M^v_{xt}(s)$ is a $P, \mathcal{F}^s$ martingale. We define next the process $w^v_{xt}(s)$ by

$$dw^v_{xt}(s) = dw - \sigma^{-1}(X_{xt}(s))g(X_{xt}(s), v(s))ds, \; w^v_{xt}(t) = 0,$$

and we consider the probability P^v_{xt} on $\mathcal{F}^s, s > t$ given by

$$\frac{dP^v_{xt}}{dP}|_{\mathcal{F}^s} = M^v_{xt}(s).$$

The process $w^v_{xt}(s)$ is for $s > t$ a $P^v_{xt}, \mathcal{F}^s$ standard Wiener process. Moreover, on $\Omega, \mathcal{F}^s, P^v_{xt}$, the process $X(s)$ appears as the solution of

$$dX = g(X, v)ds + \sigma(X)dw^v(s), \tag{14.1.4}$$

where we have omitted to write explicitly the parameters x, t. We are interested in the random field

$$\Phi^v(x,t) = E[M^v_{xt}(T)h(X_{xt}(T)) + \int_t^T M^v_{xt}(s)l(X_{xt}(s), v(s))ds|\mathcal{F}^t], \tag{14.1.5}$$

in which $l(x, v, s)$ is an adapted random field and $h(x)$ is a random variable that $\mathcal{F}^T$-measurable.

We note also that

$$\Phi^v(x,t) = E^v[h(X_{xt}(T)) + \int_t^T l(X_{xt}(s), v(s))ds|\mathcal{F}^t].$$

We next consider the strong formulation. First note that in the preceding formulation there is no difference between an open-loop control and a closed-loop control. Indeed, since $X(s)$ is a fixed process, we may consider a control given by a feedback $v(x, s)$ to be an adapted process $v(X(s), s)$. We shall compare the impact of feedbacks on the strong and weak formulations. By strong formulation of the state, we mean that the state $x^v_{xt}(s)$ depends now on the control and is the solution of the equation

$$\begin{aligned} dx &= g(x, v(x,s))ds + \sigma(x)dw, \\ x(t) &= x, \end{aligned} \tag{14.1.6}$$

and we define the random field

$$\Psi^v(x,t) = E[h(x^v_{xt}(T) + \int_t^T l(x^v_{xt}(s), v(x^v_{xt}(s), s))ds|\mathcal{F}^t]. \tag{14.1.7}$$

The two random fields $\Phi^v(x, t)$ and $\Psi^v(x, t)$ differ in general. They coincide only when they are deterministic. This is because the processes $w(s)$ with probability P in (14.1.6) and $w_{xt}^v(s)$ with probability P_{xt}^v in (14.1.4) are both standard Wiener processes. When g, σ are deterministic, $x_{xt}^v(s)$ and $X_{xt}(s)$ are identical deterministic functionals of $w(.)$ and $w_{xt}^v(s)$ respectively. When h and l are deterministic, $h(x_{xt}^v(T) + \int_t^T l(x_{xt}^v(s), v(x_{xt}^v(s), s))ds$ and $h(X_{xt}(T) + \int_t^T l(X_{xt}(s), v(X_{xt}(s), s))ds$ are identical deterministic functionals of $w(.)$ and $w_{xt}^v(s)$ respectively. Therefore, $\Phi^v(x,t)$ and $\Psi^v(x,t)$ are deterministic and coincide.

We are interested in the minimization problems

$$\Phi(x,t) = \inf_v \Phi^v(x,t), \tag{14.1.8}$$

$$\Psi(x,t) = \inf_v \Psi^v(x,t).$$

Since $\Phi^v(x,t)$ is a random field, the infimum is not the pointwise infimum, but the essential infimum. So we want a random field $\Phi(x,t)$ such that $\Phi(x,t) \le \Phi^v(x,t)$, a.s. $\forall v$, and such that it is the maximal one. We still call this random field the *value function*. The same applies to $\Psi(x,t)$.

14.2 Backward Stochastic Partial Differential Equations

We have noticed that the set of open-loop and closed-loop controls in the first problem (14.1.8) coincide, since the process $X_{xt}(s)$ is fixed. Nevertheless, when we consider a control v, we will prefer to assume that it is a random field $v(x,s)$. We claim that $\Phi^v(x,t)$ is the solution of the following BSPDE (backward stochastic partial differential equation):

$$-d_t\Phi^v + A\Phi^v dt = [l(x, v(x)) + D\Phi^v.g(x, v(x)) + K^v.\sigma^{-1}(x)g(x, v(x)) + \text{tr}(DK^v\sigma)]dt$$
$$-K^v.dw(t),\ \Phi^v(x,T) = h(x), \tag{14.2.1}$$

where we recall the definition of the operator $A: A\varphi(x) = -\sum_{i,j=1}^n a_{ij}(x)\dfrac{\partial^2\varphi}{\partial x_i \partial x_j}$ (x), with $a(x) = \dfrac{1}{2}\sigma(x)\sigma^*(x)$. In fact, this equation has two unknown random fields $\Phi^v(x,t)$ and $K^v(x,t)$. At this stage, we assume that this problem has a solution that is uniquely defined. We proceed formally, so we do not make precise the functional spaces in which the random fields are defined. When l, g, h are deterministic functions, then we have $K^v = 0$, and $\Phi^v(x,t)$ is deterministic, the solution of the linear PDE

$$-\frac{\partial\Phi^v}{\partial t} + A\Phi^v = l(x, v(x)) + D\Phi^v.g(x, v(x)),$$

$$\Phi^v(x,T) = h(x).$$

To check that the solution of (14.2.1) coincides with (14.1.5), we use a verification argument. We need an extension of Itô's formula, due to Kunita [30]. Consider a random field $\Gamma(x,t)$ such that

$$-d_t\Gamma = L(x,t)dt - K(x,t).dw. \tag{14.2.2}$$

Then the process $\Gamma(X(s),s)$ has the Itô differential

$$\begin{aligned} d\Gamma(X(s),s) &= [-L(X(s),s) - A\Gamma(X(s),s) + \mathrm{tr}(DK\,\sigma)(X(s),s)]ds + \\ &+ D\Gamma(X(s),s).\sigma(X(s))dw(s) + K(X(s),s).dw(s), \end{aligned} \tag{14.2.3}$$

in which $X(s)$ is the solution of (14.1.2). Note that there is no drift term in (14.1.2). If there were a drift, then formula (14.2.3) should incorporate it exactly as in the ordinary Itô's formula. We apply (14.2.3) to $\Phi^v(x,t)$ and obtain

$$\begin{aligned} d\Phi^v(X(s),s) &= [-l(X(s),v(X(s)) - D\Phi^v.g(X(s),v(X(s)) - K^v.\sigma^{-1}g(X(s),v(X(s)))]ds+ \\ &+ D\Phi^v(X(s),s).\sigma(X(s))dw(s) + K^v(X(s),s).dw(s). \end{aligned} \tag{14.2.4}$$

We then deduce

$$\begin{aligned} d(M^v(s)\Phi^v(X(s),s)) &= -M^v(s)l(X(s),v(X(s))ds+ \\ &+M^v(s)[\Phi(X(s),s)\sigma^{-1}g(X(s),v(X(s))) + K^v(X(s),s) \\ &+\sigma^* D\Phi^v(X(s),s)].dw(s). \end{aligned} \tag{14.2.5}$$

Integrating between t and T, and taking the conditional expectation with respect to $\mathcal{F}^t$, we obtain that $\Phi^v(x,t)$ coincides with the right-hand side of (14.1.5).

We next proceed with another characterization of $\Phi^v(x,t)$. We introduce a stochastic Fokker–Planck equation, the solution of which is denoted by $m^v_{xt}(\xi,s)$, $s>t$. It is defined as follows:

$$\begin{aligned} d_s m &= [-\mathrm{div}\,(mg(\xi,v(\xi))) + \sum_{ij=1}^n \frac{\partial^2(ma_{ij}(\xi))}{\partial\xi_i\partial\xi_j}]ds + \\ &+ -\sum_{ij=1}^n \frac{\partial}{\partial\xi_i}(m\sigma_{ij}(\xi))dw_j + m\sigma^{-1}g(\xi,v(\xi)).dw, \\ m(\xi,t) &= \delta(\xi - x). \end{aligned} \tag{14.2.6}$$

The initial condition at time t is the Dirac measure at the point x. We claim that

$$\Phi^v(x,t) = E[\int_t^T \int m^v_{xt}(\xi,s)l(\xi,v(\xi,s))d\xi ds + \int m^v_{xt}(\xi,T)h(\xi)d\xi|\mathcal{F}^t]. \tag{14.2.7}$$

Indeed, by Itô's formula, we have

$$d_s \int m^v(\xi, s)\Phi^v(\xi, s)d\xi = \int d_s m^v \Phi^v d\xi + \int m^v d_s \Phi^v d\xi + \\ + \int m^v(\text{tr}\,(DK^v\sigma) + K^v.\sigma^{-1}g)d\xi ds.$$

Replacing the differentials of m^v and Φ^v, we obtain

$$d_s \int m^v(\xi, s)\Phi^v(\xi, s)d\xi = -\int m^v(\xi, s)l(\xi, v(\xi))d\xi ds + \\ + \int m^v[D\Phi^v.\sigma dw + (\sigma^{-1}g + K^v).dw]d\xi.$$

Integrating between t and T and taking the conditional expectation with respect to $\mathcal{F}^t$, we obtain formula (14.2.7). When the functions $l(x, v)$ and $h(x)$ are deterministic, we get

$$\Phi^v(x, t) = \int_t^T \int \bar{m}^v_{xt}(\xi, s)l(\xi, v(\xi, s))d\xi ds + \int \bar{m}^v_{xt}(\xi, T)h(\xi)d\xi,$$

where $\bar{m}^v_{xt}(\xi, s) = E[m^v_{xt}(\xi, s)|\mathcal{F}^t]$, which is the solution of the deterministic Fokker–Planck equation

$$\frac{\partial \bar{m}}{\partial s} = -\text{div}\,(\bar{m}g(\xi, v(\xi))) + \sum_{ij=1}^n \frac{\partial^2(\bar{m}a_{ij}(\xi))}{\partial\xi_i\partial\xi_j},$$

$$\bar{m}(\xi, t) = \delta(\xi - x).$$

A third characterization can be given, derived from (14.2.4). Defining $\Phi^v_{xt}(s) = \Phi^v(X_{xt}(s), s)$, $v_{xt}(s) = v(X_{xt}(s), s)$, $Z^v_{xt}(s) = K^v(X_{xt}(s), s) + \sigma^*(X_{xt}(s), s)D\Phi^v(X_{xt}(s), s)$, we can rewrite (14.2.4) as follows:

$$\Phi^v_{xt}(s) = h(X_{xt}(T)) + \int_s^T [l(X_{xt}(\tau), v_{xt}(\tau)) + (\sigma^{-1}(X_{xt}(\tau)), \tau)^* Z^v_{xt}(\tau).g(X_{xt}(\tau), v_{xt}(\tau))]d\tau - \\ - \int_s^T Z^v_{xt}(\tau).dw(\tau). \tag{14.2.8}$$

This is an equation with unknowns $\Phi^v_{xt}(s)$, $Z^v_{xt}(s)$. We recover $\Phi^v(x, t) = \Phi^v_{xt}(t)$. We also check easily, from (14.2.5), that

$$\Phi^v_{xt}(s) = E^v[h(X_{xt}(T)) + \int_s^T l(X_{xt}(\tau), v_{xt}(\tau))d\tau|\mathcal{F}^s]. \tag{14.2.9}$$

Considering now the strong formulation (14.1.6), (14.1.7), we get

$$-d_t\Psi^v + A\Psi^v dt = [l(x, v(x)) + D\Psi^v.g(x, v(x)) + \text{tr}(D\chi^v\sigma)]dt - \chi^v.dw(t), \tag{14.2.10}$$

$$\Psi^v(x, T) = h(x),$$

and m^v_{xt} is changed into μ^v_{xt}, the solution of

$$d_s\mu = [-\text{div}\,(\mu g(\xi, v(\xi))) + \sum_{ij=1}^{n} \frac{\partial^2(\mu a_{ij}(\xi))}{\partial\xi_i\partial\xi_j}]ds + \tag{14.2.11}$$

$$-\sum_{ij=1}^{n} \frac{\partial}{\partial\xi_i}(\mu\sigma_{ij}(\xi))dw_j,$$

$$\mu(\xi, t) = \delta(\xi - x),$$

with

$$\Psi^v(x, t) = E[\int_t^T \int \mu^v_{xt}(\xi, s)l(\xi, v(\xi, s))d\xi ds + \int \mu^v_{xt}(\xi, T)h(\xi)d\xi|\mathcal{F}^t]. \tag{14.2.12}$$

Also, on setting $\Psi^v_{xt}(s) = \Psi^v(x^v_{xt}(s), s)$, $v_{xt}(s) = v(x^v_{xt}(s), s)$, $Z^v_{xt}(s) = \chi^v(x^v_{xt}(s), s) + \sigma^*(x^v_{xt}(s), s)D\Psi^v(x^v_{xt}(s), s)$, we obtain

$$\Psi^v_{xt}(s) = E[h(x^v_{xt}(T)) + \int_s^T l(x^v_{xt}(\tau), v_{xt}(\tau))d\tau|\mathcal{F}^s], \tag{14.2.13}$$

$$\Psi^v_{xt}(s) = h(x^v_{xt}(T)) + \int_s^T l(x^v_{xt}(\tau), v_{xt}(\tau))d\tau - \int_s^T Z^v_{xt}(\tau).dw(\tau). \tag{14.2.14}$$

For $v(.)$ fixed we can reconcile (14.2.1) and (14.2.10) by replacing w by w^v in (14.2.10). Similarly, we can reconcile (14.2.9) and (14.2.12) by replacing P by P^v and x^v by X in (14.2.12), and we can reconcile (14.2.8) and (14.2.14) by replacing w by w^v and x^v by X in (14.2.14).

14.3 Nonlinear BSPDE

We recall the definition of the Hamiltonian

$$H(x, q) = \inf_v(l(x, v) + q.g(x, v)). \tag{14.3.1}$$

We omit to put explicitly the constraints on the control, since we are not developing a rigorous approach. We want to characterize the value function $\Phi(x,t)$, defined by (14.1.8), as the solution of the stochastic nonlinear BSPDE

$$-d_t\Phi + A\Phi dt = [H(x, D\Phi + (\sigma^{-1})^* K) + \text{tr}(DK\,\sigma)]dt - K.dw \quad (14.3.2)$$
$$\Phi(x,T) = h(x).$$

There are two unknowns, $\Phi(x,t)$ and $K(x,t)$, which are adapted random fields.

If we consider instead

$$\Psi(x,t) = \inf_v \Psi^v(x,t), \quad (14.3.3)$$

then it will be the solution of a different BSPDE, namely

$$-d_t\Psi + A\Psi dt = [H(x, D\Psi) + \text{tr}(D\chi\,\sigma)]dt - \chi.dw, \quad (14.3.4)$$
$$\Psi(x,T) = h(x).$$

This last equation has been considered by Peng; see [42]. We shall check the characterization for $\Phi(x,t)$ formally. A similar argument holds for $\Psi(x,t)$. First, assume as usual that we can define a unique $\bar{v}(x,q)$ that achieves the infimum in the Hamiltonian, and consider the feedback

$$\hat{v}(x,t) = \bar{v}(x, D\Phi + (\sigma^{-1})^* K). \quad (14.3.5)$$

We see immediately, recalling (14.2.1), that

$$\Phi(x,t) = \Phi^{\hat{v}}(x,t). \quad (14.3.6)$$

On the other hand, for a feedback v, consider Φ^v defined by (14.2.1) and set $\tilde{\Phi}^v = \Phi^v - \Phi$, $\tilde{K}^v = K^v - K$. We have

$$\begin{aligned}
-d_t\tilde{\Phi}^v + A\tilde{\Phi}^v dt &= [l(x,v(x)) + D\Phi^v.g(x,v(x)) + K^v.\sigma^{-1}(x)g(x,v(x)) \\
&- H(x, D\Phi + (\sigma^{-1})^*K)]dt + \text{tr}(D\tilde{K}^v\,\sigma)dt - \tilde{K}^v.dw \\
&\geq [\,D\tilde{\Phi}^v.g(x,v(x)) + \tilde{K}^v.\sigma^{-1}(x)g(x,v(x)) + \text{tr}(D\tilde{K}^v\,\sigma)]dt - \tilde{K}^v.dw \\
&= [D\tilde{\Phi}^v.g(x,v(x)) + \text{tr}(D\tilde{K}^v\,\sigma)]dt - \tilde{K}^v.dw^v.
\end{aligned}$$

Consider $\tilde{\Phi}^v(X(s),s)$ and recall that for the probability P^v, w^v is a standard Wiener process, and $X(s)$ appears as the solution of (14.1.4). Using the Itô–Kunita formula, we obtain easily that

$$d\tilde{\Phi}^v(X(s),s) \leq (\tilde{K}^v + \sigma^* D\tilde{\Phi}^v)(X(s),s).dw^v.$$

Integrating between t and T and taking the P^v expectation with respect to $\mathcal{F}^t$, we obtain immediately $\tilde{\Phi}^v(x,t) \geq 0$, since $\tilde{\Phi}^v(x,T) = 0$.

Another way to recover (14.3.2) is to use formulas (14.2.6), (14.2.7). The approach is of the necessary condition type. Consider an optimal feedback $\hat{v}$ and let $m = m^{\hat{v}}$ be the corresponding solution of (14.2.6). We next consider another feedback v and $\hat{v} + \theta v$, where θ is a real number, which will tend to 0. We can define $m^{\hat{v}+\theta v}$ and

$$\tilde{m} = \frac{dm^{\hat{v}+\theta v}}{d\theta}|_{\theta=0}.$$

We check that $\tilde{m} = \tilde{m}_{xt}(\xi, s)$ is the solution of

$$d_s\tilde{m} = [-\text{div}\,(\tilde{m}g(\xi,\hat{v}(\xi))) + \sum_{ij=1}^{n} \frac{\partial^2(\tilde{m}a_{ij}(\xi))}{\partial\xi_i\partial\xi_j} - \text{div}(mD_v g(\xi,\hat{v}(\xi))v(\xi))]ds +$$

$$+ - \sum_{ij=1}^{n} \frac{\partial}{\partial\xi_i}(\tilde{m}\sigma_{ij}(\xi))dw_j + (\tilde{m}\sigma^{-1}g(\xi,\hat{v}(\xi)) + m\sigma^{-1}D_v g(\xi,\hat{v}(\xi))v(\xi)).dw,$$

$$\tilde{m}(\xi,t) = 0. \tag{14.3.7}$$

From formula (14.2.7) we obtain

$$\frac{d\Phi^{\hat{v}+\theta v}}{d\theta}|_{\theta=0} = E[\int_t^T \int \tilde{m}(\xi,s)l(\xi,\hat{v}(\xi))d\xi ds + \int \tilde{m}(\xi,T)h(\xi)d\xi +$$

$$+ \int_t^T \int m(\xi,s)D_v l(\xi,\hat{v}(\xi))v(\xi)d\xi ds|\mathcal{F}^t].$$

We set $\Phi(\xi,s) = \Phi^{\hat{v}}(\xi,s)$, noting that

$$\int \tilde{m}(\xi,T)h(\xi)d\xi = \int \tilde{m}(\xi,T)\Phi(\xi,T)d\xi =$$

$$= \int_t^T \int d_s(\tilde{m}(\xi,s)\Phi(\xi,s)).$$

Considering equation (14.3.7) and the equation of $\Phi(\xi,s)$ and applying Itô's calculus, we get

$$\frac{d\Phi^{\hat{v}+\theta v}}{d\theta}|_{\theta=0} = E\int_t^T \int m(\xi,s)[D_v l(\xi,\hat{v}(\xi))v(\xi) +$$

$$+ (D\Phi + (\sigma^{-1})^* K)(\xi,s).\, D_v g(\xi,\hat{v}(\xi))v(\xi)]d\xi ds|\mathcal{F}^t].$$

Assuming that $m(\xi, s)$ does not vanish, and expressing that the Gâteaux differential must vanish for all v if $\hat{v}$ is optimal, we obtain

$$D_v l(\xi, \hat{v}(\xi)) + (D\Phi + (\sigma^{-1})^* K)^*(\xi, s) D_v g(\xi, \hat{v}(\xi)) = 0,$$

which we interpret as saying that $l(\xi, v) + (D\Phi + (\sigma^{-1})^* K)(\xi).g(\xi, v)$ attains its infimum in v at the point $\hat{v}(\xi, s)$. It is then clear that $\Phi(\xi, s)$ is the solution of (14.3.2).

Consider next the process $\Phi_{xt}(s) = \Phi(X_{xt}(s), s)$. Using equation (14.3.2) and applying the Itô–Kunita formula, we can check that

$$\begin{aligned} d\Phi_{xt}(s) &= -H(X_{xt}(s), D\Phi + (\sigma^{-1})^* K(X_{xt}(s), s))ds + \\ &\quad + (K + \sigma^* D\Phi)(X_{xt}(s), s).dw(s). \end{aligned}$$

Setting $Z_{xt}(s) = (K + \sigma^* D\Phi)(X_{xt}(s), s)$, we see that the pair $\Phi_{xt}(s)$, $Z_{xt}(s)$ is the solution of

$$\begin{aligned} \Phi_{xt}(s) &= h(X_{xt}(T)) + \int_s^T H(X_{xt}(\tau), (\sigma^{-1})^*(X_{xt}(\tau)) Z_{xt}(\tau))d\tau \\ &\quad - \int_s^T Z_{xt}(\tau).dw(\tau). \end{aligned} \tag{14.3.8}$$

This equation is also given in [14]. We can check that

$$\Phi_{xt}(s) = \inf_v \Phi_{xt}^v(s). \tag{14.3.9}$$

This follows immediately from the definition of $\Phi_{xt}(s)$ and the property (14.1.8). It can also be checked directly.

EXERCISE 14.1. From equations (14.3.8) and (14.2.8), prove the property (14.3.9).

■

Consider next the strong formulation. The optimal feedback is now $\hat{v}(x, t) = \bar{v}(x, D\Psi)$. The optimal state is $x_{xt}^{\hat{v}}(s)$. To simplify the notation, we set $y(s) = x_{xt}^{\hat{v}}(s)$ and $u(s) = \hat{v}(x_{xt}^{\hat{v}}(s), s)$. So the dynamics of the optimal state is

$$\begin{aligned} dy &= g(y(s), u(s))ds + \sigma(y(s))dw(s), \\ y(t) &= x. \end{aligned} \tag{14.3.10}$$

We define a process similar to $\Phi_{xt}(s)$, namely $\Psi_{xt}(s) = \Psi(y(s), s)$, and to simplify notation, we drop the indices x, t. We obtain easily

$$d\Psi(s) = -l(y(s), u(s))ds + (\chi + \sigma^* D\Psi)(y(s), s).dw(s).$$

We can set $Z_{xt}(s) = Z(s) = (\chi + \sigma^* D\Psi)(y(s), s)$ and check the relation

$$\Psi(s) = h(y(T)) + \int_t^T l(y(s), u(s))ds - \int_t^T Z(s).dw(s). \tag{14.3.11}$$

But unlike what we had in (14.3.8), we do not obtain a self-contained equation for the pair $\Psi(s), Z(s)$, since the optimal control $u(s)$ cannot be obtained from $\Psi(s), Z(s)$. In fact, $u(s)$ satisfies

$$\begin{gathered} u(s) \text{ minimizes in } v \\ l(y(s), v) + D\Psi(y(s), s) \ . \ g(y(s), v). \end{gathered} \tag{14.3.12}$$

We need here the stochastic maximum principle. If we set $p(s) = D\Psi(y(s), s)$, we can check that $p(s)$ satisfies the adjoint equation

$$\begin{aligned} -dp(s) = [&D_x l(y(s), u(s)) + (D_x g(y(s), u(s)))^* p(s) \\ &+ \sum_{j=1}^n (D_x \sigma^j)^*(y(s), s) r^j(s)]ds - \sum_{j=1}^n r^j(s) dw_j(s), \end{aligned}$$

$$p(T) = D_x h(y(T)), \tag{14.3.13}$$

where σ^j is the vector (σ_{ij}) and $r^j = D^2\Psi\sigma^j + D\chi^j$. Then (14.3.12) becomes

$$\begin{gathered} u(s) \text{ minimizes in } v \\ l(y(s), v) + p(s) \ . \ g(y(s), v). \end{gathered} \tag{14.3.14}$$

REMARK 14.1. It is interesting to notice the difference between the weak and strong formulations. The Bellman equations are different, and also the stochastic maximum principle is different. In the weak formulation, it amounts to the BSDE (14.3.8), and in the strong formulation it boils down to the stochastic maximum principle, with the adjoint equation (14.3.13) and the necessary condition (14.3.14). In the Markov case, the Bellman equations are identical, but (14.3.8) and (14.3.11) remain different, but since $\Phi_{xt}(s) = \Phi(X_{xt}(s), s)$ and $\Psi_{xt}(s) = \Phi(y(s), s)$, they have the same probability distribution if the basic probability is P^u for the first process and P for the second process. More precisely, $\Phi_{xt}(s)$ and $\Psi_{xt}(s)$ are solutions of

$$\begin{aligned} \Phi_{xt}(s) = h(X_{xt}(T)) &+ \int_s^T H(X_{xt}(\tau), D\Phi(X_{xt}(\tau), \tau))d\tau \\ &- \int_s^T \sigma^* D\Phi(X_{xt}(\tau), \tau).dw(\tau), \end{aligned} \tag{14.3.15}$$

$$\Psi_{xt}(s) = h(y(T)) + \int_s^T l(y(s), u(s))ds - \int_s^T \sigma^* D\Phi(y(\tau), \tau).dw(\tau).$$

14.4 The Case of Two Noises

A common case leading to nonlinear BSPDEs is that in which the state dynamics depends on two noises, as follows:

$$dx = g(x, v(x, s))ds + \sigma(x)dw + \sigma^0(x)dw^0, \tag{14.4.1}$$
$$x(t) = x,$$

in which $w(s)$ and $w^0(s)$ are standard independent Wiener processes taking values respectively in R^d and R^{d_0}. We set $a(x) = \sigma(x)\sigma^*(x) + \sigma^0(x)(\sigma^0(x)^*$ and assume that $a(x)$ is invertible. The feedback $v(x, s)$ is a random field adapted to $\mathcal{F}^{0,s} = \sigma(w^0(\tau), \tau \leq s)$. The payoff is given by

$$\Phi^v(x, t) = E[h(x_{xt}^v(T)) + \int_t^T l(x_{xt}^v(\tau), v_{xt}(\tau))d\tau|\mathcal{F}^{0,t}]. \tag{14.4.2}$$

All functions $g, \sigma, l, h, \sigma^0$ are deterministic. Nevertheless, because of the conditioning, $\Phi^v(x, t)$ is a random field adapted to the filtration $\mathcal{F}^{0,t}$. Nevertheless, by techniques similar to those above, we check that $\Phi^v(x, t)$ is the solution of the BSPDE

$$-d_t\Phi^v + A\Phi^v dt = [l(x, v(x)) + D\Phi^v.g(x, v(x)) + \text{tr}(DK^v\sigma^0)]dt$$
$$-K^v.dw^0(t), \tag{14.4.3}$$

$$\Phi^v(x, T) = h(x).$$

Setting

$$\Phi(x, t) = \inf_v \Phi^v(x, t), \tag{14.4.4}$$

we have that $\Phi(x, t)$ is the solution of the nonlinear BSPDE

$$-d_t\Phi + A\Phi dt = [H(x, D\Phi + \text{tr}(DK\sigma^0))]dt - K.dw^0, \tag{14.4.5}$$
$$\Phi(x, T) = h(x).$$

If we introduce the random field $m_{xt}^v(\xi, s)$, the solution of

$$d_s m = [-\text{div}\,(mg(\xi, v(\xi))) + \sum_{ij=1}^n \frac{\partial^2(ma_{ij}(\xi))}{\partial\xi_i\partial\xi_j}]ds + \tag{14.4.6}$$

$$+ - \sum_{i=1}^n \sum_{j=1}^{d_0} \frac{\partial}{\partial\xi_i}(m\sigma_{ij}^0(\xi))dw_j^0,$$

$$m(\xi, t) = \delta(\xi - x),$$

we obtain

$$\Phi^v(x,t) = E[\int_t^T \int m_{xt}^v(\xi,s) l(\xi, v(\xi,s)) d\xi ds + \int m_{xt}^v(\xi,T) h(\xi) d\xi | \mathcal{F}^{0,t}], \tag{14.4.7}$$

so $m_{xt}^v(\xi,s)$ is the conditional probability density of the state at time s, given $\mathcal{F}^{0,s}$ and the initial condition $x(t) = x$.

14.5 The Linear–Quadratic Case

We shall consider only the strong formulation framework. To simplify, we take w to be one-dimensional, so $\sigma(x)$ is a vector in R^n. Equation (14.3.4) becomes

$$- d_t\Psi - \frac{1}{2}\sigma^* D^2\Psi\sigma dt = [H(x, D\Psi) + D\chi.\sigma)]dt - \chi dw, \tag{14.5.1}$$
$$\Psi(x,T) = h(x),$$

in which χ is a scalar random field. We assume now

$$g(x,v) = Fx + Gv,\ l(x,v) = \frac{1}{2}x^* Mx + \frac{1}{2}v^* Nv,\ h(x) = \frac{1}{2}x^* M_T x,$$
$$\sigma(x) = Cx,$$

where F, C are $n \times n$ matrices, G is an $n \times m$ matrix, and M, M_T are $n \times n$ symmetric matrices. They are, moreover, adapted stochastic processes. We compute easily

$$H(x,q) = \frac{1}{2}x^* Mx + q.Fx - \frac{1}{2}q^* GN^{-1}G^* q,$$

and we look for a solution $\Psi(x,t) = \frac{1}{2}x^* P(t)x,\ \chi(x,t) = \frac{1}{2}x^*\Gamma(t)x$, where $P(t)$ and $\Gamma(t)$ are adapted stochastic processes and symmetric matrices. We obtain easily the stochastic Riccati equation

$$- dP_t = [PF + F^* P - PGN^{-1}G^* P + C^* PC + M + \tag{14.5.2}$$
$$+ \Gamma C + C^*\Gamma]dt - \Gamma dw,$$
$$P(T) = M_T.$$

We refer to Peng [42] for a more complete treatment. We note also that there is no analogue for equation (14.3.2). Of course, we cannot assume $\sigma(x) = Cx$, since $\sigma(x)$ must be invertible, but we may think of σ as constant and invertible. We cannot obtain an explicit solution. To simplify, let us consider the case $n = 1$. Equation (14.5.1 becomes

$$
\begin{aligned}
- d_t\Psi - \frac{1}{2}\sigma^2\frac{\partial^2\Psi}{\partial x^2}dt &= [\frac{1}{2}Mx^2 + Fx\frac{\partial\Psi}{\partial x} - \frac{1}{2}\frac{G^2}{N}(\frac{\partial\Psi}{\partial x})^2 + \sigma\frac{\partial\chi}{\partial x}]dt - \chi dw, \\
\Psi(x,T) &= \frac{1}{2}M_T x^2,
\end{aligned} \tag{14.5.3}
$$

whereas equation (14.3.2) yields

$$
\begin{aligned}
-d_t\Phi - \frac{1}{2}\sigma^2\frac{\partial^2\Phi}{\partial x^2}dt &= [\frac{1}{2}Mx^2 + Fx(\frac{\partial\Phi}{\partial x} + \frac{K}{\sigma}) - \frac{1}{2}\frac{G^2}{N}(\frac{\partial\Phi}{\partial x} + \frac{K}{\sigma})^2 + \sigma\frac{\partial K}{\partial x}]dt \\
&\quad -Kdw, \\
\Phi(x,T) &= \frac{1}{2}M_T x^2.
\end{aligned} \tag{14.5.4}
$$

We can solve by the formula

$$
\begin{aligned}
\Psi(x,t) &= \frac{1}{2}P(t)x^2 + x\rho(t) + s(t), \\
\chi(x,t) &= \frac{1}{2}\Gamma(t)x^2 + x\gamma(t) + \delta(t),
\end{aligned}
$$

in which $P(t)$ is the solution of the stochastic Riccati equation

$$
\begin{aligned}
-dP(t) &= (2FP - \frac{G^2P^2}{N}M)dt - \Gamma(t)dw, \\
P(T) &= M_T, \\
-d\rho(t) &= [(F - \frac{G^2P}{N})\rho + \Gamma\sigma]dt - \gamma dw, \\
\rho(T) &= 0, \\
-ds &= (\sigma^2 P - \frac{1}{2}\frac{G^2}{N}\rho^2 + \gamma\sigma)dt - \delta dw, \\
s(T) &= 0.
\end{aligned}
$$

However, we cannot find an explicit formula for the solution of (14.5.4).

Chapter 15
Principal Agent Control Problems

15.1 Risk Sharing

15.1.1 Setting of the Model

In this chapter, we take inspiration from [17]. Our approach is more analytic and relies on dynamic programming, whereas Cvitanić and Zhang rely more on BSDEs. We consider two decision-makers, called principal and agent. The principal proposes a contract to the agent. They have their own utility functions. When they share risk, they form a team, and the utility function of the team is a combination of the two utility functions. We shall consider first a simple example, which reduces to the entrepreneur's problem, developed in the preceding chapter. We suppose that we have an underlying probability space $(\Omega, \mathcal{A}, P)$ on which is defined a scalar standard Wiener process $w(t)$. We denote by $\mathcal{F}^t$ the filtration generated by $w(t)$. There is a control $v(t)$ that is adapted to the filtration $\mathcal{F}^t$. It is the decision of the agent, and it models the selection of scale of a project. The output of the project is then given by

$$\begin{aligned} dX &= (rX + \alpha X v(t))dt + Xv(t)dw(t), \\ X(0) &= x, \end{aligned} \tag{15.1.1}$$

where $\alpha > 0$. We shall assume that x is a positive number. At some fixed horizon T, both observe the output $X(T)$ and share it. The agent gets C_T, and the principal gets the remainder $X(T) - C_T$. Here C_T is a random variable that is $\mathcal{F}^T$-measurable and a decision variable that belongs to the principal. The utility functions of the principal and of the agent are represented respectively by $U_P(x)$ and $U_A(x)$. We assume

$$\begin{aligned} &U_P, U_A : R \to R,\ C^2, \\ &U'_P, U'_A > 0 \qquad U''_P, U''_A < 0, \end{aligned} \tag{15.1.2}$$

A. Bensoussan, *Estimation and Control of Dynamical Systems*, Interdisciplinary Applied Mathematics 48, https://doi.org/10.1007/978-3-319-75456-7_15

$$U'_P, U'_A(-\infty) = +\infty \qquad U'_P, U'_A(+\infty) = 0. \tag{15.1.3}$$

So the objective of the agent is represented by $EU_A(C_T)$, and the objective of the principal is represented by $EU_P(X(T) - C_T)$. Since they form a team, they maximize the combined objective

$$\max_{C_T, v(.)} J_x(C_T, v(.)) = \max_{C_T, v(.)} E[U_P(X(T) - C_T) + \beta U_A(C_T)], \tag{15.1.4}$$

where $\beta > 0$.

15.1.2 Reduction to the Entrepreneur's Problem

We shall assume that

$$-U'_P(0) + \beta U'_A(0) > 0. \tag{15.1.5}$$

This assumption and the assumptions (15.1.2), (15.1.3) imply that for $x > 0$, the function $C \to U_P(x - C) + \beta U_A(C)$ is concave, and its derivative $-U'_P(x - C) + \beta U'_A(C)$ decreases from $+\infty$ to $-\infty$. It is 0 at a unique point $\hat{C}(x) > 0$. Defining the function

$$\begin{aligned} F(x) &= \max_C [U_P(x - C) + \beta U_A(C)] \\ &= U_P(x - \hat{C}(x)) + \beta U_A(\hat{C}(x)), \end{aligned} \tag{15.1.6}$$

it is clear that the problem reduces to

$$\max_{v(.)} EF(X(T)). \tag{15.1.7}$$

The function $F(x)$ is C^2, monotonically increasing, and concave, since

$$F'(x) = U'_P(x - \hat{C}(x)), \;\; F''(x) = \frac{\beta U''_P(x - \hat{C}(x)) U''_A(\hat{C}(x))}{U''_P(x - \hat{C}(x)) + \beta U''_A(\hat{C}(x))}. \tag{15.1.8}$$

As for the entrepreneur's problem, we can write the HJB equation for the value function $\Phi(x, t)$. We obtain easily

$$\begin{aligned} \frac{\partial \Phi}{\partial t} + rx \frac{\partial \Phi}{\partial x} - \frac{1}{2} \alpha^2 \frac{(\frac{\partial \Phi}{\partial x})^2}{\frac{\partial^2 \Phi}{\partial x^2}} &= 0, \\ \Phi(x, T) &= F(x). \end{aligned} \tag{15.1.9}$$

15.1.3 Solution

We consider the derivative $\lambda(x,t)$, which is the solution of

$$\frac{\partial \lambda}{\partial t}+rx\frac{\partial \lambda}{\partial x}+(r-\alpha^2)\lambda+\frac{1}{2}\alpha^2\lambda^2\frac{\frac{\partial^2\lambda}{\partial x^2}}{(\frac{\partial \lambda}{\partial x})^2}=0, \tag{15.1.10}$$

$$\lambda(x,T)=F'(x).$$

We solve it by the equation $x=G(\lambda,t)$, in which $G(\lambda,t)$ is the solution of

$$\frac{\partial G}{\partial t}-rG-(r-\alpha^2)\lambda\frac{\partial G}{\partial \lambda}+\frac{1}{2}\alpha^2\lambda^2\frac{\partial^2 G}{\partial \lambda^2}=0, \tag{15.1.11}$$

$$G(\lambda,T)=U_P'^{-1}(\lambda)+U_A'^{-1}(\frac{\lambda}{\beta}).$$

The final condition is obtained by writing $U'_P(x-\hat{C}(x))=\lambda$, and hence $x-\hat{C}(x)=U_P'^{-1}(\lambda)$. On the other hand, by the definition of $\hat{C}(x)$, we have

$$-U'_P(x-\hat{C}(x))+\beta U'_A(\hat{C}(x))=0,$$

whence $\lambda=\beta U'_A(\hat{C}(x))$, which implies $\hat{C}(x)=U_A'^{-1}(\frac{\lambda}{\beta})$. Note that $G(\lambda,t)$ is defined for $\lambda\geq 0$. The function $U_P'^{-1}(\lambda)+U_A'^{-1}(\frac{\lambda}{\beta})$ is decreasing from $U_P'^{-1}(0)+U_A'^{-1}(0)=+\infty$ to $U_P'^{-1}(+\infty)+U_A'^{-1}(+\infty)=-\infty$. The function $G(\lambda,t)$ is the solution of a linear PDE for which we can give an explicit solution. The best way to find it is to use the probabilistic interpretation. We consider the stochastic differential equation

$$\begin{aligned} dY(s)&=-(r-\alpha^2)Y(s)ds-\alpha Y(s)dw(s),\\ Y(t)&=\lambda, \end{aligned}$$

whose solution is denoted by $Y_{\lambda t}(s)$. If we set $\Gamma(\lambda)=U_P'^{-1}(\lambda)+U_A'^{-1}(\frac{\lambda}{\beta})$, then we show easily that

$$G(\lambda,t)=E\Gamma(Y_{\lambda t}(T))\exp -r(T-t). \tag{15.1.12}$$

We can write explicitly

$$Y_{\lambda t}(T)=\lambda\exp -[(r-\frac{\alpha^2}{2})(T-t)+X\alpha\sqrt{T-t}],$$

in which X is a Gaussian with mean 0 and variance 1. Therefore, after some easy but tedious calculations, we get

$$G(\lambda,t) = \frac{1}{\lambda\alpha\sqrt{2\pi(T-t)}}\int_0^{+\infty}\Gamma(\eta)\exp$$

$$-\frac{1}{2\alpha^2}\left(\frac{1}{\sqrt{T-t}}\log\frac{\lambda}{\eta}-(r+\frac{\alpha^2}{2})\sqrt{T-t}\right)^2 d\eta. \quad (15.1.13)$$

The function $G(\lambda,t)$ decreases from $G(0,t)=+\infty$ to $G(+\infty,t)=-\infty$. So for all x real, the equation $G(\lambda,t)=x$ has a unique positive solution $\lambda(x,t)$. We next write

$$\Phi(x,t)=\Psi(\lambda(x,t),t), \quad (15.1.14)$$

and as with the entrepreneur's problem, we check easily that Ψ is the solution of the linear equation

$$\frac{\partial\Psi}{\partial t}+\frac{1}{2}\alpha^2\lambda^2\frac{\partial^2\Psi}{\partial\lambda^2}-r\lambda\frac{\partial\Psi}{\partial\lambda}=0, \quad (15.1.15)$$

$$\Psi(\lambda,T)=U_P(U_P'^{-1}(\lambda))+\beta U_A(U_A'^{-1}(\frac{\lambda}{\beta})).$$

EXERCISE 15.1. Prove the final condition in (15.1.15).

The optimal feedback is given by

$$\hat{v}(x,t)=-\alpha\frac{\lambda(x,t)}{x\frac{\partial\lambda(x,t)}{\partial x}}=-\alpha\frac{\lambda(x,t)}{x}\frac{\partial G}{\partial\lambda}(\lambda(x,t),t), \quad (15.1.16)$$

so the optimal state $\hat{X}(t)$ is the solution of the stochastic differential equation

$$d\hat{X}=[r\hat{X}-\alpha^2\lambda(\hat{X},t)\frac{\partial G}{\partial\lambda}(\lambda(\hat{X},t),t)]dt- \quad (15.1.17)$$

$$-\alpha\lambda(\hat{X},t)\frac{\partial G}{\partial\lambda}(\lambda(\hat{X},t),t)dw,$$

$$\hat{X}(0)=x.$$

Of course, we have to prove that this equation has a solution. It turns out that it has an explicit solution. We introduce the process $Y_x(t)$, the solution of

$$dY=-rYdt-\alpha Ydw, \quad (15.1.18)$$

$$Y(0)=\lambda(x,0).$$

Then

$$\hat{X}(t) = G(Y(t), t) \tag{15.1.19}$$

is a solution. Indeed, $Y(t) = \lambda(\hat{X}, t)$, and using Itô's formula and the equation (15.1.11), we obtain immediately (15.1.17). The feedback (15.1.16) leads to the control

$$\hat{v}(t)G(Y(t), t) = -\alpha Y(t)\frac{\partial G}{\partial \lambda}(Y(t), t), \tag{15.1.20}$$

and we can check as for the entrepreneur's problem that this control is optimal. Finally, we can define the dual function

$$\begin{aligned}\tilde{\Phi}(\lambda, t) = &\quad \sup_x(\Phi(x, t) - x\lambda)\\ &= \Phi(G(\lambda, t), t) - \lambda G(\lambda, t)\end{aligned}$$

and check that it is the solution of the equation

$$\begin{gathered}\frac{\partial \tilde{\Phi}}{\partial t} + \frac{1}{2}\alpha^2\lambda^2\frac{\partial^2 \tilde{\Phi}}{\partial \lambda^2} - r\lambda\frac{\partial \tilde{\Phi}}{\partial \lambda} = 0 \\ \Psi(\lambda, T) = \tilde{F}(\lambda)\end{gathered} \tag{15.1.21}$$

$$\begin{aligned}&= U_P(U_P'^{-1}(\lambda)) - \lambda U_P'^{-1}(\lambda) + \beta[U_A(U_A'^{-1}(\frac{\lambda}{\beta})) - \frac{\lambda}{\beta}U_A'^{-1}(\frac{\lambda}{\beta})]\\ &= \tilde{U}_P(\lambda) + \beta\tilde{U}_A(\frac{\lambda}{\beta}).\end{aligned}$$

We can then obtain $\Phi(x, t)$ by the formula

$$\Phi(x, t) = \min_{\lambda \geq 0}(\tilde{\Phi}(\lambda, t) + x\lambda). \tag{15.1.22}$$

15.1.4 Example

Assume that

$$U_P(x) = -\frac{1}{\gamma_P}\exp -\gamma_P x, \;\; U_A(x) = -\frac{1}{\gamma_A}\exp -\gamma_A x. \tag{15.1.23}$$

We obtain

$$G(\lambda, T) = -(\frac{1}{\gamma_P} + \frac{1}{\gamma_A})\log \lambda + \frac{\log \beta}{\gamma_A} \tag{15.1.24}$$

and

$$G(\lambda,t) = [\frac{\log\beta}{\gamma_A} + (r - \frac{\alpha^2}{2})(\frac{1}{\gamma_P} + \frac{1}{\gamma_A})(T-t) - (\frac{1}{\gamma_P} + \frac{1}{\gamma_A})\log\lambda]\exp -r(T-t). \tag{15.1.25}$$

We have next

$$\hat{v}(x,t) = \alpha\frac{(\frac{1}{\gamma_P} + \frac{1}{\gamma_A})\exp -r(T-t)}{x},\ \hat{C}(x) = \frac{\log\beta}{\gamma_P+\gamma_A} + \frac{\gamma_P}{\gamma_P+\gamma_A}x, \tag{15.1.26}$$

$$\Psi(\lambda,T) = -\frac{\lambda}{\gamma_P+\gamma_A},\ \Psi(\lambda,t) = -\frac{\lambda\exp -r(T-t)}{\gamma_P+\gamma_A},$$

$$\begin{aligned}
\tilde{\Phi}(\lambda,T) &= \frac{\lambda\log\lambda}{\gamma_P+\gamma_A} - \lambda(\frac{1}{\gamma_P} + \frac{\log\beta+1}{\gamma_A}),\\
\tilde{\Phi}(\lambda,t) &= [\frac{\lambda\log\lambda}{\gamma_P+\gamma_A} - \lambda(\frac{1}{\gamma_P} + \frac{\log\beta+1}{\gamma_A} + (r - \frac{\alpha^2}{2})(\frac{1}{\gamma_P} + \frac{1}{\gamma_A})(T-t))]\\
&\quad \exp -r(T-t),\\
\lambda(x,t) &= \beta^{\frac{\gamma_P}{\gamma_P+\gamma_A}}\exp(r - \frac{\alpha^2}{2})(T-t)\exp -[\frac{\gamma_P\gamma_A}{\gamma_P+\gamma_A}x\ \exp r(T-t)],\\
\Phi(x,t) &= -\frac{\exp -\frac{\alpha^2}{2}(T-t)}{\gamma_P+\gamma_A}\beta^{\frac{\gamma_P}{\gamma_P+\gamma_A}}\exp -[\frac{\gamma_P\gamma_A}{\gamma_P+\gamma_A}x\ \exp r(T-t)].
\end{aligned}$$

EXERCISE 15.2. Solve the problem with utility functions

$$U_P(x) = \frac{1}{\gamma_P}x^{\gamma_P},\ U_A(x) = \frac{1}{\gamma_A}x^{\gamma_A}.$$

15.2 Implementing Contracts

15.2.1 Decentralized Solution

We indicated that the principal chooses the contract C_T and the agent chooses the scale of the project $v(.)$. What happens when they do not use the team objective? The principal has the objective $EU_P(X(T) - C_T)$, and the agent has the objective $EU_A(C_T)$. In practice, the contract is a function of the final output $C_T = \varphi(X(T))$, and the principal must decide what it is. He does that by observing the reaction of the agent and optimizing his objective functional. The agent solves a stochastic control problem identical to the team problem, except for the objective to maximize.

So the agent solves the HJB equation

$$\frac{\partial \chi}{\partial t} + rx\frac{\partial \chi}{\partial x} - \frac{1}{2}\alpha^2 \frac{(\frac{\partial \chi}{\partial x})^2}{\frac{\partial^2 \chi}{\partial x^2}} = 0, \tag{15.2.1}$$

$$\chi(x,T) = U_A(\varphi(x)).$$

So it is a functional $\chi(x,t;\varphi(.))$. Set $\mu(x,t;\varphi(.)) = \frac{\partial \chi(x,t;\varphi(.))}{\partial x}$. Then the optimal final state $\hat{X}(T)$ is defined by the equation

$$\mu(x,0;\varphi(.))Z(T) = U_A'(\varphi(\hat{X}(T)))\varphi'(\hat{X}(T)), \tag{15.2.2}$$

where $x = \hat{X}(0;\varphi(.))$, the initial budget, and $Z(T)$ is the final value of the process $Z(t)$, the solution of

$$dZ = -rZdt - \alpha Z dw,$$

$$Z(0) = 1.$$

We set $\hat{X}(T;\varphi(.))$ to denote the solution of (15.2.2). The principal tries to solve the following problem:

$$\max_{\varphi(.)} EU_P(\hat{X}(T;\varphi(.)) - \varphi(\hat{X}(T;\varphi(.)))). \tag{15.2.3}$$

This is a very complex problem, and we shall not try to solve it. Following [17], we may try to adjust $\varphi(.)$ so that the principal gets the same payoff as in solving the team problem. In solving the team problem, the principal gets $EU_P(U_P'^{-1}(\lambda(x,0)Z(T)))$. We can verify this condition if φ satisfies

$$\hat{X}(T;\varphi(.)) - \varphi(\hat{X}(T;\varphi(.))) = U_P'^{-1}(\lambda(x,0)Z(T))$$

or

$$U_P'(\hat{X}(T;\varphi(.)) - \varphi(\hat{X}(T;\varphi(.)))) = \lambda(x,0)Z(T)$$

$$= \frac{\lambda(x,0)}{\mu(x,0;\varphi(.))} U_A'(\varphi(\hat{X}(T;\varphi(.))))\varphi'(\hat{X}(T;\varphi(.))).$$

It is thus natural to require that the function $\varphi(.)$ satisfy

$$\varphi'(\xi) = \frac{U_P'(\xi - \varphi(\xi))}{U_A'(\varphi(\xi))} \frac{\mu(x,0;\varphi(.))}{\lambda(x,0)}. \tag{15.2.4}$$

We may consider this functional equation as a fixed-point problem. Indeed, set $y = \mu(x, 0; \varphi(.))$, and consider the family of differential equations

$$\varphi'(\xi) = \frac{U'_P(\xi - \varphi(\xi))}{U'_A(\varphi(\xi))} \frac{y}{\lambda(x, 0)}.$$

We suppose that we can get a function $\varphi(\xi; y)$, the solution of this differential equation, for each value y. We must then find y such that $y = \mu(x, 0; \varphi(.; y))$. In this fixed-point problem, x is simply a parameter. We assume implicitly that the function $U_A(\varphi(.; y))$ satisfies all the properties needed to solve the problem of the agent.

Can we have $\varphi(.) = \hat{C}(.)$? Since $U'_P(\xi - \hat{C}(\xi)) = \beta U'_A(\hat{C}(\xi))$, we have necessarily

$$\hat{C}'(\xi) = \beta \frac{\mu(x, 0; \hat{C}(.))}{\lambda(x, 0)}. \tag{15.2.5}$$

The function $\hat{C}(.)$ must be linear. We have seen that such is the case when the utilities are exponential. Indeed, see (15.1.26), we have $\hat{C}(x) = \frac{\log \beta}{\gamma_P + \gamma_A} + \frac{\gamma_P}{\gamma_P + \gamma_A} x$. Then we get easily

$$F(x) = -\frac{\gamma_P + \gamma_A}{\gamma_P \gamma_A} \beta^{\frac{\gamma_P}{\gamma_P + \gamma_A}} \exp -\frac{\gamma_P \gamma_A}{\gamma_P + \gamma_A} x.$$

Next, $\tilde{F}(x) = U_A(\hat{C}(x)) = -\frac{1}{\gamma_A} \beta^{-\frac{\gamma_A}{\gamma_P + \gamma_A}} \exp -\frac{\gamma_P \gamma_A}{\gamma_P + \gamma_A} x$. We see immediately that $\tilde{F}(x) = \frac{\gamma_P}{\gamma_P + \gamma_A} \frac{1}{\beta} F(x)$. Comparing equation (15.2.1) with (15.1.9), we see easily that $\chi(x, t) = \frac{\gamma_P}{\gamma_P + \gamma_A} \frac{1}{\beta} \Phi(x, t)$. Also, $\mu(x, 0; \hat{C}(.)) = \frac{\gamma_P}{\gamma_P + \gamma_A} \frac{1}{\beta} \lambda(x, 0)$, and thus $\hat{C}'(\xi) = \frac{\gamma_P}{\gamma_P + \gamma_A}$, which is indeed satisfied. For nonexponential utilities, the function $\hat{C}(x)$ cannot be used to implement the contract in a decentralized manner.

15.2.2 Implementing the Team Solution

Let us assume that we can solve the problem described in the preceding section and find a function $\varphi(.)$. It will be different from $\hat{C}(.)$, as seen in the previous section. The corresponding optimal final state $\hat{X}(T; \varphi(.))$ will be different from the state $\hat{X}(T)$ defined by (15.1.19), the team optimization. However, we have

$$\hat{X}(T) - \hat{C}(\hat{X}(T)) = \hat{X}(T; \varphi(.)) - \varphi(\hat{X}(T; \varphi(.))),$$

and the optimal objective functional of the principal is the same in both cases. The optimal objective functional of the agent is not the same, unless we have by chance

$$EU_A(\varphi(\hat{X}(T;\varphi(.)))) = EU_A(\hat{C}(\hat{X}(T))).$$

This relation does not mean that $\varphi(.) = \hat{C}(.)$, nor that $\hat{X}(T;\varphi(.)) = \hat{X}(T)$.

If the principal wants to implement the team solution exactly, in a decentralized way, he can offer the contract $C_T = X(T) - U_P'^{-1}(\lambda(x,0)Z(T))$. As we can see, this contract is not a deterministic function of the final wealth $X(T)$. There is a random term $\lambda(x,0)Z(T)$ correlated with $X(T)$. The problem of the agent becomes

$$\max_{v(.)} EU_A(X(T) - U_P'^{-1}(\lambda(x,0)Z(T))).$$

Dynamic programming is not directly applicable to this problem, at least as above. The only way to apply dynamic programming would be to consider two processes $X(t)$ and $Z(t)$. However, it is much easier to use the martingale method. We try to obtain the optimal state for this problem, directly, by solving

$$\max_{X(T)} EU_A(X(T) - U_P'^{-1}(\lambda(x,0)Z(T)))$$

subject to the constraint derived from the budget equation $EX(T)Z(T) = x$. In this formulation, $X(T)$ is now any random variable that is $\mathcal{F}^T$-measurable. We take care of the budget equation using a Lagrange multiplier η. So we solve

$$\max_{X(T)} E[U_A(X(T) - U_P'^{-1}(\lambda(x,0)Z(T))) - \eta X(T)Z(T)].$$

Writing the first-order condition, we obtain

$$\hat{X}(T) = U_P'^{-1}(\lambda(x,0)Z(T))) + U_A'^{-1}(\eta Z(T)).$$

We find η by writing the budget equation

$$EZ(T)[U_P'^{-1}(\lambda(x,0)Z(T))) + U_A'^{-1}(\eta Z(T))] = x. \tag{15.2.6}$$

We can assert that $\eta = \frac{\lambda(x,0)}{\beta}$ is a solution. Indeed, we know that the optimal state for the team problem is

$$\hat{X}(T) = U_P'^{-1}(\lambda(x,0)Z(T))) + U_A'^{-1}(\lambda(x,0)Z(T)),$$

and it satisfies the budget equation. Now, from the monotonicity of the function $U_A'^{-1}(.)$, equation (15.2.6) has a unique solution. So the optimal states coincide, and the team solution can be implemented in this way by the principal.

15.2.3 Generalizations

15.2.3.1 Effort of the Agent

In the preceding model, the agent chooses the scale of the project. Let us consider, as in the entrepreneur's problem, that he decides also about his effort. The problem is described as follows. The output of the project evolves as follows:

$$dX = (b(u(t)) + \alpha v(t))dt + v(t)dw(t), \tag{15.2.7}$$
$$X(0) = x.$$

The process $u(.)$ is adapted to the filtration $\mathcal{F}^t$. The function $b(u)$ is monotonically increasing, and u represents the effort of the agent. Note that in this model, we do not have X as a multiplicative factor. The principal proposes the contract C_T, but the agent has also an aversion to effort. He subtracts the accumulated effort $\int_0^T g(u(t))dt$ from the contract. So he applies his utility function to the difference $C_T - \int_0^T g(u(t))dt$. The function $g(u)$ is also monotonically increasing. The objective functional of the principal is $EU_P(X(T) - C_T)$, and that of the agent is $EU_A(C_T - \int_0^T g(u(t))dt)$. The team problem is

$$\max_{C_T,u(.),v(.)} EU_P(X(T) - C_T) + \beta EU_A(C_T - \int_0^T g(u(t))dt). \tag{15.2.8}$$

Because of the integral, one needs to introduce an additional process $Y(t)$, the solution of

$$\frac{dY}{dt} = g(u(t)), \tag{15.2.9}$$
$$Y(0) = 0.$$

Therefore, the objective functional of the team becomes

$$\max_{C_T,u(.),v(.)} EU_P(X(T) - C_T) + \beta EU_A(C_T - Y(T)).$$

Since C_T is $\mathcal{F}^T$-measurable, the problem reduces to

$$\max_{u(.),v(.)} EF(X(T), Y(T)), \tag{15.2.10}$$

with

$$F(x, y) = \max_C [U_P(x - C) + \beta U_A(C - y)]. \tag{15.2.11}$$

It is clear that $F(x,y) = F(x-y)$, where $F(x)$ has been defined by (15.1.6). The HJB equation is easily written as

$$\frac{\partial \Phi}{\partial t} + \max_u\left(\frac{\partial \Phi}{\partial x} b(u) + \frac{\partial \Phi}{\partial y} g(u)\right) - \frac{1}{2}\alpha^2 \frac{(\frac{\partial \Phi}{\partial x})^2}{\frac{\partial^2 \Phi}{\partial x^2}} = 0, \tag{15.2.12}$$
$$\Phi(x,T) = F(x-y).$$

Assume that

$$r = \max_u (b(u) - g(u)) > 0, \tag{15.2.13}$$

in which case, the value function $\Phi(x,y,t)$ is simply $\Phi(x-y,t)$, where $\Phi(x,t)$ is the solution of

$$\frac{\partial \Phi}{\partial t} + r\frac{\partial \Phi}{\partial x} - \frac{1}{2}\alpha^2 \frac{(\frac{\partial \Phi}{\partial x})^2}{\frac{\partial^2 \Phi}{\partial x^2}} = 0, \tag{15.2.14}$$
$$\Phi(x,T) = F(x).$$

This equation is very close to (15.1.9), except for $rx\frac{\partial \Phi}{\partial x}$, which is replaced by $r\frac{\partial \Phi}{\partial x}$. If we set $\lambda(x,t) = \frac{\partial \Phi}{\partial x}(x,t)$, we obtain the following equation for λ:

$$\frac{\partial \lambda}{\partial t} + r\frac{\partial \lambda}{\partial x} - \alpha^2\lambda + \frac{1}{2}\alpha^2\lambda^2 \frac{\frac{\partial^2 \lambda}{\partial x^2}}{(\frac{\partial \lambda}{\partial x})^2} = 0, \tag{15.2.15}$$
$$\lambda(x,T) = F'(x).$$

We solve this equation again by inversion, $G(\lambda,t) = x$. We get the following linear PDE for G:

$$\frac{\partial G}{\partial t} + \alpha^2\lambda\frac{\partial G}{\partial \lambda} + \frac{1}{2}\alpha^2\lambda^2\frac{\partial^2 G}{\partial \lambda^2} = r, \tag{15.2.16}$$
$$G(\lambda,T) = \Gamma(\lambda).$$

We get easily

$$G(\lambda,t) = -r(T-t) + \tilde{G}(\lambda,t), \tag{15.2.17}$$

with $\tilde{G}(\lambda,t)$ given by

$$\tilde{G}(\lambda,t) = \frac{1}{\lambda\alpha\sqrt{2\pi(T-t)}}\int_0^{+\infty} \Gamma(\eta)\exp$$
$$-\frac{1}{2\alpha^2}\left(\frac{1}{\sqrt{T-t}}\log\frac{\lambda}{\eta} - \frac{\alpha^2}{2}\sqrt{T-t}\right)^2 d\eta, \tag{15.2.18}$$

which is the expression (15.1.13) with $r = 0$. Finally, $\Phi(x,t) = \Psi(\lambda(x,t),t)$, with Ψ the solution of

$$\frac{\partial \Psi}{\partial t} + \frac{1}{2}\alpha^2\lambda^2\frac{\partial^2\Psi}{\partial\lambda^2} = 0, \tag{15.2.19}$$

$$\Psi(\lambda, T) = U_P(U_P'^{-1}(\lambda)) + \beta U_A(U_A'^{-1}(\frac{\lambda}{\beta})).$$

15.2.3.2 Nonlinear Volatility Selection with Exponential Utilities

We consider a more general model than that in Section 15.2.3.1, however with exponential utilities. The model is described as follows:

$$dX = (rX + b(u(t), v(t))dt + v(t)dw(t), \tag{15.2.20}$$

$$X(0) = x,$$

$$\frac{dY}{dt} = \mu X(t) + g(u(t), v(t)), \tag{15.2.21}$$

$$Y(0) = 0,$$

with the team objective

$$\max_{u(.),v(.)} EF(X(T) - Y(T)) \tag{15.2.22}$$

and

$$F(x) = \max_C[U_P(x - C) + \beta U_A(C)].$$

We assume exponential utilities

$$U_P(x) = -\frac{1}{\gamma_P}\exp -\gamma_P x,\ U_A(x) = -\frac{1}{\gamma_A}\exp -\gamma_A x.$$

An easy calculation shows that

$$F(x) = -\frac{\gamma_A + \gamma_P}{\gamma_A\gamma_P}\exp -\frac{\gamma_A\gamma_P x}{\gamma_A + \gamma_P}\beta^{\frac{\gamma_P}{\gamma_A+\gamma_P}}. \tag{15.2.23}$$

We can write the HJB given equation as follows:

$$\frac{\partial \Phi}{\partial t} + rx\frac{\partial \Phi}{\partial x} + \mu x\frac{\partial \Phi}{\partial y} + \tag{15.2.24}$$

$$+\max_{u,v}[\frac{\partial\Phi}{\partial x}b(u,v)+\frac{\partial\Phi}{\partial y}g(u,v)+\frac{1}{2}v^2\frac{\partial^2\Phi}{\partial x^2}]=0,$$

$$\Phi(x,y,T)=F(x-y).$$

We look for a solution of the form

$$\Phi(x,y,t)=-l(t)\exp-\frac{\gamma_A\gamma_P}{\gamma_A+\gamma_P}(k(t)x-y). \tag{15.2.25}$$

EXERCISE 15.3. Show that

$$k(t)=\frac{\mu}{r}+(1-\frac{\mu}{r})\exp r(T-t)$$

and that $l(t)$ is the solution of the linear differential equation

$$l'(t)+l(t)\frac{\gamma_A\gamma_P}{\gamma_A+\gamma_P}\min_{u,v}[\frac{1}{2}v^2k^2(t)\frac{\gamma_A\gamma_P}{\gamma_A+\gamma_P}-k(t)b(u,v)+g(u,v)]=0$$

$$l(T)=\frac{\gamma_A+\gamma_P}{\gamma_A\gamma_P}\beta^{\frac{\gamma_P}{\gamma_A+\gamma_P}}.$$

We assume, of course, that $\min_{u,v}[\frac{1}{2}v^2k^2(t)\frac{\gamma_A\gamma_P}{\gamma_A+\gamma_P}-k(t)b(u,v)+g(u,v)]$ is finite.

15.3 General Approach

15.3.1 Description of the Problem

We follow the presentation of [17]. The agent decides on his effort, and the principal decides on the contract arrangements. The scaling of projects is fixed to simplify the problem. We consider a Wiener process $w(t)$ defined on a probability space $\Omega, \mathcal{A}, P$, and we set $\mathcal{F}^t=\sigma(w(s), s\leq t)$ and $\mathcal{F}_t^s=\sigma(w(\tau)-w(t), s\leq\tau\leq t)$. We then define the output of the project by

$$dX=\sigma dw(s),\ \ X(t)=x. \tag{15.3.1}$$

The solution is denoted by $X_{xt}(s)$. We rely on the weak formulation to introduce the effort of the agent. Since the state $X_{xt}(s)$ is fixed, there is no difference between open-loop and closed-loop controls. So consider a feedback $v(x,s)$ that is an adapted random field. To this feedback we associate the control $v_{xt}(s)=v(X_{xt}(s),s)$. We may omit the indices xt to simplify notation. We denote it by $v(s)$ and associate a process $w_{xt}^v(s)$ by setting

$$w_{xt}^v(s)=w(s)-w(t)-\int_t^s v(\tau)d\tau. \tag{15.3.2}$$

Using Girsanov's theorem, we introduce a change of probability P^v_{xt} defined by its Radon–Nikodym derivative

$$\frac{dP^v_{xt}}{dP}|_{\mathcal{F}^s_t} = M^v_{xt}(s) = \exp\left(\int_t^s v(\tau)dw(\tau) - \frac{1}{2}\int_t^s v^2(\tau)d\tau\right). \tag{15.3.3}$$

We know that on the probability space $\Omega, \mathcal{A}, P^v_{xt}$, the process $w^v(s) = w^v_{xt}(s)$ is a standard Wiener process, and the output $X_{xt}(s)$ appears as the solution of

$$dX = v(s)\sigma ds + \sigma dw^v(s), \;\; X(t) = x. \tag{15.3.4}$$

This approach is useful in separating the impact of the agent from that of the principal. The principal decides on a contract. It consists of a cash flow to be paid continuously to the agent and a lump sum to be paid at the end of the contract, which is the horizon T. We denote by $c(s)$ and C_T respectively the cash flow and the final payment. The process $c(s)$ is adapted to the filtration $\mathcal{F}^s_t$, and the final payoff C_T is a random variable that is $\mathcal{F}^T_t$-measurable. As we shall see, it is not realistic to consider these processes feedbacks on the state. This creates major challenges. The agent considers $c(.)$ and C_T as given. His control is $v(.)$. He wants to maximize the objective functional

$$K^A_{x,t;c(.),C_T}(v(.)) = E^v[\int_t^T u_A(X_{xt}(s), c(s), v(s))ds + U_A(X_{xt}(T), C_T)|\mathcal{F}^t]. \tag{15.3.5}$$

If the agent comes with a reaction function $v(.)$ that maximizes the preceding quantity, he obtains the value

$$\Phi^A(x, t; c(.), C_T) = \sup_{v(.)} K^A_{x,t;c(.),C_T}(v(.)). \tag{15.3.6}$$

Of course, the reaction function becomes a very complex functional of $c(.), C_T$. We call it $\hat{v}(.)$. The principal uses it in computing his own objective functional

$$K^P_{xt}(c(.), C_T) = E^{\hat{v}}[\int_t^T u_P(X_{xt}(s), c(s))ds + U_P(X_{xt}(T), C_T)|\mathcal{F}^t] \tag{15.3.7}$$

and tries to achieve his value function

$$\Phi^P(x, t) = \sup_{c(.),C_T} K^P_{xt}(c(.), C_T). \tag{15.3.8}$$

15.3.2 The Agent Problem

It is clear that the problem formulated in the preceding section is quite complex. The agent's problem is a non-Markov stochastic control problem. To simplify notation, we shall write $u_A(X(s), v(s))$ and $U_A(X(T))$ instead of $u_A(X(s), c(s), v(s))$ and $U_A(X(T), C_T)$, since for the agent, $c(.)$ and C_T are external. Of course, this implies that the fields $u_A(x, v)$ and $U_A(x)$ are random for fixed x, v and adapted to $\mathcal{F}^s$. We use the techniques of Chapter 14. The random field $\Phi^A(x, t)$ satisfies the nonlinear BSPDE

$$\begin{aligned} - d_t \Phi^A &= [\sup_v (u_A(x, v) + (\frac{\partial \Phi^A}{\partial x}\sigma + K^A)v) + \\ &+ \frac{1}{2}\frac{\partial^2 \Phi^A}{\partial x^2}\sigma^2 + \sigma\frac{\partial K^A}{\partial x}]dt - K^A dw, \\ \Phi^A(x, T) &= U_A(x). \end{aligned} \tag{15.3.9}$$

Consider next $\Phi^A_{xt}(s) = \Phi^A(X_{xt}(s), s)$. Setting $Z^A_{xt}(s) = \sigma\frac{\partial \Phi^A(X_{xt}(s), s)}{\partial x} + K^A(X_{xt}(s), s)$, we obtain the BSDE

$$\Phi^A_{xt}(s) = h(X_{xt}(T)) + \int_s^T \sup_v (u_A(X_{xt}(\tau), v) + vZ^A_{xt}(\tau))d\tau - \int_s^T Z^A_{xt}(\tau)dw(\tau). \tag{15.3.10}$$

15.3.3 The Principal Problem

We recall that $u_A = u_A(x, c, u)$ and $U_A = U_A(x, C_T)$. We assume that the problem $\sup_v (u_A(x, c, v) + vz)$ has a unique solution for every pair (x, c). We denote by $\varphi_A(x, c, z)$ this solution, and we assume that it is invertible in z. In other words, the equation $\varphi_A(x, c, z) = v$ in z has a unique solution, denoted by $Z_A(x, c, u)$. For instance, if the supremum is attained at a point where the gradient is 0, we will write

$$\frac{\partial u_A(x, c, \varphi_A)}{\partial v} + z = 0$$

and

$$Z_A(x, c, v) = -\frac{\partial u_A(x, c, v)}{\partial u}.$$

In our framework, the principal will take into account a minimum value of the agent at time 0, thus a deterministic quantity, denoted by η. We can consider that this is the minimum value for the agent to accept the contract. With this guaranteed value, he will accept the contract. The contract comprises a process $c(t)$ and a random variable C_T. They are simply adapted. When they are indicated to the agent, he will react by calibrating his effort, according to the preceding section. The principal knows that and will decide appropriately and maximize his own payoff. Let us explain that the principal can proceed by deciding two processes $c(t)$, $v(t)$ and the variable C_T. He then solves the stochastic differential equation

$$\begin{aligned} d\eta &= [-u_A(X(t), c(t), v(t)) - v(t)Z_A(X(t), c(t), v(t))]dt + \\ &\quad + Z_A(X(t), c(t), v(t))dw(t), \qquad (15.3.11) \\ \eta(0) &= \eta. \end{aligned}$$

We recall that in this equation, $X(t)$ is given by

$$dX = \sigma dw(t), \;\; X(0) = x, \qquad (15.3.12)$$

so it is a given process. We set $Z_A(t) = Z_A(X(t), c(t), v(t))$. It follows that $v(t) = \varphi_A(X(t), c(t), Z_A(t))$. Therefore, the process $\eta(t)$ is the solution of the SDE

$$\begin{aligned} d\eta &= -\sup_v(u_A(X(t), c(t), v) + vZ_A(t))dt + Z_A(t)dw(t), \qquad (15.3.13) \\ \eta(0) &= \eta. \end{aligned}$$

Let $\eta(T)$ be the final value of this process. The principal decides the final contract C_T by solving the equation $\eta(T) = U_A(X(T), C_T)$, so $C_T = C_T(X(T), \eta(T))$, where the feedback $C_T(x, \eta)$ is the solution of $\eta = U_A(x, C_T)$. But then the process $\eta(t)$ appears as the solution of the backward SDE

$$\eta(t) = U_A(X(T), C_T) + \int_t^T \sup_v(u_A(X(s), c(s), v) + vZ_A(s))ds - \int_t^T Z_A(s)dw(s). \qquad (15.3.14)$$

If the agent is presented with the contract $c(.)$ and C_T, he will solve exactly problem (15.3.14) in the unknowns $\eta(t)$ and $Z_A(t)$. Since this problem has a unique solution, he will find a wealth $\eta(t)$ and a process $Z_A(t)$ as designed by the principal. He will check that the initial wealth is η, which is acceptable. So he will accept the contract and will calibrate his effort as $v(.)$, exactly as anticipated by the principal. Therefore, the principal is free to design the processes $c(t)$ and $v(t)$ as he wishes. The final contract C_T is, however, already fixed.

It is then clear that the principal will solve a stochastic control problem in which the states are $X(t)$ and $\eta(t)$, and the controls are $c(t)$ and $v(t)$. Defining $w^v(.)$ by $dw^v(t) = dw(t) - v(t)dt$, $w^v(0) = 0$, we write the state equations as

$$dX = \sigma v(t)dt + \sigma dw^v(t), \qquad (15.3.15)$$

$$d\eta = -u_A(X(t), c(t), v(t))dt + Z_A(X(t), c(t), v(t))dw^v(t),$$
$$X(0) = x, \ \eta(0) = \eta,$$

and the payoff is defined by

$$K^P_{x,\eta}(c(.), v(.)) = E^v[\int_0^T u_P(X(t), c(t))dt + U_P(X(T), C_T(X(T), \eta(T)))]. \tag{15.3.16}$$

This problem can be solved by standard dynamic programming methods. We solve the PDE

$$\frac{\partial \Psi}{\partial t} + \frac{1}{2}\sigma^2 \frac{\partial^2 \Psi}{\partial x^2} + \sup_{c,v}[\frac{\partial \Psi}{\partial x}\sigma u - \frac{\partial \Psi}{\partial \eta} u_A(x, c, v) + \tag{15.3.17}$$
$$+\frac{1}{2}\frac{\partial^2 \Psi}{\partial \eta^2} Z_A^2(x, c, v) + \frac{\partial^2 \Psi}{\partial x \partial \eta}\sigma Z_A(x, c, v) + u_P(x, c)] = 0,$$
$$\Psi(x, \eta, T) = U_P(x, C_T(x, \eta)).$$

This Bellman equation defines optimal feedbacks $\hat{c}(x, \eta)$, $\hat{v}(x, \eta)$. We obtain the optimal controls $\hat{c}(t)$ and $\hat{v}(t)$ by solving the equation

$$d\eta = [-u_A(X(t), \hat{c}(X(t), \eta), \hat{v}(X(t), \eta)) - \tag{15.3.18}$$
$$\hat{v}(X(t), \eta) Z_A(X(t), \hat{c}(X(t), \eta), \hat{v}(X(t), \eta))]dt + Z_A(X(t), \hat{c}(X(t), \eta),$$
$$\hat{v}(X(t), \eta))dw(t),$$
$$\eta(0) = \eta.$$

If the solution is denoted by $\hat{\eta}(t)$, then the optimal controls are given by $\hat{c}(t) = \hat{c}(X(t), \hat{\eta}(t))$ and $\hat{v}(t) = \hat{v}(X(t), \hat{\eta}(t))$. The value of the principal is $\Psi(x, \eta, 0)$. The value of the agent is, as guaranteed, η.

15.3.4 Generalization

In the previous model, the fact that the state $X(t)$ remained fixed played an essential role. Here we describe a situation that is different. We introduce a process $Y^v_{xyt}(s)$ depending on $v(.)$, which is a continuous adapted process, defined by

$$\frac{dY}{ds} = g(X_{xt}(s), v(s)), \tag{15.3.19}$$
$$Y(t) = y,$$

and the payoff of the agent is modified as follows:

$$\begin{aligned} K^A_{x,y,t;c(.),C_T}(v(.)) &= E^v[U_A(X_{xt}(T), Y^v_{xyt}(T), C_T) \\ &\quad + \int_t^T u_A(X_{xt}(s), c(s), v(s))ds|\mathcal{F}^t]. \end{aligned} \qquad (15.3.20)$$

Omitting to write explicitly $c(.)$, C_T, which are external for the agent, we write

$$\Phi^v(x, y, t) = E^v[U_A(X_{xt}(T), Y^v_{xyt}(T)) + \int_t^T u_A(X_{xt}(s), v(s))ds|\mathcal{F}^t]. \qquad (15.3.21)$$

In all these considerations, x, y, t are fixed parameters. Therefore, the control $v(s)$ can also depend on x, y, t, in which case we write it as $v_{xyt}(s)$. The next step is to define the random field

$$\Phi(x, y, t) = \sup_{v(.)} \Phi^v(x, y, t). \qquad (15.3.22)$$

We have omitted to denote it by $\Phi^A(x, y, t)$ to reduce notation.

We claim that this random field is the solution of the backward stochastic partial differential equation (BSPDE)

$$\begin{aligned} -d_t\Phi &= [\sup_v(u_A(x, v) + (\frac{\partial \Phi}{\partial x}\sigma + K)v + \frac{\partial \Phi}{\partial y}g(x, v)) + \\ &\quad + \frac{1}{2}\frac{\partial^2 \Phi}{\partial x^2}\sigma^2 + \sigma\frac{\partial K}{\partial x}]dt - Kdw, \\ \Phi(x, y, T) &= U_A(x, y), \end{aligned} \qquad (15.3.23)$$

which we call a stochastic HJB equation.

As usual, this BSPDE contains two unknown adapted random fields $\Phi(x, y, t)$ and $K(x, y, t)$ determined uniquely by (15.3.23). Solving it, we define an optimal random feedback $\hat{v}(x, y, t)$ by taking a point achieving the supremum in the brackets on the right-hand side of (15.3.23). We get next $\hat{Y}_{xyt}(s)$ by solving for $s > t$:

$$\begin{aligned} \frac{d\hat{Y}}{ds} &= g(X_{xt}(s), \hat{v}(X_{xt}(s), \hat{Y}, s)), \\ \hat{Y}(t) &= y. \end{aligned}$$

This leads to $\hat{v}(s) = \hat{v}_{xyt}(s) = \hat{v}(X_x(s), \hat{Y}_{xyt}(s), s)$, and finally $\Phi(x, y, t) = \Phi^{\hat{v}}(x, y, t)$. To check this last point, we use the Itô–Kunita formula. If we write (15.3.23) as

$$-d_t\Phi = \Gamma(x, y, t)dt - K(x, y, t)dw,$$

then the Itô–Kunita differentiation formula reads

$$d\Phi(X_{xt}(s), \hat{Y}_{xyt}(s), s) = \Bigg(-\Gamma(X_{xt}(s), \hat{Y}_{xyt}(s), s) + \frac{\partial K(X_{xt}(s), \hat{Y}_{xyt}(s), s)}{\partial x}\sigma +$$

$$\frac{\partial \Phi(X_{xt}(s), \hat{Y}_{xyt}(s), s)}{\partial y} g(X_{xt}(s), \hat{v}_{xyt}(s)) + \frac{1}{2}\frac{\partial^2 \Phi(X_{xt}(s), \hat{Y}_{xyt}(s), s)}{\partial x^2}\sigma^2 \Bigg) +$$

$$+ \left(\frac{\partial \Phi(X_{xt}(s), \hat{Y}_{xyt}(s), s)}{\partial x}\sigma + K(X_{xt}(s), \hat{Y}_{xyt}(s), s) \right) dw(s).$$

We note that $\hat{Y}_{xyt}(s) = Y^{\hat{v}}_{xyt}(s)$, and we set

$$Z^{\hat{v}}_{xyt}(s) = (\frac{\partial \Phi}{\partial x}\sigma + K)(X_{xt}(s), Y^{\hat{v}}_{xyt}(s), s).$$

We obtain easily

$$d\Phi(X_{xt}(s), Y^{\hat{v}}_{xyt}(s), s) = -(u_A(X_{xt}(s), Y^{\hat{v}}_{xyt}(s), s) + \hat{v}(s) Z^{\hat{v}}_{xyt}(s))ds + \\ + Z^{\hat{v}}_{xyt}(s) dw(s),$$

from which we obtain easily $\Phi(x, y, t) = \Phi^{\hat{v}}(x, y, t)$. Similarly, for an adapted process $v(.)$, we can state that

$$d\Phi(X_{xt}(s), Y^{v}_{xyt}(s), s) \leq -u_A(X_{xt}(s), Y^{v}_{xyt}(s), s)ds + \\ + (\frac{\partial \Phi}{\partial x}\sigma + K)(X_{xt}(s), Y^{v}_{xyt}(s), s)(dw - v(s)ds),$$

which implies

$$\Phi(x, y, t) \geq E^{v}[U_A(X_{xt}(T), Y^{v}_{xyt}(T)) + \int_t^T u_A(X_{xt}(s), v(s))ds|\mathcal{F}^t] \\ = \Phi^{v}(x, y, t),$$

and thus (15.3.22) is proved. We next consider the random field $\zeta(x, y, t) = \frac{\partial \Phi}{\partial y}(x, y, t)$. By setting $L(x, y, t) = \frac{\partial K}{\partial y}(x, y, t)$ and differentiating (15.3.23) in y, we obtain the equation

$$-d_t\zeta = [(\sigma\frac{\partial \zeta}{\partial x} + L)\hat{v} + \frac{\partial \zeta}{\partial y} g(x, \hat{v}) + \frac{1}{2}\sigma^2\frac{\partial^2 \zeta}{\partial x^2} + \sigma\frac{\partial L}{\partial x})]dt \quad (15.3.24) \\ -Ldw, \zeta(x, y, T) = \frac{\partial U_A(x, y)}{\partial y}.$$

Let us check that $\zeta(x, y, t)$ is uniquely defined by the feedback $\hat{v}(., ., s)$, $t \leq s \leq T$. Indeed, we define, for x, y, t, given the random field $m_{xyt}(\xi, \eta, s)$, the solution of the forward SPDE,

$$\begin{aligned} d_s m &= [-\sigma \frac{\partial}{\partial \xi}(m\hat{v}) - \frac{\partial}{\partial \eta}(mg(\xi, \hat{v})) + \frac{1}{2}\sigma^2 \frac{\partial^2 m}{\partial \xi^2}]ds + \\ &\quad + (-\sigma \frac{\partial m}{\partial \xi} + \hat{v}m)dw, \ t < s < T, \\ m(\xi, \eta, t) &= \delta(\xi - x) \otimes \delta(\eta - y), \end{aligned} \tag{15.3.25}$$

and we check easily that

$$\zeta(x, y, t) = E[\int \frac{\partial U_A(\xi, \eta)}{\partial \eta} m_{xyt}(\xi, \eta, T) d\xi d\eta | \mathcal{F}^t]. \tag{15.3.26}$$

Let us now define the adapted stochastic processes $(t \leq s)$

$$\Phi_{xyt}(s) = \Phi(X_{xt}(s), Y^{\hat{v}}_{xyt}(s), s), \quad \zeta_{xyt}(s) = \zeta(X_{xt}(s), Y^{\hat{v}}_{yt}(s), s). \tag{15.3.27}$$

From the BSPDE (15.3.23) and (15.3.24), using the Itô–Kunita differentiation rule, we check easily that

$$\begin{aligned} \Phi_{xyt}(s) &= U_A(X_{xt}(T), Y^{\hat{v}}_{xyt}(T)) + \int_s^T u_A(X_{xt}(\tau), \hat{v}_{xyt}(\tau))d\tau \\ &\quad + \int_s^T Z_{xyt}(\tau)\hat{v}_{xyt}(\tau)d\tau - \int_s^T Z_{xyt}(\tau)dw(\tau), \end{aligned} \tag{15.3.28}$$

$$\zeta_{xyt}(s) = \frac{\partial U_A}{\partial y}(X_{xt}(T), Y^{\hat{v}}_{xyt}(T)) + \int_s^T \Gamma_{xyt}(\tau)\hat{v}_{xyt}(\tau)d\tau - \int_s^T \Gamma_{xyt}(\tau)dw(\tau), \tag{15.3.29}$$

and

$$\hat{v}_{xyt}(\tau) = \varphi_A(X_{xt}(\tau), Z_{xyt}(\tau), \zeta_{xyt}(\tau)), \tag{15.3.30}$$

in which $\varphi_A(x, z, \zeta)$ achieves the maximum in v of the function $u_A(x, v) + zv + \zeta g(x, v)$. Moreover,

$$Z_{xyt}(\tau) = (\frac{\partial \Phi}{\partial x}\sigma + K)(X_{xt}(\tau), Y^{\hat{v}}_{yt}(\tau), \tau), \tag{15.3.31}$$

$$\Gamma_{xyt}(\tau) = (\frac{\partial^2 \Phi}{\partial x \partial y}\sigma + \frac{\partial K}{\partial y})(X_{xt}(\tau), Y^{\hat{v}}_{xyt}(\tau), \tau). \tag{15.3.32}$$

Also, $\Phi_{xyt}(t) = \Phi(x, y, t)$ and $\zeta_{xyt}(t) = \zeta(x, y, t)$. We note also that

$$\hat{v}_{xyt}(s) = \hat{v}(X_x(s), \hat{Y}_{xyt}(s), s),$$

and the notation is consistent. We consider next the case $\Phi(t) = \Phi_{x00}(t), \zeta(t) = \zeta_{x00}(t), Z(t) = Z_{x00}(t), \Gamma(t) = \Gamma_{x00}(t)$. The corresponding system (15.3.28), (15.3.29), (15.3.30) generalizes equation (15.3.10). In (15.3.10) there were two unknowns $\Phi(t), Z(t)$, and in (15.3.28), (15.3.29) there are four unknowns $\Phi(t), \zeta(t), Z(t), \Gamma(t)$. It is thus important that this system has a unique solution. We shall consider this issue in the next section.

We turn now to the principal problem. He will define a stochastic control problem that incorporates the agent problem. He has his own decisions $c(t), C_T$. It is convenient not to use $v(t)$ as in (15.3.15), but a pair $z(t), \gamma(t)$, reminiscent of the processes $Z(t), \Gamma(t)$. There will be two additional states $\eta(t), \zeta(t)$ reminiscent of $\Phi(t), \zeta(t)$. We also set $\tilde{w}(t) = w^v(t)$. We state the problem as follows:

$$\begin{aligned} dX &= \sigma\varphi_A(X, c, z, \zeta)dt + \sigma d\tilde{w}, \\ dY &= g(X, \varphi_A(X, c, z, \zeta))dt, \\ d\eta &= -u_A(X, c, \varphi_A(X, c, z, \zeta))dt + z(t)d\tilde{w}, \\ d\zeta &= -\gamma(t)d\tilde{w}, \end{aligned} \tag{15.3.33}$$

with initial conditions $X(0) = x, Y(0) = 0, \eta(0) = \eta, \zeta(0) = \zeta$. We then adjust the final contract C_T so that

$$\begin{aligned} \eta(T) &= U_A(X(T), Y(T), C_T), \\ \zeta(T) &= \frac{\partial U_A(X(T), Y(T), C_T)}{\partial y}. \end{aligned} \tag{15.3.34}$$

In the simpler case (15.3.14), we were requiring only the first condition, and there was not a $Y(T)$. So the contract should have more components. The system (15.3.33) contains four state variables and three control variables $c(t), z(t), \gamma(t)$. We write

$$d\tilde{w} = dw - \varphi_A(X(t), c(t), z(t), \zeta(t))dt,$$

and we assume that the map $C_T \to U_A(x, y, C_T), \frac{\partial U_A(x,y,C_T)}{\partial y}$ is invertible, for fixed parameters x, y. The inverse map is denoted by $C_T(x, y, \eta, \zeta)$. For given controls $c(t), z(t), \gamma(t)$, if we choose the final contract C_T in order to guarantee (15.3.34), then the pair $\eta(t), \zeta(t)$ together with $z(t), \gamma(t)$ satisfies (15.3.28), (15.3.29). If this system has a unique solution, the agent will have no other choice for the control $v(t)$ than to choose the quantity $\varphi_A(X(t), c(t), z(t), \zeta(t))$. It remains to write the payoff of the principal, namely

$$K^P_{x,0,\eta,0}(c(.), z(.), \gamma(.)) = \tilde{E}[\int_0^T u_P(X(t), c(t))dt + U_P(X(T), C_T(X(T), Y(T), \eta(T), \zeta T)))], \qquad (15.3.35)$$

where $\tilde{E}$ refers to the probability for which $\tilde{w}(t)$ is a Wiener process. As for the case (15.3.15), (15.3.16), we can write an HJB equation for the value function

$$\Psi(x, y, \eta, \zeta, t) = \sup_{c(.),z(.),\gamma(.)} K^P_{x,y,\eta,\zeta,t}(c(.), z(.), \gamma(.)),$$

where $K^P_{x,y,\eta,\zeta,t}(c(.), z(.), \gamma(.))$ is the obvious extension of $K^P_{x,0,\eta,0}(c(.), z(.), \gamma(.))$ starting at time t, with initial conditions x, y, η, ζ. We write it formally as

$$\frac{\partial\Psi}{\partial t} + \frac{1}{2}\sigma^2\frac{\partial^2\Psi}{\partial x^2} + \sup_{c,z,\gamma}[\frac{\partial\Psi}{\partial x}\sigma\varphi_A(x, c, z, \zeta) + \frac{\partial\Psi}{\partial y}g(x, \varphi_A(x, c, z, \zeta)) - \frac{\partial\Psi}{\partial\eta}u_A(x, c, \varphi_A(x, c, z, \zeta)) + \frac{1}{2}\frac{\partial^2\Psi}{\partial\eta^2}z^2 + \frac{1}{2}\frac{\partial^2\Psi}{\partial\zeta^2}\gamma^2 + \sigma z\frac{\partial^2\Psi}{\partial x\partial\eta} - \sigma\gamma\frac{\partial^2\Psi}{\partial x\partial\zeta} - z\gamma\frac{\partial^2\Psi}{\partial\eta\partial\zeta}] = 0, \qquad (15.3.36)$$

$$\Psi(x, y, \eta, \zeta, T) = U_P(x, C_T(x, y, \eta, \zeta)).$$

15.3.5 Study of Problems (15.3.28), (15.3.29), (15.3.30), and (15.3.23)

If we can solve (15.3.23), then the system (15.3.28), (15.3.29), (15.3.30) has a solution given by (15.3.27), (15.3.31), (15.3.32). So the important thing is uniqueness. We are going to prove uniqueness when $t \in [T-\epsilon, T]$ for ϵ fixed but sufficiently small. It will follow that $\Phi_{xy,T-\epsilon}(T-\epsilon) = \Phi(x, y, T-\epsilon)$ and $\zeta_{xy,T-\epsilon}(T-\epsilon) = \frac{\partial\Phi}{\partial y}(x, y, T-\epsilon)$. If we consider now that $t \in [T-2\epsilon, T-\epsilon]$, we can write from (15.3.28), (15.3.29), and $t \le s \le T-\epsilon$ that

$$\Phi_{xyt}(s) = \Phi(X_{xt}(T-\epsilon), Y^{\hat{u}}_{yt}(T-\epsilon), T-\epsilon) + \int_s^{T-\epsilon} u_A(X_{xt}(\tau), \hat{u}_{xyt}(\tau))d\tau + \int_s^{T-\epsilon} Z_{xyt}(\tau)\hat{u}_{xyt}(\tau)d\tau - \int_s^{T-\epsilon} Z_{xyt}(\tau)dw(\tau),$$

$$\zeta_{xyt}(s) = \frac{\partial \Phi}{\partial y}(X_{xt}(T-\epsilon), Y_{yt}^{\hat{u}}(T-\epsilon), T-\epsilon) + \int_s^{T-\epsilon} \Gamma_{xyt}(\tau)\hat{u}_{xyt}(\tau)d\tau$$
$$- \int_s^{T-\epsilon} \Gamma_{xyt}(\tau)dw(\tau),$$

and the same reasoning will lead to uniqueness in the interval $[T-2\epsilon, T-\epsilon]$. So uniqueness holds on the interval $[T-2\epsilon, T]$. By continuation, it will hold on $[0, T]$. To simplify notation we shall prove uniqueness for T fixed and sufficiently small of the system

$$\Phi(t) = U_A(X(T), Y^{\hat{u}}(T)) + \textstyle\int_t^T u_A(X(s), \hat{u}(s))dt \qquad (15.3.37)$$
$$+ \int_t^T Z(s)\hat{u}(s)dt - \int_t^T Z(s)dw(s),$$

$$\zeta(t) = \frac{\partial U_A}{\partial y}(X(T), Y^{\hat{u}}(T)) + \int_t^T \Gamma(s)\hat{u}(s)ds - \int_t^T \Gamma(s)dw(s), \qquad (15.3.38)$$

and

$$\hat{u}(s) = \varphi_A(X(s), Z(s), \zeta(s)), \qquad (15.3.39)$$

with $X(0) = x, Y(0) = 0$. We shall assume that the functions U_A and $\frac{\partial U_A}{\partial y}$ are globally Lipschitz in y, that u_A and g are globally Lipschitz in v, and that φ_A takes values in a bounded set and is globally Lipschitz in z, ζ. We shall prove uniqueness only for processes $Z(s)$ and $\Gamma(s)$ that are bounded. This unpleasant assumption will imply for the BSPDE (15.3.23) that the functions $z(x, y, t) = (\frac{\partial \Phi}{\partial x}\sigma + K)(x, y, t)$ and $\Gamma(x, y, t) = \frac{\partial z}{\partial y}(x, y, t)$ are bounded. Suppose we have two solutions Φ, ζ, Z, Γ and $\tilde{\Phi}, \tilde{\zeta}, \tilde{Z}, \tilde{\Gamma}$. We set

$$v(s) = \varphi_A(X(s), Z(s), \zeta(s)), \ \ \tilde{v}(s) = \varphi_A(X(s), \tilde{Z}(s), \tilde{\zeta}(s)).$$

The processes $v(s), \tilde{v}(s), Z(s), \tilde{Z}(s), \Gamma(s), \tilde{\Gamma}(s)$ are bounded by assumption. From Itô's calculus one obtains easily

$$[E|\tilde{\Phi}(t) - \Phi(t)|^2 + E|\tilde{\zeta}(t) - \zeta(t)|^2] \exp kt + k \int_t^T [E|\tilde{\Phi}(s) - \Phi(s)|^2$$
$$+E|\tilde{\zeta}(s) - \zeta(s)|^2] \exp ks \, ds +$$
$$+ \int_t^T [E|\tilde{Z}(s) - Z(s)|^2 + E|\tilde{\Gamma}(s) - \Gamma(s)|^2] \exp ks \, ds =$$

$$[E|U_A(X(T), Y^{\tilde{v}}(T)) - U_A(X(T), Y^v(T))|^2 + E|\frac{\partial U_A}{\partial y}(X(T), Y^{\tilde{v}}(T))$$

$$-\frac{\partial U_A}{\partial y}(X(T), Y^v(T))|^2] \exp kT +$$

$$+2\int_t^T E(\tilde{\Phi}(s) - \Phi(s))[u_A(X(s), \tilde{v}(s)) - u_A(X(s), v(s)) + \tilde{Z}(s)\tilde{v}(s)$$

$$-Z(s)v(s)] \exp ks\, ds +$$

$$+2\int_t^T E(\tilde{\zeta}(s) - \zeta(s))(\tilde{\Gamma}(s)\tilde{v}(s) - \Gamma(s)v(s)) \exp ks\, ds.$$

Next, from the assumptions, we can state

$$E|U_A(X(T), Y^{\tilde{v}}(T)) - U_A(X(T), Y^v(T))|^2 + E|\frac{\partial U_A}{\partial y}(X(T), Y^{\tilde{v}}(T))$$

$$-\frac{\partial U_A}{\partial y}(X(T), Y^v(T))|^2 \leq$$

$$CT\int_0^T |\tilde{v}(s) - v(s)|^2 ds \leq CT[\int_0^T |\tilde{Z}(s) - Z(s)|^2 ds + \int_0^T |\tilde{\zeta}(s) - \zeta(s)|^2 ds].$$

Next,

$$\int_t^T E(\tilde{\Phi}(s) - \Phi(s))[u_A(X(s), \tilde{v}(s)) - u_A(X(s), v(s)) + \tilde{Z}(s)\tilde{v}(s)$$

$$-Z(s)v(s)] \exp ks\, ds+$$

$$+\int_t^T E(\tilde{\zeta}(s) - \zeta(s))(\tilde{\Gamma}(s)\tilde{v}(s) - \Gamma(s)v(s)) \exp ks\, ds \leq C\int_t^T |\tilde{Z}(s) - Z(s)|(|\tilde{\Phi}(s)$$

$$-\Phi(s)| + |\tilde{\zeta}(s) - \zeta(s)|) \exp ks\, ds+$$

$$+C\int_t^T |\tilde{\Gamma}(s) - \Gamma(s)||\tilde{\zeta}(s) - \zeta(s)| \exp ks\, ds + C\int_t^T (|\tilde{\Phi}(s) - \Phi(s)|^2$$

$$+|\tilde{\zeta}(s) - \zeta(s)|^2) \exp ks\, ds,$$

thanks to the assumption that $Z(s)$ and $\Gamma(s)$ are bounded. We continue the inequalities

$$\leq \epsilon\int_t^T (|\tilde{Z}(s) - Z(s)|^2 + |\tilde{\zeta}(s) - \zeta(s)|^2) \exp ks\, ds + \frac{C}{\epsilon}\int_t^T (|\tilde{\Phi}(s)$$

$$-\Phi(s)|^2 + |\tilde{\zeta}(s) - \zeta(s)|^2) \exp ks\, ds.$$

By taking ϵ sufficiently small and k sufficiently large and by modifying the generic constants, we get easily from these inequalities that

$$k\int_0^T [E|\tilde{\Phi}(s)-\Phi(s)|^2+E|\tilde{\zeta}(s)-\zeta(s)|^2]\exp ks\,ds+$$

$$+\int_0^T [E|\tilde{Z}(s)-Z(s)|^2+E|\tilde{\Gamma}(s)-\Gamma(s)|^2]\exp ks\,ds \leq CT\exp kT$$

$$\int_0^T (|\tilde{Z}(s)-Z(s)|^2+|\tilde{\zeta}(s)-\zeta(s)|^2)\exp ks\,ds,$$

and for T sufficiently small, we necessarily have $\tilde{\Phi}(s) = \Phi(s), \tilde{\zeta}(s) = \zeta(s)$, $\tilde{Z}(s) = Z(s), \tilde{\Gamma}(s) = \Gamma(s)$. So the uniqueness is proven.

We turn now to the study of the BSPDE (15.3.23). This type of equation has been studied by Shige Peng [42] in the nondegenerate case, i.e., there is also a second-order term in the y argument. It would take too long to develop a full analytical proof. We simply give the formal argument, based on the optimality principle, as in Peng [42]. To be justified, it requires considerable regularity on the random field $\Phi(x, y, t)$. We require first that it have an Itô differential, so it can be written as follows:

$$\Phi(x,y,t)=U_A(x,y)+\int_t^T L(x,y,s)ds-\int_t^T K(x,y,s)dw(s), \qquad (15.3.40)$$

and we identify the random field $L(x, y, t)$. Note that an expression like (15.3.40) is necessarily uniquely defined. We proceed with the optimality principle. We can write

$$\Phi^u(x,y,t)=E[\textstyle\int_t^{t+\epsilon} M_t^v(\tau)u_A(X_{xt}(\tau),u(\tau))d\tau+M_t^v(t+\epsilon)\Phi^v(X_{xt}(t+\epsilon),$$
$$Y_{xyt}^v(t+\epsilon,t+\epsilon)|\mathcal{F}^t].$$

Since the control $v(.)$ splits between the intervals $(t, t+\epsilon)$ and $(t+\epsilon, T)$, we can write

$$\Phi(x,y,t)=\sup_u E[\textstyle\int_t^{t+\epsilon} M_t^u(\tau)u_A(X_{xt}(\tau),u(\tau))d\tau+M_t^u(t+\epsilon)\Phi(X_{xt}(t+\epsilon),$$
$$Y_{xyt}^u(t+\epsilon,t+\epsilon)|\mathcal{F}^t].$$

From the Itô–Kunita formula we derive

$$E[M_t^u(t+\epsilon)\Phi(X_{xt}(t+\epsilon),Y_{xyt}^u(t+\epsilon,t+\epsilon)|\mathcal{F}^t]=\Phi(x,y,t)+$$

$$+\int_t^{t+\epsilon} M_t^u(\tau)[-L(X_{xt}(\tau),Y_{xyt}^u(\tau),\tau)+\frac{\partial\Phi}{\partial y}(X_{xt}(\tau),Y_{xyt}^u(\tau),\tau)g(X_{xt}(\tau),u(\tau))+$$

$$+(\frac{1}{2}\sigma^2\frac{\partial^2\Phi}{\partial x^2}+\sigma\frac{\partial K}{\partial x})(X_{xt}(\tau),Y_{xyt}^u(\tau),\tau)+(\sigma\frac{\partial\Phi}{\partial x}+K)(X_{xt}(\tau),Y_{xyt}^u(\tau),\tau)u(\tau)]d\tau.$$

By canceling the term $\Phi(x, y, t)$, dividing by ϵ, and letting ϵ tend to 0, we show (formally) that $L(x, y, t)$ is the random field in brackets on the right-hand side of (15.3.23).

Finally, introducing ideas as in (15.3.25), (15.3.26), we can give a new interpretation to (15.3.23). Consider a feedback $v(x, y, t)$. We define $\Phi^v(x, y, t)$ as the solution of the BSPDE

$$
\begin{aligned}
-d_t\Phi^v &= [u_A(x, v) + (\frac{\partial \Phi^v}{\partial x}\sigma + K^v)v + \frac{\partial \Phi^v}{\partial y}g(x, v)) + \\
&\quad + \frac{1}{2}\frac{\partial^2 \Phi^v}{\partial x^2}\sigma^2 + \sigma\frac{\partial K^v}{\partial x}]dt - K^v dw \\
\Phi^v(x, y, T) &= U_A(x, y).
\end{aligned} \tag{15.3.41}
$$

Note that the symbol v must be replaced with the feedback $v(x, y, t)$. We associate to the feedback the control $v_{xyt}(s)$ as explained above for $\hat{v}_{xyt}(s)$. To save notation we shall denote in the same way the feedback v and the control $v_{xyt}(s)$ associated to it. We consider the processes $X_{xt}(s)$ and $Y^v_{xyt}(s)$. We can check that

$$
\begin{aligned}
\Phi^v(x, y, t) =& E[M^v_{xyt}(T)U_A(X_{xt}(T), Y^v_{xyt}(T)) + \int_t^T M^v_{xyt}(\tau)u_A(X_{xt}(\tau), \\
& v(X_{xt}(\tau), Y^v_{xyt}(\tau), \tau))d\tau|\mathcal{F}^t],
\end{aligned} \tag{15.3.42}
$$

and of course $M^v_{xyt}(s)$ is the solution of $dM^v_{xyt}(s) = M^v_{xyt}(s)v(X_{xt}(s), Y^v_{xyt}(s), s))dw(s)$, $M^v_{xyt}(t) = 1$. We associate to the feedback v the probability density $m^v_{xyt}(\xi, \eta, s)$, the solution of the SPDE (called the Fokker–Planck equation)

$$
\begin{aligned}
d_s m &= [-\sigma\frac{\partial}{\partial \xi}(mv) - \frac{\partial}{\partial \eta}(mg(\xi, v)) + \frac{1}{2}\sigma^2\frac{\partial^2 m}{\partial \xi^2}]ds + \\
&\quad + (-\sigma\frac{\partial m}{\partial \xi} + vm)dw,\ t < s < T \\
m(\xi, \eta, t) &= \delta(\xi - x) \otimes \delta(\eta - y),
\end{aligned} \tag{15.3.43}
$$

in which the symbol v must be replaced by $v(\xi, \eta, s)$. From (15.3.41) and (15.3.43), we deduce easily that

$$
\begin{aligned}
\Phi^v(x, y, t) =& E[\int_t^T \int\int u_A(\xi, v(\xi, \eta, s))m^v_{xyt}(\xi, \eta, s)d\xi d\eta ds \\
& + \int\int U_A(\xi, \eta)m^v_{xyt}(\xi, \eta, T)d\xi d\eta|\mathcal{F}^t].
\end{aligned} \tag{15.3.44}
$$

Let us check that $\Phi(x, y, t) = \sup_v \Phi^v(x, y, t)$ is the solution of the stochastic HJB equation (15.3.23). It is again a formal proof, which uses calculus of variations techniques. Let $\hat{v}$ be an optimal feedback. Note that it may depend on x, y, t, so it should be written as $\hat{v}_{xyt}(\xi, \eta, s)$. We omit to write x, y, t, which in this context are purely fixed parameters. The idea is to compute the Gâteaux differential of Φ^v with respect to the feedback v and write the necessary condition, namely that this Gâteaux differential vanishes at the optimal feedback $\hat{v}$. Let us compute formally the Gâteaux differential. Let $m(\xi,\eta,s)=m^{\hat{v}}_{xyt}(\xi,\eta,s)$ be the solution of the Fokker–Planck equation (15.3.43) corresponding to $v=\hat{v}$. We consider a perturbation $\hat{v}+\theta v$, and we want to compute the Gâteaux differential $\frac{d}{d\theta}\Phi^{\hat{v}+\theta v}(x,y,t)|_{\theta=0}$. We introduce $\tilde{m}(\xi,\eta,s)$ as the solution of

$$d_s\tilde{m}=[-\sigma\frac{\partial}{\partial\xi}(\tilde{m}\hat{v})-\frac{\partial}{\partial\eta}(\tilde{m}g(\xi,\hat{v}))+\frac{1}{2}\sigma^2\frac{\partial^2\tilde{m}}{\partial\xi^2}]ds+(-\sigma\frac{\partial\tilde{m}}{\partial\xi}+\hat{v}\tilde{m})dw$$
$$+(-\sigma\frac{\partial}{\partial\xi}(mv)-\frac{\partial}{\partial\eta}(m\frac{\partial}{\partial v}g(\xi,\hat{v})v))ds+vmdw$$
$$\tilde{m}(\xi,\eta,t)=0.$$

By easy calculations, we obtain

$$\frac{d}{d\theta}\Phi^{\hat{v}+\theta v}(x,y,t)|_{\theta=0}=E[\int_t^T\int\int(\frac{\partial}{\partial v}u_A(\xi,\hat{v})v\,m+u_A(\xi,\hat{v})\,\tilde{m})d\xi d\eta ds+$$
$$+\int\int U_A(\xi,\eta)\tilde{m}(\xi,\eta,T)d\xi d\eta|\mathcal{F}^t]. \tag{15.3.45}$$

By definition of $\hat{v}$, the random field $\Phi(x,y,t)$ is the solution of the BSPDE

$$-d_t\Phi=[u_A(x,\hat{v})+(\frac{\partial\Phi}{\partial x}\sigma+K)\hat{v}+\frac{\partial\Phi}{\partial y}g(x,\hat{v}))+$$
$$+\frac{1}{2}\frac{\partial^2\Phi}{\partial x^2}\sigma^2+\sigma\frac{\partial K}{\partial x}]dt-Kdw, \tag{15.3.46}$$
$$\Phi(x,y,T)=U_A(x,y),$$

where $K=K^{\hat{v}}$. Combining this BSPDE and the SPDE of $\tilde{m}$, we obtain the expression

$$\frac{d}{d\theta}\Phi^{\hat{v}+\theta v}(x,y,t)|_{\theta=0}=E[\int_t^T\int\int m[\frac{\partial}{\partial v}u_A(\xi,\hat{v})+\frac{\partial\Phi}{\partial\xi}\sigma+K$$
$$+\frac{\partial\Phi}{\partial\eta}\frac{\partial}{\partial v}g(\xi,\hat{v})]v\,d\xi d\eta ds|\mathcal{F}^t].$$

Since v is an arbitrary random field adapted to $\mathcal{F}^t$, and assuming that m does not vanish on a set of positive measure, it follows that

$$\frac{\partial}{\partial v}u_A(\xi,\hat{v})+\frac{\partial\Phi}{\partial\xi}\sigma+K+\frac{\partial\Phi}{\partial\eta}\frac{\partial}{\partial v}g(\xi,\hat{v})=0.$$

This condition is a necessary condition for $u_A(\xi, v) + (\frac{\partial \Phi}{\partial \xi}\sigma + K)v + \frac{\partial \Phi}{\partial \eta}g(\xi, v)$. If it is sufficient, we can assert that $\hat{v}$ attains the maximum of $u_A(\xi, v) + (\frac{\partial \Phi}{\partial \xi}\sigma + K)v + \frac{\partial \Phi}{\partial \eta}g(\xi, v)$, which implies that Φ is the solution of (15.3.23). We notice also that the optimal feedback $\hat{v}_{xyt}(\xi, \eta, s)$ does not depend on x, y, t.

EXERCISE 15.4. Show that the following sequence is monotone increasing:

$$\begin{aligned} -d_t\Phi^k &= [u_A(x, u^k) + (\frac{\partial \Phi^k}{\partial x}\sigma + K^k)u^k + \frac{\partial \Phi^k}{\partial y}g(x, u^k)) + \\ &+ \frac{1}{2}\frac{\partial^2 \Phi^k}{\partial x^2}\sigma^2 + \sigma\frac{\partial K^k}{\partial x}]dt - K^k dw, \qquad (15.3.47) \\ \Phi^k(x, y, T) &= U_A(x, y), \end{aligned}$$

with u^{k+1} achieving the maximum of $u_A(\xi, v) + (\frac{\partial \Phi^k}{\partial \xi}\sigma + K^k)v + \frac{\partial \Phi^k}{\partial \eta}g(\xi, v)$ in v.

15.4 Examples and Applications

15.4.1 Exponential Utilities and Lump-Sum Payments

We consider the particular case of (15.3.19), (15.3.20), (15.3.35). So u_A and u_P are 0, and

$$U_A(x, y, C_T) = -\exp -\gamma_A(C_T - y), \; U_P(x, C_T) = -\exp -\gamma_P(x - C_T) \qquad (15.4.1)$$

and $g(x, v) = \mu x + g(v)$. We begin by solving the agent problem, namely (15.3.23), which reads

$$\begin{aligned} -d_t\Phi &= [\sup_v((\frac{\partial \Phi}{\partial x}\sigma + K)v + \frac{\partial \Phi}{\partial y}(\mu x + g(v)) + \qquad (15.4.2) \\ &+ \frac{1}{2}\frac{\partial^2 \Phi}{\partial x^2}\sigma^2 + \sigma\frac{\partial K}{\partial x}]dt - K dw, \\ \Phi(x, y, T) &= -\exp -\gamma_A(C_T - y). \end{aligned}$$

We check by direct calculation that

$$\begin{aligned} \Phi(x, y, t) &= -\exp -\gamma_A(\varphi(t) - y - \mu x(T - t)), \qquad (15.4.3) \\ K(x, y, t) &= -\gamma_A\Phi(x, y, t)(\rho(t) + \sigma\mu(T - t)), \end{aligned}$$

in which the pair $\varphi(t), \rho(t)$ is the solution of the BSDE

$$d\varphi = (\frac{1}{2}\gamma_A\rho^2 + \inf_v(g(v) - \rho v))dt + (\rho(t) + \sigma\mu(T-t))dw, \quad (15.4.4)$$
$$\varphi(T) = C_T.$$

We get immediately $\zeta(x, y, t) = \gamma_A\Phi(x, y, t)$. The optimal feedback $\hat{v}(x, y, t)$ does not depend on x, y and reduces to $\hat{v}(t)$, which achieves the infimum of $g(v) - v\rho(t)$. Assuming that the derivative in v of this function vanishes at $\hat{v}(t)$, we get immediately $\hat{v}(t) = (g')^{-1}(\rho(t))$. This simplifies considerably the principal's problem. Defining the process $\tilde{w}(t)$ by $d\tilde{w}(t) = dw(t) - (g')^{-1}(\rho(t))dt$ and $\tilde{w}(0) = 0$, we consider the probability $\tilde{P}$ such that $\tilde{w}(t)$ becomes a Wiener process. Then the principal's payoff is given by $-\tilde{E}[\exp -\gamma_P(X(T) - C_T)] = -\tilde{E}[\exp -\gamma_P(X(T) - \varphi(T))]$. The principal incorporates the agent's problem as follows. He considers the dynamical system $X(t)$ and $\varphi(t)$ the solution of the SDEs

$$dX = \sigma((g')^{-1}(\rho)dt + d\tilde{w}), \quad (15.4.5)$$

$$d\varphi = (\frac{1}{2}\gamma_A\rho^2 + g((g')^{-1}(\rho)) + \sigma\mu(T-t)(g')^{-1}(\rho))dt +$$
$$+ (\rho + \sigma\mu(T-t))d\tilde{w},$$

$$X(0) = x, \ \varphi(0) = \varphi,$$

in which the initial value φ is decided by the principal. In the system (15.4.5), the adapted process $\rho(t)$ is a control. The principal maximizes the payoff

$$-\tilde{E}[\exp -\gamma_P(X(T) - \varphi(T))]. \quad (15.4.6)$$

If the principal solves this problem, he proposes the contract C_T to the agent. The agent will solve problem (15.4.4), for which necessarily the process $\rho(t)$ coincides with the principal's control, provided that the BSDE (15.4.4) has a unique solution. The agent computes his expected return on his objective, which is by definition $\Phi(x, 0, 0)$; hence from (15.4.3), we have $-\exp -\gamma_A(\varphi - \mu x T)$. The agent accepts the contract if this value is sufficiently high. So the principal fixes the value of φ in order to be sure that the corresponding contract is acceptable to the agent. This being done, it remains to solve the stochastic control problem (15.4.5), (15.4.6). This is done by solving the HJB equation for the principal's value function $\Psi(x, \varphi, t)$, namely

$$\frac{\partial\Psi}{\partial t} + \frac{1}{2}\sigma^2\frac{\partial^2\Psi}{\partial x^2} + \frac{1}{2}\sigma^2\mu^2(T-t)^2\frac{\partial^2\Psi}{\partial\varphi^2} + \sigma^2\mu(T-t)\frac{\partial^2\Psi}{\partial x\partial\varphi} + \quad (15.4.7)$$

$$+\sup_{\rho}[\sigma(g')^{-1}(\rho)\frac{\partial\Psi}{\partial x}+(\frac{1}{2}\gamma_A\rho^2+g((g')^{-1}(\rho))+\sigma\mu(T-t)(g')^{-1}(\rho))\frac{\partial\Psi}{\partial\varphi}$$

$$+(\frac{1}{2}\rho^2+\rho\sigma\mu(T-t))\frac{\partial^2\Psi}{\partial\varphi^2}\sigma\rho\frac{\partial^2\Psi}{\partial x\partial\varphi}]=0,$$

$$\Psi(x,\varphi,T)=-\exp-\gamma_P(x-\varphi).$$

We look for a solution of the form

$$\Psi(x,\varphi,t)=-\Lambda(t)\exp-\gamma_P(x-\varphi). \tag{15.4.8}$$

An easy calculation shows that $\Lambda(t)$ is the solution of the linear differential equation

$$-\Lambda'(t)-\frac{1}{2}\gamma_P^2\sigma^2(1-\mu(T-t))^2\Lambda(t)+ \tag{15.4.9}$$

$$+\gamma_P\Lambda(t)\sup_{\rho}[\sigma(1-\mu(T-t))((g')^{-1}(\rho)+\gamma_P\rho)-\frac{1}{2}(\gamma_A+\gamma_P)\rho^2$$

$$-g((g')^{-1}(\rho))]=0.$$

$$\Lambda(T)=1$$

We see that the optimal $\rho(t)$ is deterministic and achieves the maximum of the function $\sigma(1-\mu(T-t))((g')^{-1}(\rho)+\gamma_P\rho)-\frac{1}{2}(\gamma_A+\gamma_P)\rho^2-g((g')^{-1}(\rho))$, provided that this function has a unique maximum. If we take $g(v)=\frac{1}{2}v^2$, we get easily, denoting by $\hat{\rho}(t)$ the optimal control,

$$\hat{\rho}(t)=\frac{\sigma(1+\gamma_P)(1-\mu(T-t))}{1+\gamma_P+\gamma_A}.$$

We state the following result.

PROPOSITION 15.1. *We assume that the BSDE (15.4.4) has a unique solution $\varphi(t)$ and $\rho(t)$. We also assume that the function of ρ,*

$$\sigma(1-\mu(T-t))((g')^{-1}(\rho)+\gamma_P\rho)-\frac{1}{2}(\gamma_A+\gamma_P)\rho^2-g((g')^{-1}(\rho)),$$

has a unique maximum $\hat{\rho}(t)$. Then the principal chooses the contract C_T by the formula

$$\hat{C}_T=\varphi-\int_t^T(\frac{1}{2}\gamma_A\hat{\rho}^2+\inf_v(g(v)-\hat{\rho}v))ds+$$

$$-\int_t^T(\hat{\rho}(s)+\sigma\mu(T-s))dw(s),$$

with a constant φ sufficiently large to guarantee that the expected payoff of the agent $-\exp -\gamma_A(\varphi - \mu x T)$ is acceptable to him.

REMARK 15.1. The fact that the BSDE (15.4.4) has a unique solution is a difficult question, because it has quadratic growth. We refer to Cvitanić–Zhang [17] for details.

15.4.2 General Utilities, Quadratic Cost, and Lump-Sum Payments

We consider the model

$$
\begin{aligned}
U_A(x, y, C_T) &= U_A(C_T) - y, \qquad (15.4.10)\\
U_P(x, C_T) &= U_P(x - C_T),\\
u_A, u_P = 0,\ g(x, v) &= \frac{k}{2}v^2.
\end{aligned}
$$

The BSPDE solving the agent's problem reduces to

$$
\begin{aligned}
-\, d_t\Phi &= [\sup_v((\frac{\partial \Phi}{\partial x}\sigma + K)v + \frac{\partial \Phi}{\partial y}\frac{k}{2}v^2) + \qquad (15.4.11)\\
&\quad + \frac{1}{2}\frac{\partial^2 \Phi}{\partial x^2}\sigma^2 + \sigma\frac{\partial K}{\partial x}]dt - Kdw,\\
\Phi(x, y, T) &= U_A(C_T) - y.
\end{aligned}
$$

We get the solution $\Phi(x, y, t) = \varphi(t) - y$, with $\varphi(t)$, $K(t)$ the solution of the SDE

$$
\begin{aligned}
&-d\varphi - \sup_v(Kv - \frac{k}{2}v^2)\,dt - Kdw,\\
&\varphi(T) = U_A(C_T),
\end{aligned}
$$

or

$$
\begin{aligned}
-\, d\varphi &= \frac{1}{2}\frac{K^2}{k}dt - Kdw, \qquad (15.4.12)\\
\varphi(T) &= U_A(C_T).
\end{aligned}
$$

This SDE has an explicit solution. If we define $\zeta(t) = \exp\frac{\varphi(t)}{k}$, we obtain the SDE

$$
\begin{aligned}
d\zeta &= \zeta\frac{K}{k}dw,\\
\zeta(T) &= \exp\frac{U_A(C_T)}{k},
\end{aligned}
$$

so clearly $\zeta(t) = E[\exp \frac{U_A(C_T)}{k}|\mathcal{F}^t]$, and $K(t)$ is obtained by the representation of the martingale $\zeta(t)$. Consequently, $\varphi(t) = k \log \zeta(t)$. The optimal feedback of the agent is $\hat{v}(x, y, t) = \hat{v}(t) = \dfrac{K(t)}{k}$. We can then define the principal's problem. Define $\tilde{w}(t)$ by $d\tilde{w}(t) = dw(t) - \dfrac{K(t)}{k}dt$ and $\tilde{w}(0) = 0$. We consider the probability $\tilde{P}$ such that $\tilde{w}(t)$ becomes a Wiener process. Then the principal uses the dynamical system

$$dX = \sigma \frac{K(s)}{k} ds + \sigma d\tilde{w}(s),\ X(t) = x, \tag{15.4.13}$$

$$d\varphi = \frac{1}{2}\frac{K^2}{k} ds + K d\tilde{w}(s),\ \varphi(t) = \varphi,$$

where the initial condition φ is also chosen by the principal. The principal chooses the control $K(.)$ in order to maximize the payoff $\tilde{E}U_P(X_{xt}(T) - U_A^{-1}(\varphi_{\varphi t}(T))$. Define the value function

$$\Psi(x, \varphi, t) = \sup_{K(.)} \tilde{E}U_P(X_{xt}(T) - U_A^{-1}(\varphi_{\varphi t}(T)).$$

The function $\Psi(x, \varphi, t)$ is the solution of the HJB equation

$$\frac{\partial \Psi}{\partial t} + \frac{1}{2}\sigma^2 \frac{\partial^2 \Psi}{\partial x^2} + \sup_K [\sigma \frac{K}{k}\frac{\partial \Psi}{\partial x} + \frac{1}{2}\frac{K^2}{k}\frac{\partial \Psi}{\partial \varphi} + \tag{15.4.14}$$

$$+ \frac{1}{2}K^2 \frac{\partial^2 \Psi}{\partial \varphi^2} + K\sigma \frac{\partial^2 \Psi}{\partial x \partial \varphi}] = 0,$$

$$\Psi(x, \varphi, T) = U_P(x - U_A^{-1}(\varphi)).$$

Although complex, this equation can be solved, almost explicitly. We begin with a change of arguments. We set $\zeta = \exp \dfrac{\varphi}{k}$ and define

$$H(x, \zeta, t) = \zeta \Psi(x, k \log \zeta, t).$$

We check easily that H is the solution of a simpler PDE, namely

$$\frac{\partial H}{\partial t} + \frac{1}{2}\sigma^2 \frac{\partial^2 H}{\partial x^2} + \sup_v [\frac{1}{2}v^2 \frac{\partial^2 H}{\partial \zeta^2} + \sigma v \frac{\partial^2 H}{\partial x \partial \zeta}] = 0, \tag{15.4.15}$$

$$H(x, \zeta, T) = \zeta U_P(x - U_A^{-1}(k \log \zeta)).$$

We next consider the partial derivative $\lambda(x, \zeta, t) = \dfrac{\partial H}{\partial \zeta}$. After some tedious calculations we can see that $\lambda(x, \zeta, t)$ is the solution of the PDE

$$\frac{\partial \lambda}{\partial t} + \frac{1}{2}\sigma^2 \frac{\partial^2 \lambda}{\partial x^2} + \frac{1}{2}\sigma^2 \left(\frac{\frac{\partial \lambda}{\partial x}}{\frac{\partial \lambda}{\partial \zeta}}\right)^2 \frac{\partial^2 \lambda}{\partial \zeta^2} - \sigma^2 \frac{\frac{\partial \lambda}{\partial x}}{\frac{\partial \lambda}{\partial \zeta}} \frac{\partial^2 \lambda}{\partial x \partial \zeta} = 0, \tag{15.4.16}$$

$$\lambda(x, \zeta, T) = \Delta(x, \zeta) =$$

$$U_P(x - U_A^{-1}(k \log \zeta)) - k \frac{U_P'(x - U_A^{-1}(k \log \zeta))}{U_A'(U_A^{-1}(k \log \zeta))}.$$

The interesting "miracle" is that as in Chapter 13, we simplify considerably by looking at $\lambda(x, \zeta, t)$ as the solution of an equation $G(x, \lambda, t) = \zeta$. In fact, we get the linear equation

$$\frac{\partial G}{\partial t} + \frac{1}{2}\sigma^2 \frac{\partial^2 G}{\partial x^2} = 0, \tag{15.4.17}$$

$$G(x, \lambda, T) = \Gamma(x, \lambda),$$

where $\Gamma(x, \lambda)$ is obtained as the solution (in ζ) of the equation $\Delta(x, \zeta) = \lambda$. We must assume that the function $\Delta(x, \zeta)$ is monotone. Since (15.4.15) can be written as

$$\frac{\partial H}{\partial t} + \frac{1}{2}\sigma^2 \frac{\partial^2 H}{\partial x^2} - \frac{1}{2}\sigma^2 \frac{(\frac{\partial \lambda}{\partial x})^2}{\frac{\partial \lambda}{\partial \zeta}} = 0,$$

we obtain H as the solution of a linear equation.

We can give a probabilistic interpretation of the result (15.4.17). We first interpret (15.4.15) as the HJB equation of a stochastic control problem. The dynamical system is

$$dX = \sigma dw, \; X(t) = x, \tag{15.4.18}$$

$$d\zeta = v dw, \; \zeta(t) = \zeta,$$

and

$$H(x, \zeta, t) = \sup_{v(.)} E[\zeta_{\zeta t}(T) U_P(X_{xt}(T) - U_A^{-1}(k \log \zeta_{\zeta t}(T)))]. \tag{15.4.19}$$

We can use the martingale approach. We note that $E\zeta_{\zeta t}(T) = \zeta$. Because the functional in (15.4.19) does not contain $v(.)$ explicitly, we can try to find the optimal $\zeta(T)$ that maximizes $E[\zeta_{\zeta t}(T) U_P(X_{xt}(T) - U_A^{-1}(k \log \zeta_{\zeta t}(T)))]$ subject to the constraint $E\zeta_{\zeta t}(T) = \zeta$. We accommodate this scalar constraint with a Lagrange multiplier $\lambda = \lambda(x, \zeta, t)$. So we solve the problem

$$\sup_{\zeta(T)} \left(E[\zeta(T) U_P(X(T) - U_A^{-1}(k \log \zeta(T)))] - \lambda E\zeta(T) \right)$$

without constraint. Since $\zeta(T)$ and $X(T)$ are both $\mathcal{F}^T$-measurable, we can maximize pointwise. So we see that $\zeta(T)$ is the solution of the equation $\Delta(X(T), \zeta(T)) = \lambda$, or $\zeta(T) = \Gamma(X(T), \lambda)$. We next choose $\lambda = \lambda(x, \zeta, t)$ such that $\zeta = E\Gamma(X_{xt}(T), \lambda)$, which is in fact $\zeta = G(x, \lambda, t)$. This shows that our notation is coherent and that $\lambda(x, \zeta, t) = \dfrac{\partial H}{\partial \zeta}$, which is a standard interpretation of the Lagrange multiplier. Finally, the contract C_T is given by the formula

$$C_T = U_A^{-1}(k \log \Gamma(X(T), \lambda)),$$

so it is a deterministic function of $X(T)$.

15.4.3 Risk-Neutral Principal and Log-Utility Agent

First, to guarantee a positive process $X(s)$, we modify its equation to get a geometric Brownian motion

$$\begin{aligned} dX &= \sigma X dw, \\ X(t) &= x. \end{aligned} \tag{15.4.20}$$

We next assume

$$\begin{aligned} U_A(x, y, C_T) &= \log C_T - y, \\ U_P(x, C_T) &= x - C_T, \\ u_A, u_P = 0,\ g(x, v) &= \frac{1}{2} v^2. \end{aligned} \tag{15.4.21}$$

The value function of the agent $\Phi(x, y, t)$ remains $\varphi(t) - y$, with $\varphi(t), K(t)$ the solution of the SDE

$$\begin{aligned} -d\varphi &= \frac{1}{2} K^2 dt - K dw, \\ \varphi(T) &= \log C_T, \end{aligned} \tag{15.4.22}$$

and the optimal feedback of the agent is $\hat{v}(x, y, t) = K(t)$. Setting $\zeta(t) = \exp \varphi(t)$, we get easily

$$\begin{aligned} d\zeta &= \zeta K dw, \\ \zeta(T) &= C_T. \end{aligned} \tag{15.4.23}$$

The principal considers then the dynamical system

$$
\begin{aligned}
d\zeta &= \zeta K dw,\ dX = \sigma X dw, \\
\zeta(t) &= \zeta,\ X(t) = x,
\end{aligned}
\tag{15.4.24}
$$

with the payoff functional

$$\sup_{K(.)} \frac{1}{\zeta} E\zeta_{\zeta t}(T)(X_{xt}(T) - \zeta_{\zeta t}(T)). \tag{15.4.25}$$

Defining

$$H(x,\zeta,t) = \sup_{K(.)} E\zeta_{\zeta t}(T)(X_{xt}(T) - \zeta_{\zeta t}(T)),$$

one obtains easily that $H(x, \zeta, t)$ is the solution of the HJB equation

$$
\begin{gathered}
\frac{\partial H}{\partial t} + \frac{1}{2}\sigma^2 x^2 \frac{\partial^2 H}{\partial x^2} + \sup_v [\frac{1}{2} v^2 \frac{\partial^2 H}{\partial \zeta^2} + v\sigma x \frac{\partial^2 H}{\partial x \partial \zeta}] = 0, \\
H(x,\zeta,T) = \zeta(x-\zeta).
\end{gathered}
\tag{15.4.26}
$$

To solve this equation, we consider the derivative $\lambda(x, \zeta, t) = \frac{\partial H}{\partial \zeta}$, the solution of

$$
\begin{gathered}
\frac{\partial \lambda}{\partial t} + \frac{1}{2}\sigma^2 x^2 \frac{\partial^2 \lambda}{\partial x^2} + \frac{1}{2}\sigma^2 x^2 \frac{\partial^2 \lambda}{\partial \zeta^2}\left(\frac{\frac{\partial \lambda}{\partial x}}{\frac{\partial \lambda}{\partial \zeta}}\right)^2 - \sigma^2 x^2 \frac{\partial^2 \lambda}{\partial x \partial \zeta}\frac{\frac{\partial \lambda}{\partial x}}{\frac{\partial \lambda}{\partial \zeta}} = 0, \\
\lambda(x,\zeta,T) = x - 2.\zeta
\end{gathered}
\tag{15.4.27}
$$

In fact, we see that that $\lambda(x, \zeta, t) = x - 2\zeta$ is a solution. Therefore, $H(x, \zeta, t)$ is the solution of

$$
\begin{gathered}
\frac{\partial H}{\partial t} + \frac{1}{2}\sigma^2 x^2 \frac{\partial^2 H}{\partial x^2} + \frac{\sigma^2 x^2}{4} = 0, \\
H(x,\zeta,t) = \zeta(x-\zeta),
\end{gathered}
$$

which implies

$$H(x,\zeta,t) = \zeta x - \zeta^2 + \frac{x^2}{4}(\exp \sigma^2 (T-t) - 1), \tag{15.4.28}$$

so the optimal payoff of the principal is $\frac{H(x,\zeta,t)}{\zeta}$. The optimal v in the expression in square brackets in (15.4.26) is $\frac{\sigma x}{2}$. Therefore, the optimal path of (15.4.24) is $\zeta(s)-\zeta=\frac{1}{2}(X(s)-x)$, so the optimal $K(s)$ is given by

$$K(s)=\frac{\sigma X(s)}{X(s)-x+2\zeta}.$$

To avoid singularities, we assume that the initial conditions satisfy $2\zeta-x>0$. The optimal contract $C_T=\zeta(T)$ is thus given by

$$C_T=\zeta+\frac{1}{2}(X(T)-x), \tag{15.4.29}$$

and we note that it takes positive values.

15.4.4 Moral Hazard with Unobservable Effort

We describe another example taken from [17]. The agent is hired to perform a task that results in the output

$$dX=\mu dt+\sigma dw(t). \tag{15.4.30}$$

However, due to additional effort, the agent achieves a higher output

$$dX^{-v}=(\mu+\sigma v(t))dt+\sigma dw(t), \tag{15.4.31}$$

with $v(t)\geq 0$. However, he reports only $X(t)$. Defining $dw^{-v}(t)=v(t)dt+dw(t)$, we can write

$$dX=(\mu-\sigma v(t))dt+\sigma dw^{-v}(t). \tag{15.4.32}$$

The contract is composed of two parts; first a stopping time τ representing the closure of the contract, in which a lump sum is given, and second, a payment flow during the lifetime of the contract. The cumulative payment is denoted by $i(t)$, which is an adapted nondecreasing process. Lump-sum payments are possible at any time. The cash flow of the agent is $\lambda v(t)dt+di(t)$, where the first term represents the part of the additional output that the agent keeps for himself, $0<\lambda\leq 1$. Therefore, for a given contact $(\tau, i(t))$ and an additional effort $v(.)$, the payoff of the agent is

$$\varphi^v(0)=\varphi^{v,\tau,i}(0)=E^{-v}[\int_0^\tau \exp(-\gamma t)(\lambda v(t)dt+di(t))+R\exp-\gamma\tau], \tag{15.4.33}$$

where R is a lump sum given at the termination of the contract. To simplify the value, R is fixed. The payoff of the principal is expressed as

$$K^{v,\tau,i} = E^{-v}[\int_0^\tau \exp -rt((\mu - \sigma v(t))dt - di(t)) + L \exp -r\tau], \qquad (15.4.34)$$

in which L is fixed.

We begin with the agent problem. Let us define

$$\varphi^v(t) = E^{-v}[\int_t^\tau \exp -\gamma(s-t)(\lambda v(s)ds + di(s)) + R \exp -\gamma(\tau - t)|\mathcal{F}^t]. \qquad (15.4.35)$$

We see easily that $\varphi^v(t)$ is the solution of the BSDE

$$\begin{aligned} d\varphi^v &= [\gamma\varphi^v + \sigma v(Z^v - \lambda)]dt - di(t) + \sigma Z^v dw, \qquad (15.4.36) \\ \varphi^v(\tau) &= R. \end{aligned}$$

We are interested in taking the supremum in v of φ^v. We shall introduce a condition that will lead to a simple solution. We say that a contract τ, i is implementable if there exists a pair $\varphi(.), Z(.)$ such that

$$\begin{aligned} d\varphi &= \gamma\varphi dt - di + \sigma Z dw, \qquad (15.4.37) \\ \varphi(\tau) &= R, \;\; Z(t) \geq \lambda, \;\; 0 \leq t \leq \tau. \end{aligned}$$

We have the following interesting and useful result. If the contract τ, i is implementable, then $\varphi(t) \geq \varphi^v(t)$, under some assumptions on $v(.)$. In fact, we need to be able to assert that

$$E^{-v} \int_t^\tau |Z^v|^2 ds, < E^{-v} \int_t^\tau |Z|^2 ds + \infty.$$

If this is true, we claim that

$$\varphi(t) = \sup_v \varphi^v(t). \qquad (15.4.38)$$

An easy calculation shows that

$$d[(\varphi(t) - \varphi^v(t)) \exp -\gamma t] = \exp -\gamma t[-\sigma v(Z - \lambda) + \sigma(Z - Z^v)dw^v],$$

and since $v(Z - \lambda) \geq 0$, we obtain, after integrating between t and T and taking the conditional expectation with respect to $\mathcal{F}^t$, $\varphi(t) - \varphi^v(t) \geq 0$. Moreover, every $v \geq 0$ such that $v = 0$ when $Z > \lambda$ is optimal. In particular, $v = 0$ is optimal.

We turn now to the principal's problem. He has to choose the pair (τ, i) to optimize (15.4.34) when v chosen as just indicated. We propose the following approach. The principal chooses a value φ and a pair of adapted processes $Z(t)$ and $i(t)$. The second is monotonically nondecreasing. He then defines the process $\varphi(t)$ by the formula

$$d\varphi(t) = \gamma\varphi dt - di + \sigma Z dw. \tag{15.4.39}$$

In fact, for an arbitrary φ, we introduce the set

$$\mathcal{A}(\varphi) = \{Z(.), i(.)| \, Z(t) \geq \lambda, \ \tau = \inf_t\{\varphi(t) \leq R\} < +\infty\}.$$

If this set is not empty, the contract (τ, i) is implementable, and $\varphi = \sup_v \varphi^v(0)$. If the value φ is sufficiently attractive, the agent accepts the contract and chooses a positive $v(t)$ such that $v(t) = 0$ whenever $Z(t) > \lambda$. Note that if $\varphi \leq R$, we have $\tau = 0$. This is identical to "no contract." So we consider only the values $\varphi > R$ and assume that $\mathcal{A}(\varphi)$ is not empty. We shall show below that such is the case.

We set $\mathcal{A}_0(Z) = \{v(.)|v(t) = 0, \text{ if } Z(t) > \lambda\}$. If the contract (τ, i) defined by the pair $(Z(.), i(.))$ is proposed, the agent will choose any of the elements of $\mathcal{A}_0(Z)$. The corresponding payoff of the principal becomes

$$K^{v,Z,i}(\varphi) = E^{-v}[\int_0^\tau \exp -rt((\mu - \sigma v(t))dt - di(t)) + L \exp -r\tau]. \tag{15.4.40}$$

Since the principal cannot influence the choice of $v(.)$, we assume that he considers the best case $\sup_{v\in\mathcal{A}_0(Z)} K^{v,Z,i}(\varphi)$. Therefore, his value function is given by

$$\Psi(\varphi) = \sup_{Z,i\in\mathcal{A}(\varphi)} \sup_{v\in\mathcal{A}_0(Z)} K^{v,Z,i}(\varphi). \tag{15.4.41}$$

Let us check that $\mathcal{A}(\varphi)$ is not empty. We can take $Z(t) = \lambda$. It remains to find a process $i(t)$ such that if we define $\varphi(t)$ by (15.4.39), then $\tau < +\infty$ a.s. We recall that $\varphi > R$. Let $\bar{R} > \varphi > R$, and we choose $i(t)$ such that $\varphi(t)$ is a diffusion reflected at $\bar{R}$. It is well known that $E\tau = \Gamma(\varphi)$, where $\Gamma(\varphi)$ is the solution of the two-point boundary value problem

$$\gamma\varphi\Gamma'(\varphi) + \frac{1}{2}\lambda^2\sigma^2\Gamma''(\varphi) + 1 = 0, \ R < \varphi < \bar{R},$$
$$\Gamma(R) = 0, \ \Gamma'(\bar{R}) = 0,$$

and since $\Gamma(\varphi)$ is bounded, we get $E\tau$ bounded, and thus $\tau < +\infty$ a.s. This is why we have to accept processes $i(t)$ that may have jumps. So the value function is well defined. We next consider the Bellman equation associated with the value function. Clearly, if $\varphi = R$, we have $\Psi(R) = L$. Assume now that $\varphi > R$. We may implement an immediate jump $di(0) = \epsilon$, so that φ becomes $\varphi - \epsilon$. There is an immediate cost of ϵ, so we can claim that

$$\Psi(\varphi) \geq \Psi(\varphi - \epsilon) - \epsilon,$$

and since ϵ is arbitrary, assuming Ψ is smooth, we obtain $\Psi'(\varphi) \geq -1$. Another decision consists in taking $Z(t) = \lambda$ for $0 \leq t \leq \epsilon$ and $di(t) = 0$, $v(t) = 0$ on the same interval. Therefore, on $(0, \epsilon)$,

$$\varphi(t) = \tilde{\varphi}(t) = \varphi \exp \gamma t + \sigma\lambda \int_0^t \exp \gamma(t-s) dw(s).$$

Let $\tilde{\tau}$ be the first time at which $\tilde{\varphi}(t)$ hits the value R, so on the interval $(0, \tilde{\tau} \wedge \epsilon)$, the process has not yet reached the value R. From the optimality principle, we can claim that

$$\Psi(\varphi) \geq \mu E \int_0^{\tilde{\tau}\wedge\epsilon} \exp -rt\, dt + E \exp -r\tilde{\tau} \wedge \epsilon\, \Psi(\tilde{\varphi}(\tilde{\tau} \wedge \epsilon)),$$

and a standard expansion of the right-hand side yields easily the inequality

$$0 \geq \mu - r\Psi + \gamma\varphi \frac{\partial \Psi}{\partial \varphi} + \frac{1}{2}\lambda^2\sigma^2 \frac{\partial^2 \Psi}{\partial \varphi^2}.$$

In fact, we can also replace λ by any value $z \geq \lambda$. So we can assert that

$$\Psi'(\varphi) + 1 \geq 0 \tag{15.4.42}$$

$$\inf_{z\geq\lambda}(-\frac{1}{2}z^2\sigma^2\frac{\partial^2\Psi}{\partial\varphi^2}) - \gamma\varphi\frac{\partial\Psi}{\partial\varphi} + r\Psi - \mu \geq 0,$$

and in fact, we have also the complementarity slackness condition

$$(\Psi'(\varphi) + 1)[\inf_{z\geq\lambda}(-\frac{1}{2}z^2\sigma^2\frac{\partial^2\Psi}{\partial\varphi^2}) - \gamma\varphi\frac{\partial\Psi}{\partial\varphi} + r\Psi - \mu] = 0, \tag{15.4.43}$$

since one of the possibilities described above needs to occur for any initial value φ. This is a formal but standard argument. So $\Psi(\varphi)$ is the solution of a variational inequality.

We then define the following problem. Find $R^* > R$ and $\Psi(\varphi)$, $R < \varphi < R^*$, such that

$$-\frac{1}{2}\lambda^2\sigma^2\frac{\partial^2\Psi}{\partial\varphi^2} - \gamma\varphi\frac{\partial\Psi}{\partial\varphi} + r\Psi - \mu = 0, \; R < \varphi < R^*, \tag{15.4.44}$$

$$\Psi(R) = L, \; \Psi'(R^*) + 1 = 0 \quad \gamma R^* + r\Psi(R^*) = \mu.$$

We extend $\Psi(\varphi)$ for $\varphi > R^*$ by setting $\Psi(\varphi) = \Psi(R^*) - \varphi + R^*$. We get a C^2 function. We may then state the following proposition.

PROPOSITION 15.2. *We assume that there exist a solution $\Psi(\varphi)$ of the differential equation (15.4.44) and a number R^* such that all the relations hold and $\Psi(\varphi)$ is* *concave. Then $\Psi(\varphi)$ satisfies (15.4.42) and (15.4.43). Moreover, $\Psi(\varphi)$ is the value function of the principal. The optimal policy of the principal is described as follows. If $R < \varphi \leq R^*$, then he chooses $v = 0$, $Z(t) = \lambda$, and $i(t)$ is such that the process $\varphi(t)$ defined by (15.4.39) is reflected on the boundary R^* and remains below R^*. If $\varphi > R^*$, he gives an immediate lump sum $di(0) = \varphi - R^*$ to the agent, and proceeds as in the case $\varphi \leq R^*$.*

PROOF. The solution of (15.4.44) extended by the affine function beyond R^* is C^2 and concave. It follows that $\inf_{z\geq\lambda}(-\frac{1}{2}z^2\sigma^2\frac{\partial^2\Psi}{\partial\varphi^2}) = -\frac{1}{2}\lambda^2\sigma^2\frac{\partial^2\Psi}{\partial\varphi^2}$. Moreover, for $\varphi > R^*$, we have

$$-\frac{1}{2}\lambda^2\sigma^2\frac{\partial^2\Psi}{\partial\varphi^2} - \gamma\varphi\frac{\partial\Psi}{\partial\varphi} + r\Psi - \mu = \varphi\gamma + r(\Psi(R^*) - \varphi + R^*) - \mu$$
$$= (\varphi - R^*)(\gamma - r) \geq 0,$$

and thus the second inequality (15.4.42) holds. Since Ψ is concave, $\Psi'(\varphi) + 1$ is decreasing and vanishes for $\varphi > R^*$, so the first inequality (15.4.42) holds also. But (15.4.43) holds also, so $\Psi(\varphi)$ satisfies (15.4.42), (15.4.43). Let us prove that it is the value function for the principal. Consider a process $\varphi(t)$ satisfying (15.4.39) and $v(t) \geq 0$, with $v(t) = 0$ when $Z(t) > \lambda$. From Itô's formula we have

$$d(\Psi(\varphi(t))\exp -rt) = \exp -rt[(-r\Psi(\varphi(t)) + \gamma\varphi\frac{\partial\Psi(\varphi(t))}{\partial\varphi} - \frac{\partial\Psi(\varphi(t))}{\partial\varphi}di(t) +$$
$$+ \frac{\partial\Psi(\varphi(t))}{\partial\varphi}\sigma Z(t)dw + \frac{1}{2}\sigma^2 Z(t)^2\frac{\partial^2\Psi(\varphi(t))}{\partial\varphi^2}dt],$$

and using the two ineqaulities (15.4.42), it follows that

$$d(\Psi(\varphi(t))\exp -rt) \leq \exp -rt[-\mu dt - \frac{1}{2}\sigma^2\frac{\partial^2\Psi(\varphi(t))}{\partial\varphi^2}(\lambda^2 - Z(t)^2) + di(t) +$$
$$+ \frac{\partial\Psi(\varphi(t))}{\partial\varphi}\sigma Z(t)dw].$$

Using the concavity of Ψ and the fact that $Z(t) \geq \lambda > 0$, hence $\lambda^2 - Z(t)^2 < 0$, we get

$$d(\Psi(\varphi(t))\exp -rt) \leq \exp -rt[-\mu dt + di(t) + \frac{\partial\Psi(\varphi(t))}{\partial\varphi}\sigma Z(t)dw]$$
$$\leq \exp -rt[-\mu dt + di(t) + \frac{\partial\Psi(\varphi(t))}{\partial\varphi}\sigma Z(t)dw^{-u} -$$
$$- \frac{\partial\Psi(\varphi(t))}{\partial\varphi}\sigma v(t)dt]$$

$$\leq \exp -rt[-\mu dt + di(t) + \frac{\partial \Psi(\varphi(t))}{\partial \varphi}\sigma Z(t)dw^{-v} +$$
$$+ v(t)dt].$$

Integrating between 0 and τ and taking the mathematical expectation with respect to P^{-v}, we get easily $\Psi(\varphi) \geq K^{v,Z,i}(\varphi)$, so also

$$\Psi(\varphi) \geq \sup_{Z,i\in\mathcal{A}(\varphi)} \sup_{v\in\mathcal{A}_0(Z)} K^{v,Z,i}(\varphi).$$

On the other hand, assume first $R < \varphi \leq R^*$. We take $v(t) = 0$ and $Z(t) = \lambda$. Take next $i(t)$ such that the process

$$d\varphi(t) = \gamma\varphi dt - di + \sigma\lambda dw,$$
$$\varphi(0) = \varphi,$$

is reflected on R^* and remains below R^*. One obtains easily $\Psi(\varphi) = K^{0,\lambda,i}(\varphi)$. Now when $\varphi > R^*$, we modify $i(t)$ by adding an immediate payment of size $\varphi - R^*$, and we apply the previous policy for $\varphi = R^*$. We again obtain $\Psi(\varphi) = K^{0,\lambda,i}(\varphi)$. Combining results, we see that $\Psi(\varphi)$ is the value function of the principal. Moreover, we have obtained his optimal control. □

15.5 Contracting Under Hidden Agent Type

15.5.1 The Problem

We discuss another contract problem, developed in [17]. Our approach is more analytic. The output of the agent is given by

$$dX = (\sigma v(s) + \theta)ds + \sigma dw(s), \tag{15.5.1}$$
$$X(t) = x.$$

There is an additional parameter θ representing the skill of the agent. It is known to the agent but unknown to the principal. The principal has a probability distribution of this parameter given by $f(\theta)$ on a range (θ_L, θ_H). The agent decides on his effort $v(s)$. He decides also how to react to a payment contract made by the principal. The principal proposes a menu, depending on the parameter θ. The payment is a lump sum given at T. We assume that it is of the form $C_T = \Phi(X(T), \theta)$, so it is a function of the final output, indexed by the parameter θ. Of course, this function is the decision of the principal. The agent has the possibility to choose a contract $\Phi(x, \theta')$ with $\theta' \neq \theta$. In this case, he does not declare the true value θ. The principal will try to design a contract that is self-revealing, or truth-telling, so that the agent will have no interest in hiding the true value of the parameter.

Let us use the notation $X_{xt}^{\theta}(s; v(.))$ for the solution of (15.5.1). If the agent chooses the contract $\Phi(x, \theta')$, he will solve the following control problem:

$$\sup_{v(.)} E[U_A(\Phi(X_{xt}^{\theta}(T; v(.)), \theta') - \frac{k\sigma^2}{2}\int_t^T v^2(s)ds], \tag{15.5.2}$$

We denote by $V(x, t; \theta, \theta')$ this quantity, which thus represents the value function of the agent. The agent will then define θ' such that

$$V(x, t; \theta) = \sup_{\theta'} V(x, t; \theta, \theta'). \tag{15.5.3}$$

We see that the optimal θ' depends on the initial conditions x, t and of course on the true value θ. If the optimal θ' is θ, the menu of contracts is truth-telling. In solving the problem (15.5.2), the agent defines his optimal effort $\hat{v}$. This optimal effort depends on θ, θ'. Let us denote by $\hat{X}_{xt}^{\theta,\theta'}(s)$ the optimal output, the solution of (15.5.1), for the optimal effort. The principal gets the value function

$$\Gamma(x, t; \theta, \theta') = E[U_P(\hat{X}_{xt}^{\theta,\theta'}(T) - \Phi(\hat{X}_{xt}^{\theta,\theta'}(T), \theta'))], \tag{15.5.4}$$

If we denote by $\hat{\theta}$ the optimal choice of θ' made by the agent, the value function of the principal is

$$\Gamma(x, t; \theta) = \Gamma(x, t; \theta, \hat{\theta}). \tag{15.5.5}$$

The problem of the principal is to choose the function $\Phi(x, \theta)$ in order to maximize this value function.

15.5.2 Preliminaries

We begin by obtaining useful formulas for the functions $V(x, t; \theta, \theta')$ and $\Gamma(x, t; \theta, \theta')$. Consider first the agent's problem. By classical dynamic programming arguments, we see that V is the solution of Bellman equation

$$\frac{\partial V}{\partial t} + \theta\frac{\partial V}{\partial x} + \sup_v[v\sigma\frac{\partial V}{\partial x} - \frac{k\sigma^2}{2}v^2] + \frac{1}{2}\sigma^2\frac{\partial^2 V}{\partial x^2} = 0,$$
$$V(x, T; \theta, \theta') = U_A(\Phi(x, \theta')),$$

or

$$\frac{\partial V}{\partial t} + \theta\frac{\partial V}{\partial x} + \frac{1}{2k}(\frac{\partial V}{\partial x})^2 + \frac{1}{2}\sigma^2\frac{\partial^2 V}{\partial x^2} = 0,$$
$$V(x, T; \theta, \theta') = U_A(\Phi(x, \theta')).$$

Let us define the function

$$\chi(x,t;\theta,\theta') = \exp\frac{V(x,t;\theta,\theta')}{k\sigma^2},$$

An easy calculation shows that χ is the solution of a linear equation, namely

$$\begin{aligned}\frac{\partial\chi}{\partial t}+\theta\frac{\partial\chi}{\partial x}+\tfrac{1}{2}\sigma^2\tfrac{\partial^2\chi}{\partial x^2} &= 0,\\ \chi(x,T;\theta,\theta') &= \exp\frac{U_A(\Phi(x,\theta'))}{k\sigma^2}.\end{aligned} \tag{15.5.6}$$

We can give a probabilistic interpretation to the function χ. We introduce the process

$$X^{\theta}_{xt}(s) = x+\theta(s-t)+\sigma(w(s)-w(t)); \tag{15.5.7}$$

then we can assert that

$$\chi(x,t;\theta,\theta') = E\exp\frac{U_A(\Phi(X^{\theta}_{xt}(T),\theta'))}{k\sigma^2}.$$

Combining results, we can state the first important relation:

$$\exp\frac{V(x,t;\theta,\theta')}{k\sigma^2} = E\exp\frac{U_A(\Phi(X^{\theta}_{xt}(T),\theta'))}{k\sigma^2}. \tag{15.5.8}$$

Also,

$$\exp\frac{V(x,t;\theta)}{k\sigma^2} = E\exp\frac{U_A(\Phi(X^{\theta}_{xt}(T),\hat{\theta}))}{k\sigma^2}. \tag{15.5.9}$$

We turn now to the expression of $\Gamma(x,t;\theta,\theta')$. We first define the optimal $\hat{v}$ of the agent. From the Bellman equation above, the optimal effort is given by a feedback $\hat{v}(x,s) = \dfrac{1}{k\sigma}\dfrac{\partial V}{\partial x} = \dfrac{\sigma}{\chi}\dfrac{\partial\chi}{\partial x}$. So the optimal output $\hat{X}^{\theta,\theta'}_{xt}(s)$ is the solution of

$$\begin{aligned}d\hat{X} &= \left(\theta+\frac{\sigma^2}{\chi}\frac{\partial\chi}{\partial x}(\hat{X})\right)dt+\sigma dw,\\ \hat{X}(t) &= x.\end{aligned}$$

But then $\Gamma(x,t;\theta,\theta')$ is the solution of the linear equation depending on χ, namely

$$\begin{aligned}\frac{\partial\Gamma}{\partial t}+\theta\frac{\partial\Gamma}{\partial x}+\frac{\sigma^2}{\chi}\frac{\partial\chi}{\partial x}\frac{\partial\Gamma}{\partial x}+\frac{1}{2}\sigma^2\frac{\partial^2\Gamma}{\partial x^2} &= 0,\\ \Gamma(x,T;\theta,\theta') &= U_P(x-\Phi(x,\theta')).\end{aligned}$$

Again, a simple calculation shows that

$$\Gamma(x,t;\theta,\theta') = \frac{\tilde{\Gamma}(x,t;\theta,\theta')}{\chi(x,t;\theta,\theta')}, \tag{15.5.10}$$

with $\tilde{\Gamma}(x,t;\theta,\theta')$ the solution of

$$\begin{gathered}\frac{\partial \tilde{\Gamma}}{\partial t} + \theta \frac{\partial \tilde{\Gamma}}{\partial x} + \frac{1}{2}\sigma^2 \frac{\partial^2 \tilde{\Gamma}}{\partial x^2} = 0, \\ \tilde{\Gamma}(x,T;\theta,\theta') = U_P(x - \Phi(x,\theta')) \exp \frac{U_A(\Phi(x,\theta'))}{k\sigma^2},\end{gathered} \tag{15.5.11}$$

and thus we have obtained the formula

$$\begin{aligned}\Gamma(x,t;\theta,\theta') &= \exp - \frac{V(x,t;\theta,\theta')}{k\sigma^2} \\ &\quad E\left[U_P(X_{xt}^{\theta}(T) - \Phi(X_{xt}^{\theta}(T),\theta')) \exp \frac{U_A(\Phi(X_{xt}^{\theta}(T),\theta'))}{k\sigma^2}\right].\end{aligned} \tag{15.5.12}$$

The interesting feature is that it is the same process $X_{xt}^{\theta}(s)$ that enters into formulas (15.5.8) and (15.5.12). That will facilitate considerably the solution. Also, we can write from (15.5.5),

$$\begin{aligned}\Gamma(x,t;\theta) &= \exp - \frac{V(x,t;\theta)}{k\sigma^2} \\ &\quad E\left[U_P(X_{xt}^{\theta}(T) - \Phi(X_{xt}^{\theta}(T),\hat{\theta})) \exp \frac{U_A(\Phi(X_{xt}^{\theta}(T),\hat{\theta}))}{k\sigma^2}\right].\end{aligned} \tag{15.5.13}$$

A next interesting formula relates to computing $\dfrac{\partial \chi(x,t;\theta,\theta')}{\partial \theta} = \Delta(x,t;\theta,\theta')$. From equation (15.5.6), we obtain

$$\begin{gathered}\frac{\partial \Delta}{\partial t} + \theta \frac{\partial \Delta}{\partial x} + \frac{1}{2}\sigma^2 \frac{\partial^2 \Delta}{\partial x^2} = - \frac{\partial \chi(x,t;\theta,\theta')}{\partial x}, \\ \Delta(x,T;\theta,\theta') = 0.\end{gathered}$$

But then we check that

$$\Delta(x,t;\theta,\theta') = \frac{1}{\sigma^2}[\tilde{\chi}(x,t;\theta,\theta') - \chi(x,t;\theta,\theta')(x + \theta(T-t))], \tag{15.5.14}$$

in which $\tilde{\chi}$ is the solution of

$$\frac{\partial \tilde{\chi}}{\partial t} + \theta \frac{\partial \tilde{\chi}}{\partial x} + \frac{1}{2}\sigma^2 \frac{\partial^2 \tilde{\chi}}{\partial x^2} = 0, \tag{15.5.15}$$

$$\tilde{\chi}(x, T; \theta, \theta') = x \exp \frac{U_A(\Phi(x, \theta'))}{k\sigma^2}.$$

We can then provide a simple probabilistic formula for $\tilde{\chi}(x, t; \theta, \theta')$, namely

$$\tilde{\chi}(x, t; \theta, \theta') = E X_{xt}^{\theta}(T) \exp \frac{U_A(\Phi(X_{xt}^{\theta}(T), \theta'))}{k\sigma^2}.$$

But then

$$\frac{\partial \chi(x, t; \theta, \theta')}{\partial \theta} = \frac{1}{\sigma^2}[E X_{xt}^{\theta}(T) \exp \frac{U_A(\Phi(X_{xt}^{\theta}(T), \theta'))}{k\sigma^2} - \chi(x, t; \theta, \theta')(x + \theta(T - t))].$$

Now

$$\frac{1}{k\sigma^2} \exp \frac{V(x, t; \theta)}{k\sigma^2} \frac{\partial V(x, t; \theta)}{\partial \theta} = \frac{\partial \chi(x, t; \theta, \hat{\theta})}{\partial \theta},$$

and thus we have obtained the formula

$$\exp \frac{V(x, t; \theta)}{k\sigma^2}[\frac{1}{k} \frac{\partial V(x, t; \theta)}{\partial \theta} + (x + \theta(T - t))] = \tag{15.5.16}$$

$$= E X_{xt}^{\theta}(T) \exp \frac{U_A(\Phi(X_{xt}^{\theta}(T), \hat{\theta}))}{k\sigma^2}.$$

REMARK 15.2. In [17], the preceding formula is established in the case $\hat{\theta} = \theta$.

15.5.3 *The Principal's Problem*

The principal considers first the function $V(x, t; \theta)$ as given. This means that he parameterizes the value function of the agent. He tries then to maximize his value function $\Gamma(x, t; \theta)$. From formula (15.5.13), since $V(x, t; \theta)$ is fixed, he wants to adjust the function Φ in order to maximize

$$E\left[U_P(X_{xt}^{\theta}(T) - \Phi(X_{xt}^{\theta}(T), \hat{\theta})) \exp \frac{U_A(\Phi(X_{xt}^{\theta}(T), \hat{\theta}))}{k\sigma^2}\right].$$

But because $V(x, t; \theta)$ is fixed, he must take into account the relations (15.5.9) and (15.5.16), which represent constraints for the optimization problem. But these two constraints are scalar (for x, t, θ) and fixed. So one can introduce two Lagrange multipliers λ and μ, so the problem becomes an unconstrained maximization problem, namely to maximize

$$E\left[\exp\frac{U_A(\Phi(X_{xt}^\theta(T), \hat{\theta}))}{k\sigma^2}[U_P(X_{xt}^\theta(T) - \Phi(X_{xt}^\theta(T), \hat{\theta})) - \lambda X_{xt}^\theta(T) - \mu]\right]. \tag{15.5.17}$$

But this unconstrained optimization can be performed pointwise, by solving

$$\sup_\Phi \exp\frac{U_A(\Phi)}{k\sigma^2}[U_P(\xi - \Phi) - \lambda\xi - \mu].$$

Taking the derivative in Φ, we obtain the following equation: for $\Phi(\xi)$

$$k\sigma^2\frac{U_P'(\xi - \Phi)}{U_A'(\Phi)} = U_P(\xi - \Phi) - \lambda\xi - \mu. \tag{15.5.18}$$

From this equation, we derive a function $\Phi(\xi; \lambda, \mu)$. We define the Lagrange multipliers by solving the relations

$$\exp\frac{V(x, t; \theta)}{k\sigma^2} = E\exp\frac{U_A(\Phi(X_{xt}^\theta(T); \lambda, \mu))}{k\sigma^2}, \tag{15.5.19}$$

$$\begin{aligned}\exp\frac{V(x, t; \theta)}{k\sigma^2}[\frac{1}{k}\frac{\partial V(x, t; \theta)}{\partial\theta} + (x + \theta(T - t))] = \\ = EX_{xt}^\theta(T)\exp\frac{U_A(\Phi(X_{xt}^\theta(T); \lambda, \mu))}{k\sigma^2}.\end{aligned} \tag{15.5.20}$$

This defines Lagrange multipliers depending on $x, t, \theta, V, \dfrac{\partial V}{\partial\theta}$. If we insert these values in the function $\Phi(\xi; \lambda, \mu)$, and assuming $V(x, t; \theta)$ fixed, we get a function $\Phi(\xi; x, t, \theta)$. In fact, it represents $\Phi(\xi; \hat{\theta}(x, t; \theta))$. At this stage, we cannot go further without assuming the truth-telling property $\hat{\theta} = \theta$ and that the function depends parametrically on the initial conditions x, t.

So far, the principal has considered the function V a given function. In fact, in order that the agent accept the contract, he must guarantee only a sufficient value $V(x, t; \theta) \geq r(x, t; \theta)$. Recalling formula (15.5.13), we see that the principal defines V by solving the following problem (we omit to write x, t):

$$\begin{aligned}\sup_{V(\theta)\geq r(\theta)}\int_{\theta_L}^{\theta_H} f(\theta)\exp -\frac{V(\theta)}{k\sigma^2}E[\exp\frac{U_A(\Phi(X^\theta(T); \lambda, \mu)}{k\sigma^2}U_P(X^\theta(T) \\ - \Phi(X^\theta(T); \lambda, \mu))]d\theta,\end{aligned} \tag{15.5.21}$$

with the Lagrange multipliers λ, μ obtained from (15.5.19), (15.5.20).

REMARK 15.3. Note that the definition of $V(\theta)$ does not require that the contract be truth-telling, in contrast to what is mentioned in [17]. It is only in considering $\Phi(\xi; x, t, \theta)$ as the contract $\Phi(\xi; \theta)$ that we assume implicitly the truth-telling property.

15.5.4 Examples

We consider the situation $U_A(x) = U_P(x) = x$, so the situation in which the principal and the agent are risk-neutral. Equation (15.5.18) yields immediately

$$k\sigma^2 = \xi - \Phi - \lambda\xi - \mu,$$

so in fact, $\Phi(\xi; \lambda, \mu) = a + b\xi$, where the two constants have to defined by the constraints. From (15.5.19) and (15.5.20) we write

$$\exp\frac{V}{k\sigma^2} = E\exp\frac{a + bX_{xt}^{\theta}(T)}{k\sigma^2},$$

$$\exp\frac{V}{k\sigma^2}[\frac{1}{k}\frac{\partial V}{\partial\theta} + (x + \theta(T-t))] = EX_{xt}^{\theta}(T)\exp\frac{a + bX_{xt}^{\theta}(T)}{k\sigma^2}.$$

We can solve easily these two equations and obtain

$$b = \frac{V'}{T-t},\ a = V - \frac{V'}{T-t}(x + \theta(T-t)) - \frac{1}{2k}\frac{V'^2}{T-t}.$$

Also,

$$E[\exp\frac{U_A(\Phi(X^{\theta}(T))}{k\sigma^2}U_P(X^{\theta}(T) - \Phi(X^{\theta}(T)))] = \exp\frac{V}{k\sigma^2}\times$$
$$[\frac{V'}{k} + x + \theta(T-t) - V - \frac{1}{2}\frac{V'^2}{k(T-t)}].$$

The problem (15.5.21) reduces to

$$\min_{V(\theta)\geq r(\theta)}\int_{\theta_L}^{\theta_H} f(\theta)[\frac{V'}{k} - V - \frac{1}{2}\frac{V'^2}{k(T-t)}]d\theta. \tag{15.5.22}$$

This boils down to a problem of calculus of variations. Assume, for instance, that f is uniform and $r(\theta) = r_0$. It is standard to check that the function $V(\theta)$ satisfies on the interval (θ_L, θ_H) the variational inequality

$$-\frac{V''}{k(T-t)}+1 \geq 0,$$

$$V \geq r_0,$$

$$(V-r_0)(-\frac{V''}{k(T-t)}+1)=0.$$

Hence

$$V(x,t,\theta)=V(t,\theta)=r_0+\frac{k(T-t)}{2}((\theta-\theta^*)^+)^2, \tag{15.5.23}$$

$$\theta^* = \max(\theta_L, \theta_H - \frac{1}{k}).$$

It follows that

$$b = k(\theta-\theta^*)^+,\ a = r_0 - k(\theta-\theta^*)^+(x+\theta(T-t)). \tag{15.5.24}$$

Finally, the contract $\Phi(\xi,\theta)$ becomes (it depends on the initial conditions x, t, so we use ξ for the argument)

$$\Phi(\xi,\theta) = r_0 + k(\theta-\theta^*)^+(\xi - x - \theta(T-t)). \tag{15.5.25}$$

We see that the optimal contract depends on the initial conditions (x, t). From formula (15.5.8), we get easily

$$V(x,t;\theta,\theta') = r_0 + k(T-t)(\theta-\theta')(\theta'-\theta^*)^+ + \frac{k(T-t)}{2}((\theta'-\theta^*)^+)^2. \tag{15.5.26}$$

The contract is truth-telling. Indeed, if the real parameter θ is less than θ^*, we have $V(x,t,\theta) = r_0$. If $\theta' < \theta^*$, then $V(x,t;\theta,\theta') = r_0$, so choosing $\theta' < \theta^*$ does not bring any advantage. But if he chooses $\theta' > \theta^* > \theta$, the gradient satisfies $\frac{\partial V}{\partial \theta'} = \theta - \theta' < 0$, so there is no reason not to reveal. If $\theta > \theta^*$, then he will not choose $\theta' < \theta^*$. But for $\theta' > \theta^*$, the gradient vanishes at $\theta' = \theta$, which is a maximum.

15.5.5 Controlling Volatility

We consider the following model: The output is given by

$$dX = v(s)dw(s),\ X(t) = x, \tag{15.5.27}$$

in which $v(s)$ is a control. This control is defined by the principal, who fixes the level of risk. The principal also fixes the contract payable at T, depending on a parameter θ that represents some skill of the agent. If θ is the value of the parameter, the real Wiener process is $w^{\theta}(s)$ such that $dw = \theta ds + dw^{\theta}$. So the output satisfies

$$dX = v(s)(\theta ds + dw^{\theta}(s)). \tag{15.5.28}$$

The agent chooses from the menu of contracts $C_T(\theta')$, in which θ' may be different from θ. So his payoff is

$$V(x,t;\theta,\theta') = EM_t^{\theta}(T)U_A(C_T(\theta')),$$

and he chooses θ' in order to maximize this quantity in θ'. So his value function is

$$V(x,t,\theta) = \sup_{\theta'} V(x,t;\theta,\theta').$$

As usual, the principal will consider this function as given when he designs the contract and chooses the risk level. Since

$$M_t^{\theta}(T) = \exp[\theta(w(T)-w(t)) - \frac{1}{2}\theta^2(T-t)],$$

we compute easily

$$\frac{\partial V}{\partial \theta} = E\{M_t^{\theta}(T)[w(T)-w(t)-\theta(T-t)]U_A(C_T(\hat{\theta}))\},$$

in which $\hat{\theta}$ is the optimal value of θ' chosen by the agent. This can be rewritten as

$$\frac{\partial V}{\partial \theta} + \theta(T-t)V = E[M_t^{\theta}(T)(w(T)-w(t))U_A(C_T(\hat{\theta}))]. \tag{15.5.29}$$

Note that we have also

$$V = EM_t^{\theta}(T)U_A(C_T(\hat{\theta})). \tag{15.5.30}$$

The principal will consider (15.5.29) and (15.5.30) to be constraints on the contract $C_T(\hat{\theta})$. The principal has also to choose the risk level. However, this is equivalent to choosing the value of $X(T)$, with the constraint

$$EX(T) = x. \tag{15.5.31}$$

This is simply a consequence of the representation of martingales theorem. So the principal has two decisions $X(T)$ and $C_T(\hat{\theta})$, with three scalar constraints (15.5.29), (15.5.30), (15.5.31). His objective is to maximize the payoff

$EM_t^\theta(T)U_P(X(T) - C_T(\hat{\theta}))$. We will not try to find the function V as in the previous section. However, it is important to guarantee a minimum level of V, since otherwise, no contract will be acceptable to the agent. To cope with this constraint, we impose a constraint on $C_T(\hat{\theta})$, namely $C_T(\hat{\theta}) \geq L$. If we assume that

$$U_A(L) \geq r(\theta), \ \forall\theta, \tag{15.5.32}$$

then the constraint $C_T(\hat{\theta}) \geq L$ guarantees that the agent does not reject the contract. So the principal maximizes $EM_t^\theta(T)U_P(X(T)-C_T(\hat{\theta}))$ on the set $C_T(\hat{\theta}) \geq L$, with the constraints (15.5.29), (15.5.30), (15.5.31). We introduce Lagrange multipliers λ, μ, ν for the three equality constraints, so the principal solves the problem

$$\max_{X(T), C_T(\hat{\theta}) \geq L} \quad EM_t^\theta(T)U_P(X(T) - C_T(\hat{\theta})) - EM_t^\theta(T)U_A(C_T(\hat{\theta}))$$
$$(\lambda + \mu(w(T) - w(t)) - \nu EX(T). \tag{15.5.33}$$

Since $X(T)$ and $C_T(\hat{\theta})$ are $\mathcal{F}^T$- measurable, we can optimize pointwise. We first optimize in $X(T)$ by writing the first-order condition

$$M_t^\theta(T)U'_P(X(T) - C_T(\hat{\theta})) = \nu. \tag{15.5.34}$$

Assuming that U'_P has an inverse I_P, we obtain the relation

$$X(T) - C_T(\hat{\theta}) = I_P(\frac{\nu}{M_t^\theta(T)}). \tag{15.5.35}$$

Using this relation, we can eliminate $X(T)$ in the functional (15.5.33). We obtain the problem

$$\max_{C_T(\hat{\theta}) \geq L} [-EM_t^\theta(T)U_A(C_T(\hat{\theta}))(\lambda + \mu(w(T) - w(t))) - \nu EC_T(\hat{\theta})]. \tag{15.5.36}$$

Again we optimize pointwise. We obtain easily

$$C_T(\hat{\theta}) = \left| \begin{array}{ll} L & \text{if } \lambda + \mu(w(T) - w(t)) \geq 0, \\ L \vee I_A(\frac{-\nu}{M_t^\theta(T)(\lambda + \mu(w(T) - w(t)))}) & \text{if } \lambda + \mu(w(T) - w(t)) < 0, \end{array} \right. \tag{15.5.37}$$

in which I_A is the inverse of U'_A. So we have expressed the values of $X(T)$ and $C_T(\hat{\theta})$ in terms of the Lagrange multipliers λ, μ, ν. It remains to compute these Lagrange multipliers using the constraints (15.5.29), (15.5.30), (15.5.31). We can see that the contract $C_T(\hat{\theta})$ depends only on $w(T) - w(t)$ and the values $x, t, \theta, V, \dfrac{\partial V}{\partial \theta}$.

Chapter 16
Differential Games

16.1 Open-Loop Deterministic Nash Equilibrium

16.1.1 Description of the Problem

We have N players, whose decisions are functions of time $v^i(t), i = 1, \cdots, N$, with values in R^{m_i}. In fact, there will be constraints, so

$$v^i(t) \in U^i_{\text{ad}}, \text{ a.e. } t. \tag{16.1.1}$$

We shall denote by $v(t)$ the whole vector $(v^1(t), \cdots, v^N(t))$. The state equation is

$$\begin{aligned}\frac{dx}{dt} &= g(x(t), v(t)), \\ x(0) &= x_0.\end{aligned} \tag{16.1.2}$$

We shall not detail our assumptions, instead either stating them case by case or proceeding formally. Each player has a cost objective given by

$$J^i(v(.)) = \int_0^T l^i(x(t), v(t))dt + h^i(x(T)). \tag{16.1.3}$$

We shall also use the following notation: $v = (v^i, \bar{v^i})$. In this notation, $\bar{v^i}$ represents all the controls v^j except $j = i$. In other words, we describe the set emphasizing player i. A Nash equilibrium is a vector $u(.)$ such that

$$J^i(u(.)) \le J^i(v^i(.), \bar{u^i}(.)), \; \forall v^i(.), \forall i. \tag{16.1.4}$$

A. Bensoussan, *Estimation and Control of Dynamical Systems*, Interdisciplinary Applied Mathematics 48, https://doi.org/10.1007/978-3-319-75456-7_16

An important case is $N = 2$ and $l^1(x, v) = -l^2(x, v) = l(x, v)$, $h^1(x) = -h^2(x) = h(x)$. Then $J^1(v(.)) = -J^2(v(.)) = J(v(.))$. This is called a two-person **zero-sum game**. Then a Nash equilibrium $u(.)$ satisfies

$$J(u(.)) = \inf_{v^1(.)} J(v^1(.), u^2(.)) = \sup_{v^2(.)} J(u^1(.), v^2(.)). \tag{16.1.5}$$

But then also

$$J(u(.)) = \inf_{v^1(.)} \sup_{v^2(.)} J(v^1(.), v^2(.)) = \sup_{v^2(.)} \inf_{v^1(.)} J(v^1(.), v^2(.)). \tag{16.1.6}$$

In this context, the Nash equilibrium is called a **saddle point** for the functional $J(v^1(.), v^2(.))$. In general, one has

$$\sup_{v^2(.)} \inf_{v^1(.)} J(v^1(.), v^2(.)) \leq \inf_{v^1(.)} \sup_{v^2(.)} J(v^1(.), v^2(.)). \tag{16.1.7}$$

If the inequality is strict, we have a **duality gap**. In that case, there is no Nash equilibrium. A Nash equilibrium is also called a saddle point. Note that (16.1.5) is stronger than (16.1.6) and not equivalent. We may have no duality gap yet no saddle point.

16.1.2 Maximum Principle

Clearly, each player solves a control problem, parameterized by the decisions of the other players. We can write the maximum principle for each of them and obtain the following system:

$$\begin{aligned} \frac{dy}{dt} &= g(y, u), \\ -\frac{dp^i}{dt} &= D_x H^i(y, u, p^i), \\ y(0) = x \quad & p^i(T) = D_x h^i(y(T)), \\ H^i(y(t), u(t), p^i(t)) &= \inf_{v^i \in U^i_{\text{ad}}} H^i(y(t), v^i, \bar{u^i}(t), p^i(t)), \ \forall t. \end{aligned} \tag{16.1.8}$$

We recall the definition of the Hamiltonian (in the framework of the maximum principle):

$$H^i(x, v, q^i) = l^i(x, v) + q^i.g(x, v), \tag{16.1.9}$$

so

$$D_x H^i(x, v, q^i) = D_x l^i(x, v) + (D_x g)^*(x, v) q^i. \qquad (16.1.10)$$

In the case of a zero-sum game, we have $H^1(x, v, q) = -H^2(x, v, -q) = H(x, v, q)$. We have $p^1(t) = -p^2(t) = p(t)$, and the maximum principle becomes

$$\begin{aligned} \frac{dy}{dt} &= g(y, u), \qquad (16.1.11) \\ -\frac{dp}{dt} &= D_x H(y, u, p), \\ y(0) = x \quad & p(T) = D_x h(y(T)), \\ H(y(t), u(t), p(t)) &= \inf_{v^1} \sup_{v^2} H(y(t), v^1, v^2, p(t)) = \sup_{v^2} \inf_{v^1} H(y(t), v^1, v^2, p(t)). \end{aligned}$$

So there should be no duality gap for the function $H(y(t), v^1, v^2, p(t))$.

16.1.3 Example

We consider the following example ([19], Example 4.1), with $N = 2$, $n = 1$. We have

$$\begin{aligned} l^1(x, v) &= x + \frac{\alpha}{2}(v^1)^2 - v^2, \ h^1(x) = 0, \\ l^2(x, v) &= x - v^2, \ h^2(x) = 0, \\ g(x, v) &= -v^1\sqrt{x} + v^2 + 1, \\ U^1_{\text{ad}} = [0, +\infty) \quad & U^2_{\text{ad}} = [0, 1]. \end{aligned}$$

We see that the state must remain positive. This constraint will not be binding in the sequel. The maximum principle conditions can be written as follows:

$$\begin{aligned} \frac{dy}{dt} &= -u^1\sqrt{y} + u^2 + 1, \ y(0) = x_0, \qquad (16.1.12) \\ -\frac{dp^1}{dt} &= -p^1\frac{u^1}{2\sqrt{y}} + 1, \ p^1(T) = 0, \\ -\frac{dp^2}{dt} &= -p^2\frac{u^1}{2\sqrt{y}} + 1, \ p^2(T) = 0, \\ \alpha u^1 &= p^1\sqrt{y}, \\ u^2 = 1 \text{ if } p^2 - 1 < 0, \quad & u^2 = 0 \text{ if } p^2 - 1 > 0. \end{aligned}$$

Obviously, $p^1 = p^2 = p$, where p is the solution of

$$-\frac{dp}{dt} = 1 - \frac{p^2}{2\alpha},\ p(T) = 0, \tag{16.1.13}$$

which has an independent solution. We have explicitly

$$p(t) = \sqrt{2\alpha}\frac{1 - \exp -\sqrt{\frac{2}{\alpha}}(T-t)}{1 + \exp -\sqrt{\frac{2}{\alpha}}(T-t)}, \tag{16.1.14}$$

and we see that $p(t)$ is monotonically decreasing and positive. Assume that $2\alpha > 1$. There exists a time $\tau < T$ such that $p(t) > 1$ for $t < \tau$, and $p(t) < 1$ for $\tau < t < T$. An easy calculation shows that

$$T - \tau = \sqrt{\frac{\alpha}{2}} \log \frac{\sqrt{2\alpha} + 1}{\sqrt{2\alpha} - 1}. \tag{16.1.15}$$

From the last conditions (16.1.12), we can state that

$$\begin{aligned} &u^2(t) = 0,\ \forall 0 < t < \tau,\ u^2(t) = 0,\ \forall \tau < t < T, \\ &u^1(t) = \frac{p(t)}{\alpha}\sqrt{y(t)}, \end{aligned} \tag{16.1.16}$$

and the optimal state $y(t)$ is the solution of

$$\begin{aligned} \frac{dy}{dt} &= -\frac{p(t)}{\alpha}y(t) + 1,\ 0 < t < \tau, \\ \frac{dy}{dt} &= -\frac{p(t)}{\alpha}y(t) + 2,\ \tau < t < T, \\ y(0) &= x_0. \end{aligned} \tag{16.1.17}$$

We see that if $x_0 > 0$, then $y(t) > 0$, so the positivity constraint is not binding.

16.2 Closed-Loop Deterministic Nash Equilibrium

16.2.1 Setting of the Problem

Here we allow the players to use feedbacks $v^i(x, t)$ on the state. We know that in deterministic optimal control, there is no difference between open-loop and closed-loop optimal controls, because when applied to the optimal state, the value of the optimal feedback reduces to the value of the optimal open-loop control. We shall observe differences for games. The reason is related to the status of information. Although the problem is deterministic, there is still a difference in the information framework between open-loop and closed-loop differential games. In the open-loop case, a player has no information whatsoever about the opponents' controls, whereas in the closed-loop case, he has some, since he uses the state that depends on the opponents' decisions. If we write $v(x,t)$ for the aggregated controls, the state equation becomes

$$\frac{dx}{dt} = g(x(t), v(x(t), t)), \tag{16.2.1}$$
$$x(0) = x_0,$$

and we have the payoff functionals

$$J^i(v(.)) = \int_0^T l^i(x(t), v(x(t), t))dt + h^i(x(T)). \tag{16.2.2}$$

A Nash equilibrium $u(x,t)$ will satisfy the conditions

$$J^i(u(.)) \leq J^i(v^i(.), \overline{u^i(.)})\, \forall v^i(.), \forall i. \tag{16.2.3}$$

16.2.2 Dynamic Programming

Again, when we fix all opponents' feedback controls, the problem of the individual player, say i, is a control problem. Because we want an optimal feedback, we use dynamic programming. Defining Lagrangians

$$L^i(x, v^i, \overline{u^i(x)}, q^i) = l^i(x, v^i, \overline{u^i(x)}) + q^i.g(x, v^i, \overline{u^i(x)}),$$

the Bellman equation for player i will be

$$\frac{\partial \Phi^i}{\partial t} + \inf_{v^i \in U^i_{\text{ad}}} L^i(x, v^i, \overline{u^i(x)}, D\Phi^i) = 0,$$

$$\Phi^i(x, T) = h^i(x),$$

and $u^i(x,t)$ is the point of the infimum of $L^i(x, v^i, \overline{u^i(x)}, D\Phi^i)$. We can summarize the operations as follows. Consider the Lagrangians $L^i(x, v, q^i)$. We set $q = (q^1, \cdots, q^N)$. We define a Nash point $\hat{v}(x,q) = (\cdots, \hat{v}^i(x,q), \cdots)$ of the Lagrangians $L^i(x, v, q^i)$ in $v = (v^1, \cdots, v^N)$. We then set $H^i(x,q) =$

$L^i(x, \hat{v}(x,q), q^i)$. The vector $\Phi(x,t)$ of value functions $\Phi^1(x,t), \cdots \Phi^N(x,t)$ satisfies the system of equations

$$\frac{\partial \Phi^i}{\partial t} + H^i(x, D\Phi) = 0, \tag{16.2.4}$$

$$\Phi^i(x,T) = h^i(x),$$

with $D\Phi = (D\Phi^1, \cdots, D\Phi^N)$. If we can solve this system, the optimal feedbacks are obtained by the formulas $u^i(x,t) = \hat{v}^i(x, D\Phi(x,t))$.

REMARK 16.1. An open-loop control can be considered to be a degenerate closed-loop control. One can convince oneself that an open-loop Nash equilibrium is a closed-loop Nash equilibrium, of course different from the one coming from (16.2.4), but the converse is not true. ■

Consider next a two-person zero-sum game. We have $L^2(x,v,-q) = -L^1(x,v,q) = -L(x,v,q)$. To avoid difficulty of notation, let us set $\mathbf{q} = (q^1, q^2)$ and reserve q for a generic vector in R^n. So the Nash point of the Lagrangians is denoted by $\hat{v}(x,\mathbf{q}) = \hat{v}(x, q^1, q^2)$. We write $\hat{v}(x,q) = \hat{v}(x,q,-q)$. Similarly,

$$H^1(x,\mathbf{q}) = L^1(x, \hat{v}(x,\mathbf{q}), q^1) = L(x, \hat{v}(x,\mathbf{q}), q^1),$$
$$H^2(x,\mathbf{q}) = L^2(x, \hat{v}(x,\mathbf{q}), q^2) = -L(x, \hat{v}(x,\mathbf{q}), -q^2).$$

Let us define $H(x,q) = L(x, \hat{v}(x,q), q)$. Therefore, $H^1(x,q,-q) = H(x,q)$ and $H^2(x,q,-q) = -H(x,q)$. Moreover, one checks easily that

$$\begin{aligned} H(x,q) &= \inf_{v^1} \sup_{v^2} L(x, v^1, v^2, q) = \sup_{v^2} \inf_{v^1} L(x, v^1, v^2, q) \\ &= \inf_{v^1} L(x, v^1, \hat{v}^2(x,q), q) = \sup_{v^2} L(x, \hat{v}^1(x,q), v^2, q). \end{aligned} \tag{16.2.5}$$

Turning back to the system (16.2.4), we see that it reduces to one equation $\Phi^1(x,t) = \Phi(x,t) = -\Phi^2(x,t)$, namely

$$\frac{\partial \Phi}{\partial t} + H(x, D\Phi) = 0 \tag{16.2.6}$$

$$\Phi(x,t) = h(x),$$

which is called an **Isaacs equation**. Recall the inequality

$$\sup_{v^2} \inf_{v^1} L(x, v^1, v^2, q) \leq \inf_{v^1} \sup_{v^2} L(x, v^1, v^2, q).$$

The equality as written in (16.2.5) is a condition, called an *Isaacs condition*.

16.2.3 Example

We return to the example 16.1.3. We begin with the Lagrangians

$$L^1(x, v, q^1) = x + \frac{\alpha}{2}(v^1)^2 - v^2 + q^1(-v^1\sqrt{x} + v^2 + 1),$$
$$L^2(x, v, q^2) = x - v^2 + q^2(-v^1\sqrt{x} + v^2 + 1).$$

We look for a Nash point at the level of the Lagrangians. Recalling the constraints $U^1_{\text{ad}} = [0, +\infty)$, $U^2_{\text{ad}} = [0, 1]$, we get easily

$$\hat{v}^1(x, q) = \frac{(q^1)^+\sqrt{x}}{\alpha}, \; \hat{v}^2(x, q) = 1\!\!1_{\{q^2<1\}},$$

from which we obtain

$$H^1(x, q) = x + (q^1 - 1)1\!\!1_{\{q^2<1\}} + q^1 - x\frac{((q^1)^+)^2}{2\alpha},$$
$$H^2(x, q) = x - (1 - q^2)^+ + q^2(1 - \frac{(q^1)^+x}{\alpha}).$$

We can then state the system of Bellman equations

$$\frac{\partial\Phi^1}{\partial t} + (\frac{\partial\Phi^1}{\partial x} - 1)1\!\!1_{\{\frac{\partial\Phi^2}{\partial x}<1\}} + \frac{\partial\Phi^1}{\partial x} + 1 - \frac{x}{2\alpha}((\frac{\partial\Phi^1}{\partial x})^+)^2 = 0,$$
$$\frac{\partial\Phi^2}{\partial t} - \frac{x}{\alpha}\frac{\partial\Phi^2}{\partial x}(\frac{\partial\Phi^1}{\partial x})^+ + \frac{\partial\Phi^2}{\partial x} + x - (1 - \frac{\partial\Phi^2}{\partial x})^+ = 0,$$
$$\Phi^1(x, T) = 0, \; \Phi^2(x, T) = 0. \tag{16.2.7}$$

Although they look complex, these equations have very simple solutions, namely

$$\Phi^1(x, t) = xp(t) + \rho(t),$$
$$\Phi^2(x, t) = x\pi(t) + \sigma(t),$$

with $p(t)$ already appearing in the open-loop case, given by formula (16.1.14). We obtain $\pi(t)$ by solving

$$\frac{d\pi}{dt} + 1 - \frac{p(t)}{\alpha}\pi(t) = 0, \; \pi(T) = 0. \tag{16.2.8}$$

EXERCISE 16.1. Check that $0 \leq \pi(t) < p(t)$.

We obtain the optimal feedbacks by the formulas

$$u^1(x,t) = \frac{p(t)\sqrt{x}}{\alpha},\ u^2(x,t) = \mathbb{1}_{\{\pi(t)<1\}}. \tag{16.2.9}$$

We see that the second player does not use a real feedback, but an open-loop control. However, it is different from the optimal open-loop control, which was $\mathbb{1}_{\{p(t)<1\}}$.

Applying these optimal feedbacks shows that the optimal state is the solution of

$$\frac{dz}{dt} = -\frac{p}{\alpha}z + 1 + \mathbb{1}_{\{\pi(t)<1\}},\ z(0) = x_0,$$

and the trajectory differs from $y(t)$, see (16.1.17), which is the solution of

$$\frac{dy}{dt} = -\frac{p}{\alpha}y + 1 + \mathbb{1}_{\{p(t)<1\}},\ y(0) = x_0.$$

Since $\pi(t) < p(t)$, we get easily $z(t) \geq y(t)$.

16.3 Deterministic Linear–Quadratic Games

16.3.1 Open-Loop Nash Equilibrium

We consider a situation with $N = 2$ and $n = 1$ in order to simplify. The state equation is given by

$$\frac{dx}{dt} = a(t)x(t) + b^1(t)v^1(t) + b^2(t)v^2(t), \tag{16.3.1}$$
$$x(0) = x_0,$$

and the objectives of the two players are

$$J^i(v(.)) = \frac{1}{2}\int_0^T (\alpha^i(t)x^2(t) + \beta^i(t)(v^i(t))^2)dt + \frac{1}{2}\gamma^i x^2(T). \tag{16.3.2}$$

The Hamiltonians are given by (we omit the time to save notation)

$$H^i(x,v,q^i) = \frac{1}{2}(\alpha^i x^2 + \beta^i(v^i)^2) + q^i(ax + b^1v^1 + b^2v^2). \tag{16.3.3}$$

We denote by $u(t)$ the Nash equilibrium, by $y(t)$ the corresponding state, and by $p^1(t)$, $p^2(t)$ the adjoint states. We state that

$$u^i(t) = -\frac{p^i(t)b^i(t)}{\beta^i(t)}; \tag{16.3.4}$$

hence

$$\frac{dy}{dt} = a(t)y(t) - \sum_{i=1}^{2} \frac{p^i(t)(b^i(t))^2}{\beta^i(t)} \tag{16.3.5}$$

$$-\frac{dp^i(t)}{dt} = \alpha^i(t)y(t) + a(t)p^i(t),\ y(0) = x_0,\ \ p^i(T) = \gamma^i y(T).$$

The equations (16.3.4), (16.3.5) express the maximum principle. We can decouple this system by setting $p^i(t) = P^i(t)y(t)$. The functions $P^i(t)$ are solutions of the Riccati equations

$$-\frac{dP^1}{dt} - 2aP^1 + \frac{(b^1)^2}{\beta^1}(P^1)^2 + \frac{(b^2)^2}{\beta^2}P^1P^2 = \alpha^1 \tag{16.3.6}$$

$$-\frac{dP^2}{dt} - 2aP^2 + \frac{(b^2)^2}{\beta^2}(P^2)^2 + \frac{(b^1)^2}{\beta^1}P^1P^2 = \alpha^1,$$

$$P^1(T) = \gamma^1,\ \ P^2(T) = \gamma^2.$$

We can give an explicit solution of this system in the stationary case, when all parameters do not depend on time. Indeed, set $\nu^i = \dfrac{(b^i)^2}{\beta^i}$ and $Q(t) = \nu^1P^1(t) + \nu^2P^2(t)$, $\mu(t) = Q(t) - a$, $M = \nu^1\alpha^1 + \nu^2\alpha^2$, $N = \nu^1\gamma^1 + \nu^2\gamma^2$. We also set $\theta^2 = M + a^2$. Combining the two equations (16.3.6) and rearranging, we obtain easily

$$-\frac{d\mu}{dt} + \mu^2 = \theta^2, \tag{16.3.7}$$

$$\mu(T) = N - a,$$

which has an explicit solution

$$\mu(t) = Q(t) - a = \theta \frac{(N - a + \theta)\exp 2\theta(T - t) + N - a - \theta}{(N - a + \theta)\exp 2\theta(T - t) - (N - a - \theta)}. \tag{16.3.8}$$

Going back to the equations (16.3.6), we see easily that

$$-\frac{dP^i}{dt} + P^i(t)Q(t) = \alpha^i, \tag{16.3.9}$$

$$P^i(T) = \gamma^i, \tag{16.3.10}$$

which are linear decoupled equations.

REMARK 16.2. The optimal controls are defined by

$$u^i(t) = -\frac{b^i(t)P^i(t)}{\beta^i(t)}y(t). \tag{16.3.11}$$

Therefore, they are obtained by a feedback on the optimal state, although they are open-loop controls. As we shall see in the next section, these feedbacks do not form a Nash equilibrium among all feedbacks. This is quite different from deterministic optimal control. It should be emphasized that the feedbacks (16.3.11) are valid only for the optimal trajectory.

16.3.2 Closed-Loop Nash Equilibrium

We consider the same problem (16.3.1), (16.3.2), but we look for a Nash equilibrium among feedback controls. We use dynamic programming. We define a Nash point equilibrium at the level of the Lagrangians (16.3.3) and obtain the Hamiltonians

$$H^1(x,q) = \frac{1}{2}\alpha^1x^2 + q^1ax - \frac{1}{2}(q^1)^2\frac{(b^1)^2}{\beta^1} - q^1q^2\frac{(b^2)^2}{\beta^2}, \tag{16.3.12}$$

$$H^2(x,q) = \frac{1}{2}\alpha^2x^2 + q^2ax - \frac{1}{2}(q^2)^2\frac{(b^2)^2}{\beta^2} - q^1q^2\frac{(b^1)^2}{\beta^1},$$

which leads to the following system of PDEs; see (16.2.4):

$$\frac{\partial\Phi^1}{\partial t} + \frac{1}{2}\alpha^1x^2 + ax\,\frac{\partial\Phi^1}{\partial x} - \frac{1}{2}\frac{(b^1)^2}{\beta^1}(\frac{\partial\Phi^1}{\partial x})^2 - \frac{(b^2)^2}{\beta^2}\frac{\partial\Phi^1}{\partial x}\frac{\partial\Phi^2}{\partial x} = 0, \tag{16.3.13}$$

$$\frac{\partial\Phi^2}{\partial t} + \frac{1}{2}\alpha^2x^2 + ax\,\frac{\partial\Phi^2}{\partial x} - \frac{1}{2}\frac{(b^2)^2}{\beta^2}(\frac{\partial\Phi^2}{\partial x})^2 - \frac{(b^1)^2}{\beta^1}\frac{\partial\Phi^1}{\partial x}\frac{\partial\Phi^2}{\partial x} = 0,$$

$$\Phi^1(x,T) = \frac{1}{2}\gamma^1x^2,\ \Phi^2(x,T) = \frac{1}{2}\gamma^2x^2.$$

We look for a quadratic solution $\Phi^i(x,t) = \frac{1}{2}Q^i(t)x^2$. We obtain easily the system of Riccati equations

$$-\frac{dQ^1}{dt} - 2aQ^1 + \frac{(b^1)^2}{\beta^1}(Q^1)^2 + 2\frac{(b^2)^2}{\beta^2}Q^1Q^2 = \alpha^1, \tag{16.3.14}$$

$$-\frac{dQ^2}{dt} - 2aQ^2 + \frac{(b^2)^2}{\beta^2}(Q^2)^2 + 2\frac{(b^1)^2}{\beta^1}Q^1Q^2 = \alpha^2,$$

$$Q^1(T) = \gamma^1,\ Q^2(T) = \gamma^2.$$

We can see that these equations are different from those of the open-loop Nash equilibrium (16.3.6). There is a factor 2 in the coupling term. The technique used in the open-loop case breaks down because of this coefficient. Nevertheless, we will mimic the approach. Recall the definition of ν^i and the fact that the constants do not depend on time. We define

$$Q(t) = \nu^1 Q^1(t) + \nu^2 Q^2(t),\ \Gamma(t) = \nu^1 Q^1(t) - \nu^2 Q^2(t).$$

Of course, we obtain easily Q^1 and Q^2 from Q and Γ. we obtain easily the following system for Q and Γ:

$$-\frac{dQ}{dt} - 2aQ + \frac{3}{2}Q^2 = M + \frac{\Gamma^2}{2}, \tag{16.3.15}$$
$$-\frac{d\Gamma}{dt} + (Q - 2a)\Gamma = \alpha^1\nu^1 - \alpha^2\nu^2,$$

$$Q(T) = N,\ \Gamma(T) = \gamma^1\nu^1 - \gamma^2\nu^2.$$

There is some improvement in the system (16.3.15) compared to (16.3.14). Indeed, the second equation is linear. We have the following.

PROPOSITION 16.1. *There exists a solution of (16.3.15) such that* $Q(t) > 0$.

PROOF. We interpret the Riccati equation as one obtained from an optimal control problem. Namely, consider the state evolution

$$\frac{dx}{ds} = ax + \frac{3}{2}v(s),\ x(t) = x$$

and the objective

$$K_{xt}(v(.)) = \frac{1}{2}\left[\int_t^T ((M + \frac{\Gamma^2(s)}{2})x^2(s)ds + \frac{3}{2}v^2(s))ds + Nx^2(T)\right]. \tag{16.3.16}$$

We see easily that

$$\frac{1}{2}Q(t)x^2 = \inf_{v(.)} K_{xt}(v(.)).$$

It follows immediately that $Q(t) > 0$. In addition, we can write

$$\frac{1}{2}Q(t)x^2 \leq K_{xt}(0).$$

If we set $||\Gamma|| = \sup_s |\Gamma(s)|$, we get easily the inequality

$$K_{xt}(0) \leq \frac{1}{2}x^2 \exp a(T-t)\,[\frac{1}{a}(M + \frac{||\Gamma||^2}{2}) + N].$$

So

$$Q(t) \leq \exp a(T-t)\,[\frac{1}{a}(M+\frac{||\Gamma||^2}{2})+N]. \tag{16.3.17}$$

Now, Γ is the solution of a linear equation. We have the explicit formula

$$\Gamma(t) = (\gamma^1 v^1 - \gamma^2 v^2)\exp - \int_t^T (Q(s)-2a)ds +$$

$$+(\alpha^1 v^1 - \alpha^2 v^2)\int_t^T (\exp - \int_t^s (Q(\tau)-2a)d\tau)ds.$$

Using the fact that $Q(s) > 0$, we get

$$||\Gamma|| \leq \exp 2aT\,(|\gamma^1 v^1 - \gamma^2 v^2| + \frac{|\alpha^1 v^1 - \alpha^2 v^2}{2a}).$$

Recalling (16.3.17), it follows that we have obtained a priori bounds on $Q(t)$ and $\Gamma(t)$. They have been obtained on the condition that $Q(t)$ could be interpreted as the infimum of the payoff of a control problem. If we consider a local solution $Q(t), \Gamma(t)$, i.e., defined on $t \in (T-\epsilon, T]$ for ϵ sufficiently small, we obtain a positive solution for $Q(t)$, since $Q(T) > 0$. The control interpretation can be obtained on this small interval, and thus the bounds are valid on this small interval. Since the bounds do not depend on ϵ, we can extend the solution beyond $T-\epsilon$, and in fact on the interval $[0, T]$. We deduce $Q^1(t)$ and $Q^2(t)$, and the interpretation of the value functions $\Phi^1(x,t)$, $\Phi^2(x,t)$, which are positive. Hence $Q^1(t)$, $Q^2(t)$ are positive. This concludes the proof. □

REMARK 16.3. The optimal feedbacks are given by the formulas

$$u^i(x,t) = -\frac{b^i}{\beta^i}Q^i(t)x, \tag{16.3.18}$$

which are different from the formulas used in the open-loop case; see (16.3.11).

16.3.3 Two-Person Zero-Sum Differential Game: Open-Loop Controls

We study a simple case. The dynamics is given by

$$g(x,v) = ax + b^1 v^1 + b^2 v^2, \tag{16.3.19}$$

and the payoffs are

$$J(v(.)) = \frac{1}{2}\int_0^T (\alpha x^2(t) + \beta^1(v^1(t))^2 - \beta^2(v^2(t))^2)dt + \frac{1}{2}\gamma x^2(T), \tag{16.3.20}$$

$$J^1(v(.)) = -J^2(v(.)) = J(v(.)).$$

If we write the system (16.1.8), we obtain $p^1(t) = -p^2(t) = p(t)$ and the system

$$\frac{dy}{dt} = ay - (\nu^1 - \nu^2)p, \; y(0) = x_0, \tag{16.3.21}$$

$$-\frac{dp}{dt} = ap + \alpha y, \quad p(T) = \gamma y(T),$$

with the notation $\nu^i = \dfrac{(b^i)^2}{\beta^i}$. We solve the system as follows: $p(t) = P(t)y(t)$, with $P(t)$ the solution of the Riccati equation

$$-\frac{dP}{dt} - 2aP + \nu P^2 = \alpha, \tag{16.3.22}$$

$$P(T) = \gamma,$$

where we have set $\nu = \nu^1 - \nu^2$. If the Riccati equation (16.3.22) has a bounded solution, then we solve the state equation

$$\frac{dy}{dt} = (a - \nu P(t))y, \tag{16.3.23}$$

$$y(0) = x_0,$$

and we can define the pair

$$u^1(t) = -\frac{b^1}{\beta^1}P(t)y(t), \; u^2(t) = \frac{b^2}{\beta^2}P(t)y(t). \tag{16.3.24}$$

We shall need the functional

$$\mathcal{K}(v(.)) = \frac{\alpha}{2}\int_0^T z(t)^2 dt + \frac{\gamma}{2}z(T)^2 - \frac{\beta^2}{2}\int_0^T v(t)^2 dt, \tag{16.3.25}$$

$$z(t) = b^2\int_0^t \exp a(t-s)\, v(s)ds.$$

We then state our next result.

Proposition 16.2. *Assume that the Riccati equation (16.3.22) has a bounded solution. Then the pair* $u^1(.), u^2(.)$ *satisfies*

$$J(u(.)) = \sup_{v^2(.)} \inf_{v^1(.)} J(v^1(.), v^2(.)). \tag{16.3.26}$$

If $\mathcal{K}(v(.)) \leq 0, \forall v(.)$, *then the pair* $u^1(.), u^2(.)$ *is an open-loop Nash equilibrium. A sufficient condition is the existence of a bounded solution of the Riccati equation*

$$\frac{dS}{dt} + 2aS + \nu^2 S^2 + \alpha = 0, \tag{16.3.27}$$
$$S(T) = \gamma.$$

Proof. We are in a situation in which there exists a duality gap. So the existence of a solution of the Riccati equation (16.3.22) is not sufficient to eliminate the duality gap and to obtain an open-loop Nash equilibrium. One needs the concavity of the functional $\mathcal{K}(v(.))$. We proceed with the proof:

Consider two open-loop controls $v^1(.), v^2(.)$ and the corresponding trajectory $x(.)$ given by

$$\frac{dx}{dt} = ax + b^1 v^1 + b^2 v^2, \; x(0) = x_0.$$

We define

$$\tilde{v}^1(t) = v^1(t) + \frac{b^1}{\beta^1} P(t)x(t),$$
$$\tilde{v}^2(t) = v^2(t) - \frac{b^2}{\beta^2} P(t)x(t).$$

Then we can obtain the formula

$$J(v(.)) = \frac{1}{2} P(0)x_0^2 + \frac{1}{2}[\beta^1 \int_0^T (\tilde{v}^1(t))^2 dt - \beta^2 \int_0^T (\tilde{v}^2(t))^2 dt]. \tag{16.3.28}$$

Note that this inequality does not imply that we can make a change of control variables from v^1, v^2 to $\tilde{v}^1, \tilde{v}^2$. Nonetheless, the pair $(u^1(.), u^2(.))$ corresponds to $\tilde{v}^1 = 0, \; \tilde{v}^2 = 0$, and therefore, $J(u(.)) = \frac{1}{2}P(0)x_0^2$. Moreover,

$$\inf_{v^1(.)} \sup_{v^2(.)} J(v^1(.), v^2(.)) \geq \frac{1}{2} P(0)x_0^2 = J(u(.)) \geq \sup_{v^2(.)} \inf_{v^1(.)} J(v^1(.), v^2(.)). \tag{16.3.29}$$

Indeed, for all $v^1(.)$, one can find a control $v^2(.)$ such that $\tilde{v}^2(t) = 0$. The corresponding cost is represented symbolically by $J(v^1(.), \tilde{v}^2(.) = 0)$. But then from formula (16.3.28), we get $J(v^1(.), \tilde{v}^2(.) = 0) \geq \frac{1}{2}P(0)x_0^2$. Therefore, also $\sup_{v^2(.)} J(v^1(.), v^2(.)) \geq \frac{1}{2}P(0)x_0^2$. Since $v^1(.)$ is arbitrary, we get the left inequality. A similar reasoning will lead to the right inequality.

We then compute $J(v^1(.), u^2(.))$. The corresponding trajectory is

$$\frac{dx}{dt} = ax + b^1v^1 + b^2u^2,\ x(0) = x_0.$$

If we set $\tilde{x}(t) = x(t) - y(t)$, we can prove the formula

$$J(v^1(.), u^2(.)) = \frac{1}{2}P(0)x_0^2 + \frac{\alpha}{2}\int_0^T (\tilde{x}(t))^2 + \frac{\gamma}{2}(\tilde{x}(T))^2 + \frac{\beta^1}{2}\int_0^T (v^1(.) - u^1(.))^2 dt, \tag{16.3.30}$$

from which we obtain

$$\inf_{v^1(.)} J(v^1(.), u^2(.)) = J(u(.)) = \frac{1}{2}P(0)x_0^2.$$

Therefore,

$$\sup_{v^2(.)}\inf_{v^1(.)} J(v^1(.), v^2(.)) \geq \inf_{v^1(.)} J(v^1(.), u^2(.)) = \frac{1}{2}P(0)x_0^2,$$

and from the first inequality (16.3.29) we conclude immediately (16.3.26). To obtain that the duality gap vanishes, we need the assumption on $\mathcal{K}(v(.))$. Indeed, we can compute

$$J(u^1(.), v^2(.)) = \frac{1}{2}P(0)x_0^2 + \mathcal{K}(v^2(.) - u^2(.)). \tag{16.3.31}$$

Therefore,

$$\sup_{v^2(.)} J(u^1(.), v^2(.)) = \frac{1}{2}P(0)x_0^2 + \sup_{v^2(.)} \mathcal{K}(v^2(.) - u^2(.)).$$

But

$$\sup_{v^2(.)} J(u^1(.), v^2(.)) \geq \inf_{v^1(.)}\sup_{v^2(.)} J(v^1(.), v^2.(.))$$

Therefore,

$$\begin{aligned}\inf_{v^1(.)}\sup_{v^2(.)} J(v^1(.), v^2(.)) - \sup_{v^2(.)}\inf_{v^1(.)} J(v^1(.), v^2(.)) &\leq \sup_{v^2(.)} J(u^1(.), v^2(.)) - \frac{1}{2}P(0)x_0^2\\ &\leq \sup_{v^2(.)} \mathcal{K}(v^2(.) - u^2(.)),\end{aligned}$$

and if the assumption is satisfied, we have $\sup_{v^2(.)} \mathcal{K}(v^2(.) - u^2(.)) = 0$, and the duality gap vanishes. Let us check that this is satisfied when the second condition (16.3.26) holds. Indeed, an easy calculation shows that

$$\mathcal{K}(v(.)) = -\frac{\beta^2}{2}\int_0^T (v(t) - \frac{b^2}{\beta^2}S(t)z(t))^2 dt, \tag{16.3.32}$$

and the result follows. □

EXERCISE 16.2. Prove formulas (16.3.28), (16.3.30), (16.3.31), (16.3.32).

We can give an example of a case in which the duality gap is not 0 and in fact $+\infty$. We consider the following example, due to P. Zhang [56]. In the model (16.3.19), (16.3.20), we take $b^1 = b^2 = 1$, $\beta^1 = \beta^2 = 1$, $a = 1$, and $\alpha = 2, \gamma = 0, T = 1$. Consequently, $\nu^1 = \nu^2 = 1, \nu = 0$. The Riccati equation (16.3.22) reduces to

$$-\frac{dP}{dt} - 2P = 2, \ \ P(1) = 0,$$

whence

$$P(t) = \exp 2(1-t) - 1.$$

We check easily that

$$J(u(.)) == \sup_{v^2(.)} \inf_{v^1(.)} J(v^1(.), v^2(.)) = \frac{x_0^2}{2}.(\exp 2 - 1) \tag{16.3.33}$$

Next choose $v^1(.)$ arbitrary, and set

$$v_n^2 = -v^1 + n.$$

The corresponding trajectory is obtained as follows:

$$\frac{dx_n}{dt} = x_n + n, \quad x_n(0) = x_0.$$

An easy calculation shows that

$$J(v^1(.), v_n^2(.)) = n^2\frac{(\exp -1)^2}{2} + x_0 n(\exp -1)^2 + x_0^2\frac{\exp 2 - 1}{2} + n\int_0^1 v^1(t)dt.$$

Therefore,

$$\sup_{v^2(.)} J(v^1(.), v^2(.)) \geq J(v^1(.), v_n^2(.)) \to +\infty,$$

which implies

$$\inf_{v^1(.)} \sup_{v^2(.)} J(v^1(.), v^2(.)) = +\infty.$$

The Riccati equation (16.3.27) becomes

$$\frac{dS}{dt} + 2S + S^2 + 2 = 0, \;\; S(1) = 0,$$

whose solution is $S(t) = -1 + \tan(1 - t + \frac{\pi}{4})$, which has a singularity when $t = 1 - \frac{\pi}{4}$.

EXERCISE 16.3. Consider the zero-sum differential game

$$\frac{dx}{dt} = 2v^1 - v^2, \;\; x(0) = x_0,$$

$$J(v(.)) = \frac{1}{2}\int_0^T ((v^1)^2 - (v^2)^2)dt + \frac{1}{2}x(T)^2.$$

Show that the P Riccati equation has a bounded solution, and that the S Riccati equation has a solution only if $T < 1$.

16.3.4 *Two-Person Zero-Sum Differential Game: Closed-Loop Controls*

If we consider the system (16.2.4), it appears quickly that the value functions $\Phi^i(x, t)$ satisfy $\Phi^2(x, t) = -\Phi^1(x, t)$, and the function $\Phi^1(x, t) = \Phi(x, t)$ satisfies the Isaacs equation

$$\frac{\partial \Phi}{\partial t} + \frac{1}{2}\alpha x^2 - \frac{1}{2}\nu(\frac{\partial \Phi}{\partial x})^2 + ax\frac{\partial \Phi}{\partial x} = 0, \qquad (16.3.34)$$

$$\Phi(x, T) = \frac{1}{2}\gamma x^2,$$

whose solution is $\Phi(x, t) = \frac{1}{2}P(t)x^2$ with $P(t)$ the solution of the Riccati equation

$$\frac{dP}{dt} + 2aP - \nu P^2 + \alpha = 0, \qquad (16.3.35)$$

$$P(T) = \gamma,$$

and we define the optimal feedbacks by the formulas

$$u^1(x,t) = -\frac{b^1}{\beta^1}P(t)x,\ u^2(x,t) = \frac{b^2}{\beta^2}P(t)x. \tag{16.3.36}$$

If the Riccati equation has a bounded solution, these formulas define a closed-loop Nash equilibrium. This follows directly from dynamic programming arguments and the existence of a smooth solution of the Isaacs equation (see below for more general details).

Let us check that the existence of a bounded solution of (16.3.35) is sufficient but not necessary to guarantee the existence of a closed-loop Nash equilibrium, as we shall see in the following example, due to P. Bernhard [10]. The state equation is

$$\frac{dx}{dt} = (2-t)v^1(t) + tv^2(t),\ x(0) = x_0,$$

$$J(v(.)) = \frac{1}{2}x(2)^2 + \int_0^2 ((v^1(t))^2 - (v^2(t))^2)dt.$$

The Riccati equation for this problem is

$$\frac{dP}{dt} + 2P^2(t-1) = 0,\ P(2) = 1,$$

whose solution is

$$P(t) = \frac{1}{(t-1)^2},$$

which is not bounded on the interval (0, 2). On the other hand, the feedbacks

$$u^1(x,t) = -\frac{2-t}{2(1-t)^2}x,\ u^2(x,t) = \frac{t}{2(t-1)^2}x$$

form a closed-loop Nash equilibrium. Indeed, we can write

$$J(v(.)) = \frac{1}{2}x_0^2 + \int_0^2 [(v^1(x(t),t) - u^1(x,t))^2 - (v^2(x(t),t) - u^2(x,t))^2]dt$$

for every pair of feedbacks $v^1(x,t)$, $v^2(x,t)$, and the conclusion follows. Note that although the feedbacks carry a singularity, when used in the state equation, the singularity vanishes, since the optimal trajectory is $y(t) = x_0(t-1)^2$.

16.4 Optimality Principle for Two-Person Zero-Sum Game

16.4.1 More on the Isaacs Equation

We consider a general two-person zero-sum game

$$\frac{dx}{ds} = g(x, v), \qquad x(t) = x, \tag{16.4.1}$$

with $v = (v^1, v^2)$. We define

$$J_{xt}(v(.)) = J_{xt}(v^1(.), v^2(.)) = \int_t^T l(x(s), v(s))ds + h(x(T)), \tag{16.4.2}$$

$= J_{xt}^1(v(.)) = J_{xt}(v(.))$, and $J_{xt}^2(v(.)) = -J_{xt}(v(.))$. The first player is the minimizer and the second player the maximizer. The players use feedback controls $v^1(x, t), v^2(x, t)$, assuming that the state equation (16.4.1) is well defined. In the smooth case, there is a Nash point $u^1(x, t), u^2(x, t)$, obtained by solving the Isaacs equation

$$\frac{\partial \Phi}{\partial t} + H(x, t, D\Phi) = 0, \qquad \Phi(x, T) = h(x), \tag{16.4.3}$$

with the following notation: We call the function $L(x, t, v, q) = l(x, t, v) + q.g(x, v)$ the Lagrangian. The Hamiltonian is defined by

$$H(x, t, q) = \inf_{v^1} \sup_{v^2} L(x, t, v^1, v^2, q) = \sup_{v^2} \inf_{v^1} L(x, t, v^1, v^2, q), \tag{16.4.4}$$

assuming that there is no duality gap and that there is a saddle point $\hat{v}^1(x, t, q), \hat{v}^2(x, t, q)$. The Nash equilibrium is given by

$$u^1(x, t) = \hat{v}^1(x, t, D\Phi(x, t)), \; u^2(x, t) = \hat{v}^2(x, t, D\Phi(x, t)), \tag{16.4.5}$$

and it satisfies

$$J_{xt}(u^1(.,.), v^2(.,.)) \leq J_{xt}(u^1(.,.), u^2(.,.)) \leq J_{xt}(v^1(.,.), u^2(.,.)), \quad \forall \text{feedbacks } v^1(.,.), v^2(.,.), \tag{16.4.6}$$

where $(.,.)$ represents the two arguments x, t.

Suppose now that the Isaacs condition (16.4.4) is not satisfied. We can still give a meaning to

$$H^+(x,t,q) = \inf_{v^1}\sup_{v^2} L(x,t,v^1,v^2,q),\ \ H^-(x,t,q) = \sup_{v^2}\inf_{v^1} L(x,t,v^1,v^2,q).$$

So we can consider two PDEs

$$\frac{\partial \Phi^+}{\partial t} + H^+(x,t,D\Phi^+) = 0, \tag{16.4.7}$$
$$\Phi^+(x,T) = h(x),$$

$$\frac{\partial \Phi^-}{\partial t} + H^-(x,t,D\Phi^-) = 0, \tag{16.4.8}$$
$$\Phi^-(x,T) = h(x).$$

This is not a system. They are two separate equations. We can give an interpretation of these two equations as a differential game in which we are introducing a difference between the two players. To understand this, we go back to the definition of H^+ and H^-. We assume that we can define a function $\hat{v}^2(x,t,q;v^1)$ such that

$$\sup_{v^2} L(x,t,v^1,v^2,q) = L(x,t,v^1,\hat{v}^2(x,t,q;v^1),q), \tag{16.4.9}$$

which we call the response function of player 2 to player 1. Player 1 defines a feedback $\bar{v}^1(x,t,q)$ by solving

$$\inf_{v^1} L(x,t,v^1,\hat{v}^2(x,t,q;v^1),q). \tag{16.4.10}$$

Similarly, we define the best response of player 1 to player 2, called $\hat{v}^1(x,t,q;v^2)$, by

$$\inf_{v^1} L(x,t,v^1,v^2,q) = L(x,t,\hat{v}^1(x,t,q;v^2),v^2,q), \tag{16.4.11}$$

and one can define $\bar{v}^2(x,t,q)$ by solving

$$\sup_{v^2} L(x,t,\hat{v}^1(x,t,q;v^2),v^2,q). \tag{16.4.12}$$

Note that the saddle point $\hat{v}^1(x,t,q)$, $\hat{v}^2(x,t,q)$ defined above, see (16.4.4), is a fixed point for the response functions, namely

$$\hat{v}^1(x,t,q) = \hat{v}^1(x,t,q;\hat{v}^2(x,t,q)), \tag{16.4.13}$$
$$\hat{v}^2(x,t,q) = \hat{v}^2(x,t,q;\hat{v}^1(x,t,q)). \tag{16.4.14}$$

We next turn back to the PDE (16.4.7), and we take $\bar{v}^1(x,t,q)$, $q = D\Phi^+(x,t)$ in the definition of $\hat{v}^2(x,t,q;v^1)$. We obtain feedbacks

$$u^{+2}(x,t;v^1) = \hat{v}^2(x,t,D\Phi^+(x,t);v^1),\ u^{+1}(x,t) = \bar{v}^1(x,t,D\Phi^+(x,t)). \tag{16.4.15}$$

We will claim that they form an equilibrium among feedbacks of the form $v^1(x,t)$ for player 1 and $v^2(x,t,v^1)$ for player 2. More precisely, for such feedbacks we define the state equation by

$$\begin{aligned} \frac{dx}{ds} &= g(x(s), v^1(x(s),s), v^2(x(s),s,v^1(x,s))), \\ x(t) &= x, \end{aligned} \tag{16.4.16}$$

and the payoff is

$$J_{xt}(v^1(.,.), v^2(.,.,.)) = \int_t^T l(x(s), v^1(x(s),s), v^2(x(s),s,v^1(x,s)))ds + h(x(T)). \tag{16.4.17}$$

We claim that

$$\begin{aligned} J_{xt}(u^{+1}(.,.), v^2(.,.,.)) \le \Phi^+(x,t) &= J_{xt}(u^{+1}(.,.), u^{+2}(.,.;.)) \\ &\le J_{xt}(v^1(.,.), u^{+2}(.,.;.)). \end{aligned} \tag{16.4.18}$$

So we state the following result.

PROPOSITION 16.3. *Assume that (16.4.7) has a smooth solution. Then we have (16.4.18) and*

$$\begin{aligned} \Phi^+(x,t) &= \inf_{v^1(.,.)} \sup_{v^2(.,.,.)} J_{xt}(v^1(.,.), v^2(.,.,.)) \\ &= \sup_{v^2(.,.,.)} \inf_{v^1(.,.)} J_{xt}(v^1(.,.), v^2(.,.,.)). \end{aligned} \tag{16.4.19}$$

A similar formula holds for $\Phi^-(x,t)$, with the roles of players 1 and 2 interchanged.

PROOF. We prove the left inequality. Player 1 uses a feedback $u^{+1}(x,t)$, and player 2 uses a response strategy $v^2(x,t,v^1)$. Using these strategies, the state equation becomes

$$\begin{aligned} \frac{dx}{ds} &= g(x(s), u^{+1}(x(s),s), v^2(x(s),s,u^{+1}(x,s))), \\ x(t) &= x. \end{aligned} \tag{16.4.20}$$

We have

$$\frac{d\Phi^+}{ds}(x(s),s) + l(x(s),u^{+1}(x(s),s),v^2(x(s),s,u^{+1}(x,s))) = \frac{\partial\Phi^+}{\partial s}(x(s),s)+$$

$$l(x(s),u^{+1}(x(s),s),v^2(x(s),s,u^{+1}(x,s))) + D\Phi^+(x(s),s).g(x(s),$$
$$u^{+1}(x(s),s),v^2(x(s),s,u^{+1}(x,s))) \leq$$

$$\frac{\partial\Phi^+}{\partial s}(x(s),s) + \sup_{v^2} l(x(s),u^{+1}(x(s),s),v^2)$$
$$+D\Phi^+(x(s),s).g(x(s),u^{+1}(x(s),s),v^2) =$$
$$\frac{\partial\Phi^+}{\partial s}(x(s),s) + H^+(x(s),D\Phi^+(x(s),s)) = 0.$$

This last equality is a consequence of the definition of $u^{+1}(x,s)$; see (16.4.15). Integrating between t and T, we obtain the first inequality (16.4.18). We prove in a similar way the right-hand one. The assertion (16.4.19) is an immediate consequence of (16.4.18). □

16.4.2 Optimality Principle

We begin with a useful lemma.

LEMMA 16.1. *Let $\Gamma(v^1,v^2)$ be a continuous function on $R^{m_1}\times R^{m_2}$ and let U^1_{ad}, U^2_{ad} be compact subsets of R^{m_1}, respectively R^{m_2}. We denote by $v^2(.)\equiv v^2(v^1)$ a measurable map from R^{m_1} to R^{m_2}. Then*

$$\inf_{v^1\in U^1_{ad}}\sup_{v^2\in U^2_{ad}}\Gamma(v^1,v^2) = \inf_{v^1\in U^1_{ad}}\sup_{v^2(.)\in U^2_{ad}}\Gamma(v^1,v^2(v^1)) = \tag{16.4.21}$$
$$= \sup_{v^2(.)\in U^2_{ad}}\inf_{v^1\in U^1_{ad}}\Gamma(v^1,v^2(v^1)).$$

PROOF. The first equality follows simply from $\sup_{v^2\in U^2_{\text{ad}}}\Gamma(v^1,v^2) = \sup_{v^2(.)\in U^2_{\text{ad}}}\Gamma(v^1,v^2(v^1))$, $\forall v^1$, which is obvious. Let us prove the second equality. By the assumptions, we can assert that $\forall\theta>0$, there exists a measurable map $v^2_\theta(.)$ from U^1_{ad} to U^2_{ad} such that

$$\sup_{v^2\in U^2_{\text{ad}}}\Gamma(v^1,v^2) \leq \Gamma(v^1,v^2_\theta(v^1)) + \theta\;\forall\, v^1\in U^1_{\text{ad}}.$$

This is a measurable selection theorem, a proof of which can be found in [22]. But then

$$\inf_{v^1\in U^1_{ad}} \sup_{v^2\in U^2_{ad}} \Gamma(v^1, v^2) \leq \inf_{v^1\in U^1_{ad}} \Gamma(v^1, v^2_\theta(v^1)) + \theta$$

$$\leq \sup_{v^2(.)\in U^2_{ad}} \inf_{v^1\in U^1_{ad}} \Gamma(v^1, v^2(v^1)) + \theta.$$

Since θ is positive but otherwise arbitrary, it follows that

$$\inf_{v^1\in U^1_{ad}} \sup_{v^2\in U^2_{ad}} \Gamma(v^1, v^2) \leq \sup_{v^2(.)\in U^2_{ad}} \inf_{v^1\in U^1_{ad}} \Gamma(v^1, v^2(v^1)).$$

But the reverse inequality is true, whence the result. □

Similarly, we have

$$\sup_{v^2\in U^2_{ad}} \inf_{v^1\in U^1_{ad}} \Gamma(v^1, v^2) = \sup_{v^2\in U^2_{ad}} \inf_{v^1(.)\in U^1_{ad}} \Gamma(v^1(v^2), v^2) = \tag{16.4.22}$$

$$= \inf_{v^1(.)\in U^1_{ad}} \sup_{v^2\in U^2_{ad}} \Gamma(v^1(v^2), v^2).$$

REMARK 16.4. The result (16.4.21) is compatible with the existence of a duality gap; $\inf_{v^1\in U^1_{ad}} \sup_{v^2\in U^2_{ad}} \Gamma(v^1, v^2) > \sup_{v^2\in U^2_{ad}} \inf_{v^1\in U^1_{ad}} \Gamma(v^1, v^2)$. So it is a way to eliminate the duality gap, introducing an asymmetry between the players. Let us consider the following example of a finite game. We have $v^1 = (1, 2)$ and $v^2 = (1, 2, 3)$. Then

$$\Gamma(1, 1) = -6,\ \Gamma(1, 2) = 2,\ \Gamma(1, 3) = -3,$$
$$\Gamma(2, 1) = 4,\ \Gamma(2, 2) = -5,\ \Gamma(2, 3) = -4.$$

We have

$$\inf_{v^1}\sup_{v^2}\Gamma(v^1, v^2) = 2,\ \sup_{v^2}\inf_{v^1}\Gamma(v^1, v^2) = -4,$$

so there is a duality gap. On the other hand, there are nine possible functions $v^2(v^1)$, and one can verify that

$$\sup_{v^2(.)}\inf_{v^1}\Gamma(v^1, v^2(v^1)) = 2.$$

We turn now to the differential game (16.4.1) and (16.4.2). The controls will be $v^1(s)$ and $v^2(s, v^1(s))$, which we will represent by $v^1(.)$ and $v^2(., .)$. Note that we do not consider feedbacks. To some extent, the feedback is replaced by a response function of the second player to the decision of the first player. The state equation evolves as follows:

$$\frac{dx}{ds} = g(x(s), v^1(s), v^2(s, v^1(s))), \tag{16.4.23}$$
$$x(t) = x,$$

and the payoff is given by

$$J_{xt}(v^1(.), v^2(.,.)) = \int_t^T l(x(s), v^1(s), v^2(s, v^1(s)))ds + h(x(T)). \tag{16.4.24}$$

The **upper-value** function is defined by

$$\Phi^+(x,t) = \sup_{v^2(.,.)} \inf_{v^1(.)} J_{xt}(v^1(.), v^2(.,.)). \tag{16.4.25}$$

Similarly, we shall consider controls $v^2(s)$ and $v^1(s, v^2(s))$ and define the **lower-value** function

$$\Phi^-(x,t) = \inf_{v^1(.,.)} \sup_{v^2(.)} J_{xt}(v^1(.), v^2(.,.)). \tag{16.4.26}$$

We state the **optimality principle** as follows, similar to the control case.

PROPOSITION 16.4. *The upper-value function satisfies the functional equation*

$$\Phi^+(x,t) = \sup_{v^2(.,.)} \inf_{v^1(.)} [\int_t^{t+\epsilon} l(x(s), v^1(s), v^2(s, v^1(s)))ds$$
$$+\Phi^+(x(t+\epsilon), t+\epsilon)], \ \forall \epsilon | t+\epsilon \leq T. \tag{16.4.27}$$

A similar functional equation holds for the lower-value function.

PROOF. As in the control case, this result is mostly algebraic, and it requires limited assumptions. Let us denote by $W(x,t)$ the right-hand side of (16.4.27). From the definition of $W(x,t)$, for all $\theta > 0$ we can find $v_\theta^2(.,.)$ whose value depends also on x, t, ϵ such that

$$\Phi^+(x,t) - \theta \leq \inf_{v^1(.)} [\int_t^{t+\epsilon} l(x_\theta(s), v^1(s), v_\theta^2(s, v^1(s)))ds + \Phi^+(x_\theta(t+\epsilon), t+\epsilon)],$$

where $x_\theta(s)$ satisfies

$$\frac{dx_\theta}{ds} = g(x_\theta(s), v^1(s), v_\theta^2(s, v^1(s))),$$
$$x_\theta(t) = x.$$

Of course $x_\theta(s)$ depends also on the controls $v^1(s), v^2_\theta(s, v^1(s))$. Next, from the definition of $\Phi^+(y, t+\epsilon)$, there exists a response function $v^2_{\theta y}(s, v^1(s))$ such that

$$\Phi^+(y, t+\epsilon) - \theta \leq \inf_{v^1(.)}[\int_{t+\epsilon}^T l(x^\theta_{y,t+\epsilon}(s), v^1(s), v^2_{\theta y}(s, v^1(s)))ds + h(x^\theta_{y,t+\epsilon}(T))],$$

where $x^\theta_{y,t+\epsilon}(s)$ satisfies

$$\frac{dx_{\theta y}}{ds} = g(x_{\theta y}(s), v^1(s), v^2_{\theta y}(s, v^1(s))),$$
$$x_{\theta y}(t+\epsilon) = y.$$

We then combine the response functions into a single one:

$$\tilde{v}^2_\theta(s, v^1) = \left| \begin{array}{ll} v^2_\theta(s, v^1), & t < s < t+\epsilon, \\ v^2_{\theta x(t+\epsilon)}(s, v^1), & t+\epsilon < s < T, \end{array} \right.$$

Let $\tilde{x}_\theta(s)$ be the state corresponding to $v^1(s)$ and $\tilde{v}^2_\theta(s, , v^1)$. Combining results, we obtain

$$W(x,t) - 2\theta \leq \int_t^T l(\tilde{x}_\theta(s), v^1(s), \tilde{v}^2_\theta(s, v^1(s)))ds + h(\tilde{x}_\theta(T))$$
$$= J_{xt}(v^1(.), \tilde{v}^2_\theta(.,.)), \ \forall v^1(.).$$

Therefore, $W(x,t) - 2\theta \leq \Phi^+(x,t)$, and since θ is arbitrary, we can claim that $W(x,t) \leq \Phi^+(x,t)$. Let us now prove the reverse inequality. We consider an arbitrary response function $v^2(.,.)$; hence

$$W(x,t) \geq \inf_{v^1(.)}[\int_t^{t+\epsilon} l(x(s), v^1(s), v^2(s, v^1(s)))ds + \Phi^+(x(t+\epsilon), t+\epsilon)].$$

For every θ we can find $v^1_\theta(.)$ such that

$$W(x,t) + \theta \geq \int_t^{t+\epsilon} l(x_\theta(s), v^1_\theta(s), v^2(s, v^1_\theta(s)))ds + \Phi^+(x_\theta(t+\epsilon), t+\epsilon),$$

where $x_\theta(s)$ is the solution of

$$\frac{dx_\theta}{ds} = g(x_\theta(s), v^1_\theta(s), v^2(s, v^1_\theta(s))),$$
$$x_\theta(t) = x.$$

Next,

$$\Phi^+(y,t+\epsilon) \geq \inf_{v^1(.)}[\int_{t+\epsilon}^T l(x_{y,t+\epsilon}(s), v^1(s), v^2(s, v^1(s)))ds + h(x_{y,t+\epsilon}(T))],$$

and for every θ there exists $v^1_{\theta y}(s)$ such that

$$\Phi^+(y,t+\epsilon)+\theta \geq \int_{t+\epsilon}^T l(x_{y,t+\epsilon}(s), v^1_{\theta y}(s), v^2(s, v^1_{\theta y}(s)))ds + h(x_{y,t+\epsilon}(T)).$$

Applying the above with $y = x_\theta(t+\epsilon)$, and setting $v^1_{\theta y}(s)) = v^1_{\theta\epsilon}(s)$, we get

$$\Phi^+(x_\theta(t+\epsilon), t+\epsilon)+\theta \geq \int_{t+\epsilon}^T l(x_{x_\theta(t+\epsilon),t+\epsilon}(s), v^1_{\theta\epsilon}(s), v^2(s, v^1_{\theta\epsilon}(s)))ds$$
$$+ h(x_{x_\theta(t+\epsilon),t+\epsilon}(T)).$$

We combine again the two evolutions, before and after $t+\epsilon$, and consider the control

$$\tilde{v}^1_\theta(s) = \left| \begin{array}{l} v^1_\theta(s),\ t < s < t+\epsilon, \\ v^1_{\theta\epsilon}(s),\ t+\epsilon < s < T. \end{array} \right.$$

Let $\tilde{x}_\theta(s)$ be the state corresponding to the pair $\tilde{v}^1_\theta(s), v^2(s, \tilde{v}^1_\theta(s))$. We can write

$$W(x,t) + 2\theta \geq \int_t^T l(\tilde{x}_\theta(s), \tilde{v}^1_\theta(s), v^2(s, \tilde{v}^1_\theta(s)))ds + h(\tilde{x}_\theta(T)),$$

so also,

$$W(x,t) + 2\theta \geq \inf_{v^1(.)} J_{xt}(v^1(.), v^2(.,.)),$$

and since $v^2(.,.)$ is arbitrary, we obtain $W(x,t) + 2\theta \geq \Phi^+(x,t)$. Since θ is arbitrary, we get $W(x,t) \geq \Phi^+(x,t)$. This completes the proof. □

16.4.3 Viscosity Solution of Upper and Lower Isaacs Equations

We call (16.4.7) and (16.4.8) upper and lower Isaacs equations. We shall focus on the upper-value function and upper Isaacs equation. But analogous considerations will hold for the lower-value function and lower Isaacs equation. We will consider a situation in which the upper-value function satisfies

$$\Phi^+(x,t) \text{ is bounded and uniformly Lipschitz continuous on } R^n \times [0,T]. \tag{16.4.28}$$

This is a drastic assumption, which can be substantially relaxed, but it simplifies considerably the analysis. It is satisfied in the following case:

$$g, l, Dg, Dl, h, Dh \text{ bounded independent of } t. \tag{16.4.29}$$

The fact that the controls of both players are not defined in terms of feedbacks is important in obtaining the regularity property (16.4.28). We have the following.

PROPOSITION 16.5. *Under the assumption (16.4.29), the upper-value function is a viscosity solution of the Isaacs equation (16.4.7).*

PROOF. Consider a subsolution $\Psi(x,t)$ of (16.4.27), i.e., a bounded uniformly continuous function such that

$$\begin{aligned}\Psi(x,t) \leq \sup\nolimits_{v^2(.,.)} \inf\nolimits_{v^1(.)}[\int_t^{t+\epsilon} l(x(s), v^1(s), v^2(s, v^1(s)))ds \\ +\Psi(x(t+\epsilon), t+\epsilon)],\ \forall \epsilon | t+\epsilon \leq T, \\ \Psi(x,T) \leq h(x).\end{aligned} \tag{16.4.30}$$

We claim that it is a viscosity subsolution of the Isaacs equation (16.4.7). We must show that for all $q, \rho \in D^+\Psi(x,t)$, we have

$$\rho + H^+(x,t,q) \geq 0. \tag{16.4.31}$$

We know that we can construct a function $\varphi_{xt}(y,s)$ that is C^1 such that $\varphi_{xt}(y,s) \geq \Psi(y,s)\ \forall y, s$, $\varphi_{xt}(x,t) = \Psi(x,t)$ and $\rho = \dfrac{\partial \varphi_{xt}}{\partial s}(x,t)$ and $q = D\varphi_{xt}(x,t)$. From the assumptions, we have

$$0 \leq \sup_{v^2(.,.)} \inf_{v^1(.)} [\int_t^{t+\epsilon} L(x, v^1(s), v^2(s, v^1(s)), q)ds] + \rho\epsilon + o(\epsilon),$$

hence also

$$0 \leq \sup_{v^2} L(x, v^1, v^2, q) + \rho + \frac{o(\epsilon)}{\epsilon},\ \forall v^1,$$

and (16.4.31) follows immediately. Consider next a supersolution of (16.4.27), i.e., a bounded uniformly continuous function such that

$$\begin{aligned}\Psi(x,t) \geq \sup\nolimits_{v^2(.,.)} \inf\nolimits_{v^1(.)}[\int_t^{t+\epsilon} l(x(s), v^1(s), v^2(s, v^1(s)))ds \\ +\Psi(x(t+\epsilon), t+\epsilon)],\ \forall \epsilon | t+\epsilon \leq T, \\ \Psi(x,T) \geq h(x).\end{aligned} \tag{16.4.32}$$

We show that it is a viscosity supersolution of the Isaacs equation (16.4.7). We must show that for all $q, \rho \in D^- \Psi(x,t)$, we have

$$\rho + H^+(x,t,q) \leq 0, \tag{16.4.33}$$

We know that we can construct a function $\varphi_{xt}(y,s)$ that is C^1 such that $\varphi_{xt}(y,s) \leq \Psi(y,s)\ \forall y,s$, $\varphi_{xt}(x,t) = \Psi(x,t)$ and $\rho = \frac{\partial \varphi_{xt}}{\partial s}(x,t)$ and $q = D\varphi_{xt}(x,t)$. We deduce as above that

$$\frac{o(\epsilon)}{\epsilon} \geq \inf_{v^1} L(x, v^1, v^2(v^1), q) + \rho$$

for every function $v^2(v^1)$. Therefore, also

$$\frac{o(\epsilon)}{\epsilon} \geq \inf_{v^1} \sup_{v^2} L(x, v^1, v^2, q) + \rho,$$

and (16.4.33) follows. This completes the proof. □

The maximum principle, see Proposition 10.6 and Theorem 11.4, allows us to claim that viscosity subsolutions are lower than viscosity supersolutions, and consequently, the upper value is the unique viscosity solution of (16.4.7).

16.5 Stochastic Open-Loop Nash Equilibrium

16.5.1 Stochastic Maximum Principle

We consider a probability space $\Omega, \mathcal{A}, P$ equipped with a filtration $\mathcal{F}^t$ and a standard $\mathcal{F}^t$ Wiener process $w(t)$ with values in R^k. The initial state is a random variable ξ with values in R^n that is $\mathcal{F}^0$-measurable and independent of the Wiener process. We have N players with controls $v^i(t), i = 1, \cdots, N$, with values in R^{m_i}. We again set $v(.) = (v^1(.), \cdots, v^N(.))$. This time, they are processes adapted to the filtration $\mathcal{F}^t$. We use the notation $v(.) = (v^i(.), \overline{v^i}(.))$ to emphasize a particular player. The state equation is given by

$$\begin{aligned} dx &= g(x(t), v(t))dt + \sigma(x(t))dw(t), \\ x(0) &= \xi, \end{aligned} \tag{16.5.1}$$

and the objective payoffs are

$$J^i(v(.)) = E[\int_0^T l^i(x(t), v(t))dt + h^i(x(T))]. \tag{16.5.2}$$

The concept of open-loop Nash equilibrium is the following (identical to the deterministic case):

$$J^i(u(.)) \leq J^i(v^i(.), \bar{u^i}(.)), \forall v^i(.), \forall i. \tag{16.5.3}$$

As already discussed in the deterministic case, we favor the stochastic maximum principle to characterize an open-loop Nash equilibrium $u(.)$. We consider the Hamiltonians

$$H^i(x, v, q^i) = l^i(x, v) + q^i.g(x, v).$$

The stochastic maximum principle reads,

$$\begin{aligned} dy &= g(y, u)dt + \sigma(y)dw(t), \qquad (16.5.4)\\ -dp^i &= [D_x H^i(y, u, p^i) + \sum_{j=1}^{k}(D_x\sigma^j)^*(y)r^{ij}(t)]dt \\ &\quad - \sum_{j=1}^{k} r^{ij}(t)dw^j(t), \\ y(0) = x &\quad p^i(T) = D_x h^i(y(T)), \\ H^i(y(t), u(t), p^i(t)) &= \inf_{v^i \in U^i_{\mathrm{ad}}} H^i(y(t), v^i, \bar{u^i}(t), p^i(t)), \ \forall t, \end{aligned}$$

where we note that $u(t)$ are the optimal controls and $y(t)$,$p^1(t), \cdots, p^N(t)$ the optimal state and adjoint states. The processes $r^{ij}(t)$ are adapted and represent the co-unknowns of the SBDE defining $p^i(t)$; see Theorem 11.1.

16.5.2 Example

We consider the same example as in the deterministic case; see section 16.3. We have

$$\begin{aligned} g(x, v) &= ax + b^1v^1 + b^2v^2, \ \sigma(x) = \sigma x, \\ l^i(x, v) &= \frac{1}{2}(\alpha^i x^2 + \beta^i(v^i)^2), \ h^i(x) = \frac{1}{2}\gamma^i x^2. \end{aligned}$$

Recalling that $\nu^i = \dfrac{(b^i)^2}{\beta^i}$, we see that the stochastic maximum principle reduces to

$$\begin{aligned} dy &= (ay - \nu^1 p^1 - \nu^2 p^2)dt + \sigma y dw \\ y(0) &= \xi, \\ -dp^i &= (\alpha^i y + ap^i + \sigma r^i)dt - r^i dw, \\ p^i(T) &= \gamma^i y(T), \quad i = 1, 2, \end{aligned} \tag{16.5.5}$$

and

$$u^i(t) = -\frac{b^i}{\beta^i} p^i(t). \tag{16.5.6}$$

We look for a solution of the form

$$p^i(t) = P^i(t)y(t),\ r^i(t) = \rho^i(t)y(t), \tag{16.5.7}$$

where $P^i(t)$ and $\rho^i(t)$ are deterministic. We obtain easily the system

$$\begin{aligned} &\frac{dP^1}{dt} + (2a + \sigma^2)P^1 - \nu^1(P^1)^2 - \nu^2 P^1 P^2 + \alpha^1 = 0, \\ &\frac{dP^2}{dt} + (2a + \sigma^2)P^2 - \nu^2(P^2)^2 - \nu^1 P^1 P^2 + \alpha^2 = 0, \\ &P^1(T) = \gamma^1\ P^2(T) = \gamma^2. \end{aligned} \tag{16.5.8}$$

This system is almost identical to the deterministic one ($\sigma = 0$); see (16.3.6). We have also $\rho^i(t) = \sigma P^i(t)$. We can then find the optimal state $y(t)$ by solving

$$\begin{aligned} dy &= (a - \nu^1 P^1(t) - \nu^2 P^2(t))y dt + \sigma y dw, \\ y(0) &= \xi. \end{aligned}$$

The open-loop Nash equilibrium is obtained from feedbacks on the optimal state

$$u^i(t) = -\frac{b^i}{\beta^i} P^i(t)y(t).$$

16.6 Stochastic Closed-Loop Nash Equilibrium

If we return to (16.5.1), (16.5.2), we consider controls defined by feedbacks $v^i(x,t)$ and look for a Nash equilibrium. We use dynamic programming to characterize the value functions $\Phi^i(x,t)$. In fact, they are very similar to (16.2.4) except for additional second-order terms

$$-\frac{\partial \Phi^i}{\partial t} + A\Phi^i = H^i(x, D\Phi), \tag{16.6.1}$$

$$\Phi^i(x, T) = h^i(x),$$

with

$$A\varphi(x) = -\sum_{ij=1}^{n} a_{ij}(x)\frac{\partial^2 \varphi}{\partial x_i \partial x_j}.$$

We apply this system to the linear–quadratic case considered in the previous section, Section 16.5.2. We obtain the system

$$\frac{\partial \Phi^1}{\partial t} + \frac{1}{2}\alpha^1 x^2 + ax\,\frac{\partial \Phi^1}{\partial x} - \frac{1}{2}\nu^1(\frac{\partial \Phi^1}{\partial x})^2 - \nu^2\frac{\partial \Phi^1}{\partial x}\frac{\partial \Phi^2}{\partial x} + \frac{1}{2}\sigma^2 x^2\frac{\partial^2 \Phi^1}{\partial x^2} = 0, \tag{16.6.2}$$

$$\frac{\partial \Phi^2}{\partial t} + \frac{1}{2}\alpha^2 x^2 + ax\,\frac{\partial \Phi^2}{\partial x} - \frac{1}{2}\nu^2(\frac{\partial \Phi^2}{\partial x})^2 - \nu^1\frac{\partial \Phi^1}{\partial x}\frac{\partial \Phi^2}{\partial x} + \frac{1}{2}\sigma^2 x^2\frac{\partial^2 \Phi^2}{\partial x^2} = 0,$$

$$\Phi^1(x, T) = \frac{1}{2}\gamma^1 x^2,\ \Phi^2(x, T) = \frac{1}{2}\gamma^2 x^2.$$

We look for functions $\Phi^i(x,t) = \frac{1}{2}Q^i(t)x^2$. We obtain the system of Riccati equations

$$\frac{dQ^1}{dt} + (2a + \sigma^2)Q^1 - \nu^1(Q^1)^2 - 2\nu^2 Q^1 Q^2 + \alpha^1 = 0, \tag{16.6.3}$$

$$\frac{dQ^2}{dt} + (2a + \sigma^2)Q^2 - \nu^2(Q^2)^2 - 2\nu^1 Q^1 Q^2 + \alpha^2 = 0,$$

$$Q^1(T) = \gamma^1,\ Q^2(T) = \gamma^2.$$

Again we obtain a system similar to that in the deterministic case, changing a into $a + \frac{\sigma^2}{2}$.

16.7 Stochastic Open-Loop Nash Equilibrium: Weak Formulation

We proceed as in Chapter 14. We consider Ω, $\mathcal{A}$, P, and a filtration $\mathcal{F}^t$. We consider also a standard $\mathcal{F}^t$ Wiener process with values in R^n. We define the state equation by

$$dX = \sigma(X)dw, \tag{16.7.1}$$

$$X(t) = x,$$

with $\sigma(x)$ invertible. We shall denote by $X_{xt}(s)$ the solution of (16.7.1). We next consider two players in order to simplify, and they choose controls that are adapted processes $v^1(t), v^2(t)$ with values in R^{m_1}, R^{m_2}. These controls can be defined by feedbacks on $X(s)$, since this is a fixed process. We set $v(t) = (v^1(t), v^2(t))$. We introduce the drift $g(X(s), v(s))$ and the process $w^{v(.)}(s)$ defined by

$$dw_{xt}^{v(.)} = dw - \sigma^{-1}(X_{xt}(s))g(X_{xt}(s), v(s))ds, \tag{16.7.2}$$

with $w_{xt}^{v(.)}(t) = w(t)$ to fix ideas. The process $X_{xt}(s)$ can be viewed as the solution of

$$dX = g(X(s), v(s))ds + \sigma(X(s))dw^{v(.)}. \tag{16.7.3}$$

We next define the martingale $M_{xt}^{v(.)}(s)$ by

$$\begin{aligned} dM_{xt}^{v(.)}(s) &= M_{xt}^{v(.)}(s)\sigma^{-1}(X_{xt}(s))g(X_{xt}(s), v(s))dw(s), \\ M_{xt}^{v(.)}(t) &= 1. \end{aligned} \tag{16.7.4}$$

We define the payoffs of players $i = 1, 2$ by

$$\Phi^{i,v(.)}(x,t) = E[M_{xt}^{v(.)}(T)(h^i(X_{xt}(T)) + \int_t^T l^i(X_{xt}(s), v(s))ds|\mathcal{F}^t] \tag{16.7.5}$$

and we define the new probability $P_{xt}^{v(.)}$ on $\Omega, \mathcal{A}$ by

$$\frac{dP_{xt}^{v(.)}}{dP}|\mathcal{F}^s = M_{xt}^{v(.)}(s).$$

We have also

$$\Phi^{i,v(.)}(x,t) = E^{v(.)}[h^i(X_{xt}(T)) + \int_t^T l^i(X_{xt}(s), v(s))ds|\mathcal{F}^t]. \tag{16.7.6}$$

Consider the Lagrangians

$$L^i(x, v, q^i) = l^i(x, v) + q^i.g(x, v).$$

We denote by $\bar{v}^i(x, q)$, with $q = (q^1, q^2)$, a Nash equilibrium for the Lagrangians and set

$$H^i(x, q) = L^i(x, \bar{v}(x, q), q^i). \tag{16.7.7}$$

We note that

$$\bar{v}^1(x,q) = \bar{v}^1(x, \bar{v}^2(x,q), q^1),$$
$$\bar{v}^2(x,q) = \bar{v}^2(x, \bar{v}^1(x,q), q^2),$$

in which the functions $\bar{v}^1(x, v^2, q^1)$ and $\bar{v}^2(x, v^1, q^2)$ achieve respectively the infima of $L^1(x, v^1, v^2, q^1)$ in v^1 and $L^2(x, v^1, v^2, q^2)$ in v^2, the other arguments being fixed.

Let $u(.) = (u^1(.), u^2(.))$ be an open-loop Nash equilibrium for the functionals $\Phi^{i,v(.)}(x,t)$. We set

$$\Phi^i(x,t) = \Phi^{i,u(.)}(x,t).$$

Each player must solve an open-loop non-Markov stochastic control problem, as in Chapter 14.

If we set

$$H^1(x, u^2(t), q^1) = \inf_{v^1} L^1(x, v^1, u^2(t), q^1),$$
$$H^2(x, u^1(t), q^2) = \inf_{v^2} L^2(x, u^1(t), v^2, q^2),$$

we can apply (14.3.2) to claim that $\Phi^1(x,t), \Phi^2(x,t)$ satisfies the system of BSPDEs

$$-d_t\Phi^1 + A\Phi^1 dt = [H^1(x, u^2(t), D\Phi^1 + (\sigma^{-1}(x))^* K^1) + \text{tr}\,(DK^1\sigma)]dt - K^1 dw,$$
$$-d_t\Phi^2 + A\Phi^2 dt = [H^2(x, u^1(t), D\Phi^2 + (\sigma^{-1}(x))^* K^2) + \text{tr}\,(DK^2\sigma)]dt - K^2 dw,$$

$$\Phi^1(x,T) = h^1(x), \;\; \Phi^2(x,T) = h^2(x). \tag{16.7.8}$$

Consider next $\Phi^i_{xt}(s) = \Phi^i(X_{xt}(s), s)$ and $Z^i_{xt}(s) = (K^i + \sigma^* D\Phi^i)(X_{xt}(s), s)$. We can write

$$d\Phi^i_{xt}(s) = d_s\Phi^i(X_{xt}(s), s) + D_x\Phi^i(X_{xt}(s), s)\sigma(X_{xt}(s))dw(s)$$
$$-A\Phi^i(X_{xt}(s), s)ds + \text{tr}(DK^i\sigma(X_{xt}(s), s))ds,$$

and using (16.7.8), we obtain

$$d\Phi^1_{xt}(s) = -H^1(X_{xt}(s), u^2(s), (\sigma^{-1}(X_{xt}(s)))^* Z^1_{xt}(s))ds + Z^1_{xt}(s)dw(s),$$
$$d\Phi^2_{xt}(s) = -H^2(X_{xt}(s), u^1(s), (\sigma^{-1}(X_{xt}(s)))^* Z^2_{xt}(s))ds + Z^2_{xt}(s)dw(s).$$

By definition,

$$u^1(s) = \bar{v}^1(X_{xt}(s), u^2(s), (\sigma^{-1}(X_{xt}(s))^* Z^1_{xt}(s)),$$
$$u^2(s) = \bar{v}^2(X_{xt}(s), u^1(s), (\sigma^{-1}(X_{xt}(s)))^* Z^2_{xt}(s)).$$

Therefore,

$$H^1(X_{xt}(s), u^2(s), (\sigma^{-1}(X_{xt}(s)))^* Z^1_{xt}(s)) = H^1(X_{xt}(s)),$$
$$(\sigma^{-1}(X_{xt}(s)))^* Z^1_{xt}(s), (\sigma^{-1}(X_{xt}(s)))^* Z^2_{xt}(s),$$
$$H^2(X_{xt}(s), u^1(s), (\sigma^{-1}(X_{xt}(s)))^* Z^2_{xt}(s)) = H^2(X_{xt}(s)),$$
$$(\sigma^{-1}(X_{xt}(s)))^* Z^1_{xt}(s), (\sigma^{-1}(X_{xt}(s)))^* Z^2_{xt}(s).$$

It follows that the processes $\Phi^1_{xt}(s)$, $\Phi^2_{xt}(s)$ are solutions of the system of BSDEs

$$\Phi^1_{xt}(s) = h^1(X_{xt}(T)) + \int_s^T H^1(X_{xt}(\tau), (\sigma^{-1}(X_{xt}(\tau)))^* Z^1_{xt}(\tau),$$
$$(\sigma^{-1}(X_{xt}(\tau)))^* Z^2_{xt}(\tau))d\tau - \int_s^T Z^1_{xt}(\tau)dw(\tau),$$
$$\Phi^2_{xt}(s) = h^2(X_{xt}(T)) + \int_s^T H^2(X_{xt}(\tau), (\sigma^{-1}(X_{xt}(\tau)))^* Z^1_{xt}(\tau),$$
$$(\sigma^{-1}(X_{xt}(\tau)))^* Z^2_{xt}(\tau)d\tau - \int_s^T Z^2_{xt}(\tau)dw(\tau)$$
$$\forall t \leq s \leq T. \tag{16.7.9}$$

These equations are given in [14]. They reduce to a pair of independent BSDEs in the following example:

$$g(x, v) = g(x) + v^1 + v^2, \tag{16.7.10}$$
$$l^i(x, v) = l^i(x) + v^1.v^2.$$

In this case, one checks easily that

$$\bar{v}^1(x, q) = -q^2, \ \bar{v}^2(x, q) = -q^1,$$
$$H^i(x, q) = l^i(x) + q^i.g(x) - (q^i)^2.$$

Hence (16.7.9) reduces to the pair of decoupled equations

$$\Phi^i_{xt}(s) = h^i(X_{xt}(T)) + \int_s^T [l^i(X_{xt}(\tau)) + (\sigma^{-1}(X_{xt}(\tau)))^* Z^i_{xt}(\tau).g(X_{xt}(\tau))$$
$$-(Z^i_{xt}(\tau))^2]d\tau - \int_s^T Z^i_{xt}(\tau)dw(\tau). \tag{16.7.11}$$

Chapter 17
Stackelberg Differential Games

17.1 Open-Loop Deterministic Stackelberg Games

17.1.1 Generalities and Model

In Stackelberg games, the players are not on an equal footing. There is a leader and a follower. This is a generalization of the two-player zero-sum differential games considered in Section 16.4, in particular. We have the dynamical system

$$\frac{dx}{dt} = g(x(t), v^1(t), v^2(t)), \tag{17.1.1}$$
$$x(0) = x_0.$$

The payoffs are defined by

$$J^i(v^1(.), v^2(.)) = \int_0^T l^i(x(t), v^1(t), v^2(t))dt + h^i(x(T)),\ i = 1, 2. \tag{17.1.2}$$

Player 1 is the leader. She announces her control, which is a function of t, denoted by $v^1(.)$. Player 2, the follower, solves the following control problem:

$$\inf_{v^2(.)} J^2(v^1(.), v^2(.)).$$

Let $\widehat{v^2}(t; v^1(.))$ be the optimal control, called the best response function, or simply the response function for short. It depends on the whole control $v^1(.)$. The leader's problem is then to choose $v^1(.)$ in order to minimize

$$\inf_{v^1(.)} J^1(v^1(.), \widehat{v^2}(.; v^1(.))).$$

A. Bensoussan, *Estimation and Control of Dynamical Systems*, Interdisciplinary Applied Mathematics 48, https://doi.org/10.1007/978-3-319-75456-7_17

Note that if $J^1(v^1(.), v^2(.)) = -J^2(v^1(.), v^2(.)) = J(v^1(.), v^2(.))$, then this problem reduces to $\inf_{v^1(.)} \sup_{v^2(.)} J(v^1(.), v^2(.))$.

17.1.2 Maximum Principle

Consider the Hamiltonians

$$H^i(x, v^1, v^2, q^i) = l^i(x, v^1, v^2) + q^i.g(x, v^1, v^2).$$

We state the leader's maximum principle. Indeed, it is the maximum principle for the leader's control problem. Denote by $u^1(t)$ the leader's optimal control. Since the follower applies the best response function, he uses the control $u^2(t) = \widehat{v^2}(t; u^1(.))$. We also need to introduce the function $\overline{v^2}(x, q^2, v^1)$, called the reaction function, which minimizes the Hamiltonian $H^2(x, v^1, v^2, q^2)$ in v^2. It satisfies

$$D_{v^2} H^2(x, v^1, v^2, q^2) = 0. \tag{17.1.3}$$

The corresponding optimal state is denoted by $y(t)$. There are three and not two adjoint variables, denoted by $p^1(t), p^2(t), \pi(t)$. The following system holds:

$$\begin{aligned} u^2(t) &= \overline{v^2}(y(t), p^2(t), u^1(t)), \\ \frac{dy}{dt} &= g(y(t), u^1(t), u^2(t)), \ y(0) = x_0, \end{aligned} \tag{17.1.4}$$

$$-\frac{dp^1}{dt} = D_x \Big(H^1(y(t), u^1(t), \overline{v^2}(y(t), p^2(t), u^1(t)), p^1(t))$$
$$+\pi(t).D_x H^2(y(t), u^1(t), \overline{v^2}(y(t), p^2(t), u^1(t)), p^2(t)) \Big),$$
$$-\frac{dp^2}{dt} = D_x H^2(y(t), u^1(t), u^2(t), p^2(t)),$$
$$p^1(T) = D_x h^1(y(T)) + D_x^2 h^2(y(T))\pi(T), \ p^2(T) = D_x h^2(y(T)),$$
$$\frac{d\pi}{dt} = D_{q^2} \Big(H^1(y(t), u^1(t), \overline{v^2}(y(t), p^2(t), u^1(t)), p^1(t))$$
$$+\pi(t).D_x H^2(y(t), u^1(t), \overline{v^2}(y(t), p^2(t), u^1(t)), p^2(t)) \Big),$$
$$\pi(0) = 0,$$
$$D_{v^1} \Big(H^1(y(t), u^1(t), \overline{v^2}(y(t), p^2(t), u^1(t)), p^1(t))$$
$$+\pi(t).D_x H^2(y(t), u^1(t), \overline{v^2}(y(t), p^2(t), u^1(t)), p^2(t)) \Big) = 0.$$

It is important to have in mind that in this complicated set of equations, we cannot use indifferently $u^2(t)$ and $\overline{v^2}(y(t), p^2(t), u^1(t))$. We use the second expression when we have to differentiate with respect to the arguments. To understand the origin of this system, we note that for a given control $v^1(.)$, the follower computes the response function by solving his maximum principle, which can be written as follows:

$$\frac{dx}{dt} = g(x(t), v^1(t), \overline{v^2}(x(t), q^2(t), v^1(t)))\, x(0) = x_0, \tag{17.1.5}$$

$$-\frac{dq^2}{dt} = D_x H^2(x(t), v^1(t), \overline{v^2}(x(t), q^2(t), v^1(t)), q^2(t)),\;\; q^2(T) = D_x h^2(x(T)).$$

This system can be interpreted as the state equation for the leader. She has to consider the pair $x(t), q^2(t)$. So she needs two adjoint variables, denoted by $p^1(t)$ and $\pi(t)$. She needs to use the Hamiltonian (which we may call the augmented Hamiltonian of the leader)

$$H^1(x, v^1, \overline{v^2}(x, q^2, v^1), q^1) + \pi.D_x(H^2(x, v^1, \overline{v^2}(x, q^2, v^1), q^2))$$

to write her adjoint system and the necessary condition.

If we compare with the maximum principle for the open-loop deterministic Nash point, see (16.1.8), we replace the function $\overline{v^2}(y(t), p^2(t), u^1(t))$ by a fixed function $u^2(t)$. So there is no difference with respect to the arguments. it follows that $\pi(t) = 0$ and $p^1(t), p^2(t)$ reduce to the adjoint variables of the Nash equilibrium.

17.1.3 Examples

We take our examples from [19]. We first consider Example 5.1. We have

$$l^i(x, v^1, v^2) = \frac{1}{2}(x^2 + (v^i)^2) - v^i,$$
$$g(x, v^1, v^2) = v^1 + v^2,\;\; h^i(x) = 0.$$

So

$$H^i(x, v^1, v^2, q^i) = \frac{1}{2}(x^2 + (v^i)^2) - v^i + q^i(v^1 + v^2)$$

and

$$\overline{v^2}(x, q^2, v^1) = 1 - q^2.$$

So the augmented Hamiltonian of the leader is

$$\frac{1}{2}(x^2 + (v^1)^2) - v^1 + q^1(v^1 + 1 - q^2) + \pi x.$$

It follows that the maximum principle for the leader is given as follows:

$$\begin{aligned}
\frac{dy}{dt} &= u^1 + 1 - p^2 \; y(0) = x_0,\\
-\frac{dp^1}{dt} &= y(t) + \pi(t), \; p^1(T) = 0,\\
-\frac{dp^2}{dt} &= y(t), \; p^2(T) = 0,\\
\frac{d\pi}{dt} &= -p^1(t), \; \pi(0) = 0,\\
u^1(t) &= 1 - p^1(t).
\end{aligned}$$

We replace $u^1(t)$ in the first equation and define the vectors

$$\xi(t) = \begin{pmatrix} y(t) \\ \pi(t) \end{pmatrix} \quad \eta(t) = \begin{pmatrix} p^2(t) \\ p^1(t) \end{pmatrix}.$$

The system is equivalent to

$$\begin{aligned}
\frac{d\xi}{dt} &= -K^*\eta(t) + b,\\
-\frac{d\eta}{dt} &= K\xi(t),
\end{aligned}$$

$$\xi(0) = \xi_0 = \begin{pmatrix} x_0 \\ 0 \end{pmatrix} \; \eta(T) = 0,$$

with

$$K = \begin{pmatrix} 1 \; 0 \\ 1 \; 1 \end{pmatrix} \; b = \begin{pmatrix} 2 \\ 0 \end{pmatrix}.$$

We solve the system by a decoupling argument. We write

$$\eta(t) = P(t)(K^*)^{-1}\xi(t) + r(t), \; \eta(T) = 0,$$

and performing the identification, we obtain

$$-\frac{dP}{dt} + P^2 = KK^*, \; P(T) = 0,$$
$$-\frac{dr}{dt} + Pr = P(K^*)^{-1}b, \; r(T) = 0.$$

To solve the Riccati equation we consider the eigenvalues of KK^*, which are $\frac{3 \pm \sqrt{5}}{2}$. We consider the matrix of eigenvectors

$$W = \begin{pmatrix} 1 & 1 \\ \frac{1+\sqrt{5}}{2} & \frac{1-\sqrt{5}}{2} \end{pmatrix}$$

and can write

$$KK^* = W \begin{pmatrix} s_1^2 & 0 \\ 0 & s_2^2 \end{pmatrix} W^{-1},$$

with $s_1^2 = \dfrac{3+\sqrt{5}}{2}$, $s_2^2 = \dfrac{3-\sqrt{5}}{2}$. We can then diagonalize the Riccati equation by setting $P = WQW^{-1}$, where Q is a diagonal matrix $Q = \begin{pmatrix} \gamma & 0 \\ 0 & \delta \end{pmatrix}$ with γ, δ solutions of the scalar Riccati equations

$$-\frac{d\gamma}{dt} + \gamma^2 = s_1^2, \quad \gamma(T) = 0,$$
$$-\frac{d\delta}{dt} + \delta^2 = s_2^2, \; \delta(T) = 0.$$

EXERCISE 17.1. Complete the solution of the maximum principle for the leader.

We consider another example, Example 5-2 [19]. We modify the payoff of the follower, not that of the leader. We take

$$l^2(x, v^1, v^2) = x + \frac{1}{2}(v^2)^2 - v^2.$$

But then we see immediately that the payoff of the follower is the sum of a functional of $v^1(.)$ and a functional of $v^2(.)$. Therefore, the leader cannot influence the follower. The follower will solve an optimal control problem that is independent of the leader, who incorporates the follower's decision and solves a standard control problem.

EXERCISE 17.2. Write the maximum principle for the leader and check the preceding assertion.

17.1.4 Linear–Quadratic Stackelberg Games

We consider the model already studied for Nash differential games; see Section 16.3 This time, player 1 is the leader and player 2 the follower. We recall that

$$g(x, v^1, v^2) = ax + b^1 v^1 + b^2 v^2, \tag{17.1.6}$$

$$l^i(x, v^1, v^2) = \frac{1}{2}(\alpha^i x^2 + \beta^i (v^i)^2), \; h^i(x) = \frac{1}{2}\gamma^i x^2. \tag{17.1.7}$$

From the follower's Hamiltonian

$$H^2(x, v^1, v^2, q^2) = \frac{1}{2}(\alpha^2 x^2 + \beta^2 (v^2)^2) + q^2.(ax + b^1 v^1 + b^2 v^2),$$

we derive the reaction function

$$\overline{v^2}(x, q^2, v^1) = -\frac{b^2}{\beta^2} q^2.$$

Therefore, the augmented Hamiltonian of the leader becomes

$$H^1(x, v^1, \overline{v^2}(x, q^2, v^1), q^1) + \pi.D_x(H^2(x, v^1, \overline{v^2}(x, q^2, v^1), q^2)) =$$

$$\frac{1}{2}(\alpha^1 x^2 + \beta^1 (v^1)^2) + q^1.(ax + b^1 v^1 - \frac{(b^2)^2}{\beta^2} q^2) + \pi.(\alpha^2 x + a q^2).$$

We derive the maximum principle for the leader. We set $\nu^i = \frac{(b^i)^2}{\beta^i}$. We get the system

$$\frac{dy}{dt} = ay - \nu^1 p^1 - \nu^2 p^2, \; y(0) = x_0 \tag{17.1.8}$$

$$-\frac{dp^1}{dt} = p^1 a + \alpha^1 y + \alpha^2 \pi, \quad p^1(T) = \gamma^1 y(T) + \gamma^2 \pi(T),$$

$$-\frac{dp^2}{dt} = p^2 a + \alpha^2 y, \quad p^2(T) = \gamma^2 y(T)$$

$$\frac{d\pi}{dt} = a\pi - \nu^2 p^1, \quad \pi(0) = 0,$$

and $u^1(t) = -\frac{b^1}{\beta^1} p^1(t)$.

To get a structure similar to that of the preceding example, we make a transformation as follows. We set

$$y(t)=\sqrt{\nu^1\nu^2}\tilde{y}(t),\ \pi(t)=\sqrt{\frac{\alpha^1}{\alpha^2}}\nu^2\tilde{\pi}(t),\ p^1(t)=\sqrt{\alpha^1\nu^2}\tilde{p}^1(t),$$
$$p^2(t)=\sqrt{\alpha^2\nu^1}\tilde{p}^2(t),$$

and the system becomes

$$\frac{d\tilde{y}}{dt}=a\tilde{y}(t)-\sqrt{\alpha^1\nu^1}\tilde{p}^1(t)-\sqrt{\alpha^2\nu^2}\tilde{p}^2(t),\ \tilde{y}(0)=\frac{x_0}{\sqrt{\nu^1\nu^2}},$$
$$\frac{d\tilde{\pi}}{dt}=a\tilde{\pi}(t)-\sqrt{\alpha^2\nu^2}\tilde{p}^1(t),\ \tilde{\pi}(0)=0,$$
$$-\frac{d\tilde{p}^1}{dt}=a\tilde{p}^1(t)+\sqrt{\alpha^1\nu^1}\tilde{y}(t)+\sqrt{\alpha^2\nu^2}\tilde{\pi}(t),$$
$$\tilde{p}^1(T)=\gamma^1\sqrt{\frac{\nu^1}{\alpha^1}}\tilde{y}(T)+\gamma^2\sqrt{\frac{\nu^2}{\alpha^2}}\tilde{\pi}(T),$$
$$-\frac{d\tilde{p}^2}{dt}=a\tilde{p}^2(t)+\sqrt{\alpha^2\nu^2}\tilde{y}(t),\ \tilde{p}^2(T)=\gamma^2\sqrt{\frac{\nu^2}{\alpha^2}}\tilde{y}(T).$$

If we write $\xi(t)=\begin{pmatrix}\tilde{y}(t)\\ \tilde{\pi}(t)\end{pmatrix}$, $\eta(t)=\begin{pmatrix}\tilde{p}^2(t)\\ \tilde{p}^1(t)\end{pmatrix}$, we get the relations

$$\frac{d\xi}{dt}=a\xi(t)-K^*\eta(t),\ \xi(0)=\xi_0=\begin{pmatrix}\dfrac{x_0}{\sqrt{\nu^1\nu^2}}\\ 0\end{pmatrix},$$

$$-\frac{d\eta}{dt}=a\eta(t)+K\xi(t),\ \eta(T)=\begin{pmatrix}\gamma^2\sqrt{\dfrac{\nu^2}{\alpha^2}} & 0\\ \gamma^1\sqrt{\dfrac{\nu^1}{\alpha^1}} & \gamma^2\sqrt{\frac{\nu^2}{\alpha^2}}\end{pmatrix}\xi(T),$$

with $K=\begin{pmatrix}\sqrt{\alpha^2\nu^2} & 0\\ \sqrt{\alpha^1\nu^1} & \sqrt{\alpha^2\nu^2}\end{pmatrix}$. We can then proceed as in the previous section. We try a decoupling solution as follows: $\eta(t)=P(t)(K^*)^{-1}\xi(t)$. We obtain the Riccati equation

$$\frac{dP}{dt}+2aP-P^2+KK^*=0. \tag{17.1.9}$$

However, we get a final condition

$$P(T) = \begin{pmatrix} \gamma^2\nu^2 & \gamma^2\sqrt{\nu^1\nu^2\dfrac{\alpha^1}{\alpha^2}} \\ \gamma^1\sqrt{\nu^1\nu^2\dfrac{\alpha^2}{\alpha^1}} & \gamma^1\nu^1+\gamma^2\nu^2 \end{pmatrix},$$

which is not a symmetric matrix. If we assume

$$\frac{\gamma^2}{\gamma^1} = \frac{\alpha^2}{\alpha^1}, \tag{17.1.10}$$

then it becomes symmetric, and the Riccati equation can be solved. Otherwise, the existence of a solution is not guaranteed.

17.2 Feedback Stackelberg Equilibrium for Deterministic Games

17.2.1 Setting of the Problem

In contrast to Nash games, which are by construction equilbria among equal players, Stackelberg games are sequential control problems. Nevertheless, it is possible to introduce a concept of equilibrium while maintaining the distinction between a leader and a follower. Players will use feedbacks on the state. We use the terminology "feedback Stackelberg equilibrium," since we shall also define below a concept of closed-loop Stackelberg game, which will not be an equilibrium but a sequence of control problems. The leader uses a feedback $v^1(x,t)$, and the follower uses a reaction function $v^2(x,t,v^1)$. So the leader knows the state of the system and announces a strategy (feedback). The follower can use the state of the system and the current value of the leader's control. We omit the argument t to save notation. The important aspect is that we limit the choice to a single pair $u^1(x)$ and $u^2(x,v^1)$ to define an equilibrium. The cost functionals are computed as follows: First, the state evolution is obtained by solving the differential equation

$$\begin{aligned} \frac{dx}{dt} &= g(x, v^1(x), v^2(x, v^1(x))), \\ x(0) &= x_0, \end{aligned} \tag{17.2.1}$$

and we set

$$J^i(v^1(.), v^2(.,.)) = \int_0^T l^i(x(t), v^1(x(t)), v^2(x(t), v^1(x(t))))dt + h^i(x(T)). \tag{17.2.2}$$

The concept of equilibrium is defined as follows:

$$J^1(u^1(.), u^2(.,.)) \leq J^1(v^1(.), u^2(.,.)), \tag{17.2.3}$$
$$J^2(u^1(.), u^2(.,.)) \leq J^2(u^1(.), v^2(.,.)).$$

This concept is very similar to that of Nash equilibrium except for the presence of a reaction function.

17.2.2 Dynamic Programming Approach

Not surprisingly, we shall use a dynamic programming approach. So, we first embed the problem into a family of problems indexed by the initial conditions x, t. So we have

$$\frac{dx}{ds} = g(x, v^1(x), v^2(x, v^1(x))), \tag{17.2.4}$$
$$x(t) = x,$$

and

$$J^i_{xt}(v^1(.), v^2(.,.)) = \int_t^T l^i(x(s), v^1(x(s)), v^2(x(s), v^1(x(s))))ds + h^i(x(T)). \tag{17.2.5}$$

A feedback Stackelberg equilibrium will be a pair $u^1(x),\ u^2(x, v^1)$ such that

$$J^1_{xt}(u^1(.), u^2(.,.)) \leq J^1_{xt}(v^1(.), u^2(.,.)),\ \forall v^1(.), \tag{17.2.6}$$
$$J^2_{xt}(u^1(.), u^2(.,.)) \leq J^2_{xt}(u^1(.), v^2(.,.)),\ \forall v^2(.,.).$$

Since we are in the context of dynamic programming, we use Lagrangians:

$$L^i(x, v^1, v^2, q^i) = l^i(x, v^1, v^2) + q^i.g(x, v^1, v^2). \tag{17.2.7}$$

We have already introduced the reaction function of the follower. It is the function $\overline{v^2}(x, q^2, v^1)$ that minimizes $L^2(x, v^1, v^2, q^2)$ in v^2. We next consider $L^1(x, v^1, \overline{v^2}(x, q^2, v^1), q^1)$, which we minimize in v^1. Setting $\mathbf{q} = (q^1, q^2)$, we denote by $\hat{v}^1(x, \mathbf{q})$ the point of minimum of $L^1(x, v^1, \overline{v^2}(x, q^2, v^1), q^1)$ and $\hat{v}^2(x, \mathbf{q}) = \overline{v^2}(x, q^2, \hat{v}^1(x, \mathbf{q}))$. We next set

$$H^i(x, \mathbf{q}) = L^i(x, \hat{v}^1(x, \mathbf{q}), \hat{v}^2(x, \mathbf{q}), q^i), \tag{17.2.8}$$

and we consider the system of PDEs

$$\frac{\partial \Phi^i}{\partial t} + H^i(x, D\Phi) = 0 \tag{17.2.9}$$
$$\Phi^i(x, T) = h^i(x),$$

where Φ represents the pair (Φ^1, Φ^2). We assume that this system has sufficiently smooth solutions, and we define

$$u^1(x,t) = \hat{v}^1(x, D\Phi), \qquad (17.2.10)$$
$$u^2(x,t,v^1) = \overline{v^2}(x, D\Phi^2, v^1).$$

PROPOSITION 17.1. *We assume that the system (17.2.9) has a smooth solution. Then the functions defined by (17.2.10) form a feedback Stackelberg equilibrium for the problem (17.2.6) and*

$$\Phi^i(x,t) = J^i_{xt}(u^1(.), u^2(.,.)). \qquad (17.2.11)$$

PROOF. We use a verification argument. To fix the ideas, consider that the leader uses $u^1(x)$, and the follower a strategy $v^2(x, v^1) \neq u^2(x, v^1)$. Let us check that

$$\Phi^2(x,t) \leq J^2_{xt}(u^1(.), v^2(.,.)). \qquad (17.2.12)$$

We recall that

$$H^2(x, D\Phi) = L^2(x, \hat{v}^1(x, D\Phi), \overline{v^2}(x, D\Phi^2, \hat{v}^1(x, D\Phi)), D\Phi^2)$$
$$\leq L^2(x, \hat{v}^1(x, D\Phi), v^2(x, \hat{v}^1(x, D\Phi)), D\Phi^2)$$

from the definition of $\overline{v^2}(x, q^2, v^1)$. Since $\hat{v}^1(x, D\Phi) = u^1(x)$, this reads

$$H^2(x, D\Phi) \leq L^2(x, u^1(x), v^2(x, u^1(x)), D\Phi^2)$$
$$= l^2(x, u^1(x), v^2(x, u^1(x))) + D\Phi^2.g(x, u^1(x), v^2(x, u^1(x))).$$

When the leader uses $u^1(x)$, and the follower the strategy $v^2(x, v^1)$, the resulting trajectory is given by

$$\frac{dx}{ds} = g(x, u^1(x), v^2(x, u^1(x))),$$
$$x(t) = x.$$

Therefore,

$$\frac{d}{ds}\Phi^2(x(s), s) = \frac{\partial \Phi^2}{\partial s} + D\Phi^2.g(x(s), u^1(x(s)), v^2(x(s), u^1(x(s))))$$
$$\geq -l^2(x(s), u^1(x(s)), v^2(x(s), u^1(x(s)))).$$

Integrating between t and T, we obtain immediately (17.2.12). By similar reasoning we prove (17.2.11) and $\Phi^1(x,t) \leq J^1_{xt}(v^1(.), u^2(.,.))$. This completes the proof. □

17.2.3 Linear–Quadratic Feedback Stackelberg Equilibrium

If we consider the model (17.1.6), (17.1.7), then we will not see any difference between Nash equilibrium and Stackelberg equilibrium. The reason is that the reaction function $\overline{v^2}(x, q^2, v^1)$ does not depend on v^1. Indeed,

$$\overline{v^2}(x, q^2, v^1) = -\frac{b^2}{\beta^2} q^2$$

and

$$L^1(x, v^1, \overline{v^2}(x, q^2, v^1), q^1) = \frac{1}{2}(\alpha^1 x^2 + \beta^1 (v^1)^2) + q^1(ax + b^1 v^1 - v^2 q_2).$$

The minimum in v^1 is attained at $\hat{v}^1(x, \mathbf{q}) = -\frac{b^1}{\beta^1} q^1$. Also, $\hat{v}^2(x, \mathbf{q}) = \overline{v^2}(x, q^2, \hat{v}^1(x, \mathbf{q})) = -\frac{b^2}{\beta^2} q^2$. But these values are also those obtained by writing Nash equilibrium conditions. Therefore, the Hamiltonians for Nash closed-loop equilibrium and feedback Stackelberg equilibrium are identical. To see a difference, we have to introduce a coupling between the controls. So we consider

$$l^i(x, v^1, v^2) = \frac{1}{2}(\alpha^i x^2 + \beta^i (v^i)^2) + \theta v^1 v^2.$$

The reaction function becomes

$$\overline{v^2}(x, q^2, v^1) = -\frac{b^2 q^2 + \theta v^1}{\beta^2}$$

and

$$L^1(x, v^1, \overline{v^2}(x, q^2, v^1), q^1) = \frac{1}{2}(\alpha^1 x^2 + \beta^1 (v^1)^2) + q^1(ax + b^1 v^1) \\ -(\theta v^1 + q^1 b^2)\frac{b^2 q^2 + \theta v^1}{}.$$

It follows that

$$\hat{v}^1(x, \mathbf{q}) = \frac{q^1(b^1 \beta^2 - \theta b^2) - \theta b^2 q^2}{2\theta^2 - \beta^1 \beta^2}.$$

We then have

$$\hat{v}^2(x, \mathbf{q}) = \overline{v^2}(x, q^2, \hat{v}^1(x, \mathbf{q})) = \\ = \frac{b^2 q^2 (\theta^2 - \beta^1 \beta^2) + \theta q^1 (b^1 \beta^2 - \theta b^2)}{\beta^2 (\beta^1 \beta^2 - 2\theta^2)}.$$

The corresponding formulas for Nash equilibrium would be

$$\hat{v}^1(x,\mathbf{q}) = \frac{\beta^2 q^1 b^1 - \theta q^2 b^2}{\theta^2 - \beta^1\beta^2},\ \hat{v}^2(x,\mathbf{q}) = \frac{\beta^1 q^2 b^2 - \theta q^1 b^1}{\theta^2 - \beta^1\beta^2}.$$

The formulas coincide when $\theta = 0$. In general, they are different. To simplify a little, we take $b^1 = b^2 = \beta^1 = \beta^2 = 1$. Then we get

$$\hat{v}^1(x,\mathbf{q}) = \frac{q^1(1-\theta) - \theta q^2}{2\theta^2 - 1},\ \hat{v}^2(x,\mathbf{q}) = (1-\theta)\frac{q^2(1+\theta) - \theta q^1}{2\theta^2 - 1}.$$

We note the relation

$$\hat{v}^2 + \theta\hat{v}^1 + q^2 = 0,$$

which originates from the definition of feedback Stackelberg equilibrium. We first write

$$L^1(x,\hat{v}^1,\hat{v}^2,q^1) = \frac{1}{2}\alpha^1 x^2 + \frac{1}{2}\hat{v}^1 + \theta\hat{v}^1\hat{v}^2 + q^1(ax + \hat{v}^1 + \hat{v}^2).$$

Replacing $\hat{v}^2$ and making some rearrangements, we obtain

$$H^1(x,\mathbf{q}) = \frac{1}{2}\alpha^1 x^2 + aq^1 x - q^1 q^2 - \frac{1}{2(1-2\theta^2)}|q^1(1-\theta) - \theta q^2|^2. \tag{17.2.13}$$

Similarly, we can check that

$$\begin{aligned} H^2(x,\mathbf{q}) = \tfrac{1}{2}\alpha^2 x^2 + aq^2 x + \tfrac{q^2}{1-2\theta^2}(\theta q^2 - (1-\theta)q^1) - \frac{(1-\theta)^2}{2(1-2\theta^2)^2}| \\ \theta q^1 - (1+\theta)q^2|^2. \end{aligned} \tag{17.2.14}$$

For the sake of comparison, the Hamiltonians in the case of Nash equilibrium reduce to

$$H^1(x,\mathbf{q}) = \frac{1}{2}\alpha^1 x^2 + aq^1 x + q^1\frac{q^2 - \theta q^1}{\theta^2 - 1} - \frac{|q^1 - \theta q^2|^2}{2(\theta^2 - 1)^2},$$

$$H^2(x,\mathbf{q}) = \frac{1}{2}\alpha^2 x^2 + aq^2 x + q^2\frac{q^1 - \theta q^2}{\theta^2 - 1} - \frac{|q^2 - \theta q^1|^2}{2(\theta^2 - 1)^2},$$

which coincide with (17.2.13), (17.2.14) for $\theta = 0$. For $\theta = 1$, the Nash equilibrium will not exist, since the Hamiltonians are not defined. On the other hand, for the feedback Stackelberg equilibrium, we get

$$H^1(x, \mathbf{q}) = \frac{1}{2}\alpha^1 x^2 + aq^1 x - q^1 q^2 + \frac{1}{2}|q^2|^2,$$

$$H^2(x, \mathbf{q}) = \frac{1}{2}\alpha^2 x^2 + aq^2 x - |q^2|^2.$$

We see that the second Hamiltonian does not depend on q^1. Therefore, the system (17.2.9) is decoupled. The function $\Phi^2(x,t)$ is the solution of a self-contained equation, and $\Phi^1(x,t)$ is the solution of a linear equation.

17.3 Closed-Loop Memoryless Strategies For Deterministic Stackelberg Games

17.3.1 General Theory

In the definition of the feedback Stackelberg equilibrium, the follower does not use his best response. He is strictly limited to feedbacks of the form $v^2(x, v^1)$. Because of the concept of equilibrium, he will use this feedback only against the optimal leader's feedback, which is $u^1(x)$. It is natural therefore, to consider also the concept of best response. However, the best response to an open-loop control and the best response to a feedback are not the same. This is due to the fact that a feedback control carries more information than an open-loop control. The leader will then optimize her feedback, knowing that the follower will use his best response. This problem has been studied by G. Papavassilopoulos and J. B. Cruz [38]. To clarify the fact that this problem is different from that of feedback Stackelberg equilibrium, it is customary to use the terminology of closed-loop memoryless strategy. Memoryless stands for the idea that the follower does not use any past values of the state except for the initial conditions, which are known even in the case of open-loop controls. Facing a feedback control $v^1(x) = v^1(x,t)$, the follower will build his response function by solving a maximum principle set of equations that differs from (17.1.5) as follows:

$$\frac{dx}{dt} = g(x, v^1(x), \overline{v^2}(x, q^2, v^1(x))), \tag{17.3.1}$$

$$-\frac{dq^2}{dt} = D_x H^2(x, v^1(x), \overline{v^2}(x, q^2, v^1(x)), q^2) + (D_x v^1)^* D_{v^1} H^2(x, v^1(x),$$
$$\overline{v^2}(x, q^2, v^1(x)), q^2),$$
$$x(0) = x_0, \; q^2(T) = D_x h^2(x(T)).$$

There is an additional term arising from $D_x v^1$ in the adjoint equation. Once we solve (17.3.1), we obtain a pair of functions $x(t; v^1(.))$ and $q^2(t; v^1(.))$. This pair is injected into the leader's payoff to yield

$$J^1(v^1(.)) = \int_0^T l^1(x, v^1(x), \overline{v^2}(x, q^2, v^1(x)))dt + h^1(x(T)). \tag{17.3.2}$$

In fact, a useful observation is that it is sufficient to consider a feedback $v^1(x)$ that is linear, i.e., of the form

$$v^1(x,t) = v^1(t) + \tilde{v}^1(t)x. \tag{17.3.3}$$

So we need to have two open-loop controls $v^1(t)$ and $\tilde{v}^1(t)$ instead of a general feedback. Indeed, we consider the control problem with a pair of open-loop controls $v^1(t), \tilde{v}^1(t)$,

$$\frac{dx}{dt} = g(x, v^1(t) + \tilde{v}^1(t)x, \overline{v^2}(x, q^2, v^1(t) + \tilde{v}^1(t)x)), \tag{17.3.4}$$

$$-\frac{dq^2}{dt} = D_x H^2(x, v^1(t) + \tilde{v}^1(t)x, \overline{v^2}(x, q^2, v^1(t) + \tilde{v}^1(t)x), q^2) +$$

$$+ (\tilde{v}^1(t))^* D_{v^1} H^2(x, v^1(t) + \tilde{v}^1(t)x, \overline{v^2}(x, q^2, v^1(t) + \tilde{v}^1(t)x), q^2),$$

$$x(0) = x_0, \; q^2(T) = D_x h^2(x(T)),$$

$$J^1(v^1(.), \tilde{v}^1(.)) = \int_0^T l^1(x, v^1(t)+\tilde{v}^1(t)x, \overline{v^2}(x, q^2, v^1(t)+\tilde{v}^1(t)x))dt + h^i(x(T)). \tag{17.3.5}$$

We have to be careful that in (17.3.5), $v^1(.), \tilde{v}^1(.)$ refer to functions $v^1(t), \tilde{v}^1(t)$, whereas in (17.3.2), $v^1(.)$ is a feedback. We want to minimize $J^1(v^1(.), \tilde{v}^1(.))$. If we now consider an optimal feedback $\bar{v}^1(x)$ for the problem (17.3.1), (17.3.2) and the corresponding trajectory $\bar{x}(t)$, then on setting $\bar{v}^1(t) = \bar{v}^1(\bar{x}(t)) - D_x\bar{v}^1(\bar{x}(t))\bar{x}(t)$, $\bar{\tilde{v}}^1(t) = D_x\bar{v}^1(\bar{x}(t))$, we see that the cost $J^1(\bar{v}^1(.), \bar{\tilde{v}}^1(.))$ coincides with $\inf_{v^1(.)} J^1(v^1(.))$. This is an obvious consequence of the definition of $\bar{v}^1(t), \bar{\tilde{v}}^1(t)$. But then

$$\inf_{v^1(.)} J^1(v^1(.)) \geq \inf_{v^1(.), \tilde{v}^1(.)} J^1(v^1(.), \tilde{v}^1(.)).$$

But the converse is also true. Indeed, if $\bar{v}^1(t), \bar{\tilde{v}}^1(t)$ is now an optimal pair for the payoff $J^1(v^1(.), \tilde{v}^1(.))$ and $\bar{x}(t)$ is the corresponding trajectory, then the linear feedback

$$\bar{v}^1(x,t) = \bar{v}^1(t) + (\bar{\tilde{v}}^1(t))^*(x - \bar{x}(t))$$

used in computing $J^1(\bar{v}^1(.))$ leads to $J^1(\bar{v}^1(.), \bar{\tilde{v}}^1(.))$. This implies

$$\inf_{v^1(.),\tilde{v}^1(.)} J^1(v^1(.), \tilde{v}^1(.)) \geq \inf_{v^1(.)} J^1(v^1(.)),$$

and the two are equal. We can then write the maximum principle for the problem (17.3.4), (17.3.5). The optimal controls are denoted by $u^1(t), \tilde{u}^1(t), u^2(t)$, the optimal trajectory by $y(t)$; the adjoints are denoted by $p^1(t), p^2(t), \pi(t)$. We then write the set of equations

$$\begin{aligned} u^2(t) &= \overline{v^2}(y(t), p^2(t), u^1(t) + \tilde{u}^1(t)y(t)),\\ \frac{dy}{dt} &= g(y, u^1 + \tilde{u}^1 y, u^2),\ \ y(0) = x_0, \end{aligned} \tag{17.3.6}$$

$$\begin{aligned} -\frac{dp^1}{dt} &= D_x\Big(H^1(y, u^1 + \tilde{u}^1 y, \overline{v^2}(y, p^2, u^1 + \tilde{u}^1 y), p^1)\\ &\quad +\pi.D_x H^2(y, u^1 + \tilde{u}^1 y, \overline{v^2}(y, p^2, u^1 + \tilde{u}^1 y), p^2)\Big) +\\ &\quad + D_x\left(D_{v^1} H^2(y, u^1 + \tilde{u}^1 y, \overline{v^2}(y, p^2, u^1 + \tilde{u}^1 y), p^2).\tilde{u}^1\pi\right), \end{aligned}$$

$$\begin{aligned} -\frac{dp^2}{dt} &= D_x H^2(y, u^1 + \tilde{u}^1 y, u^2, p^2) +\\ &\quad + \tilde{u}^1 D_{v^1} H^2(x, u^1 + \tilde{u}^1 y, u^2, p^2), \end{aligned}$$

$$p^1(T) = D_x h^1(y(T)) + D_x^2 h^2(y(T))\pi(T),\ \ p^2(T) = D_x h^2(y(T)),$$

$$\begin{aligned} \frac{d\pi}{dt} &= D_{q^2}\Big(H^1(y, u^1 + \tilde{u}^1 y, \overline{v^2}(y, p^2, u^1 + \tilde{u}^1 y), p^1)\\ &\quad + \pi.D_x H^2(y, u^1 + \tilde{u}^1 y, \overline{v^2}(y, p^2, u^1 + \tilde{u}^1 y), p^2)\Big) +\\ &\quad + D_{q^2}\left(D_{v^1} H^2(y, u^1 + \tilde{u}^1 y, \overline{v^2}(y, p^2, u^1 + \tilde{u}^1 y), p^2).\tilde{u}^1\pi\right),\\ &\qquad \pi(0) = 0, \end{aligned}$$

$$\begin{aligned} &H^1(y, u^1 + \tilde{u}^1 y, \overline{v^2}(y, p^2, u^1 + \tilde{u}^1 y), p^1) + \pi.D_x H^2(y, u^1 + \tilde{u}^1 y, \overline{v^2}\\ &\qquad (y, p^2, u^1 + \tilde{u}^1 y), p^2) +\\ &+D_{v^1} H^2(y, u^1 + \tilde{u}^1 y, \overline{v^2}(y, p^2, u^1 + \tilde{u}^1 y)p^2).\tilde{u}^1\pi = \end{aligned}$$

$$\inf_{v^1,\tilde{v}^1}\Big(H^1(y,v^1+\tilde{v}^1y,\overline{v^2}(y,p^2,v^1+\tilde{v}^1y),p^1)$$

$$+\pi.D_xH^2(y,v^1+\tilde{v}^1y,\overline{v^2}(y,p^2,v^1+\tilde{v}^1y),p^2)$$

$$+D_{v^1}H^2(y,v^1+\tilde{v}^1y,\overline{v^2}(y,p^2,v^1+\tilde{v}^1y),p^2).\tilde{v}^1\pi\Big).$$

The last condition is written as an infimum instead of the gradient being set to be 0. The reason concerns mainly the second control $\tilde{v}^1$. Indeed, the function to minimize may not have a minimum over all values of $\tilde{v}^1$. We may need to introduce artificial constraints, such as that $\tilde{v}^1$ is bounded. This will correspond in the original problem (17.3.1) (17.3.2) to introducing bounds on the gradient of $v^1(x)$. Note that on the right-hand side of the equations for $\frac{dp^1}{dt}$ and $\frac{d\pi}{dt}$ we have written $\overline{v^2}(y,p^2,u^1+\tilde{u}^1y)$ instead of u^2, which is used on the right-hand side of $\frac{dy}{dt}$ and $\frac{dp^2}{dt}$. This is because we have to first differentiate in x and q^2, which is not the case for $\frac{dy}{dt}$ and $\frac{dp^2}{dt}$. A similar remark holds for the minimization in $v^1,\tilde{v}^1$.

17.3.2 Example

Consider the example

$$l^i(x,v^1,v^2)=\frac{1}{2}(\alpha^ix^2+(v^i)^2)+\theta v^1v^2,$$

$$g(x,v^1,v^2)=ax+v^1+v^2,$$

$$h^i(x)=\frac{1}{2}\gamma^ix^2.$$

Then $\overline{v^2}(x,q^2,v^1)=-(\theta v^1+q^2)$ and

$$H^1(x,v^1,\overline{v^2}(x,q^2,v^1),q^1)=\frac{1}{2}\alpha^1x^2+q^1ax-q^1q^2$$

$$+(\frac{1}{2}-\theta^2)(v^1)^2+v^1(-\theta q^2+(1-\theta)q^1),$$

$$D_xH^2(x,v^1,\overline{v^2}(x,q^2,v^1),q^2)=\alpha^2x+aq^2,$$

$$D_{v^1}H^2(x,v^1,\overline{v^2}(x,q^2,v^1),q^2)=-\theta^2v^1+(1-\theta)q^2.$$

The augmented Hamiltonian for the leader is given by

$$\frac{1}{2}\alpha^1 x^2 + q^1 ax - q^1 q^2 + (\frac{1}{2} - \theta^2)(v^1 + \tilde{v}^1 x)^2 + (v^1 + \tilde{v}^1 x)(-\theta q^2 + (1-\theta)q^1) + \\ + \pi(\alpha^2 x + aq^2) + \tilde{v}^1\pi(-\theta^2(v^1 + \tilde{v}^1 x) + (1-\theta)q^2).$$

Therefore, the maximum principle for the leader becomes

$$\frac{dy}{dt} = ay - p^2 + (1-\theta)(u^1 + \tilde{u}^1 y), \tag{17.3.7}$$

$$-\frac{dp^1}{dt} = ap^1 + \alpha^1 y + \alpha^2\pi + (1 - 2\theta^2)y(\tilde{u}^1)^2 + \\ + \tilde{u}^1((1-2\theta^2)u^1 + (1-\theta)p^1 - \theta p^2),$$

$$-\frac{dp^2}{dt} = ap^2 + \alpha^2 y + \tilde{u}^1(-\theta^2(u^1 + \tilde{u}^1 y) + (1-\theta)p^2),$$

$$\frac{d\pi}{dt} = a\pi - p^1 - \theta u^1 - \theta\tilde{u}^1 y + (1-\theta)\pi\tilde{u}^1,\ \pi(0) = 0,$$

$$\Big((\frac{1}{2} - \theta^2)(u^1 + \tilde{u}^1 y)^2 + (u^1 + \tilde{u}^1 y)(-\theta p^2 + (1-\theta)p^1) + \\ + \tilde{u}^1\pi(-\theta^2(u^1 + \tilde{u}^1 y) + (1-\theta)p^2)\Big) =$$

$$\inf_{v^1,\tilde{v}^1}\Big((\frac{1}{2} - \theta^2)(v^1 + \tilde{v}^1 y)^2 + (v^1 + \tilde{v}^1 y)(-\theta p^2 + (1-\theta)p^1) + \\ + \tilde{v}^1\pi(-\theta^2(v^1 + \tilde{v}^1 y) + (1-\theta)p^2)\Big),$$

$$y(0) = x_0,\ p^1(T) = \gamma^1 y(T) + \gamma^2\pi(T),\ p^2(T) = \gamma^2 y(T),\ \pi(0) = 0.$$

The global infimum is $-\infty$, as is readily seen. So we need to introduce constraints on $\tilde{v}^1$ to obtain a solution. For instance, we consider bounds on $\tilde{v}^1$. This is equivalent to introducing bounds on the gradient of $v^1(x)$ in the original problem. Another possibility will be to penalize the constraint. We refer to [38] for further details.

17.4 Open-Loop Stochastic Stackelberg Games

17.4.1 Stochastic Maximum Principle

We define a probability space $\Omega, \mathcal{A}, P$ equipped with a filtration $\mathcal{F}^t$. We consider an $\mathcal{F}^t$ Wiener process $w(t)$ with values in R^k and a random variable ξ that is $\mathcal{F}^0$-measurable. In an open-loop context, the controls of the two players are stochastic processes $v^1(t)$, $v^2(t)$ adapted to $\mathcal{F}^t$ with values in R^{m_1}, R^{m_2} respectively. The state equation is defined by the SDE

$$dx = g(x, v^1(t), v^2(t))dt + \sigma(x)dw(t), \tag{17.4.1}$$
$$x(0) = \xi,$$

and the payoffs are defined by the formulas

$$J^i(v^1(.), v^2(.)) = E[\int_0^T l^i(x(t), v^1(t), v^2(t))dt + h^i(x(T))]. \tag{17.4.2}$$

The second player, the follower, defines his optimal response $\widehat{v^2}(t; v^1(.))$ by solving a stochastic control problem with random coefficients. As done in the deterministic case, see Section 17.1, the follower will use the maximum principle. Stating

$$H^i(x, v^1, v^2, q^i) = l^i(x, v^1, v^2) + q^i.g(x, v^1, v^2),$$

we use the reaction function $\overline{v^2}(x, q^2, v^1)$, which attains the minimum of $H^2(x, v^1, v^2, q^2)$ in v^2 for given values of x, q^2, v^1. So the leader sees a pair of stochastic processes $x(t), q^2(t)$, which are adapted and solutions of

$$dx = g(x, v^1(t), \overline{v^2}(x, q^2, v^1(t)))dt + \sum_{j=1}^k \sigma^j(x)dw^j, \tag{17.4.3}$$

$$-dq^2 = [D_x H^2(x, v^1(t), \overline{v^2}(x, q^2, v^1(t)), q^2) + \sum_{j=1}^k (D_x\sigma^j(x))^* r^{2j}]dt$$

$$-\sum_{j=1}^k r^{2j} dw^j,$$

$$x(0) = \xi, \; q^2(T) = D_x h^2(x(T)),$$

and she wants to minimize the payoff

$$J^1(v^1(.)) = E[\int_0^T l^1(x, v^1(t), \overline{v^2}(x, q^2, v^1(t))dt + h^1(x(T))].$$

We can then write the following stochastic maximum principle for the leader (open-loop controls):

$$dy = g(y, u^1(t), \overline{v^2}(y, p^2, u^1(t)))dt + \sum_{j=1}^{k} \sigma^j(x)dw^j, \tag{17.4.4}$$

$$-dp^1 = D_x \Bigg(H^1(y, u^1(t), \overline{v^2}(y, p^2, u^1(t)), p^1)$$

$$+ \pi.[D_x H^2(y, u^1(t), \overline{v^2}(y, p^2, u^1(t)), p^2) + \sum_{j=1}^{k} (D_x\sigma^j(y))^* r^{2j}] \Bigg) dt +$$

$$+ \sum_{j=1}^{k} (D_x\sigma^j(y))^* r^{1j} dt - \sum_{j=1}^{k} r^{1j} dw^j,$$

$$-dp^2 = [D_x H^2(y, u^1(t), \overline{v^2}(y, p^2, u^1(t)), p^2) + \sum_{j=1}^{k} (D_x\sigma^j(y))^* r^{2j}]dt - \sum_{j=1}^{k} r^{2j} dw^j,$$

$$d\pi = D_{q^2}\Big(H^1(y, u^1(t), \overline{v^2}(y, p^2, u^1(t)), p^1) + \pi.D_x H^2(y, u^1(t),$$

$$\overline{v^2}(y, p^2, u^1(t)), p^2)\Big) + \sum_{j=1}^{k} D_x\sigma^j(y)\pi dw^j,$$

$$y(0) = \xi, \ \pi(0) = 0,$$

$$p^1(T) = D_x h^1(y(T)) + D_x^2 h^2(y(T))\pi(T), \ p^2(T) = D_x h^2(y(T)),$$

$$D_{v^1}\Big(H^1(y(t), u^1(t), \overline{v^2}(y(t), p^2(t), u^1(t)), p^1(t)) + \pi(t).D_x H^2(y(t), u^1(t),$$

$$\overline{v^2}(y(t), p^2(t), u^1(t)), p^2(t))\Big) = 0.$$

17.4.2 Example

We consider Example 16.5.2. We obtain the system

$$
\begin{aligned}
dy &= (ay - \nu^1 p^1 - \nu^2 p^2)dt + \sigma dw, \\
-dp^1 &= (ap^1 + \alpha^1 y + \alpha^2 \pi + \sigma r^1)dt - r^1 dw, \\
-dp^2 &= (ap^2 + \alpha^2 y + \sigma r^2)dt - r^2 dw, \\
d\pi &= (a\pi - \nu^2 p^1)dt + \sigma \pi dw, \\
y(0) &= \xi \pi(0) = 0, \\
p^1(T) &= \gamma^1 y(T) + \gamma^2 \pi(T) p^2(T) = \gamma^2 y(T).
\end{aligned}
\tag{17.4.5}
$$

We can solve this system as in Section 17.1.4. We set

$$
y(t) = \sqrt{\nu^1\nu^2}\tilde{y}(t),\ \pi(t) = \sqrt{\frac{\alpha^1}{\alpha^2}}\nu^2\tilde{\pi}(t),\ p^1(t) = \sqrt{\alpha^1\nu^2}\tilde{p}^1(t),
$$

$$
p^2(t) = \sqrt{\alpha^2\nu^1}\tilde{p}^2(t),
$$

$$
\tilde{r}^1(t) = \frac{r^1(t)}{\sqrt{\alpha^1\nu^2}},\ \tilde{r}^2(t) = \frac{r^2(t)}{\sqrt{\alpha^2\nu^1}},
$$

and

$$
\xi(t) = \begin{pmatrix} \tilde{y}(t) \\ \tilde{\pi}(t) \end{pmatrix},\ \eta(t) = \begin{pmatrix} \tilde{p}^2(t) \\ \tilde{p}^1(t) \end{pmatrix},\ \rho(t) = \begin{pmatrix} \tilde{r}^2(t) \\ \tilde{r}^1(t) \end{pmatrix}.
$$

We obtain a system

$$
\begin{aligned}
d\xi &= (a\xi - K^*\eta)dt + \sigma\xi dw, \\
-d\eta &= (a\eta + K\xi + \sigma\rho)dt - \rho dw,
\end{aligned}
\tag{17.4.6}
$$

$$
\xi(0) = \xi_0 = \begin{pmatrix} \dfrac{\xi}{\sqrt{\nu^1\nu^2}} \\ 0 \end{pmatrix},\qquad \eta(T) = \begin{pmatrix} \gamma^2\sqrt{\dfrac{\nu^2}{\alpha^2}} & 0 \\ \gamma^1\sqrt{\dfrac{\nu^1}{\alpha^1}} & \gamma^2\sqrt{\frac{\nu^2}{\alpha^2}} \end{pmatrix}\xi(T),
$$

with $K = \begin{pmatrix} \sqrt{\alpha^2\nu^2} & 0 \\ \sqrt{\alpha^1\nu^1} & \sqrt{\alpha^2\nu^2} \end{pmatrix}$. We obtain a solution of the form $\eta(t) = P(t)(K^*)^{-1}\xi(t)$, $\rho(t) = \sigma P(t)(K^*)^{-1}\xi(t)$, with $P(t)$ the solution of the Riccati equation

$$\frac{dP}{dt} + (2a + \sigma^2)P - P^2 + KK^* = 0, \tag{17.4.7}$$

$$P(T) = \begin{pmatrix} \gamma^2\nu^2 & \gamma^2\sqrt{\nu^1\nu^2\dfrac{\alpha^1}{\alpha^2}} \\ \gamma^1\sqrt{\nu^1\nu^2\dfrac{\alpha^2}{\alpha^1}} & \gamma^1\nu^1 + \gamma^2\nu^2 \end{pmatrix},$$

which we can solve assuming $\dfrac{\gamma^2}{\gamma^1} = \dfrac{\alpha^2}{\alpha^1}$.

17.5 Feedback Stackelberg Equilibrium for Stochastic Games

The situation is similar to that of the deterministic one; see Section 17.2. The problem is stated as follows. The state evolution is described by the diffusion

$$\begin{aligned} dx &= g(x, v^1(x), v^2(x, v^1(x)))ds + \sigma(x)dw, \\ x(t) &= x, \end{aligned} \tag{17.5.1}$$

in which the leader uses a feedback $v^1(x, s)$ that is a deterministic function, and the follower uses a reaction function $v^2(x, v^1, s)$. The respective payoffs are defined by

$$J^i_{xt}(v^1(.), v^2(., .)) = E[\int_t^T l^i(x(s), v^1(x(s)), v^2(x(s), v^1(x(s))))ds + h^i(x(T))]. \tag{17.5.2}$$

A feedback Stackelberg equilibrium will be a pair $u^1(x),\ u^2(x, v^1)$ such that

$$\begin{aligned} J^1_{xt}(u^1(.), u^2(., .)) &\le J^1_{xt}(v^1(.), u^2(., .)),\ \forall v^1(.), \\ J^2_{xt}(u^1(.), u^2(., .)) &\le J^2_{xt}(u^1(.), v^2(., .)),\ \forall v^2(., .). \end{aligned} \tag{17.5.3}$$

An equilibrium can be obtained by the dynamic programming approach. We recall the Lagrangians

$$L^i(x, v^1, v^2, q^i) = l^i(x, v^1, v^2) + q^i.g(x, v^1, v^2) \tag{17.5.4}$$

and the reaction function of the follower. It is the function $\overline{v^2}(x, q^2, v^1)$ that minimizes $L^2(x, v^1, v^2, q^2)$, in v^2. We next consider $L^1(x, v^1, \overline{v^2}(x, q^2, v^1), q^1)$, which we minimize in v^1. Setting $\mathbf{q} = (q^1, q^2)$, we denote by $\hat{v}^1(x, \mathbf{q})$ the point of minimum of $L^1(x, v^1, \overline{v^2}(x, q^2, v^1), q^1)$, and $\hat{v}^2(x, \mathbf{q}) = \overline{v^2}(x, q^2, \hat{v}^1(x, \mathbf{q}))$. We next set

$$H^i(x, \mathbf{q}) = L^i(x, \hat{v}^1(x, \mathbf{q}), \hat{v}^2(x, \mathbf{q}), q^i), \tag{17.5.5}$$

and we consider the system of PDEs

$$\frac{\partial \Phi^i}{\partial t} + A\Phi^i + H^i(x, D\Phi) = 0, \tag{17.5.6}$$
$$\Phi^i(x, T) = h^i(x),$$

where the second-order differential operator A is given by

$$A\varphi(x) = -\sum_{i,j=1}^{n} a_{ij}(x) \frac{\partial^2 \varphi}{\partial x_i \partial x_j}(x),$$

with $a = \frac{1}{2}\sigma\sigma^*$. In (17.5.6), Φ represents the pair (Φ^1, Φ^2). We assume that this system has sufficiently smooth solutions (the situation is of course more favorable than in the deterministic case), and we define

$$u^1(x) = \hat{v}^1(x, D\Phi), \tag{17.5.7}$$
$$u^2(x, v^1) = \overline{v^2}(x, D\Phi^2, v^1).$$

We obtain a result similar to that of Proposition 17.1, namely

$$\Phi^i(x, t) = J^i_{xt}(u^1(.), u^2(., .)), \tag{17.5.8}$$

and (17.5.3) holds. The proof is similar to that of Proposition 17.1 and uses a verification argument.

17.6 Stochastic Open-Loop Stackelberg Games: Weak Formulation

17.6.1 Setting of the Problem

We proceed as in the case of Nash equilibrium; see Section 16.7. We consider $\Omega, \mathcal{A}, P$ and a filtration $\mathcal{F}^s$. We consider also a standard $\mathcal{F}^s$ Wiener process with values in R^n. We define the state equation by

$$dX = \sigma(X)dw, \tag{17.6.1}$$
$$X(t) = x,$$

with $\sigma(x)$ invertible. We shall denote by $X_{xt}(s)$ the solution of (17.6.1). The leader chooses a control that is an adapted processes $v^1(s)$ with values in R^{m_1}. The follower chooses a strategy $v^2(s, v^1)$. The important remark is that these controls can be defined by feedbacks on $X(s)$, since it is a fixed process. To simplify notation, we write $v(.) = (v^1(.), v^2(., .))$. We introduce the drift $g(X(s), v^1(s), v^2(s, v^1(s)))$ and the process $w^{v(.)}(s)$ defined by

$$dw_{xt}^{v(.)} = dw - \sigma^{-1}(X_{xt}(s))g(X_{xt}(s), v^1(s), v^2(s, v^1, (s)))ds \tag{17.6.2}$$

with $w_{xt}^{v(.)}(t) = w(t)$ to fix ideas. The process $X_{xt}(s)$ can be viewed as the solution of

$$dX = g(X(s), v(s))ds + \sigma(X(s))dw^{v(.)}. \tag{17.6.3}$$

We next define the martingale $M_{xt}^{v(.)}(s)$ by

$$\begin{aligned} dM_{xt}^{v(.)}(s) &= M_{xt}^{v(.)}(s)\sigma^{-1}(X_{xt}(s))g(X_{xt}(s), v(s))dw(s), \\ M_{xt}^{v(.)}(t) &= 1. \end{aligned} \tag{17.6.4}$$

We define the payoffs of players $i = 1, 2$ by

$$\Phi^{i,v(.)}(x, t) = E[M_{xt}^{v(.)}(T)(h^i(X_{xt}(T)) + \int_t^T l^i(X_{xt}(s), v(s))ds|\mathcal{F}^t], \tag{17.6.5}$$

and defining the new probability $P_{xt}^{v(.)}$ on $\Omega, \mathcal{A}$ by

$$\frac{dP_{xt}^{v(.)}}{dP}|\mathcal{F}^s = M_{xt}^{v(.)}(s),$$

we have also

$$\Phi^{i,v(.)}(x, t) = E^{v(.)}[h^i(X_{xt}(T)) + \int_t^T l^i(X_{xt}(s), v(s))ds|\mathcal{F}^t]. \tag{17.6.6}$$

An open-loop Stackelberg equilibrium $u^1(s), u^2(s, v^1)$ satisfies

$$\begin{aligned} \Phi^{1,u^1(.),u^2(.,.)}(x, t) &\leq \Phi^{1,v^1(.),u^2(.,.)}(x, t), \ \forall v^1(.), \\ \Phi^{2,u^1(.),u^2(.,.)}(x, t) &\leq \Phi^{2,u^1(.),v^2(.,.)}(x, t), \ \forall v^2(., .). \end{aligned} \tag{17.6.7}$$

17.6.2 Stackelberg Equilibrium

Consider the Lagrangians

$$L^i(x, v^1, v^2, q^i) = l^i(x, v^1, v^2) + q^i.g(x, v^1, v^2).$$

We first define the reaction function $\overline{v^2}(x, q^2, v^1)$ of the follower, which minimizes $L^2(x, v^1, v^2, q^2)$ in v^2. We next consider $L^1(x, v^1, \overline{v^2}(x, q^2, v^1), q^1)$, which we minimize in v^1. We denote by $\hat{v}^1(x, q^1, q^2)$ the point of minimum of $L^1(x, v^1, \overline{v^2}(x, q^2, v^1), q^1)$. We set

$$\begin{aligned} H^1(x, q^1, q^2) &= \inf_{v^1} L^1(x, v^1, \overline{v^2}(x, q^2, v^1), q^1) = L^1(x, \hat{v}^1(x, q^1, q^2), \\ &\quad \overline{v^2}(x, q^2, \hat{v}^1(x, q^1, q^2)), q^1), \qquad (17.6.8) \\ H^2(x, q^1, q^2) &= L^2(x, \hat{v}^1(x, q^1, q^2), \overline{v^2}(x, q^2, \hat{v}^1(x, q^1, q^2)), q^2). \end{aligned}$$

Let $u(.) = (u^1(.), u^2(., .))$ be an open-loop Stackelberg equilibrium for the functionals $\Phi^{i,v(.)}(x, t)$. We begin with player 2, the follower. He chooses a strategy $u^2(s, v^1)$ to minimize $\Phi^{2,u^1(.),v^2(.,.)}(x, t)$. In fact, what matters is only the value of $v^2(s, u^1(s))$. So the follower has to solve an open-loop non-Markov stochastic control problem, as in Chapter 14. We set

$$\begin{aligned} H^2(x, u^1(s), q^2) &= \inf_{v^2} L^2(x, u^1(s), v^2, q^2) \\ &= L^2(x, u^1(s), \overline{v^2}(x, q^2, u^1(s)) \end{aligned}$$

and consider $\Phi^2(x, t)$, the solution of the BSPDE

$$-d_t\Phi^2 + A\Phi^2 dt = [H^2(x, u^1(t), D\Phi^2 + (\sigma^{-1}(x))^* K^2) + \text{tr}\,(DK^2\sigma)]dt - K^2 dw,$$

$$\Phi^2(x, T) = h^2(x). \qquad (17.6.9)$$

We then define $\Phi^2_{xt}(s) = \Phi^2(X_{xt}(s), s)$ and $Z^2_{xt}(s) = (K^2 + \sigma^* D\Phi^2)(X_{xt}(s), s)$, and we can write

$$\begin{aligned} d\Phi^2_{xt}(s) &= d_s\Phi^2(X_{xt}(s), s) + D_x\Phi^2(X_{xt}(s), s)\sigma(X_{xt}(s))dw(s) \\ &\quad - A\Phi^2(X_{xt}(s), s)ds + \text{tr}(DK^2\sigma(X_{xt}(s), s))ds, \end{aligned}$$

and using (17.6.9), we obtain

$$d\Phi^2_{xt}(s) = -H^2(X_{xt}(s), u^1(s), (\sigma^{-1}(X_{xt}(s)))^* Z^2_{xt}(s))ds + Z^2_{xt}(s)dw(s).$$

We have $\Phi^2_{xt}(t) = \Phi^{2,u^1(.),u^2(.,.)}(x,t)$. Also,

$$u^2(s,u^1(s)) = \overline{v2}(X_{xt}(s),(\sigma^{-1}(X_{xt}(s)))^* Z^2_{xt}(s),u^1(s)). \tag{17.6.10}$$

Note that the process $Z^2_{xt}(s)$ depends on $u^1(.)$.

We turn to the leader. She can assume that the follower uses a strategy $u^2(s,v^1)$. So she needs also to solve an open-loop non-Markov stochastic control problem. We set

$$H^1(x,u^2(s,.),q^1) = \inf_{v^1} L^1(x,v^1,u^2(s,v^1),q^1)$$

and consider the BSPDE

$$-d_t\Phi^1 + A\Phi^1 dt = [H^1(x,u^2(t,.)),D\Phi^1 + (\sigma^{-1}(x))^* K^1) + \text{tr}\,(DK^1\sigma)]dt - K^1 dw,$$

$$\Phi^1(x,T) = h^1(x). \tag{17.6.11}$$

We then define $\Phi^1_{xt}(s) = \Phi^1(X_{xt}(s),s)$ and $Z^1_{xt}(s) = (K^1 + \sigma^* D\Phi^1)(X_{xt}(s),s)$. We can write

$$\begin{aligned} d\Phi^1_{xt}(s) &= d_s\Phi^1(X_{xt}(s),s) + D_x\Phi^1(X_{xt}(s),s)\sigma(X_{xt}(s))dw(s) \\ &\quad -A\Phi^1(X_{xt}(s),s)ds + \text{tr}(DK^1\sigma(X_{xt}(s),s))ds, \end{aligned}$$

and using (17.6.11), we obtain

$$d\Phi^1_{xt}(s) = -H^1(X_{xt}(s),u^2(s,.),(\sigma^{-1}(X_{xt}(s)))^* Z^1_{xt}(s))ds + Z^1_{xt}(s)dw(s).$$

Again $Z^1_{xt}(s)$ depends on $u^2(.,.)$. Also, $\Phi^1_{xt}(t) = \Phi^{1,u^1(.),u^2(.,.)}(x,t)$.

Therefore, $u^1(s)$ is defined by

$$\begin{aligned} H^1(X_{xt}(s),u^2(s,.),(\sigma^{-1}(X_{xt}(s)))^* Z^1_{xt}(s)) &= L^1(X_{xt}(s),u^1(s),u^2(s,u^1(s)), \\ &\qquad (\sigma^{-1}(X_{xt}(s)))^* Z^1_{xt}(s)), \end{aligned}$$

and from (17.6.10), we have

$$\begin{aligned} &= L^1(X_{xt}(s),u^1(s),\overline{v2}(X_{xt}(s),(\sigma^{-1}(X_{xt}(s)))^* Z^2_{xt}(s),u^1(s)),(\sigma^{-1}(X_{xt}(s)))^* Z^1_{xt}(s)) = \\ &= \inf_{v^1} L^1(X_{xt}(s),v^1,\overline{v2}(X_{xt}(s),(\sigma^{-1}(X_{xt}(s)))^* Z^2_{xt}(s),v^1),(\sigma^{-1}(X_{xt}(s)))^* Z^1_{xt}(s)) = \\ &= H^1(X_{xt}(s),(\sigma^{-1}(X_{xt}(s)))^* Z^1_{xt}(s),(\sigma^{-1}(X_{xt}(s)))^* Z^2_{xt}(s)). \end{aligned}$$

Therefore,

$$u^1(s) = \hat{v}^1(X_{xt}(s), (\sigma^{-1}(X_{xt}(s)))^* Z^1_{xt}(s), (\sigma^{-1}(X_{xt}(s)))^* Z^2_{xt}(s)). \qquad (17.6.12)$$

Moreover,

$$H^2(X_{xt}(s), u^1(s), (\sigma^{-1}(X_{xt}(s)))^* Z^2_{xt}(s)) = H^2(X_{xt}(s), (\sigma^{-1}(X_{xt}(s)))^* Z^1_{xt}(s), (\sigma^{-1}(X_{xt}(s)))^* Z^2_{xt}(s)).$$

So the pair $\Phi^1_{xt}(s)$, $Z^1_{xt}(s)$ and $\Phi^2_{xt}(s)$, $Z^2_{xt}(s)$ is the solution of the system

$$\begin{aligned}
&\Phi^1_{xt}(s) = h^1(X_{xt}(T)) + \int_s^T H^1(X_{xt}(\tau), (\sigma^{-1}(X_{xt}(\tau)))^* Z^1_{xt}(\tau),\\
&(\sigma^{-1}(X_{xt}(\tau)))^* Z^2_{xt}(\tau))d\tau - \int_s^T Z^1_{xt}(\tau)dw(\tau),\\
&\Phi^2_{xt}(s) = h^2(X_{xt}(T)) + \int_s^T H^2(X_{xt}(\tau), (\sigma^{-1}(X_{xt}(\tau)))^* Z^1_{xt}(\tau),\\
&(\sigma^{-1}(X_{xt}(\tau)))^* Z^2_{xt}(\tau))d\tau - \int_s^T Z^2_{xt}(\tau)dw(\tau)\\
&\forall t \le s \le T, \qquad (17.6.13)
\end{aligned}$$

with the Hamiltonians defined by (17.6.8). We have a situation similar to that of Nash point equilibrium, except of course that the Hamiltonians are defined through a Stackelberg equilibrium at the level of the Lagrangians instead of a Nash equilibrium at the level of the Lagrangians. We can state that

$$u^2(s, u^1(s)) = \overline{v^2}(X_{xt}(s), (\sigma^{-1}(X_{xt}(s)))^* Z^2_{xt}(s), u^1(s)). \qquad (17.6.14)$$

In the definition of $u^2(s, v^1)$ it suffices to know the value of the argument $v^1 = u^1(s)$, which is the control that the leader will use. It is naturally possible to define $u^2(s, v^1)$ for any value of the argument by solving a BSPDE

$$\Phi^2_{xt}(s) = h^2(X_{xt}(T)) + \int_s^T H^2(X_{xt}(\tau), v^1(s), (\sigma^{-1}(X_{xt}(\tau)))^* Z^2_{xt}(\tau))d\tau - \int_s^T Z^2_{xt}(\tau)dw(\tau),$$

and

$$u^2(s, v^1(s)) = \overline{v^2}(X_{xt}(s), (\sigma^{-1}(X_{xt}(s)))^* Z^2_{xt}(s), v^1(s)).$$

Note that as already mentioned, $Z^2_{xt}(s)$ depends on $v^1(.)$.

17.7 Closed-Loop Memoryless Strategies for Stochastic Stackelberg Games

The memoryless strategies discussed in the deterministic case, see Section 17.3, can be extended to the stochastic case. It has been done by A. Bensoussan, S. Chen, and S.P. Sethi [7]. We just state how to extend the deterministic case; see Section 17.3. The leader faces the following problem:

$$dx = g(x, v^1(x), \overline{v^2}(x, q^2, v^1(x)))dt + \sigma(x)dw, \tag{17.7.1}$$

$$\begin{aligned}-dq^2 = [&D_x H^2(x, v^1(x), \overline{v^2}(x, q^2, v^1(x)), q^2) + (D_x v^1)^* D_{v^1} H^2(x, v^1(x),\\ &\overline{v^2}(x, q^2, v^1(x)), q^2)]dt\\ &+\sum_{j=1}^{k}(D_x\sigma^j(x))^* r^{2j}dt - \sum_{j=1}^{k} r^{2j}dw^j,\\ &x(0) = x_0,\ q^2(T) = D_x h^2(x(T)).\end{aligned}$$

We assume that x_0 is deterministic, and we denote by $\mathcal{F}^t$ the filtration generated by the Wiener process $w(t)$. The control is the feedback $v^1(x, t)$, which is an adapted random field, and the state is the pair $x(t), q^2(t)$, the solution of a forward backward stochastic differential equation (FBSDE). For the solvability of the system (17.7.1) we refer to [36, 41, 43, 54]. The payoff is given by

$$J^1(v^1(.)) = E[\int_0^T l^1(x, v^1(x), \overline{v^2}(x, q^2, v^1(x)))dt + h^1(x(T))]. \tag{17.7.2}$$

As in the deterministic case, one can check that one can reduce the stochastic control problem (17.7.1), (17.7.2) to a problem with a linear feedback $v^1(t)+\tilde{v}^1(t)x$, where $v^1(t)$ and $\tilde{v}^1(t)$ are adapted processes. The new problem is then stated as follows:

$$dx = g(x, v^1(t) + \tilde{v}^1(t)x, \overline{v^2}(x, q^2, v^1(t) + \tilde{v}^1(t)x))dt + \sigma(x)dw, \tag{17.7.3}$$

$$\begin{aligned}-dq^2 = [\ &D_x H^2(x, v^1(t) + \tilde{v}^1(t)x, \overline{v^2}(x, q^2, v^1(t) + \tilde{v}^1(t)x), q^2)+\\ &+(\tilde{v}^1(t))^* D_{v^1} H^2(x, v^1(t) + \tilde{v}^1(t)x, \overline{v^2}(x, q^2, v^1(t) + \tilde{v}^1(t)x), q^2)\\ &+\sum_{j=1}^{k}(D_x\sigma^j(x))^* r^{2j}dt] - \sum_{j=1}^{k} r^{2j}dw^j,\\ &x(0) = x_0,\ q^2(T) = D_x h^2(x(T)),\end{aligned}$$

$$J^1(v^1(.),\tilde{v}^1(.))=E[\int_0^T l^1(x,v^1(t)+\tilde{v}^1(t)x,\overline{v^2}(x,q^2,v^1(t)+\tilde{v}^1(t)x))dt+h^i(x(T))]. \tag{17.7.4}$$

We can then write the stochastic maximum principle for the problem (17.7.3), (17.7.4):

$$dy=g(y,u^1+\tilde{u}^1y,\overline{v^2}(x,p^2,u^1(t)+\tilde{u}^1(t)y))dt+\sigma(y)dw, \tag{17.7.5}$$

$$-dp^1=\Big[D_x\Big(H^1(y,u^1+\tilde{u}^1y,\overline{v^2}(y,p^2,u^1+\tilde{u}^1y),p^1)+\pi.D_xH^2 (y,u^1+\tilde{u}^1y,\overline{v^2}(y,p^2,u^1+\tilde{u}^1y),p^2)\Big)+$$

$$+D_x\left(D_{v^1}H^2(y,u^1+\tilde{u}^1y,\overline{v^2}(y,p^2,u^1+\tilde{u}^1y),p^2).\tilde{u}^1\pi+\pi.\sum_{j=1}^k(D_x\sigma^j(x))^*r^{2j}\right)\Bigg]dt+$$

$$+\sum_{j=1}^k(D_x\sigma^j(y))^*r^{1j}dt-\sum_{j=1}^k r^{1j}dw^j,$$

$$-dp^2=[\ D_xH^2(y,u^1+\tilde{u}^1y,\overline{v^2}(y,p^2,u^1+\tilde{u}^1y),p^2)+$$

$$+(\tilde{u}^1)^*D_{v^1}H^2(y,u^1+\tilde{u}^1y,\overline{v^2}(y,p^2,u^1+\tilde{u}^1y,p^2)$$

$$+\sum_{j=1}^k(D_x\sigma^j(y))^*r^{2j}dt]-\sum_{j=1}^k r^{2j}dw^j,$$

$$p^1(T)=D_xh^1(y(T))+D_x^2h^2(y(T))\pi(T),\ \ p^2(T)=D_xh^2(y(T)),$$

$$d\pi\ =[D_{q^2}\Big(H^1(y,u^1+\tilde{u}^1y,\overline{v^2}(y,p^2,u^1+\tilde{u}^1y),p^1) +\pi.D_xH^2(y,u^1+\tilde{u}^1y,\overline{v^2}(y,p^2,u^1+\tilde{u}^1y),p^2)\Big)+$$

$$+D_{q^2}\Big(D_{v^1}H^2(y,u^1+\tilde{u}^1y,\overline{v^2}(y,p^2,u^1+\tilde{u}^1y),p^2).\tilde{u}^1\pi\Big)]dt$$

$$+\sum_{j=1}^k D_x\sigma^j(y)\pi dw^j,$$

$$\pi(0)=0,$$

$$
\begin{aligned}
&H^1(y, u^1 + \tilde{u}^1 y, \overline{v^2}(y, p^2, u^1 + \tilde{u}^1 y), p^1) \\
&+\pi.D_x H^2(y, u^1 + \tilde{u}^1 y, \overline{v^2}(y, p^2, u^1 + \tilde{u}^1 y), p^2) + \\
&+D_{v^1} H^2(y, u^1 + \tilde{u}^1 y, \overline{v^2}(y, p^2, u^1 + \tilde{u}^1 y)p^2).\tilde{u}^1 \pi = \\
&\inf_{v^1, \tilde{v}^1} \Big(H^1(y, v^1 + \tilde{v}^1 y, \overline{v^2}(y, p^2, v^1 + \tilde{v}^1 y), p^1) \\
&+\pi.D_x H^2(y, v^1 + \tilde{v}^1 y, \overline{v^2}(y, p^2, v^1 + \tilde{v}^1 y), p^2) \\
&+ D_{v^1} H^2(y, v^1 + \tilde{v}^1 y, \overline{v^2}(y, p^2, v^1 + \tilde{v}^1 y), p^2).\tilde{v}^1 \pi \Big) .
\end{aligned}
$$

Chapter 18
Target Problems

18.1 Superreplication of Contingent Claims

18.1.1 Setting of the Problem

We follow N. Touzi [49]. We consider a financial market characterized by n assets, whose prices $X_i(s)$ follow the system of equations

$$dX_i = X_i(s) \sum_{j=1}^{n} \sigma_{ij} dw_j(s),$$
$$X_i(t) = x_i.$$

We assume that

$$\sigma = (\sigma_{ij}) \text{ is invertible.} \tag{18.1.1}$$

We denote by $X_{xt}(s)$ the vector of prices, with value x at time t. A portfolio is a process $\pi(s)$ with values in R^n, which is adapted to the filtration $\mathcal{F}^s$ generated by the Wiener process $w(s)$. The wealth process associated with an initial endowment y at time t and a portfolio $\pi(s)$ is given by

$$dY(s) = \pi(s).\sigma dw(s), \tag{18.1.2}$$
$$Y(t) = y.$$

We denote the wealth process by $Y_{yt}(s; \pi(.))$. In this setup, we do not have a nonrisky asset. We next consider a contingent claim defined by $g(X_{xt}(T))$, where g is a continuously differentiable function from R^n to R, with bounded gradient. We assume

A. Bensoussan, *Estimation and Control of Dynamical Systems*, Interdisciplinary Applied Mathematics 48, https://doi.org/10.1007/978-3-319-75456-7_18

$$0 \leq g(x) \leq C(1 + |x|). \tag{18.1.3}$$

A wealth process superreplicates the contingent claim $g(X_{xt}(T))$ if one has

$$Y_{yt}(T; \pi(.)) \geq g(X_{xt}(T)), \text{ a.s.} \tag{18.1.4}$$

We define the value function $V(x, t)$ by

$$V(x, t) = \inf\{y \geq 0 | \exists \pi(.), \text{ such that } Y_{yt}(T; \pi(.)) \geq g(X_{xt}(T)), \text{ a.s.}\} \tag{18.1.5}$$

We can replace y in this definition by any random variable that is $\mathcal{F}^t$-measurable. If there are no constraints on the portfolio, the function $V(x, t)$ is simply

$$V(x, t) = Eg(X_{xt}(T)). \tag{18.1.6}$$

We shall use the following notation in the sequel. If ξ, η are two vectors in R^n, we define the vector $\xi * \eta$ in R^n by its coordinates $(\xi * \eta)_i = \xi_i \eta_i$. We claim that the optimal portfolio corresponding to the initial endowment $V(x, t)$ is

$$\hat{\pi}_{xt}(s) = DV(X_{xt}(s), s) * X_{xt}(s). \tag{18.1.7}$$

Indeed, if there exists a wealth process $Y_{yt}(s; \pi(.))$ that superreplicates $g(X_{xt}(T))$, then necessarily $y \geq Eg(X_{xt}(T))$. So $V(x, t) \geq Eg(X_{xt}(T))$. On the other hand, if $W(x, t) = Eg(X_{xt}(T))$ and $\hat{\pi}_{xt}(s) = DW(X_{xt}(s), s) * X_{xt}(s)$, then

$$g(X_{xt}(T)) = Y_{W(x,t),t}(T; \hat{\pi}_{xt}(.)),$$

so the wealth process $Y_{W(x,t),t}(s; \hat{\pi}_{xt}(.))$ replicates the contingent claim. Therefore, $W(x, t) = Eg(X_{xt}(T)) \geq V(x, t)$, whence the equality (18.1.6). We next impose constraints on the portfolio.

18.1.2 Constraints

Let U be a closed convex set of R^n such that int $(U) \neq \emptyset$ and $0 \in U$. We define the support function of the set U by

$$\delta_U(\xi) = \sup_{x \in U} x.\xi.$$

Since $0 \in U$, we have $\delta_U(\xi) \geq 0$. So the effective domain of the function δ_U is

$$\tilde{U} := \text{dom } (\delta_U) = \{\xi | \, \delta_U(\xi) < +\infty\}.$$

The supremum is attained for every point $\xi \in \tilde{U}$. We denote by $\hat{x}(\xi) \in U$ the point at which the supremum is attained.

Moreover, following Rockafellar [47], we have the characterization

$$x \in U \Leftrightarrow \delta_U(\xi) - x.\xi \geq 0,\ \forall \xi \in \tilde{U}. \qquad (18.1.8)$$

Setting $\tilde{U}_1 = \{\xi \in \tilde{U} \mid |\xi| = 1\}$, we have also

$$x \in U \Leftrightarrow \delta_U(\xi) - x.\xi \geq 0,\ \forall \xi \in \tilde{U}_1. \qquad (18.1.9)$$

Finally,

$$x \in \text{int}(U) \Leftrightarrow \inf_{\xi \in \tilde{U}_1} (\delta_U(\xi) - x.\xi) > 0. \qquad (18.1.10)$$

We now consider the target problem (18.1.5) with constraints on the portfolio $\pi(s) \in U, \forall s$. We define

$$V(x,t) = \inf\{y \geq 0 | \exists \pi(.),\ \text{such that},\ \pi(s) \in U, \forall s;\ Y_{yt}(T;\pi(.)) \geq g(X_{xt}(T)),\ \text{a.s.}\}. \qquad (18.1.11)$$

18.1.3 Main Result for Constraints

We introduce the function

$$\hat{g}(x) = \sup_{\xi \in \tilde{U}} [g(x * \exp \xi) - \delta_U(\xi)], \qquad (18.1.12)$$

where the vector $\exp \xi$ is defined by its coordinates $\exp \xi_i,\ i = 1, \cdots, n$. We note that $\hat{g}(x) \geq g(x)$. We assume that

$$\text{the supremum is attained at } \hat{\xi}(x) \in \tilde{U}. \qquad (18.1.13)$$

It follows that

$$D\hat{g}(x) * x = Dg(x * \exp \hat{\xi}) * (x * \exp \hat{\xi}) = \hat{x}(\hat{\xi}) \in U, \forall x. \qquad (18.1.14)$$

From the characterization (18.1.8) we can assert

$$\delta_U(\xi) - \xi.(D\hat{g}(x) * x) \geq 0, \forall \xi \in \tilde{U}, \forall x, \qquad (18.1.15)$$

which is equivalent to saying that

$$H(\lambda; x, \xi) = \lambda \delta_U(\xi) - \hat{g}(x * \exp \lambda \xi) \qquad (18.1.16)$$

is nondecreasing in $\lambda, \forall x, \xi \in \tilde{U}$. Consider next the function

$$\hat{V}(x,t) = E\hat{g}(X_{xt}(T)), \tag{18.1.17}$$

which is $C^{2,1}(R^n \times (0,T))$. We observe that

$$\begin{aligned}\lambda\delta_U(\xi) - \hat{V}(x * \exp\lambda\xi, t) &= E[\lambda\delta_U(\xi) - \hat{g}(X_{x*\exp\lambda\xi,t}(T)] = \\ &= E[\lambda\delta_U(\xi) - \hat{g}(X_{xt}(T) * \exp\lambda\xi)],\end{aligned}$$

which is thus nondecreasing in λ. By reasoning the reverse to that done for $H(\lambda; x, \xi)$, we can assert that

$$D\hat{V}(x,t) * x \in U, \forall x. \tag{18.1.18}$$

Using the PDE for $\hat{V}(x,t)$ and Itô's formula, we can then write

$$\hat{g}(X_{xt}(T)) = \hat{V}(x,t) + \int_t^T \hat{\pi}_{xt}(s).\sigma dw,$$

with

$$\hat{\pi}_{xt}(s) = D\hat{V}(X_{xt}(s), s) * X_{xt}(s) \in U.$$

Since $\hat{g}(x) \geq g(x)$, we obtain immediately that the wealth associated with $\hat{V}(x,t)$ and $\hat{\pi}_{xt}(s)$ superreplicates the contingent claim $g(X_{xt}(T))$. Therefore, we have obtained

$$\hat{V}(x,t) \geq V(x,t). \tag{18.1.19}$$

We then state the following.

PROPOSITION 18.1. *Assume (18.1.3), (18.1.13) and that the value function $V(x,t)$ defined by (18.1.11) is $C^{2,1}(R^n \times (0,T))$. Then*

$$V(x,t) = \hat{V}(x,t). \tag{18.1.20}$$

PROOF. We begin with an optimality principle. We need first to consider random initial conditions. Let θ be an $\mathcal{F}^s$ stopping time. We recall that $\mathcal{F}^\theta$ is defined by

$$A \in \mathcal{F}^\theta \Rightarrow A \cap \{\theta \leq s\} \in \mathcal{F}^s, \forall s.$$

Let $\xi \in R^n$, which is $\mathcal{F}^\theta$-measurable. We assume that $\theta \leq T$ a.s. We consider wealth processes $Y_{\eta\theta}(s; \pi(.))$, where η is also $\mathcal{F}^\theta$-measurable. So the wealth satisfies

$$dY_{\eta\theta}(s) = \pi(s).\sigma dw(s), \; s > \theta,$$
$$Y_{\eta\theta}(\theta) = \eta.$$

The function $V(x,t)$ defined by (18.1.5) satisfies also

$$V(\xi,\theta) = \inf\{\eta\, \mathcal{F}^\theta \text{ measurable}|\exists\pi(.) \text{ adapted to } \mathcal{F}^s, \text{ with values in } U,$$
$$Y_{\eta\theta}(T;\pi(.)) \geq g(X_{\xi\theta}(T)), \text{ a.s.}\}, \text{ a.s.} \tag{18.1.21}$$

This is because the variables $Y_{\eta\theta}(T;\pi(.))$ and $X_{\xi\theta}(T)$ are simply the variables $Y_{yt}(T;\pi(.))$ and $X_{xt}(T)$ in which x is substituted for ξ, y for η.

Let x, t be given. For every $\epsilon > 0$, there exists an admissible portfolio $\pi^\epsilon_{xt}(.)$ such that $\pi^\epsilon_{xt}(s) \in U$ and

$$V(x,t) + \epsilon + \int_t^T \pi^\epsilon_{xt}(s).\sigma dw(s) \geq g(X_{xt}(T)). \tag{18.1.22}$$

Let θ be an $\mathcal{F}^s$ stopping time, with $t \leq \theta \leq T$. Then

$$V(x,t) + \epsilon + \int_t^\theta \pi^\epsilon_{xt}(s).\sigma dw(s) + \int_\theta^T \pi^\epsilon_{xt}(s).\sigma dw(s)$$
$$\geq g(X_{xt}(T)) \text{ a.s.} = g(X_{X_{xt}(\theta),\theta}(T)). \tag{18.1.23}$$

Therefore, we can assert from (18.1.21) that

$$V(x,t) + \epsilon + \int_t^\theta \pi^\epsilon_{xt}(s).\sigma dw(s) \geq V(X_{xt}(\theta),\theta). \tag{18.1.24}$$

Since V is smooth when $t < T$, we can use Itô's formula to obtain, assuming $\theta < T$,

$$\epsilon + \int_t^\theta \pi^\epsilon_{xt}(s).\sigma dw(s) \geq \int_t^\theta DV(X_{xt}(s),s) * X_{xt}(s).\sigma dw +$$
$$+\frac{1}{2}\int_t^\theta \sum_{ijk} \frac{\partial^2 V}{\partial x_i \partial x_j}(X_{xt}(s),s) X_{xt,i}(s) X_{xt,j}(s)\sigma_{ik}\sigma_{jk} ds.$$

Since this holds a.s., it implies

$$\pi^\epsilon_{xt}(s) = DV(X_{xt}(s),s) * X_{xt}(s), \; t < s < \theta,$$

so $\pi^\epsilon_{xt}(s)$ is fixed and $DV(X_{xt}(s),s) * X_{xt}(s) \in U$. By letting s tend to t, we obtain

$$DV(x,t) * x \in U, \forall x, t < T. \tag{18.1.25}$$

Also,

$$\sum_{ijk} \frac{\partial^2 V}{\partial x_i \partial x_j}(x,t)\sigma_{ik}\sigma_{jk} \leq 0. \tag{18.1.26}$$

But then from (18.1.25), we can assert that

$$K(\lambda; x, \xi) = \lambda \delta_U(\xi) - V(x * \exp \lambda\xi, t)$$

is nondecreasing for all $t < T$. From $K(1; x, \xi) \geq K(0, x, \xi)$ we obtain

$$V(x,t) \geq V(x * \exp \xi, t) - \delta_U(\xi), \tag{18.1.27}$$

which implies

$$V(x,t) = \sup_{\xi \in \tilde{U}} (V(x * \exp \xi, t) - \delta_U(\xi)), \ \forall t < T.$$

We cannot take $t = T$ in this relation. However, from (18.1.27), using the continuity of $V(x,t)$, we write

$$\begin{aligned} V(x,T) &\geq V(x * \exp \xi, T) - \delta_U(\xi) \\ &\geq g(x * \exp \xi) - \delta_U(\xi). \end{aligned}$$

Therefore, $V(x,T) \geq \hat{g}(x)$. But using (18.1.26), we can write

$$\begin{aligned} -\frac{\partial(V - \hat{V})}{\partial t} - \sum_{ijk} \frac{\partial^2 (V - \hat{V})}{\partial x_i \partial x_j}(x,t)\sigma_{ik}\sigma_{jk} &\geq 0, \\ V(x,T) - \hat{V}(x,T) &\geq 0, \end{aligned}$$

which implies $V(x,t) - \hat{V}(x,t) \geq 0$. From (18.1.19), we conclude the property (18.1.20), which completes the proof. □

18.2 Superreplication in Probability

18.2.1 Statement of the Problem

Following Touzi [49], we now consider the same problem, asking for superreplication in probability only. Our approach is different from Touzi's. In particular, we manage to avoid the use of viscosity theory, and proceed rather as in Chapter 13. We begin as in the previous section. The financial market is described by

$$dX_i = X_i(s) \sum_{j=1}^{n} \sigma_{ij} dw_j(s), \tag{18.2.1}$$

$$X_i(t) = x_i,$$

and we use the notation $X_{xt}(s)$ for the vector of prices, with value x at time t. A portfolio is a process $\pi(s)$ with values in R^n, which is adapted to the filtration $\mathcal{F}_t^s$ generated by the Wiener process $w(s)-w(t)$. The wealth process associated with an initial endowment y at time t and a portfolio $\pi(s)$, using the standard self-financing assumption, is given by

$$dY(s) = \pi(s).\sigma dw(s), \tag{18.2.2}$$

$$Y(t) = y.$$

We consider next a contingent claim $g(X_{xt}(T))$ at time T. We are again interested in superreplicating $g(X_{xt}(T))$. We do not introduce constraints on the portfolio, so if we set

$$V(x,t) = \inf\{y \geq 0 | \exists \pi(.) \text{ such that } Y_{yt}(T; \pi(.)) \geq g(X_{xt}(T)), \text{ a.s.}\}, \tag{18.2.3}$$

we know that $V(x,t) = Eg(X_{xt}(T))$. In fact, recalling the notation $a * b$ for the vector of components $a_i b_i$, in which a_i and b_i are the components of vectors a and b, we have

$$g(X_{xt}(T)) = V(x,t) + \int_t^T D_x V(X_{xt}(s)) * X_{xt}(s).\sigma dw(s),$$

and we can exactly replicate the contingent claim. We now change the superreplication condition as follows: Introduce a number $p \in [0, 1]$ and set

$$V(x,p,t) = \inf\{y \geq 0 | \exists \pi(.) \text{ such that } P(Y_{yt}(T; \pi(.)) \geq g(X_{xt}(T))) \geq p\}. \tag{18.2.4}$$

We clearly have $V(x,1,t) = V(x,t)$ and $V(x,0,t) = 0$. Next, for all $q(.)$ adapted to $\mathcal{F}^s$, we introduce a new process $P_{pt}(s; q(.))$ such that

$$dP_{pt}(s) = q(s).dw(s), \tag{18.2.5}$$

$$P_{pt}(t) = p,$$

where $q(s)$ is adapted to $\mathcal{F}^s$. We set

$$v(x,p,t) = \inf\{y \geq 0 | \exists \pi(.), \; q(.) \text{ such that} \tag{18.2.6}$$

$$\mathbb{1}_{\{Y_{yt}(T;\pi(.)) \geq g(X_{xt}(T))\}} \geq P_{pt}(T; q(.))\}.$$

We claim that the following holds.

LEMMA 18.1. *We have* $V(x, p, t) = v(x, p, t)$.

PROOF. We first show that $V(x, p, t) \geq v(x, p, t)$. For $y > V(x, p, t)$, $\exists \pi(.)$ such that

$$p_y = P(Y_{yt}(T; \pi(.)) \geq g(X_{xt}(T))) \geq p.$$

By the martingale representation theorem, we can find $q(.)$ adapted such that

$$\begin{aligned}\mathbb{1}_{\{Y_{yt}(T;\pi(.))\geq g(X_{xt}(T))\}} &= p_y + \int_t^T q(s).dw(s) \\ &\geq p + \int_t^T q(s).dw(s) = P_{pt}(T; q(.)),\end{aligned}$$

with $P_{pt}(s)$ defined by (18.2.5). This implies $y \geq v(x, p, t)$, and therefore, since y is arbitrary with $y > V(x, p, t)$, we have $V(x, p, t) \geq v(x, p, t)$. Let us prove the opposite inequality. Indeed, for $y > v(x, p, t)$, there exists a pair $\pi(.), q(.)$ such that on defining $Y_{yt}(s)$, $P_{pt}(s)$ by (18.2.2), (18.2.5), we have

$$\mathbb{1}_{\{Y_{yt}(T;\pi(.))\geq g(X_{xt}(T))\}} \geq P_{pt}(T; q(.)), \text{ a.s.,}$$

which implies

$$P\{Y_{yt}(T; \pi(.)) \geq g(X_{xt}(T))\} \geq p$$

and necessarily $y \geq V(x, p, t)$. Again, since y is arbitrarily greater than $v(x, p, t)$, we obtain $v(x, p, t) \geq V(x, p, t)$. This concludes the proof. □

We shall use the characterization (18.2.6) for $V(x, p, t)$. We next state and prove a lemma.

LEMMA 18.2. *We have*

$$V(x, p, T) \geq pg(x). \tag{18.2.7}$$

PROOF. By definition and the preceding lemma,

$$V(x, p, T) = \inf\{y \geq 0 | \mathbb{1}_{\{y\geq g(x)\}} \geq p\},$$

so if $V(x, p, T) < g(x)$, then $\exists\, y < g(x)$ such that $\mathbb{1}_{\{y\geq g(x)\}} \geq p$, but this is possible only with $p = 0$. So if $V(x, p, T) < g(x)$, the inequality (18.2.7) is satisfied. But if $V(x, p, T) \geq g(x)$, it is satisfied trivially. This completes the proof. □

18.2.2 Control Problem

We next introduce a control problem. The state is the pair $X_{xt}(s)$ and $P_{pt}(s; q(.))$ with

$$dX_{xt}(s) = X_{xt}(s) * \sigma dw(s), \; X(t) = x, \tag{18.2.8}$$
$$dP_{pt}(s) = q(s).dw(s), \; P_{pt}(t) = p,$$

and $q(.)$ is the control. So $X_{xt}(s)$ is not affected by any control. The payoff is defined by

$$J_{xpt}(q(.)) = E[g(X_{xt}(T)P_{pt}(T)\mathbb{1}_{T \le \tau} + V(X_{xt}(\tau), \tau)\mathbb{1}_{\tau < T}\mathbb{1}_{P_{pt}(\tau)=1}], \tag{18.2.9}$$
$$\tau = \inf\{s \ge t |\, P_{pt}(s) \notin [0, 1]\},$$

and we define the value function

$$\hat{V}(x, p, t) = \inf_{q(.)} J_{xpt}(q(.)). \tag{18.2.10}$$

We claim the following result.

LEMMA 18.3. *We have* $V(x, p, t) \ge \hat{V}(x, p, t)$.

PROOF. By definition of $V(x, p, t)$, for every ϵ, there exist $y^{\epsilon}_{xpt}, \pi^{\epsilon}_{xpt}, q^{\epsilon}_{xpt}$ such that $V(x, p, t) \le y^{\epsilon}_{xpt} \le V(x, p, t) + \epsilon$, and setting

$$Y^{\epsilon}_{xpt}(s) = Y_{y^{\epsilon}_{xpt},t}(s; \pi^{\epsilon}_{xpt}) = y^{\epsilon}_{xpt} + \int_t^s \pi^{\epsilon}_{xpt}(r).\sigma dw,$$

$$P^{\epsilon}_{xpt}(s) = P_{pt}(s; q^{\epsilon}_{xpt}) = p + \int_t^s q^{\epsilon}_{xpt}(r).dw,$$

we have also

$$\mathbb{1}_{\{Y^{\epsilon}_{xpt}(T) \ge g(X_{xt}(T))\}} \ge P^{\epsilon}_{xpt}(T).$$

Let θ be such that $t < \theta < T$. Then we note that

$$Y^{\epsilon}_{xpt}(T) = y^{\epsilon}_{xpt} + \int_t^{\theta} \pi^{\epsilon}_{xpt}(r).\sigma dw + \int_{\theta}^{T} \pi^{\epsilon}_{xpt}(r).\sigma dw,$$

$$P^{\epsilon}_{xpt}(T) = P^{\epsilon}_{xpt}(\theta) + \int_{\theta}^{T} q^{\epsilon}_{xpt}(r).dw, \; X_{xt}(T) = X_{X_{xt}(\theta),\theta}(T).$$

Therefore, from the definition of $V(x, p, t)$, we can write

$$y^{\epsilon}_{xpt} + \int_t^{\theta} \pi^{\epsilon}_{xpt}(r).\sigma dw \geq V(X_{xt}(\theta), P^{\epsilon}_{xpt}(\theta), \theta)$$

and also

$$V(x, p, t) + \epsilon + \int_t^{\theta} \pi^{\epsilon}_{xpt}(r).\sigma dw \geq V(X_{xt}(\theta), P^{\epsilon}_{xpt}(\theta), \theta). \tag{18.2.11}$$

Let

$$\tau^{\epsilon} = \inf\{s \geq t | \; P^{\epsilon}_{xpt}(s) \notin [0, 1]\}.$$

Applying (18.2.11) with $\theta = \tau^{\epsilon}$ yields

$$V(x, p, t) + \epsilon + \int_t^{\tau^{\epsilon}} \pi^{\epsilon}_{xpt}(r).\sigma dw \geq V(X_{xt}(T), P^{\epsilon}_{xpt}(T), T)1\!\!1_{T \leq \tau^{\epsilon}} +$$
$$V(X_{xt}(\tau^{\epsilon}), 1, \tau^{\epsilon})1\!\!1_{\tau^{\epsilon} < T}1\!\!1_{P^{\epsilon}_{xpt}(\tau^{\epsilon})=1} + V(X_{xt}(\tau^{\epsilon}), 0, \tau^{\epsilon})1\!\!1_{\tau^{\epsilon} < T}1\!\!1_{P^{\epsilon}_{xpt}(\tau^{\epsilon})=0}.$$

From the definition (18.2.4) we have immediately $V(x, 1, t) = V(x, t)$ and $V(x, 0, t) = 0$. Using then (18.2.7), we get immediately

$$V(x, p, t) + \epsilon + \int_t^{\tau^{\epsilon}} \pi^{\epsilon}_{xpt}(r).\sigma dw \geq g(X_{xt}(T))P^{\epsilon}_{xpt}(T)1\!\!1_{T \leq \tau^{\epsilon}} +$$
$$+ V(X_{xt}(\tau^{\epsilon}), \tau^{\epsilon})1\!\!1_{\tau^{\epsilon} < T}1\!\!1_{P^{\epsilon}_{xpt}(\tau^{\epsilon})=1}.$$

Taking the expectation, it follows that

$$V(x, p, t) + \epsilon \geq E[g(X_{xt}(T))P^{\epsilon}_{xpt}(T)1\!\!1_{T \leq \tau^{\epsilon}} +$$
$$+ V(X_{xt}(\tau^{\epsilon}), \tau^{\epsilon})1\!\!1_{\tau^{\epsilon} < T}1\!\!1_{P^{\epsilon}_{xpt}(\tau^{\epsilon})=1}],$$

and from definition (18.2.9) we obtain

$$V(x, p, t) + \epsilon \geq J_{xpt}(q^{\epsilon}_{xpt}(.)) \geq \hat{V}(x, p, t),$$

and since ϵ is arbitrary, the result follows immediately. □

18.2.3 Solution of the Control Problem

Our objective is to prove the following explicit formula.

THEOREM 18.1. *Assume (18.1.3) and g continuous. We also assume that*

$$x_1 \to g(x_1, x_{-1}) \text{ is invertible}, \ \forall x_{-1} = (x_2, \cdots, x_n), \tag{18.2.12}$$

in which the first component of x can be replaced by any component. We then have the formula, for $t < T$,

$$\hat{V}(x, p, t) = Eg(X_{xt}(T))\mathbb{1}_{g(X_{xt}(T))<\lambda(x,p,t)}, \tag{18.2.13}$$

with

$$E\mathbb{1}_{g(X_{xt}(T))<\lambda(x,p,t)} = p \tag{18.2.14}$$

and $\hat{V}(x, p, T) = pg(x)$. Moreover,

$$V(x, p, t) = Eg(X_{xt}(T))\mathbb{1}_{g(X_{xt}(T))<\lambda(x,p,t)}. \tag{18.2.15}$$

PROOF. We begin with a series of comments. We know from the definition (18.2.9) that $J_{x0t}(q(.)) = 0$, $J_{x1t}(q(.)) = V(x,t)$ for $t < T$. Also, $J_{xpT}(q(.)) = pg(x)$, $\forall p \in (0,1)$. Therefore, $\hat{V}(x, 0, t) = 0$, $\hat{V}(x, 1, t) = V(x,t)$, and $\hat{V}(x, p, T) = pg(x)$. Since $V(x, T) = g(x)$, these formulas are compatible. Let us set $W(x, p, t) = Eg(X_{xt}(T))\mathbb{1}_{g(X_{xt}(T))<\lambda(x,p,t)}$, for $t < T$.

From (18.2.14) we check easily that $\mathbb{1}_{g(X_{xt}(T))<\lambda(x,0,t)} = 0$, $\mathbb{1}_{g(X_{xt}(T))<\lambda(x,1,t)} = 1$. If $g(x) > 0$ and $g(+\infty) = \infty$, we necessarily have $\lambda(x, 0, t) = 0$ and $\lambda(x, 1, t) = +\infty$. At any rate, we recover $W(x, 0, t) = 0$, $W(x, 1, t) = V(x,t)$. Moreover,

$$W(x, p, t) = g(x)p + E(g(X_{xt}(T)) - g(x))\mathbb{1}_{g(X_{xt}(T))<\lambda(x,p,t)}$$

and

$$|E(g(X_{xt}(T)) - g(x))\mathbb{1}_{g(X_{xt}(T))<\lambda(x,p,t)}| \le E|g(X_{xt}(T)) - g(x)| \to 0, \text{ as } t \to T.$$

Therefore, also $W(x, p, t) \to pg(x)$, as $t \to T$. Hence, defining $W(x, p, T) = pg(x)$, we have $W = \hat{V}$ on the boundary $p = 0, 1$ and at time T. We now set

$$G(x, \lambda, t) = E\mathbb{1}_{g(X_{xt}(T))<\lambda}. \tag{18.2.16}$$

Clearly, the function $G(x, \lambda, t)$ is monotonically increasing in λ and remains in the interval $[0, 1]$. For $t < T$, it is continuous in λ. This follows from the assumption (18.2.12) and the properties of the Wiener process. So the function $\lambda(x, p, t)$, which is its inverse, is well defined and monotonically increasing in p. The function $G(x, \lambda, t)$ is the solution of

$$\frac{\partial G}{\partial t} + \sum_{ij} a_{ij} x_i x_j \frac{\partial^2 G}{\partial x_i \partial x_j} = 0, \tag{18.2.17}$$

$$G(x, \lambda, T) = \mathbb{1}_{g(x)<\lambda},$$

with $a_{ij} = \frac{1}{2} \sum_k \sigma_{ik} \sigma_{jk}$. If we introduce the function $\Psi(x, \lambda, t)$, the solution of

$$\frac{\partial \Psi}{\partial t} + \sum_{ij} a_{ij} x_i x_j \frac{\partial^2 \Psi}{\partial x_i \partial x_j} = 0, \tag{18.2.18}$$

$$\Psi(x, \lambda, T) = \quad g(x)\mathbb{1}_{g(x)<\lambda},$$

$$\Psi(x, \lambda, t) = \quad Eg(X_{xt}(T))\mathbb{1}_{g(X_{xt}(T))<\lambda},$$

then we have, for $t < T$,

$$W(x, p, t) = \Psi(x, \lambda(x, p, t), t). \tag{18.2.19}$$

We then set

$$Y_{xpt}(T) = g(X_{xt}(T))\mathbb{1}_{g(X_{xt}(T))<\lambda(x,p,t)}.$$

From the theorem of representation of martingales, we can assert that

$$Y_{xpt}(T) = W(x, p, t) + \int_t^T \pi_{xpt}(s).\sigma dw(s)$$

and

$$P(Y_{xyt}(T; \pi_{xpt}(.)) \geq g(X_{xt}(T))) = E\mathbb{1}_{g(X_{xt}(T))<\lambda(x,p,t)} = p.$$

Therefore, from the definition (18.2.4) we can assert that $W(x, p, t) \geq V(x, p, t)$. So if we prove that

$$W(x, p, t) = \hat{V}(x, p, t), \tag{18.2.20}$$

then, recalling Lemma 18.3, we will obtain the claim (18.2.13), and also (18.2.15). This is valid only for $t < T$. We have not defined $\lambda(x, p, t)$ for $t = T$. We have seen above that the equality is true for $t = T$, independently of the behavior of $\lambda(x, p, t)$ as $t \to T$. We shall, however, prove that $\lambda(x, p, t) \to g(x)$ as $t \to T$.

To prove (18.2.20), we proceed as in Chapter 13 and show that $W(x, p, t)$ satisfies the HJB equation for the control problem (18.2.8), (18.2.9). The difficulty is the lack of smoothness of $W(x, p, t)$, because of the final condition. So we first proceed formally, assuming the smoothness of $W(x, p, t)$ for $t < T$. We shall afterward proceed with a regularization of the final condition, to recover smoothness, and use a limit argument.

The first important relation is

$$\frac{\partial \Psi}{\partial \lambda}(x, \lambda, t) = \lambda \frac{\partial G}{\partial \lambda}(x, \lambda, t). \tag{18.2.21}$$

In view of the equations satisfied by Ψ and G, see (18.2.18), (18.2.17). it is sufficient to satisfy it at T. Unfortunately, this is not possible for lack of differentiability in λ at time T. So we shall accept this relation. In the next section, we shall give a formal argument to prove it. It then follows that

$$\frac{\partial W}{\partial p}(x, p, t) = \frac{\partial \Psi}{\partial \lambda}(x, \lambda(x, p, t), t)\frac{\partial \lambda}{\partial p}(x, p, t) =$$

$$= \frac{\frac{\partial \Psi}{\partial \lambda}(x, \lambda(x, p, t), t)}{\frac{\partial G}{\partial \lambda}(x, \lambda(x, p, t), t)} = \lambda(x, p, t), \tag{18.2.22}$$

thanks to (18.2.21). We note that for $t = T$, this relation implies

$$\lambda(x, p, T) = g(x). \tag{18.2.23}$$

We next compute the derivatives of λ from the derivatives of G, using the identity $G(x, \lambda(x, p, t), t) = p$. After easy calculations and using equation (18.2.17), we obtain

$$\frac{\partial \lambda}{\partial t} + \sum_{ij} a_{ij} x_i x_j \frac{\partial^2 \lambda}{\partial x_i \partial x_j} + \frac{\frac{\partial^2 \lambda}{\partial p^2}}{(\frac{\partial \lambda}{\partial p})^2} \sum_{ij} a_{ij} x_i x_j \frac{\partial \lambda}{\partial x_i}\frac{\partial \lambda}{\partial x_j} + \tag{18.2.24}$$

$$-\frac{2}{\frac{\partial \lambda}{\partial p}} \sum_{ij} a_{ij} x_i x_j \frac{\partial \lambda}{\partial x_j}\frac{\partial^2 \lambda}{\partial x_i \partial p} = 0.$$

We next proceed for $W(x, p, t)$ using (18.2.19). Proceeding in the same way, by computing the derivatives of the function W, we obtain

$$\frac{\partial W}{\partial t} + \sum_{ij} a_{ij} x_i x_j \frac{\partial^2 W}{\partial x_i \partial x_j} = -\sum_{ij} a_{ij} x_i x_j \frac{\frac{\partial \lambda}{\partial x_i}\frac{\partial \lambda}{\partial x_j}}{(\frac{\partial \lambda}{\partial p})^2}\frac{\partial^2 W}{\partial p^2} +$$

$$+ 2\frac{\sum_{ij} a_{ij} x_i x_j \frac{\partial \lambda}{\partial x_i}\frac{\partial^2 W}{\partial x_j \partial p}}{\frac{\partial \lambda}{\partial p}}.$$

But using (18.2.22), we can write

$$\frac{\partial W}{\partial t} + \sum_{ij} a_{ij} x_i x_j \frac{\partial^2 W}{\partial x_i \partial x_j} - \sum_{ij} a_{ij} x_i x_j \frac{\frac{\partial^2 W}{\partial x_i \partial p}\frac{\partial^2 W}{\partial x_j \partial p}}{\frac{\partial^2 W}{\partial p^2}} = 0, \tag{18.2.25}$$

with the final condition $W(x, p, T) = pg(x)$. But if we consider the minimization problem

$$\inf_q \left[\frac{1}{2} \frac{\partial^2 W}{\partial p^2} |q|^2 + \sum_{ik} \frac{\partial^2 W}{\partial x_i \partial p} x_i \sigma_{ik} q_k \right],$$

then the minimum is attained at

$$\hat{q}(x, p, t) = -\frac{\sigma^*(D_x \frac{\partial W}{\partial p} * x)}{\frac{\partial^2 W}{\partial p^2}},$$

and computing the infimum, we obtain exactly the nonlinear term in equation (18.2.25). It follows that (18.2.25) coincides with the Bellman equation, corresponding to the control problem((18.2.8), (18.2.9). Therefore, $W(x, p, t) = \hat{V}(x, p, t)$, the value function defined in (18.2.10). This concludes the formal proof.

To prove the result rigorously, we use an approximation. We first define the function

$$A(\xi) = \left| \begin{array}{c} 0 \text{ if } \xi \leq 0, \\ 6\xi^5 - 15\xi^4 + 10\xi^2 \text{ if } 0 < \xi \leq 1, \\ 1 \text{ if } \xi \geq 1, \end{array} \right.$$

which is C^2. We have that $A(\frac{\xi}{\epsilon}) \uparrow \mathbb{1}_{\xi > 0}$ as $\epsilon \downarrow 0$. We next define

$$G_\epsilon(x, \lambda, t) = E\, A(\frac{\lambda - g(X_{xt}(T))}{\epsilon}), \tag{18.2.26}$$

$$\Psi_\epsilon(x, \lambda, t) = E\, g(X_{xt}(T) A(\frac{\lambda - g(X_{xt}(T))}{\epsilon})| + \tag{18.2.27}$$
$$+ \epsilon E \int_0^{\frac{\lambda - g(X_{xt}(T))}{\epsilon}} \nu A'(\nu) d\nu.$$

In particular,

$$G_\epsilon(x, \lambda, T) = A(\frac{\lambda - g(x)}{\epsilon}),$$
$$\Psi_\epsilon(x, \lambda, T) = g(x) A(\frac{\lambda - g(x)}{\epsilon}) + \epsilon \int_0^{\frac{\lambda - g(x)}{\epsilon}} \nu A'(\nu) d\nu, \tag{18.2.28}$$
$$G_\epsilon(x, 0, t) = 0, \; G_\epsilon(x, +\infty, t) = 1,$$

$$\Psi_\epsilon(x,0,t)=0,\ \ \Psi_\epsilon(x,+\infty,t)=V(x,t)+\epsilon\int_0^1\nu A'(\nu)d\nu,\qquad(18.2.29)$$

and

$$G_\epsilon(x,\lambda,t)=EG_\epsilon(X_{xt}(T),\lambda,T),\ \ \Psi_\epsilon(x,\lambda,t)=E\Psi_\epsilon(X_{xt}(T),\lambda,T).$$

The functions $G_\epsilon(x,\lambda,t),\ \Psi_\epsilon(x,\lambda,t)$ are smooth in x,t for $t<T$ and C^2 in λ. We have also

$$\begin{aligned}G_\epsilon(x,\lambda,T)\ &\uparrow\ G(x,\lambda,T),\ \text{as } \epsilon\downarrow 0,\\ \Psi_\epsilon(x,\lambda,T)\ &\to\ \Psi(x,\lambda,T),\end{aligned}\qquad(18.2.30)$$

$$\begin{aligned}G_\epsilon(x,\lambda,t)\ &\uparrow\ G(x,\lambda,t),\ \text{as } \epsilon\downarrow 0,\\ \Psi_\epsilon(x,\lambda,t)\ &\to\ \Psi(x,\lambda,t),\end{aligned}\qquad(18.2.31)$$

where $G(x,\lambda,t),\ \Psi(x,\lambda,t)$ are given by (18.2.16), (18.2.18).

We note that

$$\begin{aligned}\frac{\partial G_\epsilon}{\partial\lambda}&=\frac{1}{\epsilon}E\,A'(\frac{\lambda-g(X_{xt}(T))}{\epsilon})>0,\\ G_\epsilon(x,0,t)&=0,\\ G_\epsilon(x,+\infty,t)&=1,\end{aligned}$$

where we have used the positivity of g and the properties of the function $A(\xi)$. It follows that we can uniquely define $\lambda_\epsilon(x,p,t)$ as the inverse of $G_\epsilon(x,\lambda,t)$:

$$G_\epsilon(x,\lambda_\epsilon(x,p,t),t)-p$$

for $p\in[0,1]$. This is also true for $t=T$. We can state that

$$\lambda_\epsilon(x,p,T)=g(x)+\epsilon A^{-1}(p),\qquad(18.2.32)$$

where $A^{-1}(p)$ is the inverse function of $A(p)$. Moreover,

$$\lambda_\epsilon(x,p,t)\to\lambda(x,p,t),\ \text{as } \epsilon\to 0.\qquad(18.2.33)$$

Indeed, $\lambda_\epsilon(x,p,t)$ remains bounded as ϵ tends to 0. If such were not the case, there would exist a subsequence that tends to $+\infty$. For this subsequence, still denoted by $\lambda_\epsilon(x,p,t)$, we would have $\dfrac{\lambda_\epsilon(x,p,t)-g(X_{xt}(T))}{\epsilon}\to+\infty$ a.s. Therefore, we would obtain

$$E\,A(\frac{\lambda_\epsilon(x,p,t)-g(X_{xt}(T))}{\epsilon})\to 1,$$

which is a contradiction. If $\lambda_\epsilon(x, p, t)$ remains bounded, we can extract a subsequence that converges to $\tilde{\lambda}$. If $\tilde{\lambda} > g(X_{xt}(T))$, we have $\dfrac{\lambda_\epsilon(x, p, t) - g(X_{xt}(T))}{\epsilon} \to +\infty$, hence $A(\frac{\lambda_\epsilon(x,p,t)-g(X_{xt}(T))}{\epsilon}) \to 1$, and similarly, if $\tilde{\lambda} < g(X_{xt}(T))$, we have $A(\frac{\lambda_\epsilon(x,p,t)-g(X_{xt}(T))}{\epsilon}) \to 0$. Since the set $\tilde{\lambda} = g(X_{xt}(T))$ has probability 0, we can assert that

$$A(\frac{\lambda_\epsilon(x, p, t) - g(X_{xt}(T))}{\epsilon}) \to 1\!\!1_{\tilde{\lambda}>g(X_{xt}(T))}, \text{ a.s.}$$

It follows that $E 1\!\!1_{\tilde{\lambda}>g(X_{xt}(T))} = p$, which implies $\tilde{\lambda} = \lambda(x, p, t)$. Since the limit is unique, the assertion (18.2.33) follows. We also have $\lambda_\epsilon(x, 0, t) = 0$, $\lambda_\epsilon(x, 1, t) = +\infty$.

We next set

$$W_\epsilon(x, p, t) = \Psi_\epsilon(x, \lambda_\epsilon(x, p, t), t)$$

Recalling (18.2.27) and

$$A(\frac{\lambda_\epsilon(x, p, t) - g(X_{xt}(T))}{\epsilon}) \to 1\!\!1_{\tilde{\lambda}>g(X_{xt}(T))}, \text{ a.s., } A'(\xi) = 0, \ \forall \xi \notin (0, 1),$$

we see immediately that

$$W_\epsilon(x, p, t) \to W(x, p, t), \text{ as } \epsilon \to 0. \tag{18.2.34}$$

Using (18.2.28) and (18.2.32), we obtain easily

$$W_\epsilon(x, p, T) = pg(x) + \epsilon \int_0^{A^{-1}(p)} \nu A'(\nu) d\nu, \tag{18.2.35}$$

and thus (18.2.34) is also true for $t = T$. We also have

$$W_\epsilon(x, 0, t) = 0, \ W_\epsilon(x, 1, t) = V(x, t) + \epsilon \int_0^1 \nu A'(\nu) d\nu. \tag{18.2.36}$$

The next step is to check that the calculations made formally for $W(x, p, t)$ are valid for the smooth function $W_\epsilon(x, p, t)$. We have in particular

$$\frac{\partial \Psi_\epsilon}{\partial \lambda}(x, \lambda, t) = \lambda \frac{\partial \Psi_\epsilon}{\partial \lambda}(x, \lambda, t),$$

and $\Psi_\epsilon(x, \lambda, t)$, $G_\epsilon(x, \lambda, t)$ satisfy the PDE (18.2.18), (18.2.17) with different final conditions, given by (18.2.28). Therefore,

$$\lambda_\epsilon(x, p, t) = \frac{\partial W_\epsilon}{\partial p}(x, p, t),$$

and $\lambda_\epsilon(x,p,t)$ is the solution of the nonlinear PDE (18.2.24), and $W_\epsilon(x,p,t)$ is the solution of the Bellman equation (18.2.25), with again different final conditions (18.2.32) and (18.2.35). Therefore, $W_\epsilon(x,p,t)$ is the value function of a stochastic control problem, where the state is the pair $X_{xt}(s)$, $P_{pt}(s)$, the solution of (18.2.8) and the objective functional

$$J^\epsilon_{xpt}(q(.)) = J_{xpt}(q(.)) + \epsilon E[\int_0^{A^{-1}(P_{pt}(T))} \nu A'(\nu)d\nu \, \mathbb{1}_{T\leq\tau}] \quad (18.2.37)$$

$$+\epsilon \int_0^1 \nu A'(\nu)d\nu \, E[\mathbb{1}_{\tau<T}\mathbb{1}_{P_{pt}(\tau)=1}].$$

Therefore,

$$|J^\epsilon_{xpt}(q(.)) - J_{xpt}(q(.))| \leq \epsilon \int_0^1 \nu A'(\nu)d\nu,$$

from which it follows immediately that

$$|W_\epsilon(x,p,t) - \hat{V}(x,p,t)| \leq \epsilon \int_0^1 \nu A'(\nu)d\nu,$$

and thus $W(x,p,t) = \hat{V}(x,p,t)$, which is the desired result (18.2.13). Then (18.2.15) follows. This concludes the proof. □

18.3 Extension

We present here an extension of the model of the preceding section. Again, it is inspired by [49], but our proof is different and is the same as in the preceding section. The financial market is composed of n assets, whose prices are characterized by a vector $X_{xt}(s)$, the solution of

$$dX = X * \sigma(\mu ds + dw(s)), \quad (18.3.1)$$
$$X(t) = x.$$

So with respect to (18.2.1), we are allowing a drift μ. For simplicity, we take μ and σ constants. They can depend on X. The wealth process associated with an endowment y at initial time t is given by

$$dY = \pi(s).\sigma(\mu ds + dw(s)), \quad (18.3.2)$$
$$Y(t) = y,$$

in which the adapted process $\pi(s)$ represents the portfolio of investments in the assets, following the standard self-financing assumption. We first consider

$$V(x,t) = \inf\{y \geq 0 | \exists \pi(.) \text{ such that } Y_{yt}(T;\pi(.)) \geq g(X_{xt}(T)), \text{ a.s.}\}. \tag{18.3.3}$$

We can reduce this problem to the case $\mu = 0$ by Girsanov's theorem. We introduce the martingale $Q_t(s)$, the solution of

$$\begin{aligned} dQ_t(s) &= -\,Q_t(s)\,\mu.dw(s), \\ Q_t(t) &= 1. \end{aligned} \tag{18.3.4}$$

Then formula (18.1.6) extends as

$$V(x,t) = EQ_t(T)g(X_{xt}(T). \tag{18.3.5}$$

Let us define a new probability on $\Omega, \mathcal{F}_t^T$ using the martingale $Q_t(s)$. We denote it by P^{Q_t}, reserving P for the original probability. If A is an event in $\mathcal{F}_t^T$, then $E^{Q_t}\mathbb{1}_A = EQ_t(T)\mathbb{1}_A$. Therefore, we have also $V(x,t) = E^{Q_t}g(X_{xt}(T)$. We know from Girsanov's theorem that the process

$$w^{Q_t}(s) = w(s) - w(t) + \int_t^s \mu d\theta$$

is a Wiener process on the probability space $\Omega, \mathcal{F}_t^T, P^{Q_t}$. We next consider the hedging problem in probability,

$$V(x,p,t) = \inf\{y \geq 0 | \exists \pi(.) \text{ such that } P(Y_{yt}(T;\pi(.)) \geq g(X_{xt}(T))) \geq p\}, \tag{18.3.6}$$

and we have immediately

$$V(x,0,t) = 0, \;\; V(x,1,t) = V(x,t), \tag{18.3.7}$$

Since by the definition, we have

$$V(x,p,T) = \inf\{y \geq 0 | \; \mathbb{1}_{y \geq g(x)} \geq p\},$$

we can state

$$V(x,p,T) \geq pg(x). \tag{18.3.8}$$

We next consider a family of processes: For every $q(.)$ adapted to $\mathcal{F}^s$, we introduce a new process $P_{pt}(s;q(.))$ such that

$$\begin{aligned} dP_{pt}(s) &= q(s).dw(s), \\ P_{pt}(t) &= p, \end{aligned} \tag{18.3.9}$$

where $q(s)$ is adapted to $\mathcal{F}^s$. As in Lemma 18.1, we have the characterization

$$V(x,p,t) = \inf\{y \geq 0 | \exists \pi(.),\ q(.) \text{ such that} \tag{18.3.10}$$

$$\mathbb{1}_{\{Y_{yt}(T;\pi(.)) \geq g(X_{xt}(T))\}} \geq P_{pt}(T; q(.))\}.$$

We next introduce the control problem similar to (18.2.8), (18.2.9), (18.2.10),

$$dX_{xt}(s) = X_{xt}(s) * \sigma dw^{Q_t}(s),\ X(t) = x,$$

$$dP_{pt}(s) = -q(s).\mu ds + q(s).dw^{Q_t}(s),\ P_{pt}(t) = p,$$

in which $q(.)$ is the control. The payoff is defined by

$$J_{xpt}(q(.)) = E^{Q_t}[g(X_{xt}(T)P_{pt}(T)\mathbb{1}_{T \leq \tau} + V(X_{xt}(\tau), \tau)\mathbb{1}_{\tau < T}\mathbb{1}_{P_{pt}(\tau)=1}],$$

$$\tau = \inf\{s \geq t |\ P_{pt}(s) \notin [0,1]\}, \tag{18.3.11}$$

and we define the value function

$$\hat{V}(x,p,t) = \inf_{q(.)} J_{xpt}(q(.)). \tag{18.3.12}$$

We then state a result similar to Theorem 18.1. Define the functions

$$G(x,\lambda,t) = E\mathbb{1}_{\lambda > Q_t(T)g(X_{xt}(T))}, \tag{18.3.13}$$

$$\Psi(x,\lambda,t) = EQ_t(T)g(X_{xt}(T))\mathbb{1}_{\lambda > Q_t(T)g(X_{xt}(T))}.$$

Note that the expectation is taken with respect to the initial probability P. We then have the following theorem.

THEOREM 18.2. *We make the assumptions of Theorem 18.1. We define for $t < T$ the function $\lambda(x,p,t)$ by solving the equation*

$$G(x,\lambda,t) = p. \tag{18.3.14}$$

Then we have

$$V(x,p,t) = \hat{V}(x,p,t) = \Psi(x,\lambda(x,p,t),t) \tag{18.3.15}$$

and $\lambda(x,p,t) = \frac{\partial V}{\partial p}(x,p,t)$. Moreover, $\lambda(x,p,t) \to g(x)$ as $t \to T$ and $\Psi(x,\lambda(x,p,t),t) \to pg(x)$ as $t \to T$.

PROOF. We proceed formally. A rigorous proof can be given by the approximation procedure of the indicator function as in Theorem 18.1. We first give the equations for $G(x,\lambda,t)$ and $\Psi(x,\lambda,t)$. If we define the process $\lambda_{\lambda t}(s) = \dfrac{\lambda}{Q_t(s)}$, then an immediate application of Itô' s lemma shows that $\lambda_{\lambda t}(s)$ is the solution of the stochastic differential equation

$$d\lambda = \lambda(s)(|\mu|^2 ds + \mu.dw),$$
$$\lambda_{\lambda t}(t) = \lambda.$$

But the first equation (18.3.13) can be written as follows:

$$G(x, \lambda, t) = E\mathbb{1}_{\lambda_{\lambda t}(T) > g(X_{xt}(T))},$$

where we recall that the components $X_i(s)$ satisfy

$$dX_i = X_i \sum_j \sigma_{ij}(\mu_j ds + dw_j),$$
$$X_i(t) = x_i.$$

The equation of $G(x, \lambda, t)$ is obtained immediately:

$$\frac{\partial G}{\partial t} + \sum_{ij}(a_{ij}x_i x_j \frac{\partial^2 G}{\partial x_i \partial x_j} + \lambda\sigma_{ij}\mu_j x_i \frac{\partial^2 G}{\partial x_i \partial \lambda}) + \tag{18.3.16}$$
$$+\frac{1}{2}|\mu|^2\lambda^2 \frac{\partial^2 G}{\partial \lambda^2} + |\mu|^2\lambda\frac{\partial G}{\partial \lambda} + \sum_{ij}\sigma_{ij}\mu_j x_i \frac{\partial G}{\partial x_i} = 0,$$
$$G(x, \lambda, T) = \mathbb{1}_{\lambda > g(x)}.$$

Next, from the second formula (18.3.13) and the definition of $\lambda_{\lambda t}(s)$, we can write

$$\Psi(x, \lambda, t) = E^{Q_t} g(X_{xt}(T))\mathbb{1}_{\lambda_{\lambda t}(T) > g(X_{xt}(T))}.$$

But we can also write

$$dX(s) = X(s) * \sigma dw^{Q_t},$$
$$d\lambda = \lambda\mu.dw^{Q_t},$$
$$X(t) = x \qquad \lambda(t) = \lambda.$$

The PDE for Ψ follows readily:

$$\frac{\partial \Psi}{\partial t} + \sum_{ij}(a_{ij}x_i x_j \frac{\partial^2 \Psi}{\partial x_i \partial x_j} + \lambda\sigma_{ij}\mu_j x_i \frac{\partial^2 \Psi}{\partial x_i \partial \lambda}) + \frac{1}{2}|\mu|^2\lambda^2 \frac{\partial^2 \Psi}{\partial \lambda^2} = 0, \tag{18.3.17}$$
$$\Psi(x, \lambda, T) = g(x)\mathbb{1}_{\lambda > g(x)}.$$

We claim that

$$\frac{\partial \Psi}{\partial \lambda} = \lambda \frac{\partial G}{\partial \lambda}. \tag{18.3.18}$$

Indeed, let $\tilde{G}(x, \lambda, t) = \lambda G(x, \lambda, t)$. From the definitions (18.3.13), we see that

$$\tilde{G}(x, \lambda, t) - \Psi(x, \lambda, t) = E(\lambda - Q_t(T)g(X_{xt}(T)))^+.$$

Therefore,

$$\frac{\partial}{\partial \lambda}(\tilde{G}(x, \lambda, t) - \Psi(x, \lambda, t)) = G(x, \lambda, t),$$

which implies (18.3.18). At time T we can interpret (18.3.18) in the sense of distributions, since

$$\begin{aligned}\frac{\partial \Psi}{\partial \lambda}(x, \lambda, T) &= g(x)\delta(\lambda - g(x)) = \lambda\delta(\lambda - g(x)) \\ &= \lambda \frac{\partial G}{\partial \lambda}(x, \lambda, T).\end{aligned}$$

We have $G(x, 0, t) = 0$, and $G(x, +\infty, t) = 1$. The function $G(x, \lambda, t)$ is continuous in λ for $t < T$, monotonically increasing from 0 to $+\infty$. Therefore, its inverse $\lambda(x, p, t)$ is well defined on the interval [0, 1] and monotonically increasing with $\lambda(x, 0, t) = 0$ and $\lambda(x, 1, t) = +\infty$. We next define

$$W(x, p, t) = \Psi(x, \lambda(x, p, t), t).$$

Then from results already proved, we have

$$\frac{\partial W}{\partial p}(x, p, t) = \lambda(x, p, t). \tag{18.3.19}$$

Also,

$$\begin{aligned}W(x, p, t) &= pg(x) + E[(Q_t(T)g(X_{xt}(T) - g(x)))\mathbb{1}_{\lambda > Q_t(T)g(X_{xt}(T))}] \\ &\to pg(x), \text{ as } t \to T,\end{aligned}$$

and $W(x, 0, t) = \Psi(x, 0, t) = 0$ and $W(x, 1, t) = \Psi(x, +\infty, t) = V(x, t)$. It remains to check that $W(x, p, t)$ is the solution of the Bellman equation of the control problem (18.3.11). We use

$$\frac{\partial W}{\partial t} = \frac{\partial \Psi}{\partial t} - \lambda \frac{\partial G}{\partial t}, \; \frac{\partial W}{\partial x_i} = \frac{\partial \Psi}{\partial x_i} - \lambda \frac{\partial G}{\partial x_i},$$

$$\frac{\partial^2 W}{\partial x_i \partial x_j} = \frac{\partial^2 \Psi}{\partial x_i \partial x_j} - \lambda \frac{\partial^2 G}{\partial x_i \partial x_j} + \frac{\dfrac{\partial G}{\partial x_i}\dfrac{\partial G}{\partial x_j}}{\dfrac{\partial G}{\partial \lambda}},$$

$$\frac{\partial G}{\partial \lambda} = \frac{1}{\dfrac{\partial^2 W}{\partial p^2}}, \; D_x G = -\frac{D_x \dfrac{\partial W}{\partial p}}{\dfrac{\partial^2 W}{\partial p^2}}.$$

Collecting results, it follows that

$$\frac{\partial W}{\partial t} + \sum_{ij} a_{ij} x_i x_j \frac{\partial^2 W}{\partial x_i \partial x_j} - \frac{1}{2} \frac{|\dfrac{\partial W}{\partial p}\mu - \sigma^*(x * D_x \dfrac{\partial W}{\partial p})|^2}{\dfrac{\partial^2 W}{\partial p^2}} = 0, \tag{18.3.20}$$

which is the Bellman equation of the control problem (18.3.11). Therefore, $W(x, p, t) = \hat{V}(x, p, t)$. We show as in Lemma 18.3 that $V(x, p, t) \geq \hat{V}(x, p, t)$. Next,

$$W(x, p, t) = E^{Q_t} g(X_{xt}(T)) \mathbb{1}_{\lambda(x,p,t) > Q_t(T) g(X_{xt}(T))},$$

so there exists an adapted process $\pi(s)$ such that

$$\begin{aligned} g(X_{xt}(T)) \mathbb{1}_{\lambda(x,p,t) > Q_t(T) g(X_{xt}(T))} &= W(x, p, t) + \int_t^T \pi(s).\sigma dw^{Q_t} \\ &= Y_{W(x,p,t),t}(T; \pi(.)) \end{aligned}$$

and

$$\{Y_{W(x,p,t),t}(T; \pi(.)) \geq g(X_{xt}(T))\} \equiv \{\lambda(x, p, t) > Q_t(T) g(X_{xt}(T))\},$$

so

$$P(\{Y_{W(x,p,t),t}(T; \pi(.)) \geq g(X_{xt}(T))\}) = p,$$

which implies from the definition of $V(x, p, t)$, see (18.3.6), that $W(x, p, t) \geq V(x, p, t)$. Comparing results, we obtain $V(x, p, t) = \hat{V}(x, p, t) = W(x, p, t)$. This completes the proof. □

Bibliography

1. Bailleul, J., Samad, T., *Encyclopedia of Systems and Control*, Springer, (2015)
2. Bayraktar, E., Sîrbu, M., Stochastic Perron's Method for Hamilton-Jacobi-Bellman Equations, SIAM J. Con. and Optimization, 51 (6) (2013)
3. Bellman, R., *Dynamic Programming,* Princeton University Press, (1957).
4. Bensoussan, A., *Perturbation Methods in Optimal Control,* Dunod-Gauthier-Villars, (1988).
5. Bensoussan, A., *Stochastic Control of Partially Observable Systems,* Cambridge University Press, (1992)
6. Bensoussan, A., Cadenillas, A., Koo, H.K., Entrepreneurial Decisions on Effort and Project with a Non-Concave Objective Function, Mathematics of Operations Research, (2015)
7. Bensoussan, A., Chen, S., Sethi, S.P., The Maximum Principle for Global Solutions of Stochastic Stackelberg Differential Games, SIAM Control and Optimization, (2016)
8. Bensoussan, A., Mookerjee, R., Mookerjee, V., Yue, W.T., Maintaining Diagnostic Knowledge-based Systems: A Control Theoretic Approach 38 Management Science, 55(2), (2009)
9. Bensoussan, A., Van Schuppen, J.H., Optimal Control of partially observable control systems with an exponential-of-integral performance index, SIAM J. Con. 23(4), (1985)
10. Bernhard, P., Linear-Quadratic two person zero-sum Differential Games, JOTA 27, (1979)
11. Bismut, J.M., On optimal control of linear stochastic equations with a linear-quadratic criterion, SIAM J. Con. and Optimization 15 (1), (1977)
12. Briand P., Hu, Y., Quadratic BSDEs with Convex Generators and unbounded terminal Conditions, Probability Theory and related Fields, 141, (2008)
13. Brockett, R.W., *Finite Dimensional Linear Systems,* Wiley, N.Y. (1970)
14. Carmona, R., *Lectures on BSDEs, Stochastic Control and Stochastic Differential Games,* (2014)
15. Cox, J., Huang, C.F., Optimal Consumption and Portfolio Policies when Asset Prices follow a Diffusion Process, Journal of Economic Theory, 49, (1989)
16. Crandall, M.G., Ishii, H., Lions, P-L., User's Guide to Viscosity Solutions of Second Order Partial Differential Equations, Bulletin of the American Mathematical Society, Vol 27, n° 1, (1992)
17. Cvitanić, J., Zhang, J., *Contract Theory in Continuous-Time Models*, Springer-Verlag, Berlin, Heidelberg, (2013)
18. Da Lio, F., Ley, O., Uniqueness Results for Second Order Bellman-Isaacs Equations under Quadratic Growth Assumptions and Applications, SIAM Journal on Control and Optimization, 45 (1), (2006)
19. Dockner, E., Jørgensen, S., Van Long, N., Sorger, G., *Differential Games in Economics and Management Science,* Cambridge University Press, Cambridge, (2000)

A. Bensoussan, *Estimation and Control of Dynamical Systems*, Interdisciplinary Applied Mathematics 48, https://doi.org/10.1007/978-3-319-75456-7

20. El Karoui, N., Hamadene, S., Matoussi, A., Backward Stochastic Differential Equations and Applications, Princeton BSDE Lecture Notes (2009)
21. Fahri, E., Panageas, S., Saving and Investing for Early Retirement: A Theoretical Analysis, Journal of Financial Analysis, 83 (2007)
22. Fleming, W.H., Soner, H.M., *Controlled Markov Processes and Viscosity Solutions,* Springer, New York, (2006)
23. Holt, C.C., Forecasting Trends and Seasonal by Exponentially Weighted Averages. Office of Naval Research Memorandum 52 (1957) reprinted in Holt, C.C., Forecasting Trends and Seasonal by Exponentially Weighted Averages. International Journal of Forecasting 20 (1), (2004).
24. Ibramovic, I.A., Has'minskii, R.Z., (1981) *Statistical Estimation,* Springer-Verlag, New York
25. Kalman, R.E., Falb, P.L., Arbib, M.A., *Topics in Mathematical System Theory,* McGraw-Hill (1969)
26. Kalman, R.E., On a new characterization of linear passive systems, Proc. 1st Allerton Conf. on Circuit and System Theory, Univ. Illinois, Urbana, (1963)
27. Karatzas, I., Lehoczky, J.P., Shreve, S.E., Optimal Portfolio and Consumption Decisions for a Small Investor on a Finite Horizon, SIAM Control and Optimization, 25, (1987)
28. Kobylanski, M., Backward Stochastic Differential Equations and Partial Differential Equations with Quadratic Growth, The Annals of Probability, Vol 28, pp. 558–602, (2000)
29. Krylov, N.V., *Controlled Diffusion Processes,* Springer-Verlag (1980)
30. Kunita, H., École d'Été de Probabilités de Saint-Flour XII, Springer-Verlag, Berlin, New York, (1982)
31. Kunita, H., Watanabe, S. On square integrable martingales, Nagoya Math J. 30, (1967)
32. Lions, P.L., *Generalized Solutions of Hamilton-Jacobi Equations,* Pitman, (1982)
33. Lions, P.L., Optimal Control of Diffusion Processes and Hamilton-Jacobi Equations, Communications in Partial Diff. Equations, (1983)
34. Liptser, R.S., Shiryaev, A.N., *Statistics of Random Processes, I-II,* Springer-Verlag, (1977)
35. Loève, M., *Probability Theory,* Springer-Verlag, New York, (1978)
36. Ma, J., Protter, P., Yong, J., Solving forward-backward stochastic differential equations explicitly-a four step scheme, Probability Theory and Related Fields, 98 (1994)
37. Nisio, M., *Stochastic Control Theory, Dynamic Programming Principle,* Springer, Tokyo, (2015)
38. Papavassilopoulos, G.P., Cruz Jr., J.B., Nonclassical Control Problems and Stackelberg Games, IEEE Transactions on Automatic Control, 24, (1979)
39. Pardoux, E., Peng, S., Adapted solution of a backward stochastic differential equation, Syst. Control Lett. 14, (1990)
40. Pardoux, E., Rascanu, A., *Stochastic Differential Equations, Backward SDEs, Partial Differential Equations,* Springer (2014)
41. Pardoux, E., Tang, S., Forward-backward stochastic differential equations and quasilinear parabolic PDEs, Probability Theory and Related Fields 114, (1999)
42. Peng, S., Stochastic Hamilton-Jacobi-Bellman Equations, SIAM J. Control and Optimization, Vol 30, No. 2, (1992)
43. Peng, S., Wu, Z., Fully coupled forward-backward stochastic differential equations and applications to optimal control, SIAM J. Control and Optimization 37, (1999)
44. Pham, H., *Continuous-time Stochastic Control Problem and Optimization with Financial Applications,* Springer (2009)
45. Pontryagin, L.S., Boltyansky, V.G., Gamgrelidze, R.V., Mishenko, E.F., *Mathematical Theory of Optimal Processes,* Nauka, Moscow, (1969)
46. Popov, V.M., *L'hyperstabilité des systèmes automatiques,* Dunod, (1973)
47. Rockafellar, R.T. *Convex Analysis,* Princeton University Press, (1970)
48. da-Silva, C.Q., Migon, H.S., Correia, L.T., Dynamic Bayesian Beta Models, Computational Statistics and Data Analysis, 55, (2011)
49. Touzi N., *Optimal Stochastic Control, Stochastic Target Problems and Backward SDE,* Springer, (2013)

50. West M., Harrison, P.J., Migon H.S., Dynamic Generalized Linear Models and Bayesian Forecasting, Journal of the American Statistical Association, 80:389, (1985)
51. Whittle, P., *Risk-sensitive Optimal Control,* Wiley, (1990)
52. Winters, P.R., Forecasting Sales by Exponentially Weighted Moving Averages, Management Science 6 (3), (1960)
53. Wonham, W.M., On a matrix Riccati equation of stochastic control, SIAM J. Con. and Optimization 6, (1968)
54. Yong, J., Forward-backward stochastic differential equations with mixed initial-terminal conditions, Trans. Amer. Math. Soc. 362, (2010)
55. Yong, J., Zhou, X.Y., *Stochastic Controls, Hamiltonian Systems and HJB Equations,* Springer, New York (1999)
56. Zhang, P. Some Results on two-person Zero-sum Linear Quadratic Differential Games, SIAM J. Con. and Optimization 43, (2005)

MIX
Papier aus verantwortungsvollen Quellen
Paper from responsible sources
FSC® C105338

If you have any concerns about our products,
you can contact us on
ProductSafety@springernature.com

In case Publisher is established outside the EU,
the EU authorized representative is:
Springer Nature Customer Service Center GmbH
Europaplatz 3, 69115 Heidelberg, Germany

Printed by Libri Plureos GmbH
in Hamburg, Germany